Food Packaging

THE AUTHOR

Dr. (Mrs.) Neelam Khetarpaul presently working as Emeritus Professor (ICAR) and former Dean and HOD (Foods and Nutrition), College of Home Science at CCS Haryana Agricultural University, Hisar is the recipient of many awards *viz.*, Distinguished Teacher Award-2000, Young Investigator Award, Ms. Manju Utreja Gold Medal and cash award for doing the best research work in the University and Best Research Paper Award by AFST (I) Hisar Chapter. She is the recipient of various Visiting Fellowships abroad funded by different national and international agencies and visited USA, Australia, UK, Hungary and Netherlands for academic pursuits. She was the country Coordinator of Indo-Netherlands Tailoring Biotechnologies project programme being funded by the Department of Biotechnology, Ministry of Science and Technology, Govt. of India and the Wageningen University, The Netherlands. She has published more than 260 research papers in various journals of national and international repute and 23 books in the discipline of Food Science and Human Nutrition.

Food Packaging

2nd Revised & Enlarged Edition

— Author —

Prof. Neelam Khetarpaul

Ph.D. (Foods and Nutrition), FICN
Emeritus Professor (ICAR)
Former Dean and HOD (Foods and Nutrition)
College of Home Science
CCS Haryana Agricultural University
Hisar – 125 004 (Haryana), India

2025

Daya Publishing House®
A Division of
Astral International Pvt. Ltd.
New Delhi – 110 002

First Edition, 2008
Second Revised and Enlarged Edition, 2025

ISBN: 978-93-5461-982-3 (HB)

Publisher's Note:

Every possible effort has been made to ensure that the information contained in this book is accurate at the time of going to press and the publisher and author cannot accept responsibility for any errors or omissions, however caused. No responsibility for loss or damage occasioned to any person acting, or refraining from action, as a result of the material in this publication can be accepted by the editor, the publisher or the author. The Publisher is not associated with any product or vendor mentioned in the book. The contents of this work are intended to further general scientific research, understanding and discussion only. Readers should consult with a specialist where appropriate.

Every effort has been made to trace the owners of copyright material used in this book, if any. The author and the publisher will be grateful for any omission brought to their notice for acknowledgement in the future editions of the book.

Published by : **Daya Publishing House®**
A Division of
Astral International Pvt. Ltd.
– ISO 9001:2015 Certified Company –
4736/23, Ansari Road, Darya Ganj
New Delhi-110 002
Ph. 011-43549197, 23278134
E-mail: info@astralint.com
Website: www.astralint.com

Digitally Printed at : **Replika Press Pvt. Ltd.**

Preface

India is a major producer of agricultural products but it processes and packages just a fraction of the output. Packaging is at the heart of the marketing strategy in terms of functionality, shelf appeal and convenience. Packaging is becoming an essential part of the value chain analysis, regarding food safety, organoleptic characteristics, ergonomics and flexibility.

Life style changes are also happening at a fast pace. Traditional meals that "grandmothers cooked" are going out of fashion and people are increasingly opting for packed ready to eat products. Convenience appears to be the single biggest driver in the ready meal market. Single person households, working women and those with higher disposable income have become increasingly influential in determining food production trends. Hence, packaging is of great importance in the final choice the consumer will make, because it directly involves convenience, appeal, information and branding.

In a very broad sense, the food industry is discovering the food service channel as a new distribution alternative. Alliances between food producers and food service operators will be the only means to develop successful solutions. The food industry has to "pull its packaging act together" if it hopes to exploit the international markets. The importance of packaging hardly needs stressing, since it is almost impossible to think of more than a handful of foods which are sold in an unpackaged state. Today packaging is essential and pervasive.

Food manufacturers know the importance of great food packaging. The food packaging could greatly affect the marketability of the product. But apart from serving as containers, food packaging also serves the purpose of opening up the avenues of communication between the manufacturer and the consumer. Food packaging helps make sure that the food is protected from tampering. It's also the means by which manufacturers transport and promote their goods. Apart from this, and probably the most important out of all the purposes of food packaging, is that it preserves the food and keeps it fresh for as long as possible. An important aspect of food packaging is **food packaging labelling.** The manufacturers are held accountable for any health hazards that the consumers may encounter because of poor packaging or poor food preservation techniques employed by the manufacturer. Furthermore, the fact that on average around 25 per cent of the ex-factory cost of consumer foods is for their packaging provides the incentive and the challenge for food packaging technologists to design and develop functional packages at minimum cost. Food packaging is the major subject which draws on several disciplines; it is a multidisciplinary speciality. This book includes 22 chapters related to history, functions, classification of materials used for food packaging, packaging methods, mechanical and chemical pulps, the kinds of deteriorative reactions, use of nanotechnology in food packaging, food packaging metals and their corrosion, packaging of foods in metal containers, use of glass in food packaging, thermoplastic polymers, important plastics processing methods, the packaging of cereals, dairy products, fruits and vegetables and meat and meat products, sterilization of packaging material and shelf life of packaged foods, vision of future food packaging *etc.* Readers with an interest in food packaging will find the information given in various chapters to be timely, representative of some of the best work in the field of food packaging, and of great value.

We hope that this revised edition of book shall be very useful for the students doing under graduation and post graduation in the disciplines of Food Science and Technology, Food Processing and Quality Control and Foods and Nutrition, research scholars and teachers. We shall regard our efforts amply rewarded if the book is received with enthusiasm.

We wish to express our sincere thanks to the publisher and especially to Mr. Anil Mittal for his continued interest in bringing out this edition.

Neelam Khetarpaul

Contents

Chapter 1

Introduction

The concept of packaged food came into existence owing to factors such as easy cooking, consumption, handling, and safety from external tampering. Changing lifestyle, convenience of consumption and increased health awareness are the main factors driving the growth of the packaged food market. Packaging materials used for food should be convenient for carrying, displaying, opening and closing.

Packaging is first impression of product and company to remember and keep the taste for long time. Food packaging is an area with which, sooner or later, every practicing food scientist and technologist becomes involved. The importance of packaging hardly needs stressing, since it is almost impossible to think of more than a handful of foods which are sold in an unpackaged state. Furthermore, the fact that on average around 25 per cent of the ex-factory cost of consumer foods is for their packaging and provides the incentive and the challenge for food packaging technologists to design and develop functional packages at minimum cost.

Food Packaging Scenario

The food processing and packaging industry is an important global business earning multi-trillion dollars. This is because of the rise in the living standards of the residents of developing countries. Moreover, people are getting more aware and conscious about their health. That is why, they are opting for packaged food. As a result, the need to follow food safety guidelines becomes stronger and the rules become more rigid. Worldwide, the sales of processed foods are estimated to be $2 trillion whereas that of packaged food range around $1 trillion. This is due to the enhancement in the incomes in developing countries which has resulted in rise in standard of living for their population. Further, people are opting for processed foods rather than foods like rice and barley. It surrounds, enhances and protects the goods we buy, from processing and manufacturing through handling and storage to the final consumer.

1. Global Overview Scenario

Food packaging is very important not only as a protection of food but also as a means of marketing the food product. Based on the latest advanced technology, food manufacturers, retailers, packaging products suppliers are working more closely than ever before to produce packaging products designed for modern lifestyles. This has led to the emergence of varied hi-fi packaging designs and packaging products, especially for food packaging, in the market. There has been explosion of ready to eat meals in the market. All food packaging highlights freshness, health and traceability of the products. Packaging has become an important part of the value chain analysis, regarding food safety.

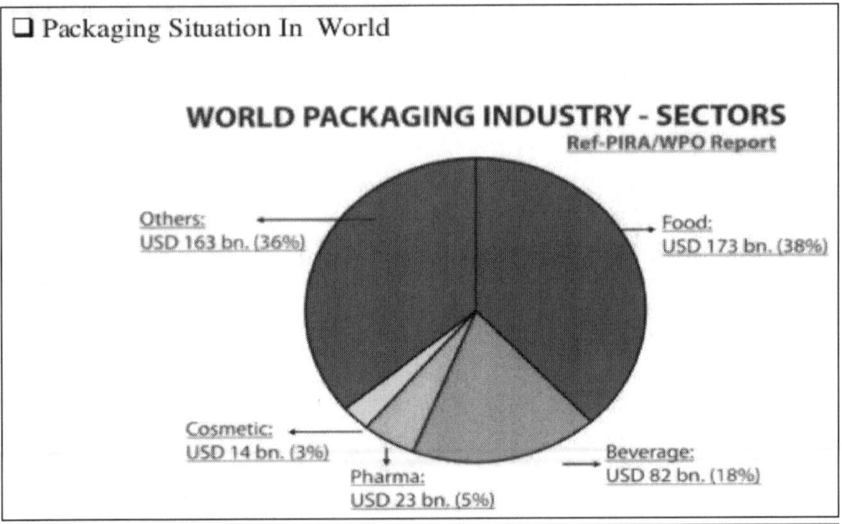

❑ Packaging Situation In World

- Consumers prefer such packaging that requires minimum time and effort to open, carry, and store such as zippers enclosure, tear notches, peel-off lids, hang hole features, and microwavable pouches

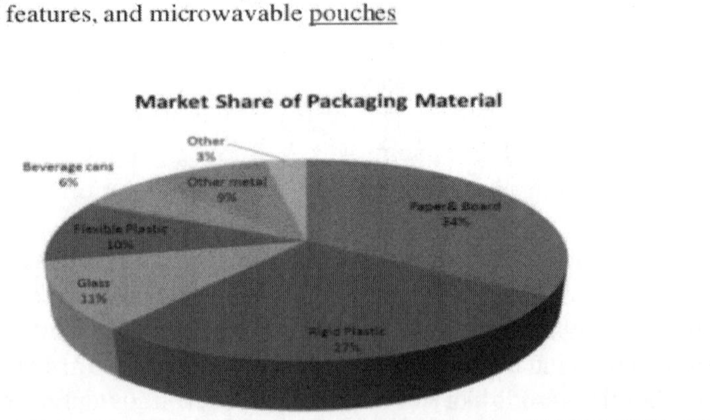

In 2017, the total value of the packaging globally was approximately $851 billion according to the latest exclusive study. The global food packaging market reached a value of US$ 302.2 billion in 2019. **Packaged Food Market** size was expected to garner $3.03 trillion by 2020, registering a CAGR of 4.5 per cent during the forecast period 2015 - 2020.The Future of Global Packaging to 2022, shows that packaging demand will grow steadily at 2.9 per cent to reach $980 billion in 2022. Customer convenience is the top priority for the vendors in the global packaging market.

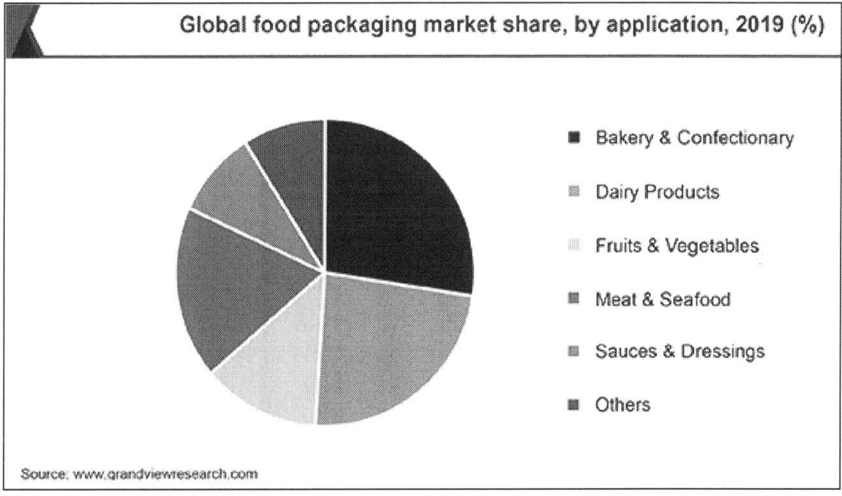

Several packaging companies are focusing on new packaging designs or redesigning the existing packaging especially in consumer goods. Consumers prefer such packaging that requires minimum time and effort to open, carry, and store such as zippers enclosure, tear notches, peel-off lids, hang hole features, and microwavable pouches.

The global packaged food market is segmented on the basis of product type and geography. Based on product type, the packaged food market share analysis is categorized into sixteen major food products namely - ready meals, baked foods, breakfast cereals, soup, baby food, potato chips, nuts, instant noodles, pasta, biscuits, chocolate confectionery, cheese, yogurt, ice creams, 'sauces, dressings and condiments', and non-alcoholic drinks. The baby food market is expected to grow at the fastest CAGR of 6.4 per cent during the forecast period. The growth is imminent due to factors such as increase in the number of working parents, and safer baby food options for infants and toddlers. Geographically, North America is the largest market in the current scenario followed by Europe, APAC and LAMEA.

The global food packaging market is mainly driven by numerous factors that include the rising urban population and the changing consumption pattern of consumers, especially in the developing economies driven by the rapid change in the lifestyle. Additionally, the strong demand for ready-to-eat and frozen food products

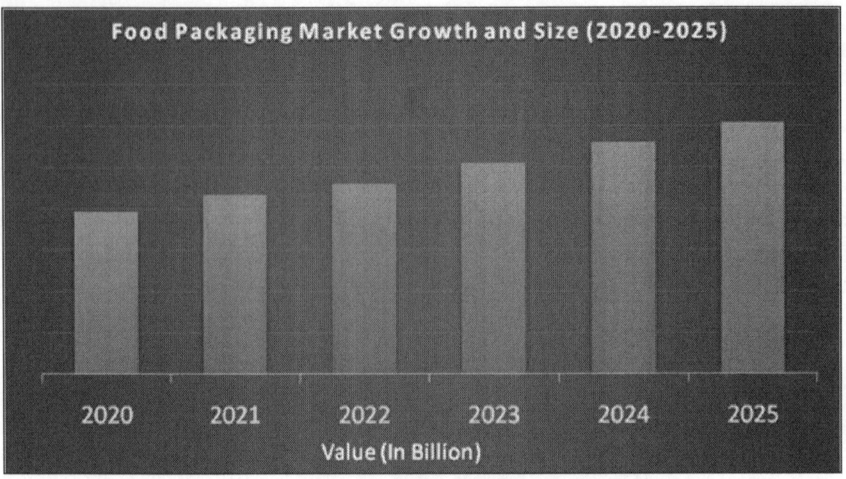

as a result of the growing working population, hectic life styles, rising disposable income and the food retail sector's expansion across the globe, will further surge the demand of the market in the coming years. Owing to this, the leading companies are investing in research and development (R and D) operations to enhance product shelf-life and develop anti-microbial packaging materials. They are also introducing lightweight solutions to reduce transportation costs and related emissions.

The technological advancement in the food packaging industry such as the advent of micro packaging, smart packaging wherein the packaging changes color if the food is contaminated and several research and development activities for bringing innovations in the packaging method will attract a large number of consumer base to the market. The food packaging solution offers remarkable

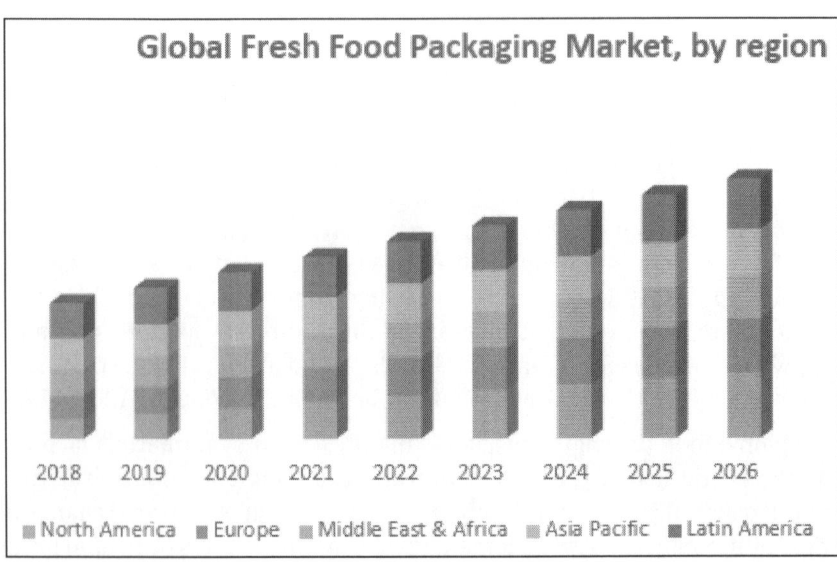

advantages such as enhanced safety level of food products, the extended shelf life of the products, and increased convenience. In parallel, the inclination of consumers towards sustainable and environment-friendly packaging and rising concerns for food safety and food contamination is thriving innumerable growth opportunities for the market. The factors that are hindering the growth of the global food packaging market include the adverse effects of the packaging materials and rising concerns related to the generation of waste.

Looking forward, IMARC Group expects the global food packaging market to exhibit moderate growth during the next five years.

II. Indian Overview Scenario

Packaged food is the fastest-growing segment in the Indian packaging industry. It is expected to fuel the demand for plastic packaging, as it ensures food quality, safety, and long shelf life. India is one of the top five markets for packaged food in the world and the second-largest in Asia. The per capita packaging consumption in India is quite low at 4.3 kgs. The Indian packaging industry constitutes about 4 per cent of the global packaging industry. The rapid growth of the market is primarily driven by the pharmaceuticals and foods and beverages industries. According to Agriculture and Agri-food Canada, the sales of packaged food in India amounted to USD 76,284.2 million in 2018, and it was expected to register a CAGR of 18 per cent during the coming years.

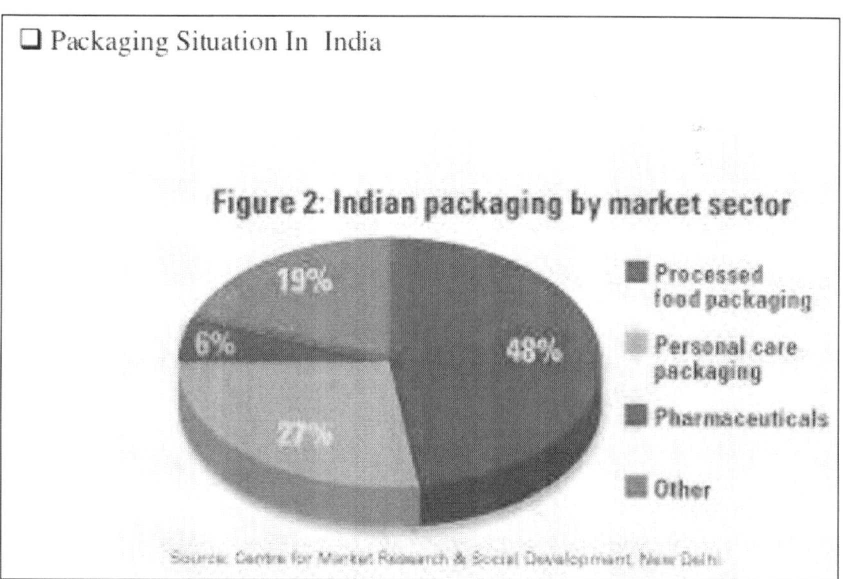

According to recent data released by the Associated Chambers of Commerce and Industry of India, the market size for organized packaged food is expected to reach INR 780 million, in 2020, from INR 530 million, in 2016. The growth of the packaging market in India is largely driven by various logistic applications, technological advancements, and the development of the packaging sector across the country.

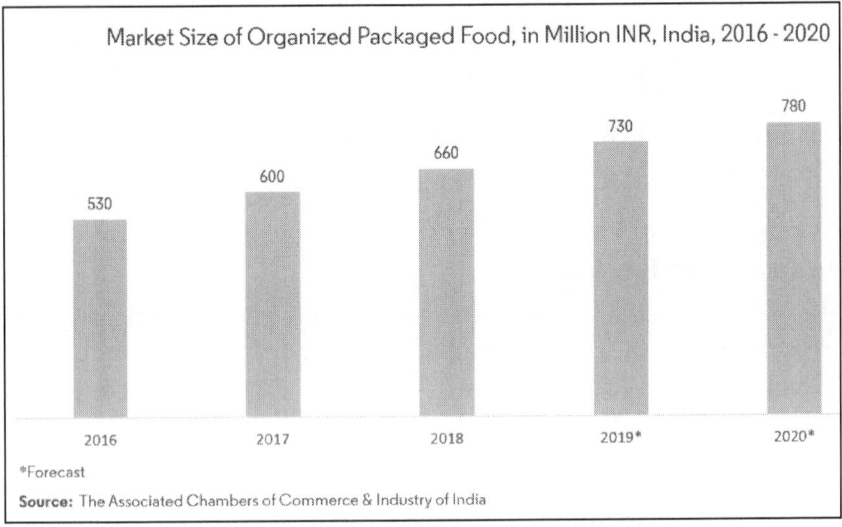

The packaging industry in India is fragmented in nature, owing to the several players competing to improve their market share. With the rising demand for packaging applications, along with technological advancement across the Indian economy, many companies are increasing their market presence by expanding their business footprint across various end-user markets. Major players include WestRock India Pvt. Ltd, OJI India Packaging Pvt. Ltd, Deccan Cans and Printers Pvt. Ltd, among others.

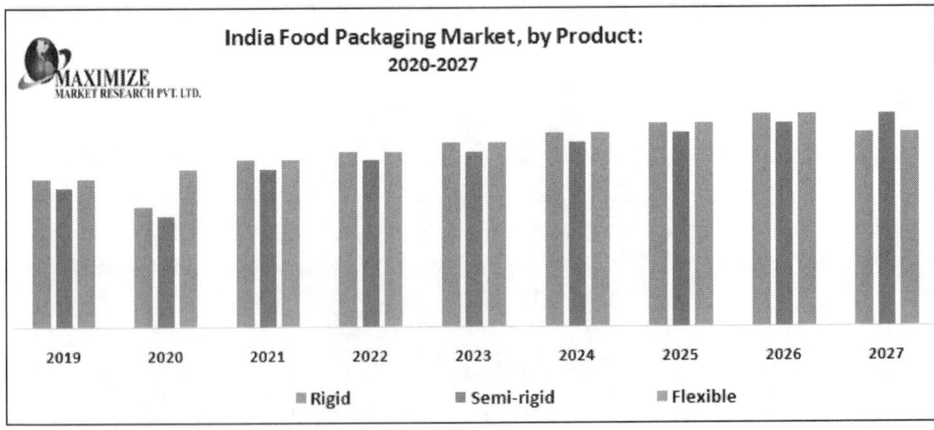

The food packaging industry in India has witnessed major innovations, in terms of branding and packaging. For instance, iD Fresh has introduced Smart Sip Tender Coconut and iD Grated Coconut, in innovative eco-friendly packaging. Many popular brands, like Amul, Mother Dairy, Patanjali, and others, use UV-protected films, color-changing films, and holographic films to reduce the chance of adulterating their products.

The Indian packaging market was valued at USD 50.5 billion in 2019, and it is expected to reach USD 204.81 billion by 2025, registering a CAGR of 26.7 per cent during the period of 2020-2025. The rising trend in food packaging is due to the following factors:

- According to the Federation of Indian Chambers of Commerce and Industry (FICCI), the expenditure on packaged foods is increasing (at inflection point), due to an increase in per capita income, urbanization, and an increase in the number of working women.
- Owing to the rising population, changing lifestyles, increased media penetration through the internet, television, and growing economy, the demand for packaging is growing. Moreover, it is one of the strongest growing sectors in the country. According to Care Ratings, a prominent credit rating company in India, more than 49 per cent of the paper produced in the country is used for packaging purposes.
- The rise of the Indian middle class, the rapid expansion of organized retail, growth of exports, and India's rising e-commerce sector is further facilitating growth. According to the Indian Institute of Packaging (IIP), the packaging consumption in India has increased by 200 per cent in the past decade, rising from 4.3 kg per person per annum (pppa) to 8.6 kg pppa.
- The food delivery and services market is fostering further growth in the country, with players, like Bigbasket, Grofers, Zomato, Swiggy, Scootsy, registering a rapid increase in sales. According to a study conducted by the Ministry of corporate affairs (India), the revenue generated by Swiggy and Zomato in India in FY 2018 was INR 4.42 billion and INR 4.66 billion, respectively.
- With the recent outbreak of COVID 19, the packaging manufacturers are facing supply chain disruption along with decreasing manufacturing at the site in many parts of the world. To ensure the smooth flow of supply chain, Global Food Safety Initiative (GFSI) certification programs are providing six-month certificate extensions by conducting a remote audit and risk assessment pertaining to COVID-19 such that the company can approve a new supplier location without an on-site audit to meet the demand.

Definitions and Levels of Food Packaging

Packaging has been defined in a number of ways reflecting different emphases:

- A populist reference source defines packaging as an industrial and marketing technique for containing, protecting, identifying and facilitating the sale and distribution of agricultural, industrial and consumer products.

- The packaging is defined as the protection of the products for the distribution, sale, and use of it whereas, the packaging industry refers to the process of designing, producing and evaluating the packages.
- Packaging can be defined as a method to protect and contain foods with the aim of minimizing the environmental impact of our consumption.
- Packaging can be defined as a method to protect and contain foods with the aim of minimizing the environmental impact of our consumption.
- The technology or instrument which protects goods or products for transportation, storage, distribution, sale and use is called as packaging. Packing may also be called as the process of design and innovation for safety of products as well as production of packages.
- Packaging can also be described as mode of communication about handling, storage information. Packaging been used as coordinate systems of finished or logistical goods.
- As per Kolter, packaging is "all the activities of designing and producing the container for a product", it can be defined as wrapping product as per customer need or as per customer point of attraction. Packaging is nothing but outer part of product which helps to sale the product readily available as well as to protect the product from damage.
- As per Business dictionary the definition of Packaging is processes such as cleaning, drying, preserving and materials such as metal, glass, paper or paperboard, plastics employed to contain, handle, protect, transport an article; role of packaging is broadening as well as it may include functions such as to attract attention, assist in product promotion and provide machine identification (barcodes, *etc.*), impart essential or additional information and help in utilization, and help in.
- The wrapping material around a consumer item that serves to contain, identify, display, describe, promote, protect, and otherwise make the product marketable and keep it clean. Packaging is much more than the product design face.
- The Packaging Institute International (PII) defines packaging as the enclosure of products, items or packages in a wrapped pouch, bag, box, cup, tray, can, tube, bottle or other container form to perform one or more of the following functions: containment, protection, preservation, communication, utility and performance. If the device or container performs one or more of these functions, it is considered a package.

The UK Institute of Packaging provides three definitions of packaging:

1. A coordinated system of preparing goods for transport, distribution, storage, retailing and end-use;

2. A means of ensuring safe delivery to the ultimate consumer in sound condition at minimum cost;
3. A techno-economic function aimed at minimizing costs of delivery while maximizing sales (and hence profits).

It is important to distinguish between packaging as defined above, and packing which can be defined as the enclosing of an individual item (or several items) in a container, usually for shipping or delivery.

Different Types of Food Packaging

A distinction is usually made between the various "levels" of packaging. A primary package, the most essential for of packaging is one which is in direct contact with the contained product. It provides the initial and usually the major protective barrier. It preserves the products and ensures protection from external bacteria or particles. Examples of primary packages include metal cans, glass bottles, and plastic pouches. It is frequently only the primary package, which the consumer purchases at retail outlets.

A secondary package, used for branding and as an extra measure of protection for the food product, contains a number of primary packages, *e.g.*, a corrugated case. It is the physical distribution carrier and is sometimes designed so that it can be used in retail outlets for the display of primary packages. In many cases, food suppliers use primary packaging to package small amounts of food and secondary packaging is then used contain the smaller primary packages together. It is a great way to bulk-sell smaller portions of food products.

A tertiary package is made up of a number of secondary packages, the most common example being a stretch-wrapped pallet of corrugated cases. In interstate and international trade, a quaternary package is frequently used to facilitate the handling of tertiary packages. This is generally a metal container up to 12 m in length which can hold many pallets and is inter-modal in nature. That is, it can be transferred to or from ships, trains, and flat-bed trucks by giant cranes. Certain designs are also able to have their temperature, humidity and gas atmosphere controlled and this is necessary in particular situations such as for the transportation of frozen foods or fresh fruits and vegetables.

Although the definitions given above cover in essence the basic role and form of packaging, it is necessary to discuss in more detail the functions of packaging and the environments where the package must perform those functions; it shall be taken up further in next chapter.

Now let us learn about the deterioration of packaged foods.

The principal aim of this chapter is to provide a brief overview of the major biochemical, chemical, physical and biological changes that occur in foods during processing and storage, and show how these combine to affect food quality. Knowledge of such changes is essential before a sensible choice of packaging materials can be made, since the rate and/or magnitude of such changes can often be minimized by selection of the correct packaging materials.

The deterioration of packaged foods (and this includes virtually all foods since today very few foods are sold without some form of packaging) depends largely on transfers that may occur between the internal environment inside the package, and the external environment which is exposed to the hazards of storage and distribution. For example, there may be transfer of moisture vapor from a humid atmosphere into a dried product, or transfer of an undesirable odor from the external atmosphere into a high fat product. In addition to the ability of packaging materials to protect and preserve foods by minimizing or preventing the transfers referred to, packaging materials must also protect the product from mechanical damage, and prevent or minimize misuse by consumers (including tampering).

Although certain types of deterioration will occur even if there is no transfer of mass (or heat, since some packaging materials can act as efficient insulators against fluctuations in ambient temperatures) between the package and its environment, it is possible in many instances to prolong the shelf life of the food through the use of packaging.

It is important that food packaging not be considered in isolation from food processing and preservation, or indeed from food marketing and distribution: all interact in a complex way, and concentrating on only one aspect to the detriment of the others is a sure-fire recipe for commercial failure.

The development of an analytical approach to food packaging is strongly recommended, and to achieve this successfully, the packaging technologist must have a good understanding of food safety and quality. The more important of these is without question food safety, that is the freedom from harmful chemical or microbial contaminants at the time of consumption. Packaging is directly related to food safety in two ways.

If the packaging material does not provide a suitable barrier around the food, microorganisms can contaminate the food and make it unsafe. However, microbial contamination can also arise if the packaging material permits the transfer of, for example, moisture or oxygen from the atmosphere into the package. In this situation, microorganisms present in the food but presenting no risk because of the initial absence of moisture or oxygen may then be able to grow and present a risk to the consumer.

Today aseptic packaging is the very common process.

Aseptic packaging can be defined as the filling of a commercially sterile product into sterile containers under aseptic conditions and sealing of the containers so that reinfection is prevented, *i.e.*, so that they are hermetically sealed. The term "aseptic" implies the absence or exclusion of any unwanted organisms from the product, package or other specific areas, while the term "hermetic" (strictly "air tight") is used to indicate suitable mechanical properties to exclude the entrance of microorganisms into a package and gas or water vapor into or from the package. The term "commercially sterile" has been defined as the absence of microorganisms capable of reproducing in the food under non-refrigerated conditions of storage and distribution, thus implying that the absolute absence of all microorganism need not be achieved.

Currently there are two specific fields of application or aseptic packaging:

1. Packaging of presterilized and sterile products, and
2. Packaging of a non-sterile product to avoid infection by microorganisms.

Examples of the first application include milk and dairy products, puddings, desserts, fruit and vegetable juices, soups, sauces, and products with particulates. Examples of the second application include fresh products such as fermented dairy products like yogurt.

There are three major reasons for the use of aseptic packaging:

1. To enable containers to be used that are unsuitable for in-package sterilization;
2. To take advantage of high-temperature-short-time (HTST) sterilization processes which are thermally efficient and generally give rise to products of a superior quality compared to those processed at lower temperatures for longer times, and
3. To extend the shelf life of products at normal temperatures by packaging them aseptically.

The first aseptic packaging of food (specifically milk in metal cans) was carried out in Denmark by Nielsen prior to 1913 and a patent for this process (termed aseptic conservation) was granted in 1921. In 1917, Dunkley in the USA patented sterilization with saturated steam of cans and lids and subsequent filling of a presterilized product. In 1923 aseptically packaged milk from South Africa reached a trade fair in London, England in perfect condition. The American Can Company developed a filling machine in 1933 called the Heat-Cool-Fill (HCF) system which used saturated steam under pressure to sterilize the cans and ends. The sterile cans were filled with sterile product and the ends sealed on in a closed chamber which was kept pressurized with steam or a mixture of steam and air. Three commercial plants were built and operated on this principle until 1945.

In the 1940s Martin developed a process in which empty metal cans were sterilized by treatment with superheated steam at 210°C prior to being filled with cold, sterile product. In 1950 the Dole Company bought the first commercial aseptic filling plant on the market.

At the end of the 1940s a dairy enterprise and machinery manufacturer in Switzerland (Alpura AG, Bern, and Sulzer AG, Winterthur, respectively) combined to develop UHT-sterilized, aseptically canned milk which was marketed in Switzerland in 1953. However, this system was not economical, mainly because of the cost of the cans, and Alpura, in collaboration with Tetra Pak AB of Sweden developed an aseptic cartoning system. Long shelf life milk packaged in this manner was first sold in Switzerland in October 1961.

Today packaging of horticultural products has become an important process.

Fresh fruits and vegetables are essential components of the human diet as they contain a number of nutritionally important compounds such as vitamins which cannot be synthesized by the human body. A fruit or vegetable is a living, respiring, edible tissue which has been detached from the parent plant. Fruits and vegetables are perishable products with active metabolism during the postharvest period. The shelf life of fruits and vegetables can be extended by, in simple terms, retarding the physiological, pathological and physical deteriorative processes (generally referred to as postharvest handling) or by inactivating the physiological processes (generally referred to as food preservation). Packaging has an important role to play in maximizing the shelf life of both the handling and preservation approaches to shelf life extension; this chapter will focus most attention on the former since the packaging requirements of the preservation processes are not unique to horticultural products.

Difficulties arise when attempts are made to draw a clear line between fruits and vegetables. Fruits and vegetables cannot be described clearly botanically or morphologically as they encompass numerous organs in vegetative or reproductive stages and belong to a large number of botanical families. Fruits tend to be restricted to reproductive organs arising from the development of floral tissues with or without fertilization. Vegetables, on the other hand, consist simply of edible plant tissues. Thus, some fruits are included as vegetables and *vice-versa*.

All fruits and vegetables continue to lose water through transpiration after they are harvested, and this loss of water is one of the main processes that affects their commercial and physiological deterioration. If transpiration is not retarded, it induces wilting, shrinkage, and loss of firmness, crispness and succulence, with concomitant deterioration in appearance, texture, and flavor. Most fruits and vegetables lose their freshness when the water loss is 3–10 per cent of their initial weight. As well as loss of weight and freshness, transpiration induces water stress which has been shown to accelerate senescence of fruits and vegetables.

Packaging of dairy products also the most important process is taking an important role in the preservation of dairy products.

For much of the world, particularly the West, milk from cattle (*Bas taurus*) accounts for nearly all the milk produced for human consumption. The composition of milk reflects the fact that it is the sole source of food for the very young mammal. Hence, it is composed of a complex mixture of lipids, proteins, carbohydrates, vitamins and minerals. The water phase carries some of the constituents in suspension while others are in solution. The fat is suspended in very small droplets as an oil in water emulsion and rises slowly to the surface on standing, a process often termed "creaming." For a detailed discussion of the chemistry of milk components, the reader is referred to a standard text.

Milk is processed into a variety of products, all having different and varying packaging requirements. The simplest product is pasteurized milk where, after a mild heat treatment, the milk is filled into a variety of packaging media and distributed. The shelf life of such a product varies from 2–10 days. UHT milk is subjected to a more complex process and the packaging is also more sophisticated; its shelf life can be up to 8 months. Cream is typically processed and packaged in a similar way to fluid milk and has a similar shelf life. Fermented dairy products, although being subjected to more complex processing operations, are also typically packaged in an analogous manner to fluid milk.

The other dairy products (butter, cheese and powders) are quite different in nature to fluid milk and their packaging requirements are therefore also quite different. Each of these groups of dairy products will be discussed in turn.

Milk for liquid consumption is often standardized with respect to fat content, and homogenized to retard the natural tendency for the fat globules to coalesce and rise to the surface. In response to consumer needs, a range of non-standard fluid milk products has been developed in recent years, such products having varying (reduced) fat levels (the term skim milk is used for such products) and additives such as calcium or other nutrients, *e.g.,* the fat-soluble vitamins. From a packaging point of view, these non-standard milks can be treated analogously to standard milks.

This function of packaging is so obvious as to be overlooked by many, but it is probably the basic function of packaging. With the exception of large, discrete products, all other products must be contained before they can be moved from one place to another. The "package," whether it be a milk bottle or a bulk cement rail wagon, must contain the product to function successfully. Without containment, pollution would be widespread.

The containment function of packaging make a huge contribution to protecting the environment from the myriad of products which are moved from one place to another on numerous occasions each day in any modern society. Faulty packaging (or underpackaging) would result in major pollution of the environment.

This is often regarded as the primary function of the package: to protect its contents from outside environmental effects, be they water, moisture vapor, gases, odors, microorganisms, dust, shocks, vibrations, compressive forces *etc.*, and to protect the environment from the product. This is especially important for those products such as toxic chemicals which may seriously damage the environment.

In the case of the majority of food products, the protection afforded by the package is an essential part of the preservation process. For example, aseptically packaged milk and fruit juices in cartons only remain aseptic for as long as the package provides protection; vacuum-packaged meat will not achieve its desired shelf life if the package permits oxygen to enter. In general, once the integrity of the package is breached, the product is no longer preserved.

Packaging also protects or conserves much of the energy expended during the production and processing of the product. For example, to produce, transport, sell and store 1 kg of bread requires 15.8 mega joules (MJ) of energy. This energy is required in the form of transport fuel, heat, power and refrigeration in farming and milling the wheat, baking and retailing the bread, and in distributing both the raw materials and the finished product. To produce the polyethylene bag to package a 1 kg loaf of bread requires 1.4 MJ of energy. This means that each unit of energy in the packaging protects eleven units of energy in the product. While eliminating the packaging might save 1.4 MJ of energy, it would also lead to spoilage of the bread and a consequent loss of 15.8 MJ of energy.

Modern industrialized societies have brought about tremendous changes in life styles and the packaging industry has had to respond to those changes. One of the major changes has been in the nature of the family and the role of women. Now an ever-increasing number of households are single-person; many couples either delay having children or opt not to at all; there is a greater percentage than ever before of women in the work force.

All these changes, as well as other factors such as the trend towards "grazing" (*i.e.*, eating snack type meals frequently but on-the-run rather than regular meals), the demand for a wide variety of food and drink at outdoor functions such as sports events, and increased leisure time, have created a demand for greater convenience in household products: foods which are pre-prepared and can be cooked or reheated in a very short time, preferably without removing them from their primary package; condiments that can be applied simply through aerosol or pump action packages; dispensers for sauces or dressings which minimize mess, *etc.* Thus, packaging plays an important role in allowing products to be used conveniently.

Two other aspects of convenience are important in package design. One of these can best be described as the apportionment function of packaging. In this context, the package functions by reducing the output from industrial production to a manageable, desirable "consumer" size. Thus, a vat of wine is " apportioned" by filling into bottles; a churn of butter is "apportioned" by packaging into 25 gram minipats; a batch of ice cream is "apportioned" by filling into 2 liter plastic tubs.

Put simply, the large scale production of products which characterizes a modern society could not succeed without the apportionment function of packaging. The relative cheapness of consumer products is largely because of their production on an enormous scale and the associated savings which result. But as the scale of production has increased, so too has the need for effective methods of apportioning the product into consumer-sized dimensions.

An associated aspect is the shape (relative proportions) of the primary package in relation to convenience in use by consumers (*e.g.*, easy to hold, open and pour as appropriate) and efficiency in building into secondary and tertiary packages.

In the movement of packaged goods in interstate and international trade, it is clearly inefficient to handle each primary package individually. Here packaging plays another very important role in permitting primary packages to be unitized into secondary packages (*e.g.*, placed inside a corrugated case) and then for these secondary packages to be unitized into a tertiary package (*e.g.*, a stretch-wrapped pallet). This unitizing activity can be carried a further stage to produce a quaternary package (*e.g.*, a container which is loaded with several pallets). If the dimensions of the primary and secondary packages are optimal, then the maximum space available on the pallet can be used. As a consequence of this unitizing function, materials handling is optimized since only a minimal number of discrete packages or loads need to be handled.

There is an old saying that "a package must protect what it sells and sell what it protects". It may be old, but it is still true; a package functions as a "silent salesman." The modern methods of consumer marketing would fail were it not for the messages communicated by the package. The ability of consumers to instantly recognize products through distinctive branding and labeling enables supermarkets to function on a self-service basis. Without this communication function (*i.e.,* if there were only plain packs and standard package sizes), the weekly shopping expedition to the supermarket would become a lengthy, frustrating nightmare as consumers attempted to make purchasing decisions without the numerous clues provided by the graphics and the distinctive shapes of the packaging.

Other communication functions of the package are equally important. Today the widespread use of modern scanning equipment at retail checkouts relies on all packages displaying a UPC (Universal Product Code) that can be read accurately and rapidly. Nutritional information on the outside of food packages has becomes mandatory in many countries.

But it is not only in the supermarket that the communication function of packaging is important. Warehouses and distribution centers would (and sometimes do) become symbolic if secondary and tertiary packages lacked labels or carried incomplete details. UPCs are also frequently used in warehouses where hand-held barcode readers linked to a computer make stock-taking quick and efficient. When international trade is involved and different languages are spoken, the use of unambiguous, readily understood symbols on the package is imperative.

Some of the major hindrances to the market growth are due to food contamination, government regulations and emergence of local brands. he companies in this market are actively gauging the customers' insights and preferences to understand the market requirements and gain significant packaged food market share. Nestle was among the first companies to facilitate this process through its R and D centre in India. Moreover, in the processing food division, Tyson Foods had a tie-up with The Hillshire Brand Company, and collaborated with Godrej Foods India to develop new products. Kraft foods and the global ketchup giant, Heinz entered into a merger to expand their prospects in the packed food market.

Top Impacting Factors

Factors such as nutritious food, and innovative packaging technology affect the growth of the packaged food market positively. The top factors, which have an impact on the packaged food market, have been elaborated as follows in the descending order of their impact:

Nutritious Food

Nowadays, consumers prefer food products, which have high nutrition value, less preservatives, low cholesterol and low fat. Major players in this industry are coming up with new food products that fulfil these requirements of the customers. There is a marked growth in the adoption of organic food. Some of the food products manufactured by the companies are positioned as 99 per cent lactose and gluten free. Therefore, awareness towards proper nutrition has led to an increased consumer inclination towards packaged food.

Innovative Packing

A variety of packaging materials are available in the packaged food market in the form of paper packets, wooden packaging and stand-up pouches. Packaging is an integral method adopted by the leading manufacturers to ensure full protection of food products from being contaminated by the presence of external agents such as temperature, humidity, and microorganisms among others. Packaging also ensures that perishable food products are not damaged due to transportation. The materials, which are used for packaging the food, should be unreactive with the food product, and durable, capable of withstanding the effects of external factors. With companies trying to fulfil the consumer needs, the packaged food market is expected to witness a surge in years to come.

Multiple Food Options

Packaged food manufacturing companies offer a variety of options, from breakfast solutions to ready meals. The breakfast solutions include different types of breakfast cereals. The awareness among the consumers regarding the benefit and need of a healthy breakfast is affecting the multiple food options. In addition, companies are offering various snacks item for consumers such as Tortilla chips by General mills, which is marketed not as a dessert but as a healthy snack. The multiplicity of option for the consumers has therefore increased the target customer base in the packaged food market.

Social Media

The impact of social media is another factor driving the packaged food industry. The social media also plays a vital role behind the growth of packaged food industry. Manufacturers are investing huge sum of money on digital marketing for promoting their products on the web. The digital marketing is keeping track on the customer's preferences and type of interest on the web. Through social media, companies are making a network of customers having same preference and taste. This helps the companies in building customer base and increase product awareness.

Packaged Food Market Key Benefits

- The study provides an in-depth analysis of the World packaged food market with current and future trends to elucidate the imminent investment pockets in the market
- Current and future trends have been outlined to determine the overall attractiveness and single out profitable trends to gain a stronger foothold in the market
- The report provides information regarding key drivers, restraints and opportunities with impact analysis
- Quantitative analysis of the current market and estimations during the period of 2014-2020 have been provided to highlight the financial appetency of the market
- Porter's five forces model and SWOT analysis of the packaged food industry illustrate the potency of the buyers and suppliers participating in the market
- Value chain analysis in the report provides a clear understanding of the roles of stakeholders involved in the value chain

Packaged Food Market Key Segmentation

The market is segmented on the basis of product type and geography.

By Product Type

- Ready meals
- Baked foods
- Breakfast cereals
- Soups
- Baby Food
- Potato Chips
- Nuts
- Instant Noodles
- Pasta
- Biscuits
- Chocolate Confectionery
- Cheese
- Yogurt
- Ice Creams

- Sauces, Dressings and Condiments
- Non-alcoholic drinks

By Geography

- North America
- Europe
- Asia-Pacific
- LAMEA

Key Players

- Nestle S.A.
- Tyson Foods, Inc.
- Kraft Foods Group, Inc.
- General Mills, Inc.
- ConAgra, Inc.
- Smithfield Food, Inc.
- Kellogg's
- JBS Foods S.A.
- Hormel Foods Corporation
- Maple Leaf Foods, Inc.

Chapter 2

History of Food Packaging

Packaging is an essential part of the human history. Packaging is not a recent phenomenon. Packaging is an activity closely associated with the evolution of society and can be, traced back to human beginnings. Concept of packaging keeps on changing as per customer and goods requirement. The nature, degree and amount of packaging at any stage of a society's growth reflect the needs, cultural patterns, material availability and technology of that society.

A study of packaging's changing roles and forms over the centuries is a study of the growth of civilization. Social changes are inevitably reflected in the way we package, deliver and consume goods. We do not know what the first package was, but we can certainly speculate. Primitive humans: nomadic hunters/gatherers, lived off the land and social groupings were restricted to family units. Definitely, they would have been subject to the geographical migrations of animals and the seasonal availability of plant food. Such an extreme nomadic existence does not encourage property accumulation beyond what can be carried on one's back or head. Primitive people needed containment and carrying devices, and out of this need came the first "package"; a wrap of leaves, an animal skin, the shell of a nut or gourd, a naturally hollow piece of wood, the fire-bearer and the "packaging" of fire.

Changes in Primitive Packaging due to Social Structure Changes

Many societal changes led to the corresponding changes in packaging; mostly the quality and quantity of existing packaging practices. Since 5000 B.C., man demonstrated his ability to domesticate plants and animals. There was an area of reasonable food supply in a given vicinity; this evolutionary stage supported larger social groups, gave birth to small tribal villages. As such, storage and transport containers were needed for milk, honey, seed grains, nuts and dried meat. Villages with access to different resources traded with their neighbours, requiring transport containers. Hence, packaging necessity can be seen from very past era. About 250 B.C., the Greek city-state period, enacted law that affected packaging. By the time, early packaging materials consisted of fabricated sacks, baskets and bags, made from materials of plant or animal origin, a clay bowl, the fire-dried clay pots. Packaging which started from natural material such as leaves then after moved to wooden material and glass. The invention of glass blowpipe and wood barrels came to existence. Wood boxes replaced hollow logs. The first wooden barrel appeared possibly in the Alpine regions of Europe and was one of the first common packaging forms for many centuries.

The discovery of glass by 2500 B.C. brought about a hard inert substance in the fire's remains. It was melted limestone, sand, soda and silicate and shaped as glass packaging; the earliest hollow glass objects like glass beads and figures disappeared in Mesopotamia (Today's Iraq) and Egypt. Glass packaging initial begun to use in 1500 B.C. in Egypt. The Romans invented the glass blow pipe in about 50 B.C. The blowpipe's invention brought glass out of the noble households and temples.

In China, Ts'ai Lun is credited with making the first true paper from the inner bark of mulberry trees. The name 'paper' given to Chinese invention was made of matted plant fibers. Ancient printing was first seen in 768 with the oldest existing printed objects (Japanese/Buddhist charms). In 868, the oldest existing book (the Diamond Sutra) was printed in Turkistan. One of the oldest record found and understand paper were used to pack the dates. During 1035 a Parsain traveler observed during his global travel that vegetables, spice and hardware were wrapped in paper for customers.

With the advent of the Industrial Revolution which started in England in about 1700, the concept spread rapidly through Europe and North America. The Industrial Revolution was or is a concept of change that transformed people with peasant occupations and local markets into an industrial society with world-wide connections. This new type of society made great use of machinery and manufactured goods on a large scale for general consumption. The changes increased the demand for barrels, boxes, baskets and bags to transport the new consumer commodities and to bring great quantities of food into the cities. The fledgling packaging industry itself had to mechanize. It became necessary to devise ways of preserving food beyond its natural biological life.

During the Industrial Revolution, the evolution of selling and information had vital packaging roles. Bulk packaging was the rule, with barrel being the workhorse of the packaging industry. Flour, apples, biscuits, molasses, gunpowder, whiskey and whale oil were transported in barrels. Packaging served primarily to contain and protect. By this time, individual packaging was of little importance until the Industrial Revolution spurred the growth of cities. The first packaged retail products were medicines, cosmetics, tea, liquors and other expensive products; 'a paper of pins'.

The first artificial plastic was prepared by Alexander Parker in 1838 and then plastic packaging has begun to be used widely after 1950 and the growing period of plastic start end of 1970. The first packaging made of plastic (based on cellulose), was created in 1856. The package from time immemorial had motivational and informational role. The package had to inform the purchaser as well as the package had to sell the product. Metal was developed as the newest part of packaging and it was made in 1866 in USA. The first canned food came out in 1959.

Brand and Labelling

In early day packaging is mandatory to protect the goods and ease to carry from one point to another but as on when we move a head uses and need of packing is getting advanced. Packaging is being known as mode of marketing the product. And further it is acting as silent salesmen. A packaging milestone in 1877, brought about the concept of 'persona'. The Quaker personage, or 'persona', gave a description of the package or product as if it were a person. Between 1890 and 1920, decoration followed the art nouveaustyle, this being followed by a period of art deco graphics and designs. The evolving printing and decorating arts applied to 'upscale' packages, many early decorations was based on works of art or national symbols or images. Early labels included pictures of pastoral life, barnyards, fruits, and the gold medals.

Demographics which is considered, the study of population structure and trends, universally became an important factor in designing products and packages. Soon the packaging industry discovered that another criteria was fast food and other institutional markets. Fast food appeared and created a demand for disposable single service packaging. This introduced, two factors to influence packaging: public health care and a rapidly growing trend toward eating out rather than at home. More concern was given to the hospital, restaurant and institutional market as well as Petroleum-derived plastics which was now added to the package designer's selection of packaging materials.

Changing Needs and New Roles

All historical changes have had an impact on the way products are bought, consumed and packaged. The packaging professionals must always turn their attention to the needs, markets and conditions of tomorrow. It has been observed that most of the goods, not essential to survival, constitute 'the good life'.

In the second half of the 20th century, the proliferation of goods was so high that packaging was forced into an entirely new role: providing the major purchase motivation rather than presenting the goods itself; the only method of differentiating was the package itself. The trend towards more intensive marketing gave marketers the objective to aim at lifestyles, emotional values, subliminal images, features, and advantages beyond the basic product itself. Today the package has become the product, and occasionally packaging has become entertainment.

Globalization and the provision of increased tonnages of high quality food to massive city at affordable prices challenges packagers. A new concern is the removal of debris generated by a consumer society and the impact that these consumption rates have on the planet's ecology.

Today, products often are wrapped in multiple layers of packaging to get them safely from the point of manufacture to consumers' cupboards and refrigerators. Food packaging can improve food safety by alleviating bacterial contamination. It has been proposed that increased use of packaging for fresh produce could prevent contamination with *Salmonella* spp., a leading cause of foodborne diseases.[3]

Disposable food packaging is in great use in food service institutions. We appreciate that in modern era, food service packaging is very popular now a days because of the convenience it affords for our busy lifestyles, but often we forget about the main benefit *i.e.* sanitation. That is why single-use products were invented over 100 years ago to help stop the spread of contagious diseases.

Packaging and the Modern Industrial Society

Food is organic in nature. One characteristic of such organic matter is that it has a limited natural biological life. Packaging brings about freedom from geographical and seasonal food production. It is because of the fact that most food is geographically and seasonally specific. In a world without packaging, we would need to live at the point of harvest to enjoy the products, and our enjoyment of them would be restricted to the natural biological life span of each. Thus, it is by proper storage, packaging and transport techniques that we are able to deliver and enjoy fresh potatoes and apples derived from them, throughout the year and throughout the country. We are no longer restricted in our choice of where to live. Man is therefore free of the natural cycles of feast and famine that are typical of societies dependent on natural regional food producing cycles. All these comes from the art of organizing a central processing and the pre-packaged concept. The advantages of a central processing and pre-packaged food concept are varied and include the following:

- ☆ Central processing allows value recovery from what would be normally wasted.
- ☆ By – products of the processed food industry form the basis of other sub-industries.

World Packaging

Humankind's global progress is such that virtually every stage in the development of society, packaging is present somewhere in it.

1. **Packaging in developed nations:** Nations in this category are not just interested on the product but are equally preoccupied with packages and graphics. To agonize over choice of package type, hire expensive

marketing groups to develop images to entice the targeted buyer and spend lavishly on graphic.

2. **Packaging in less developed nations:** At the extreme, consumers will bring their own packages or will consume food on the spot, just as they did 2000 years ago. Packagers from the more developed countries sometimes have difficulty working with less developed nations. Firstly, they fail to understand that their respective and packaging priorities are completely different. Secondly, developing nations trying to sell goods to North American markets cannot understand their preoccupation with package and graphics.

Packaging and Environmental Issues: The Sources of Waste Material

A general perception is that, if only the packaging industry would stop doing something, or, conversely, start doing something, all our landfill and pollution problems would go away. Ample evidence suggests that good packaging reduces waste. The consumer sees packaging as that part of the shopping trip that gets thrown away. Hence, packaging is garbage after product purchase. The University of Tennesssee provided the following breakdown of total landfill waste; Residential waste: 37.4 per cent; Industrial waste, 29.3 per cent; Commercial waste, 27.3 per cent; and other sources, 6 per cent.

The four Rs hierarchy and what means

- **Reduce:** Use the minimum amount of material consistent with fulfilling its basic function.
- **Reuse:** Containers or packaging components should be reused.
- **Recycle:** Packaging should be collected and the materials recycled for further use.
- **Recover:** To possible recover other value from the waste before consigning packaging to a landfill.

Modern Packaging Industry

The highly sophisticated packaging industries which characterize modern society today are far removed from the simple packaging activities of earlier times. Globalization has played a significant role in packaging. Requirement of packaging is changing rapidly as per customer desire and affordability. And as per current requirement, packaging is vital to sale the product and retain the brand in market for longer time. In today's world, 21st century is all about information technology, *i.e.* it is the world for online shopping to attract online or mall culture shopping and packaging is more than products now. The display of product work more than the actual uses and ingredients of the product.

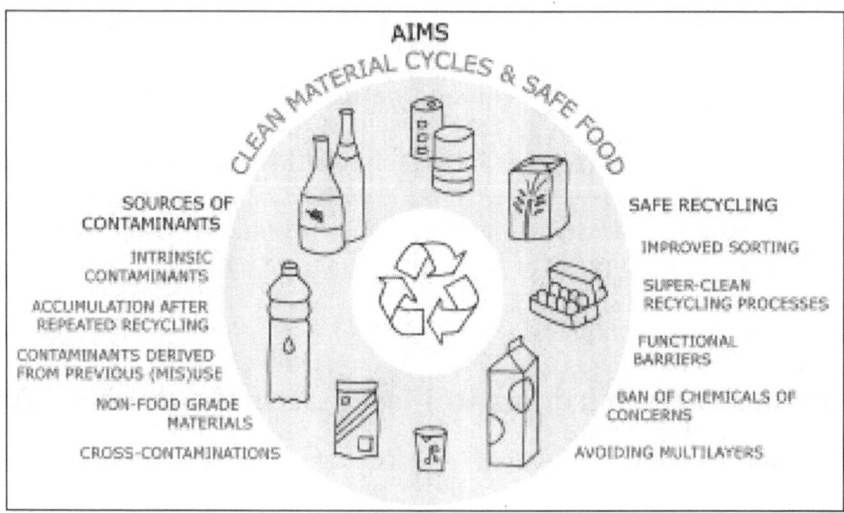

The modern packaging industry plays host to several emerging actors who influence the packaging industry in one or the other. It is in this modern age, that we use concepts such as converters in the packaging industry. Other concept worth noting include package users and suppliers.

Converters and Users: This is a new division in the packaging industry. Its objective is to take various raw materials and convert them into useful packaging materials or physical packages (cans, bottles, wraps). To this point, packaging becomes a materials application science. The company forming the physical package will also print or decorate the package.

Package 'users': It is a concept that describes the fact that, the firms that package products, are also regarded as part of the packaging industry, divided into a number of categories and each of these can be further subdivided.

The 'supplier': The supplier or manufacturer of machines for the user sector and suppliers of ancillary services, such as marketing, consumer testing and graphic design, are also important sectors of the packaging industry.

Professional Packaging Associations

There are a number of professional packaging associations whose objective is to design policies and regulate the nature, structure and composition of materials for packaging products in different countries across the globe. Some of them are:

- IoPP-Institute of Packaging Professionals
- PAC-Packaging Association of Canada
- PMMI- Packaging Machinery Manufacturers Institute
- FPA-Flexible Packaging Association
- WPO-World Packaging Organization

- ✰ Other organizations having a major impact on packaging activities are:
- ✰ ISO-International Organization for Standards
- ✰ ASTM-American Society for Testing and Materials
- ✰ TAPPI- Technical Association of the Pulp and Paper Industry
- ✰ ISTA-International Safe Transit Association

Health Implications

Concern over the health implications of packaging materials is on the rise. The latest fears cropped after Swiss scientists revealed a potential health risk in the mineral oils found in European recycled paperboard; exposure to such oils has been linked to inflammation of internal organs and cancer. While the researchers stress that exposure to such chemicals would have to occur over many years before health risks came to bear, UK cereal brand Jordan's has stopped using recycleboard, and Kellogg says it is working to minimize the mineral oil levels in its European packaging. Mineral oils are not a problem in US recycled board, according to Deborah White of the Recycled Paperboard Technical Association.

Its operation has a devastating effect on both the present and future. Thus, the need for sustainable packaging is there. The criteria for a sustainable packaging will therefore, include; recycled/renewable materials, compostable/biodegradable reusables format, source optimized, source reduced and consumer education.

Its operations has a devastating effect on both the present and future. Thus the need for sustainable packaging. The criteria for a sustainable package will therefore include; recycled/renewable materials, compostable/biodegradable,reusable format, source optimized, source reduced and consumer education. Sustainable packaging describes the following Remove, Reduce, Recycle, Renew, Re-use. These are five of the original7 R's of Sustainable Packaging revealed by Wal-Mart when it introduced its Packaging Scorecard. Sustainable packaging is the development and use of packaging which results in improved sustainability. This involves increased use of life cycle inventory (LCI) and life cycle assessment (LCA) to help guide the use of packaging which reduces the environmental impact and ecological footprint.

Chapter 3

Functions of Food Packaging

For millennia, humans stored their food in containers they found in nature *i.e.* dried gourds, shells, hollow logs, leaves as well as baskets and pottery. By the first century BC, the Chinese were wrapping foods with treated tree bark and other forerunners of paper. Centuries later, Napoleon Bonaparte used some of the first mass-produced canned food to feed his troops in the Franco-Austrian War of 1809. Plastics were discovered in the decades following that innovation but were not used beyond military purposes until well into the twentieth century.

In modern era, lot of improvement and innovations have taken place in the area of art, science and technology related to food packaging *e.g.* packaging safe for microwave cooking, disposable food packaging for ready-to serve foods, food packaging suitable for different food products based on animal or plant origin *etc.* have been evolved keeping into mind the globalization, modernization, urbanization, changing and busy life styles of families and food habits of the population. Moreover, people are getting more aware and conscious about their health. That is why, they are opting for packaged food. Now a days, the food processing and packaging industry has become an important global business earning multi-trillion dollars because of the rise in the living standards of the residents of developing countries.

Therefore, in today's society, packaging is pervasive and essential. It surrounds, enhances and protects the goods we buy, from processing and manufacturing through handling and storage to the final consumer. Without packaging, materials handling would be a messy, inefficient and costly exercise, and modern consumer marketing would be virtually impossible.

Packaging is best described as a coordinated system of preparing goods for transport, distribution, storage, retailing, and use of the goods. It is a complex, dynamic, scientific, artistic, and controversial business function. Fundamental functions of packaging: contain, protect/preserves transports and informs/sells. Packaging functions range from technical ones to marketing oriented ones.

Functions of Food Packages

Technical Functions		Marketing Functions	
Contain	Measure	Communicate	Promote
Protect	Dispense	Display	Sell
Preserve	Store and Convenience	Inform	Motivate
Technical packaging professionals need science and engineering skills.		Marketing professionals need artistic and motivational understanding	

Different packaging materials offer different advantages. Glass preserves taste well and is chemically inert. Paper and paperboard are economic to produce and easy to print on. They are also lightweight, which reduces the fuel used for the transport of goods.[7] Steel and aluminum offer the advantages of malleability, impermeability, and ease of recycling. Aluminum can also be bound to paper or plastic films for more versatility in the types of packaging that can be produced. And plastics have re

Food Packaging serves several functions, which are described as under:

Protection

This is often regarded as the primary objective and function of the food package. The protection function of package represents preventing all outside forces to intervene with the product inside. Packaging helps to protect the contents of the food package from various outside environmental effects like vibration, shock, compression, temperature, infestation *etc.* which can cause food spoilage or deterioration as a result of physical damage, chemical changes or biological damage. Certain products also require special temperature or humidity levels to preserve their shape or purpose, therefore package should protect the contents from changing their nature.

Food package keeps the product intact and protects the food contents from moisture and toxins, prevents their spillage and tampering, and helps in retaining their shape and quality. Packaging is required to provide a barrier from oxygen, dust, water vapor *etc.* It prevents mechanical damage due to the hazards of distribution. For example, the products like biscuit, jam, chips, *etc.*, need to be protected from environmental contact. In order to provide sufficient package that will protect the products, it is necessary to understand the product's characteristics and address potential hazards that would tamper those. It often proves efficient to use a secondary package when protecting larger quantities of products together and transporting them. Hence, keeping the contents of the food fresh, clean, wholesome, and safe for the required shelf life is a basic function of the packaging.

Containment

The role of containment is to conceal the product and its parts and prevent them from spillage and loss, starting from the packing line through transportation phases until it arrives to customer's home. Hence, all products must be contained before they can be moved from one place to another. The "package", whether it is a bottle of cola or a bulk cement rail wagon, must contain the product to function successfully.

Food packaging plays a vital role in preserving and transporting food items to their desired location without affecting the taste or quality till its use. Small food items are typically grouped together in one package for efficiency. Containment depends on the product's physical form and nature. Some goods can have dangerous consequences when not contained properly inside a package especially liquids and chemicals. For Example, a hygroscopic free-flowing powder or a viscous and acidic tomato concentrate need different types of containments. Also, if a product which consists of multiple parts does not contain all of them or the instruction for use when customer opens it, it degrades the product proving it unusable. Every type of package has its sensitive areas which may get damaged and cause loss of containment and therefore, producers enforce protocols and testing of package to ensure that the contents are properly contained.

The containment function contributes to protect the environment. Without containment, product loss and pollution would be wide spread. Faulty packaging could result in major pollution of the environment.

Preservation

This function is not universal for all the goods, it should be considered especially when dealing with food products, pharmaceutics and other perishable products. The importance of preservation is to keep products in controlled environment, prevent or inhibit chemical changes, biochemical changes and microbiological spoilage so that food product remains safe to use for longer time period.

The key to correct preservation is to package the product while it is in safe condition and it is imperial to understand how can this state be sustained inside the package by defining the product-spoilage mechanisms. The package attributes therefore have to address potential causes of spoilage in order for the product to maintain in desired state. Preservation is necessary for products that need to maintain certain levels of oxygen, moisture, volatiles or they are light sensitive.

Communication

A package functions as a "silent salesman" due to any special instructions or information provided by it to the consumer. It acts as a modern marketing medium for sellers to attract consumers worldwide. The modem methods of consumer marketing would fail were it not for the messages communicated by the package. The ability of consumers to instantly recognize products through distinctive branding and labeling regarding material type, shape, size, colour, merchandising display enables supermarkets to function on a self-service basis.

Marketing

Packaging is an important tool for advertisement and marketing. Based on the latest advanced technology, food manufacturers, retailers and packaging products suppliers are working more closely than ever before to produce packaging products designed for modern lifestyles. This has led to the emergence of varied hi-fi packaging designs and packaging products, especially for food packaging, in the market. There has been explosion of ready to eat meals in the market. All food packaging highlights freshness, health and traceability of the products. Packaging has become an important part of the value chain analysis, regarding food safety.

Proper food packaging with proper information plays a vital role in encouraging potential customers to buy the product. In this case, the food packaging design plays an equally important role. The more attractive the design is, the more is its chances of buying.

Packaging protects the interests of consumers. The information on the packaging includes quantity, price, additives, ingredients, inventory levels, lot number, size and weight *etc.* which is very important for merchandising.

The use of food packaging is a socioeconomic indicator of increased spending ability of the population, an increase in the gross domestic product or an increase in food availability. Packaging technology must balance food protection with other social and environment issues, including energy and material costs, heightened social and environmental consciousness, and strict regulations on pollutants and disposal of municipal solid waste.

Convenience

Based on a fact that package is a tool which helps goods remain in desired condition when reaching consumers, it also should be convenient for them to carry, transport and open the package while maintaining it safe. Food packages provide convenience to the pack handlers and user(s) throughout the packaging chain. They have certain features which add convenience in access, handling, stacking, distribution, sale, display, re-closing, opening, use, and reuse, product visibility, resealability and ease of disposal.

As known from retail assortment theory, customers make decisions while shopping and the fundamental choice is which size of product they will buy according to their needs and convenience. Certain food companies provide retailing solution based on sophisticated data from local customer researches. They also offer consultancy services for large retailers or supermarkets where they assort the goods based on psychology theory of customer behaviour. One of decisions customers tend to make while shopping is choosing the amount and also size of package either by price or their preference. To portray an example, when a person needs to buy eggs for a 6-member family that appreciates omelette, it is unlikely that they would pick the package of 6 eggs. Another available option, package of 12 or more eggs is much more relevant and convenient. Another example of convenient package is when customer purchases drink inside tin can for immediate consumption, however he would rather choose a bottled drink which can be re-opened when planning to continue drinking in a period of several hours. Different packages offer variety of uses. These examples represent common situations when convenience is truly important and therefore, majority of manufacturers offer their products in various sizes and shapes to satisfy needs of their customers.

The convenience function extends also to secondary package, the pallets or boxes used when transporting goods are often determined by equipment and the ability to fit primary packages into secondary one without wasting space based on resource utilization and easy handling. To improve food quality, intelligent packaging provides optimization and enhancement of certain foods. the intelligent packaging can include gas absorption technology or gas emitting to extend shelf life or improve food quality. Another example of intelligent packaging is including metal structures in package which interact with microwave activity to ensure the food would become crispy.

Security

Proper packaging can play an essential role in reducing the security risks of shipment. With tamper evident features, there is no chances of the food packages being destroyed or contaminated during shipment.

Portion Control

Lastly, food packaging provides information on the exact portion of contents available in that particular product.

Label Information

All food products packages require labels which provide information on the product ingredients, nutrition facts, how to use, dispose, date of manufacturing and expiry of the package or product, product legal requirements *etc.*

Information

There are several categories of information included in every package:

- Tracking information or traceability
- Product information
- Marketing and brand information

Traceability is very important not only for food safety but also for the authenticity of the product as one can track any food through all stages of production, processing and distribution. It is considered to be the entire information for each step in the process chain. For this, barcoding and radio frequency identification (RFID) systems are used; however, they might be also triple dimensional or chips with radio frequencies. It includes the metadata of product. Radio frequency identification can easily monitor and track the temperature, humidity and so on and manual scanning is not required in it. This information is not available for end-users of product, on the contrary it is essential in order for a product to reach customer. Special equipment such as scanners or detectors are needed to access the information which consists of manufacturer, basic description, package dimensions and other data regarding product that are not likely to change over longer period of time. Codes tend to vary across the world, for example UPC code is usually used in USA while EAN code is used primarily in European retail. Considering that radio frequency technology is known for more than hundred years, the use in identifying dates to Mr Charles A. Walton who patented a radio frequency identifier in 1980. It still remains quite pricy, even though the price of tag decreased from $1 (in 2003) to around 10 cents (Forbes, 2017). The RFID tags are mostly used for high value items while the warehouses and transportation companies are required to possess complex technology to access the information coded into chip which contains extensive amount of data compared to bar codes. The universal tags also diminish language barriers across supply chain and they revolutionize it.

Product information consists of data describing ingredients and nutrition volumes of food products or information with technical details for electronic goods. There are several regulations on product information by government in majority of countries. One of essential information on each product is the country of origin, manufacturer and distributor. Each product requires different information based on its nature and purpose. During recent years, society became intensively interested about products, the content of each products and materials, moreover, people became more dependent on other person's opinions about products called reviews which are shared online on websites and forums. Therefore, manufacturers chose to meet the new demand for information by attaching and printing smart codes (QR codes) which can be scanned by smart phones and then detailed product's description would appear on screen. This technology enables consumers to access more data about goods they seek without assistance of store workers.

Obviously, every manufacturer wants consumers to continue shopping products from their assortment and that is the reason why every package in retail store has a distinctive brand logo. Logos, product names, brand slogans, these are all tools used by companies to signal to customers which product they buy. In every retail store, people can choose from tens of different brands in every product category, which makes the competition among brands to impress customer even tougher. Sometimes, even the package itself can be brand-specific so customers can recognize it quicker. A good example provides a company manufacturing potato crisps Pringles. Pringles, unlike their competitors, chose to package crisps neatly into carton tubes instead of traditional plastic or foil bags. Another important communication distinction on the package is colour. Brands tend to follow strict colour code for their product lines for easier recognition. In addition, colour is one of tools to describe a sub category of products as is common in milk section: blue package means low fat milk, while red package suggests the full fat.

Other Functions

Besides traceability, tamper indication is also said to be one of the secondary functions of increasing importance.

Food tampering *i.e.* the intentional contamination of a food product, with intent to cause harm to the consumer or to a private company, is easily visible to the consumer. There are several measures to detect tampering, including banding, special membranes, breakaway closures, special printing on bottle liners or composite cans such as graphics or text that irreversibly changes upon opening and special printing that cannot be easily duplicated.

Packaging is the cornerstone of the food processing industry. The processing and packaging industries transform the food from one stage to another by using different types of techniques and methods. Hence, it must be driven by health and hygienic factors, food safety, quality, fresh tasting and balanced nutrients. Beside these, it must include toxin removal, marketing and distribution, maintenance of taste, year-round availability and long shelf life.

Despite the importance and key role which packaging plays, it is often regarded as a necessary evil or an unnecessary cost. Furthermore, in the view of many consumers packaging is, at best, somewhat superfluous, and, at worst, a serious waste of resources and an environmental menace. Such a viewpoint arises because the functions which packaging has to perform are either unknown or not considered in full. By the time most consumers come into contact with a package, its job in many cases is almost over, and it is perhaps understandable that the view that excessive packaging has been used, has gained some credence.

Packaging provides four basic functions (Brody *et al.,* 2008; Marsh and Bugusu, 2007; Robertson, 2006):

1. **Containment.** The containment function involves the ability of the packaging to maintain its integrity during the handling involved in filling, sealing, processing, transportation, and dispensing of the food (Krochta, 2007).
2. **Protection.** The need for protection depends on the food product but generally includes prevention of three major classes of external influences: chemical, biological, and physical. Chemical protection minimizes compositional changes triggered by environmental influences such as exposure to gases, typically oxygen, which can allow oxidation of lipids, flavors, colors, vitamins, *etc.*; moisture (loss or gain, which affects microbial growth, oxidation rates, and food texture), aroma (loss or gain), heat, and light (Brody *et al.,* 2008; Marsh and Bugusu, 2007). Biological protection provides a barrier to microorganisms (pathogens and spoiling

agents), insects, rodents, and other animals, thereby preventing I. INTRODUCTION 6 disease and spoilage. Physical protection shields food from mechanical damage and includes cushioning against the shock and vibration encountered during distribution, abrasion, fracture, and/or crushing (Krochta, 2007; Marsh and Bugusu, 2007).

3. **Communication.** The information that a package provides involves meeting both legal requirements and marketing objectives (Krochta, 2007). Food labels are required to provide information on the food processor, ingredients (including possible allergens), net content, nutrient content, and country of origin. Package graphics are intended to communicate product quality, branding, and sales promotion, and thus sell the product. Bar codes and radio-frequency identification tags (RFID) allow rapid check-out and tracking of inventory.
4. **Convenience.** Providing convenience (sometimes referred to as utility of use or functionality) to consumers has become a desired function of packaging (Krochta, 2007). Range of sizes; easy handling and disposal, easy opening and dispensing; portion control; reclosability, product visibility; and food preparation in the package (ovenability/microwavability) are examples of packaging designs that provide convenience to the consumer (Han, 2014; Marsh and Bugusu, 2007).

In recent years, in order to fulfil the requirements of the modern food industry, demanding consumers, and new legislation related to food safety and the environment, novel food packaging technologies have been developed. Active and intelligent packaging systems, modified atmosphere packaging, and edible films and coatings are among the most promising ones.

Levels of Packaging

1. Primary package: It is in direct contact with the product. For example, a product in can (canned food).
2. Secondary package: It contains number of primary packages. For example, a fibre board box containing several of cans.
3. Tertiary package: It is made up of number of secondary packages. For example, a stretch-wrapped pallet of fibre board boxes of canned food.
4. Quaternary package: It is used to facilitate handling (during transportation, shipping) of tertiary packages. For example, large container containing number of stretch-wrapped pallets of fibre board boxes of canned food.

Packaging Chain (Distribution Chain)

Food production → Packaging operation (primary packaging, secondary packaging) → Palletizing and warehousing → Transportation for distribution and delivery to market (distributor → wholesaler→ retailer) → Home or consumer → Product use/consumption

Hazards

Food during the time elapsed between production and ultimate consumption is exposed to various hazards (agents or conditions) which make the food unavailable for consumption or significantly affect the quality of food (by damage and any other harm).

Type, Factors and Effects of Hazard

1. **Physical Hazard**
   - Operations: handling during warehousing, transportation, distribution
   - Actions: stacking, loading and unloading in/from transport vehicles, carrying
   - Factors and effects: dropping, tearing, crushing by rope by tying, compression, stack impact (force) resulting in breakage, leakage, burst, bruising, distortion (deformation), crushing, spilling and ultimately loss of food

2. **Mechanical Hazard**
   - Operations: handling during packing, warehousing, transportation, distribution
   - Actions: stacking, lifting, loading, unloading, transportation
   - Factors and effects: piercing, puncturing and tearing by hooks, straps, nails; dropping, bumping, bounching, vibration (engine of truck, rail, aeroplane) and impact resulting in breakage, distortion, spilling and exposure to adverse climate leading ultimately to loss of food.

3. **Biological Hazard**
   - Pests (rats, rodents, moths, *etc.*) and microorganisms (molds, *etc.*) are the causing factors
   - Operations: handling during warehousing, transportation, distribution
   - Effects: different types of spoilage, insect infestation and loss of food.

4. **Climatic Hazard**
   - Operations: handling during warehousing, transportation, distribution
   - Factors: rain, sunlight, oxygen and other gases, high and low temperature, high and low humidity, high and low pressure, chemical pollutants like sulphites, chlorides and acids
   - Effects: extreme climatic conditions produce stress and results in stress crack and other kind of damage

5. Miscellaneous Hazards

Caused by fire, floods, tampering, pilferage, foreign odours, corrosive chemicals, contamination by other commodities stored together

Functions of Packaging

The primary functions of packaging are:

- ☆ Containment: The package contains its content secured within itself.
- ☆ Protection: The package provides protection to the contained food from physical, mechanical, biological, climatic and miscellaneous hazards during warehousing, loading, unloading, transportation and distribution.
- ☆ Preservation: The package prevents spoilage of food by post-process contamination of microbiological agents, prevents microbial growth by not providing suitable conditions for growth, checks abiotic deterioration caused or accelerated by agents such as oxygen, humidity, light *etc.* (lipid oxidation, oxidation of food constituents, maillard reaction, texture changes, *etc.*).
- ☆ Information (communication): The package provides information to consumer about the product. The information includes the net weight, ingredients used, nutritional composition, price, ideal storage conditions, name and address of manufacturer, date of manufacture and batch number, minimum and maximum storage or shelf life *etc.*
- ☆ Convenience: Packaging provides convenience (comfort) in terms of carrying, storing and handling of foods.
- ☆ Sales promotion: An attractive package (attractive design, colour, graphics) is by itself an advertisement of the product and enhances product sales.

Desirable Features of Packaging

- ☆ Product-package compatibility: The package serves all necessary purposes, no undesirable changes occur in food due to the reaction between the product and package
- ☆ Disposability and environment-friendly: Reuse and recycle as much as possible
- ☆ Machinability: Good performance in filling and closing in high speed machines
- ☆ Convenience to use in terms of opening, dispensing and re-closing

Packaging Design

Designing of successful packaging requires following prerequisite knowledge and must be considered:

1. **Product assessment**
   - ☆ Nature and characteristics (solid, liquid, paste, crystal, powder *etc.*), chemical composition, size, shape, weight and density of product
   - ☆ Nature and extent of deterioration (physical breakdown, seepage, sensitivity to sunlight, moisture, high and low temperature, oxygen, insect infestation *etc.*)

2. **Hazards of Distribution**
   - ☆ Method (road, rail, sea, air) and duration of transport
   - ☆ Type and intensity of hazard during distribution
   - ☆ Kind of impact on product and package (compression, drop, vibration, puncturing *etc.*)

3. **Marketing Requirement**
   - ☆ Product size, quantity per package, weight and shape for sale
   - ☆ Package for enhancing product image (self advertisement)
   - ☆ Convenience of carrying, opening, dispensing, storing *etc.*
   - ☆ Price

4. **Selection of Packaging Materials and Machinery Considerations**
   - ☆ Properties of packaging materials (barrier properties, strength properties, machinability, product and package compatibility)
   - ☆ Product characteristics
   - ☆ Market requirement and economic considerations
   - ☆ Suitability of machinery to handle the packaging materials
   - ☆ Efficiency of packaging line to cope with the designated size and shape of package
   - ☆ Efficiency of printing process

Containment

Containment means simply, to contain products to enable them to be moved or stored. It is easily overlooked. However, containment is a key factor for all other packaging functions. All products must be contained for delivery from their point of production to their ultimate destination. Without packaging, products are likely to be lost or contaminated by the environment. For this reason, we have actually used packaging for ages. Early packaging such as animal skins, baskets, or leaves from trees were used to contain liquids, powders, grains, *etc.* The containment function significantly contributes to protecting and preserving products during their distribution (Joongmin S, 2014).

Food products must be contained before they can be moved from one place to another. The containment function of packaging makes a huge contribution to protecting the environment from the myriad of products that are moved from one place to another on numerous occasions each day

Protection/Preservation

There are two broad types of damage that fresh and processed foods sustain during storage and transportation. One is physical damage such as shock, vibration, compressive forces, *etc.* The other is environmental damage that occurs due to exposure to water, light, gases, odors, microorganisms, *etc.* A good packaging system will protect or reduce these types of damage to the package contents. For example, an essential aroma or flavor in coffee or juice may easily be evaporated or oxidized without optimum barrier packaging. A shelf-stable food in a can or pouch may maintain its stability (especially against microorganisms) as long as the package provides protection. However, in the case of fresh food products, the idea protection is usually hard to achieve with packaging alone. Since temperature is a major influence on the degradation of food, it is more economical to control temperature through supply chain modification (refrigeration, freezing, *etc.*). However, packaging can also add a certain level of protection to slow down temperature change.

Freedom from harmful microbial contaminants at the time of consumption can also be influenced by the package. First if the packaging material does not provide a suitable barrier around the food, microorganisms can contaminate the food and make it unsafe. Microbial contamination can also arise if the packaging material permits the transfer moisture or oxygen (O_2) from the atmosphere into the package. In this situation, microorganisms present in the food but posing no risk because of the initial absence of moisture or O_2 may then be able to grow and present a risk to the consumer.

Communication

There is an old saying that "a package must protect what it sells and sell what it protects"; that is, the package functions as a "silent salesman." The modern methods of consumer marketing would fail if messages communicated by the package through distinctive branding and labeling, enabling supermarkets to function on a self-service basis is not correctly done. Consumers make purchasing decisions using the numerous clues provided by the graphics and the distinctive shapes of the packaging. Other communication functions of the package include a Universal Product Code (UPC) that can be read accurately and rapidly using modern scanning equipment at retail checkouts, nutritional and ingredient information (including E-numbers for additives), and country of origin.

Utility

The utility function encompasses all the packaging attributes that provide added value and convenience to the users of the product and/or package. For example, an important social trend is the growing number of mothers in the workforce and smaller households (people living alone and married couples without children). Unquestionably, food products that offer simplification and convenience have grown in popularity with this group; examples include microwavable entrees, steam-in-pouch vegetables, oven-safe meat pouches, pump-action condiments, and so on.

Chapter 4

Classification of Materials for Food Packaging

In recent decades, packaging and containers have become an essential element in food purchases. The food is **packaged and packed** with the aim of being transported and stored. That is, not only is it a container, but the container must protect what it sells and sell what it protects. There is a wide scope of packaging materials and packaging formats that are usually used in processed food industry to handle, store, and distribute processed food products, they play an important role in how these products reach the consumers in a safe and wholesome form without compromising quality. From the business perspective, the appearance of packaging is particularly important because it identifies the product in the distribution chain and differentiates it when it reaches the consumer.

Types of Packaging Materials and Formats Used in the Food Industry

The relationship between the food and contact with the packaging material continuously interact and contribute to changes that can occur over time in these products. It is therefore important that several factors are considered when choosing the right package for a particular food product. Selecting the suitable material for packaging a certain type of food depends on the functions that the package is supposed to fulfil. These functions include shielding the foods against moisture, temperature variations, oxygen, light, and biological microorganisms. Also, damage protection, permeability, food identification, and chemical and optical properties play a significant role in the material selection. Materials used in food package design and construction play a significant role in determining the shelf life of a food product also. The right selection of packaging materials and technologies maintains product quality and freshness during distribution and storage. Therefore, the material used depends on the nature of the food product because different packaging materials possess a range of performance characteristics that exert significant impacts on shelf-life. A wide range of packaging materials and packaging formats are used in the fresh and processed food industry to handle, store, and distribute fresh and processed food products, from farm to the consumer.

Materials that have been used in food packaging include wood, glass, metals (aluminum, foils and laminates, tinplate, and tin-free steel), paper and paperboards, and plastics. Today's food packages often combine several materials to exploit each material's functional or aesthetic properties. Moreover, a wider variety of plastics have been introduced in both rigid and flexible forms. Bottles and glass jars are often used for packaging liquid food stuff while solid food products are mostly packed on plastics and cardboards. Processed fruit and vegetables are usually packed in airtight metal containers to prevent oxygen transmission that might lead to spoilage of the product through microbial growth and oxidation of lipids. According to the WPO (2008), the most important consumer packaging are made of paper and board (38 per cent), followed by plastic (30 per cent) with rigid plastics alone taking an 18 per cent share, metal (19 per cent), glass (8 per cent), and others (5 per cent). Moreover, approximately 70 per cent of overall consumer packaging are used in food industry where 48 per cent of all the packaging are made from paperboard.

1. Traditional Packaging Materials

These materials have been used since the earliest times for domestic storage and local sales of foods. However, with the exception of glazed pottery, they have poor barrier properties and are only used to contain foods and keep them clean. They are also unsuited to the needs of commercial production processes and are considered by many customers to be less attractive than the newer 'industrial' materials' which have also been described below.

Leaves

Leaves are cheap and readily available, and are used as wrappers for products such as cooked foods that are quickly consumed. **Banana or plantain leaves** are used for wrapping traditional cheese and fruit confectionery such as guava cheese. **Cornhusk** is used to wrap corn paste or block brown sugar, and cooked foods of all sorts are wrapped into leaves. **'Pan' leaves** are used for wrapping spices in India; they are an excellent solution for products that are quickly consumed, as they are cheap and readily available.

Other examples are green coconut palm, papyrus leaves and bamboo and rattan fibres, which are woven into bags or baskets and used for carrying meat and vegetables in many parts of the world. **Bamboo and rattan** are widely used materials for basket making. Bamboo pots, cut out of the bamboo stem are also found. Palmyra palm leaves are used to weave boxes in which cooked foods are transported, and small banana leaf bags are used to contain coffee beans that are a traditional gift in some parts of Africa. Some of these have the potential to be developed as niche packaged products for tourist markets.

Coconut Palm

Green coconut palm and papyrus leaves are frequently woven into bags or baskets, which are used for carrying meat and vegetables in many parts of the world. Palyra palm leaves are used to weave boxes in which items such as cooked foods are transported.

Vegetable Fibres

They can also be spun into a yarn that is fine enough to make coarse canvas. These natural raw materials are converted into fibres to produce the yarn, string or cord for packaging materials. Fibres from kenaf and sisal plants are mainly used for making ropes, cord and string, which can be made into net bags to transport hard fruits. Other examples of textile containers are woven jute sacks, which are used to transport a wide variety of bulk foods including grain, flour, sugar and salt. Plant fibre sacks are flexible, lightweight and resistant to tearing, have good durability, and may be chemically treated to prevent them rotting. Their rough surfaces are non-slip, which makes stacking easier compared to synthetic fibre sacks, and they are bio-degradable. Most textile sacks can be re-used several times after washing and they are easily marked to indicate the contents. They are still widely used to transport fresh or dried crops, but they are being replaced as shipping containers by woven polypropylene or multi-walled paper sacks. Calico is a closely woven, strong cotton fabric that can be made into bags for flour, grains, legumes, coffee beans and sugar.

Such materials, although categorized by the nature of the constituent fibre, have certain common characteristics. They are very flexible, to some extent resistant to tearing and permeable to water and water vapour. Their lightweight is an advantage in handling and transport. The rough surface makes stacking easier in comparison to man-made fibre sacks, which slide due to their smooth surface. Another difference with man-made fibres is that the natural raw materials are bio-degradable when left in their pure state. However, they rot when moist limiting the number of times that they can be re-used.

Treated Skins

Leather has been used for many centuries as a flexible, lightweight and non-breakable container or bottle. Water and wine are frequently stored and transported in leather containers (camel, pig and kid goat hides). Manioc flour and solidified sugar were also packed in leather cases and pouches, but the use of leather has now ceased for most commercial food applications.

Earthenware

Earthenware is used worldwide for storage of liquids and solid foods such as curd, yoghurt, beer, dried food, honey, *etc.* Corks, wooden lids, leaves, wax, plastic sheets, or combinations of these are used to seal the pots. If well sealed, it is a gas, moisture and lightproof container.

Unglazed Earthenware

Unglazed earthenware bowls or pots are porous and the evaporating moisture makes them suitable for products that need cooling *e.g.* curd. They are still used for local sales of curd or yoghurt in parts of Asia. Glass or plastic containers have largely replaced pottery because of its high weight, fragility, variability in volume when hand-made, and the difficulty of adequately cleaning pottery containers for re-use.

Glazed Pots

They are better for storing liquids *e.g.* oils, wine, as they are moisture proof and airtight, if properly sealed. All are lightproof and if clean, restrict the entry and growth of micro-organisms, insects and rodents. One should ensure that the glazing of the earthenware does not contain lead. Most traditionally glazed pots do have lead glazings which, although they are not really harmful for serving coffee or soup, should not be used for acid drinks and other products which are to be stored for a long time.

Wood Containers

Wooden shipping containers have traditionally been used for a wide range of solid and liquid foods including fruits, vegetables, tea and beer. Wood offers good mechanical protection, good stacking characteristics and a high weight-to-strength ratio. However, plastic containers have a lower cost and have largely replaced wood in many applications. The use of wood continues for some wines and spirits because the transfer of flavour compounds from the wooden barrels improves the quality of the product; but in other applications have been replaced by aluminum, coated steel or plastic barrels.

Wooden tea chests are produced more cheaply than other containers in tea producing countries and are still widely used. Small wooden boxes are used to pack tea or spices for tourist markets now also in some countries. Wooden crates are imported for transporting fresh fruit and vegetables, fish *etc.* They are used to hold foods together and protect them from crushing, but otherwise offer little protection. However, in most applications, plastic containers have a lower cost, are more easily cleaned for reuse, do not risk contaminating foods with splinters.

Wooden Crates and Lugs

Wooden crates, once extensively used for apples, stone fruit, and potatoes have been almost totally replaced by other types of containers. The relative expense of the container, a greater concern for tare weight, and advances in material handling have reduced their use to a few specialty items, such as expensive tropical fruit. The 15-, 20-, and 25-pound wooden lugs still used for bunch grapes and some specialty crops are being gradually replaced with less costly alternatives.

Wooden Baskets and Hampers

Wire-reinforced wood veneer baskets and hampers of different sizes were once used for a wide variety of crops from strawberries to sweet potatoes. They are durable and may be nested for efficient transport when empty. However, cost, disposal problems, and difficulty in efficient palletization have severely limited their use to mostly local grower markets where they may be re-used many times.

2. Glass

Glass is defined as "an amorphous inorganic product of fusion that has been cooled to a rigid condition without crystallizing". Glass has an extremely long history in food packaging; the 1st glass objects for holding food are believed to have appeared around 3000 BC. Glass is made principally of silica, derived from sand or sandstone. For most glass, silica is combined with other raw materials in various proportions. For example, soda-lime glass, the glass typically used for food packaging, contains silica (68–73 per cent), limestone (10–13 per cent), soda ash (12–15 per cent), and alumina (1.5–2 per cent). The production of glass containers involves heating a mixture of silica (the glass former), sodium carbonate (the melting agent), and limestone/calcium carbonate and alumina (stabilizers) to high temperatures until the materials melt into a thick liquid mass that is then poured into molds. Recycled broken glass (cullet) is also used in glass manufacture and may account for as much as 60 per cent of all raw materials. Glass containers used in food packaging are often surface-coated to provide lubrication in the production line and eliminate scratching or surface abrasion and line jams. Glass coatings also increase and preserve the strength of the bottle to reduce breakage. Improved break resistance allows manufacturers to use thinner glass, which reduces weight and is better for disposal and transportation. For food packaging, bottles or jars are the types of glass packaging mostly used, mostly for packaging processed foods especially where moisture and oxygen barrier is of great importance. Glass is inactive to a wide variety of food and non-food products, very rigid and strong against pressure, transparent and has excellent barrier properties. Carbonated beverage drinks contain dissolved carbon dioxide creating pressure within the package, and glass is often the suitable packaging capable of withstanding carbon dioxide pressure.

Because it is odorless and chemically inert with virtually all food products, glass has several advantages for food-packaging applications:

- Glass possesses very good barrier properties. It is impermeable to gases, odours and vapors, so it maintains product freshness for a long period of time without impairing taste or flavor.
- They are inert and do not react with or migrate into food products.

- ✰ They are transparent to display the contents.
- ✰ They are rigid, to allow stacking without container damage.
- ✰ The ability to withstand high processing temperatures makes glass useful for heat sterilization of both low- acid and high-acid foods.
- ✰ Glass is rigid, provides good insulation, and can be produced in numerous different shapes. The transparency of glass allows consumers to see the product, yet variations in glass color can protect light-sensitive contents.
- ✰ Finally, glass packaging benefits the environment because it is reusable and recyclable.

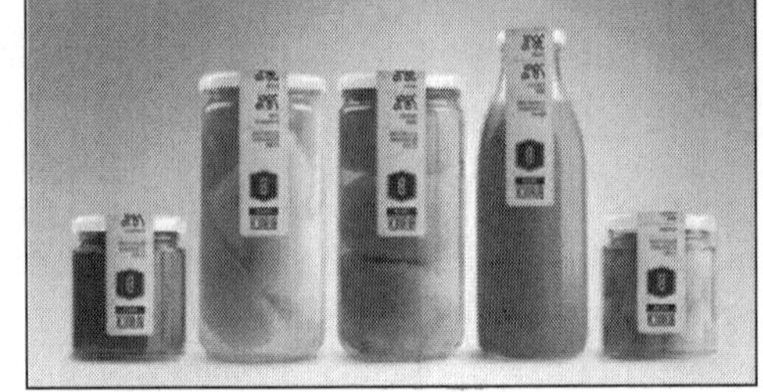

Like any material, glass has certain disadvantages too due to its heavy weight and delicacy. Despite efforts to use thinner glass, it is heavy, brittle and non-degradable. Its heavy weight adds to transportation costs. Another concern is its brittleness and susceptibility to breakage from internal pressure, impact, or thermal shock. For food packaging, glass being fragile has caused some safety concerns such as the possibility of the presence of chipped glass in food products. Glass for food packaging has declined over the last three decades, with glass losing market share to metal cans and increasingly to plastics.

3. Metals (Steel, Tin, Aluminum)

Metal is the most versatile of all packaging forms and is used in packaging in a variety of applications, from rack systems to tuna cans. It offers a combination of excellent physical protection and barrier properties, formability and decorative potential, recyclability, and consumer acceptance and therefore, is widely preferred in many food applications. It is an opaque material that provides an advantage for food that is sensitive to light. The main use of these metals is the protection and preservation of canned foods and beverages. Metal cans are convenient for ambient storage and presentation. They are tamper-proof. Metal containers are vacuum-sealed and thermally sterilized under low oxygen pressure. The decomposition of nutrients is kept to a minimum in metal containers, since metals are a perfect barrier to oxygen, light and moisture. The major limitations of metal containers are cost, the weight of the containers and the fact that they are difficult to crush.

For food packaging, four types of metal are commonly used: steel, aluminum, tin, and chromium. Metals (steel and aluminum) have very good barrier properties, relatively expensive, can be recycled (reprocessing needs high amount of energy). They are mainly used for can production for drinks and processed foods such as beans and peas, (food in cans be pasteurized).

Steel

Steel is used for cans, containers, caps, and closures. Steel, being a permanent material, can be recycled open-endedly while retaining its quality. The high mechanical strength and low weight of steel makes it relatively easy to store and ship food; metallic trays and forms for ready-to-cook food are resistant for high and low temperatures, thus can be applied for frozen and heated meal; metallic are also bottles caps and jar lids. Steel can be recycled many times without quality loss and its cost is significantly lower than aluminum hence, it is highly used in packaging systems. Steel tends to oxidize when it is exposed to moisture and oxygen, producing rust. Therefore, tin and chromium are used as protective layers for steel.

Tinplate

It has been used for preserving food for well over a hundred years. Tinplate produced from low-carbon steel (that is, blackplate), is the result of coating both sides of blackplate with thin layers of tin. The coating is achieved by dipping sheets of steel in molten tin (hot-dipped tinplate) or by the electro-deposition of tin on the steel sheet (electrolytic tinplate) to minimize corrosion of steel. Although tin provides steel with some corrosion resistance, tinplate containers are often lacquered to provide an inert barrier between the metal and the food product. Commonly used lacquers are materials in the epoxy phenolic and oleoresinous groups and vinyl resins. In addition to its excellent barrier properties to gases, water vapor, light, and odors, tinplate can be heat-treated and sealed hermetically, making it suitable for sterile products. Because it has good ductility and formability, tinplate can be used for containers of many different shapes. Thus, tinplate is widely used to form cans for drinks, processed foods, and aerosols; containers for powdered foods and sugar- or flour-based confections; and as package closures. Tinplate is an excellent substrate for modern metal coating and lithoprinting technology, enabling outstanding graphical decoration. Its relatively low weight and high mechanical strength make it easy to ship and store. Finally, tinplate is easily recycled many times without loss of quality and is significantly lower in cost than aluminum. Tin cans, made of steel sheet coated with tin as a measure of protection against corrosion of steel, especially contain products with low pH.

Chromium

If chromium is used to provide corrosion protection instead of tin, the resulting material is called electrolytic chromium-coated steel (ECCS) or tin-free steel (TFS) or chrome oxide coated steel. ECCS is less resistant to corrosion than tinplate but has better heat resistance. Even though the chrome/chrome oxide makes tin-free steel unsuitable for welding, this property makes it excellent for adhesion of coatings such as paints, lacquers, and inks. Like tinplate, tin-free steel has good formability and strength, but it is marginally less expensive than tinplate. Food cans, can ends, trays, bottle caps, and closures can all be made from tin-free steel. In addition, it can also be used to make large containers (such as drums) for bulk sale and bulk storage of ingredients or finished goods.

Aluminum

It is a lightweight, silvery white metal derived from bauxite ore, where it exists in combination with oxygen as alumina. Magnesium and manganese are often added to aluminum to improve its strength properties. Aluminum is increasingly used for canning due to its lightness, low cost and capacity to be recycled. Unlike many metals, aluminum is highly resistant to most forms of corrosion; its natural coating of aluminum oxide provides a highly effective barrier to the effects of air, temperature, moisture, and chemical attack. Therefore, aluminum has the same barrier properties as steel but with the advantage of being resistant to corrosion. Besides providing an excellent barrier to moisture, air, odors, light, and microorganisms, aluminum has good flexibility and surface resilience, excellent malleability and formability, and outstanding embossing potential. It has good resistance to temperature fluctuations and acts as an excellent gas barrier, which extends the food's shelf-life. It is also an ideal material for recycling because it is easy to reclaim and process into new products. It can be found in packaging, bottle closures and wraps and laminates. Beverage cans, foils, tubes, trays, pouches, coffee capsules and laminated papers are usually produced from aluminum. Alloying elements, such as magnesium and manganese, are sometimes added to aluminium to enhance its strength. Aluminium can be used in rigid, flexible, and semi-flexible packaging. It helps maintain the freshness and aroma of the foods, and is good for protection from radiation, oxygen, moisture, oils, and microorganisms. Soft-drinks, seafood, and pet-food are commonly enclosed by aluminium packages.

The main disadvantages of aluminum are its high cost compared to other metals (for example, steel) in food packaging systems and its inability to be welded, which renders it useful only for making seamless containers.

Aluminum Foil

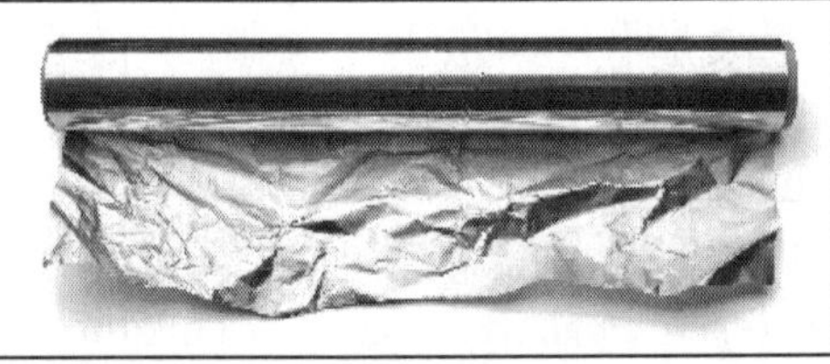

It is made by rolling pure aluminum metal into very thin sheets, followed by annealing to achieve dead-folding properties (a crease or fold made in the film will stay in place), which allows it to be folded tightly. It is a highly flexible product that allows to preserve or protect food in the domestic environment. Aluminum foil is used for aseptic cartons, pouches, wrappings, bottle capsules, push - through blisters, laminated tubes, lids, trays and containers. Aluminum foil is available in a wide range of thicknesses, with thinner foils used to wrap foods such as sweets, cheese, coffee, tea, *etc.* and thicker foils used for trays. Thin-walled aluminum cans are suitable for carbonated beverages, while wide-walled cans are suitable for steam sterilization. Like all aluminum packaging, foil provides an excellent barrier to moisture, air, odors, light, and microorganisms. It is inert to acidic foods and does not require lacquer or other protection. The material is light but strong, can be formed and converted into complex shapes, has a high thermal and electrical conductivity, and can be recycled without decrease in quality. Although aluminum is easily recyclable, foils cannot be made from recycled aluminum without pinhole formation in the thin sheets. It is difficult to use it in modern fast packaging equipment due to wrinkles, rips and marks.

Lamination of packaging involves the binding of aluminum foil to paper or plastic film to improve barrier properties. Thin gauges facilitate application. Although lamination to plastic enables heat sealability, the seal does not completely bar moisture and air. Because laminated aluminum is relatively expensive, it is typically used to package high value foods such as dried soups, herbs, and spices.

Metallized Film

A less expensive alternative to laminated packaging is metallized film. Metallized films are plastics containing a thin layer of aluminum metal. These films have improved barrier properties to moisture, oils, air, and odors, and the highly reflective surface of the aluminum is attractive to consumers. More flexible than laminated films, metallized films are mainly used to package snacks. Although the individual components of laminates and metallized films are technically recyclable, the difficulty in sorting and separating the material precludes economically feasible recycling.

Packaging Material	*Package Form and Applications*
Paper and paperboard	Wrapping paper, bag, box, sachet, pouch, case, carton, fitments in board, body of composite container, fittings in case, multiwall sack, fibre drum, tray, tub
	Take out foods(French fries, popcorn) Labels, corrugated boxes, packets
Metal	Can, crate, box, pallet, collapsible tube, closure, drum, metal strapping and banding, foil, laminate and label (aluminium)
	Tin foods and beverages(tomatoes, sweetcorn, sausages, milk, milo) *etc.*
Glass	Bottle, jar, tray, tube, ampoule, vial
	Fruit juices, jams, some pickles and chutneys, some sauces, Wines, vinegars, milk, some pickles and sauces.
Plastics (including cellulose and rubber)	Film, laminate and sheet, bag, pouch and sachet, sack (film and woven tape), bottle, jar, pot, tray, pot, blister and fitment, cushioning material and fitting, cap and closure, drum, crate, box, tub
	Plastic films (packed fruits and vegetables, cereals), bottles (milk, milk-based drinks, soups, fruit juices, and fruit drinks beverages, yogurt, rice
Timber (including plywood)	Box, crate, cask and keg, case, drum, pallet, basket and punnet
	Fruits and Vegetables

4. Papers, Paperboard and Cardboard

Products derived from wood are widely used in the packaging of food in the form of paper, paperboard or cardboard. The use of paper and paperboards for food packaging dates back to the 17th century with accelerated usage in the later part of the 19th century. Paper, paperboard and cardboard are sheet materials made from an interlaced network of cellulose fibers derived from wood by using sulfate and sulfite. Making pulp is the initial stage in making paper or cardboard, and the quality of the paper is closely related to the quality of the pulp. Pulping can be done using mechanical, chemical, or a combination process. Mechanical pulping produces papers that are characterized by relatively high bulk and low strength as well as relatively low cost. Their use in packaging is very limited. Chemical pulping produces stronger and higher quality paper and is also more expensive. Combination processes are intermediate in cost and properties. Hence, paper products are produced after pulping and/or bleaching, treating with chemicals such as slimicides and strengthening agents.

Paper is a very cheap, lightweight product with excellent printing capacity. Plain paper is used for temporary food containment and protection. It is not used to protect foods for long periods of time because it is very sensitive to moisture, has poor barrier properties and is not heat sealable. When used as primary packaging (that is, in contact with food), paper is almost always treated, coated, laminated, or impregnated with materials such as waxes, plastic, paraffin, resins, or lacquers to improve functional and protective properties. Packaging material based on paper has an advantage due to its high recyclability at relatively low cost.

Paperboard is a term that includes boxboard, chipboard and corrugated or solid fibreboards. Typically, paperboard has the following structure:

1. A top layer of white material to give surface strength and printability.
2. Middle layers of grey/brown lower grade material.
3. An under-layer of white material to stop the colour of the middle layer showing through.
4. A back layer if strength or printability are required. All layers are glued together with adhesive.

Hence, paperboard is a paper-based material that is lightweight, yet strong. It can be easily cut and manipulated to create custom shapes and structures. These characteristics make it ideal to be used in personalized packaging. Paperboard packaging comes in various grades, each suitable for different packaging requirements.

Paper and cardboards are commonly used in corrugated boxes, milk cartons, paper plates and cups, folding cartons, bags and sacks, and wrapping paper. Tissue paper, paper plates, and cups are other examples of paper and cardboard products.

Properties of Major Packaging Materials

Packaging Material	*Properties*	
	Advantage	*Disadvantage*
Paper	Strength and rigidity Printability	Opacity
Aluminium foil	Negligible permeability to gases, odours and water vapour Dimensional stability Grease resistance Brilliant appearance Dead folding characteristics	Opacity High barrier to gases
Glass	Formability and rigidity Transparency and UV protection due to color variation	Higher barrier to gases Heavy weight adds to transport cost
Cellulose film (coated)	Attractive appearance Strength Grease resistance Low permeability to water vapour, gases, and odours	Low permeability barrier
Polyethylene	Durability and flexibility Heat sealability Good moisture barrier Chemical resistance Good low-temperature performance	HDPE; poor clarity LLDPE; heat sensitive
Polystyrene	Hight tensile strength Excellent transparency	Poor barrier to gas and water vapour
Polyamide (nylon-6)	Strong Moderate oxygen barrier, and excellent odour and flavour barrier Good chemical resistance Thermal and mechanical properties similar to PET High temperature performance	Poor water vapour barrier

Source: Page *et al.* (2003), Marsh and Bugusu (2007), Mangaraj *et al.* (2009).

Paper and paperboard are mostly used for packaging dry foods but upon coating or waxing, their applications extend to packaging and serving of wet and fatty foods. SBS (or solid bleached sulfate) paperboard can be used for packing cosmetics, medicines, milk and juice, cosmetics, frozen food and more.

Cardboard is a material made up of several superimposed layers of paper, making it thicker, harder and more resistant than paper. It generally consists of three layers of paper, an outside liner, an inside liner and a corrugated medium (also known as fluting). The corrugated medium that gives it strength and rigidity. Its main use is for packaging and containers in the form of boxes which are known as corrugated or cardboard boxes. It is commonly used to make containers for shipping, such as boxes, cartons and trays and is seldom used for direct food contact. Cardboard packaging such as carton are the most widely used packaging in the horticultural industry. In recent years, paper and cardboard manufacturers are paying special attention to issues related to health and the environment by working with recycled products that increase the useful life of these raw materials.

Many different types of paper used in food packaging are as follows:

Kraft paper: It is produced by a sulfate treatment process and is available in several forms such as natural brown, unbleached, heavy duty, and bleached white. The natural kraft is the strongest of all papers and is commonly used for bags and wrapping. It is also used to package flour, sugar, and dried fruits and vegetables.

Sulfite paper: It is lighter and weaker than kraft paper. It is glazed to improve its appearance and to increase its wet strength and oil resistance. It can be coated for higher print quality and is also used in laminates with plastic or foil. It is used to make small bags or wrappers for packaging biscuits and confectionery.

Greaseproof paper: It is made through a process known as beating, in which the cellulose fibers undergo a longer than normal hydration period that causes the fibers to break up and become gelatinous. These fine fibers then pack densely to provide a surface that is resistant to oils but not wet agents. Greaseproof paper is used to wrap snack foods, cookies, candy bars, and other oily foods, a use that is being replaced by plastic films.

Glassine: It is greaseproof paper taken to an extreme (further hydration) to produce a very dense sheet with a highly smooth and glossy finish. It is used as a liner for biscuits, cooking fats, fast foods, and baked goods.

Parchment paper: It is made from acid-treated pulp (passed through a sulfuric acid bath). The acid modifies the cellulose to make it smoother and impervious to water and oil, which adds some wet strength. It does not provide a good barrier to air and moisture, is not heat sealable, and is used to package fats such as butter and lard.

Cardboard or Paperboard: It is thicker than paper with a higher weight per unit area and often made in multiple layers. It is widely used as secondary packaging. It is commonly used to make containers for shipping such as boxes, cartons, and trays and seldom used for direct food contact. The various types of paperboard are as follows:

i) **White board:** It is made from several thin layers of bleached chemical pulp; it is advised for primary packaging. It is typically used as the inner layer of a carton. White board may be coated with wax or laminated with polyethylene for heat sealability, and it is the only form of paperboard recommended for direct food contact.

ii) **Solid board:** It possesses strength and durability. It has multiple layers of bleached sulfate board. When laminated with polyethylene, it is used to create liquid cartons (known as milk board). Solid board is also used to package fruit juices and soft drinks.

iii) **Chipboard:** It is made from recycled paper and often contains blemishes and impurities from the original paper, which makes it unsuitable for direct contact with food, printing, and folding. It is often lined with white board to improve both appearance and strength. The least expensive form of paperboard, chipboard is used to make the outer layers of cartons for foods such as tea and cereals.

iv) **Fiberboard:** It can be solid or corrugated. The solid type has an inner white board layer and outer kraft layer and provides good protection against impact and compression. When laminated with plastics or aluminum, solid fiberboard can improve barrier properties and is used to package dry products such as coffee and milk powder.

v) **The corrugated type:** It is also known as corrugated board and is made with three layers of kraft paper with a central corrugating (or fluting) material. The main raw material that is used to construct the corrugated board is mostly recycled paper, made on large high-precision machinery known as corrugators. The board is cut and folded into different sizes and shapes to become corrugated packaging. These type of boards can be reused and recycled again and again as a source of pulp fiber. Corrugated boards are of different types, single faced, double faced (single wall), twin wall, and triple wall. They can be used to make packaging with different characteristics, performances, and strength. Fiberboard's resistance to impact abrasion and crushing damage makes it widely used for shipping bulk food and case packing of retail food products. Other applications of corrugated board packaging include retail packaging, pizza delivery boxes, small consumer goods packages, and so forth.

vi) **Paper laminates:** Paper laminates are coated or uncoated papers based on kraft and sulfite pulp. They can be laminated with plastic or aluminum to improve various properties. For example, paper can be laminated with polyethylene to make it heat sealable and to improve gas and moisture barrier properties. However, lamination substantially increases the cost of paper. Laminated paper is used to package dried products such as soups, herbs, and spices.

5. Plastic

Plastics are a special group of organic polymers that can be formed into a wide variety of shapes using controlled heat and pressure at relatively low temperatures, compared to metals and glass. Plastics are actually a subcategory of polymers, but in packaging the terms tend to be used interchangeably. There are two types of plastics:

Synthetic Plastic

There are different types of synthetic polymers such as polyvinyl chloride (PVC), polyethylene and its varieties (PET, HDPE, LDPE) and polystyrene (PS) but only few polymers are often used for food packaging. PVC is very resistant to humidity, fats and gases. The development of the PET range has revolutionized the packaging industry, allowing plastic to compete directly with glass bottles. PS is the plastic of choice for thermoforming due to its strength, malleability and low cost. The most widely used plastics in packaging includes: low density polyethylene (LDPE), laminated aluminium foil (LAF), high density polyethylene (HDPE), polypropylene (PP), polyethylene (PE), they have various applications ranging from plastic films, nylons, bottles *etc.* Each plastic has its own unique properties, based on its chemical composition. The performance and interaction with a variety of foods are different for each material. Thus, the plastic material for the packaging of a specific food is selected to function well within the parameters of the application. Milk, juice, and water bottles, grocery, retail, and garbage bags, and bread and frozen food bags are some of the uses of polyethylene. Polypropylene is used when heat resistance is needed. Yogurt containers and margarine tubs are applications of polypropylene.

The lightness and versatility of plastics have been confirmed over decades in the processing and packaging of food. They are the most common and most wide-ranging materials used for food packaging. Some of their widespread uses are bottles, trays, bags, foils, cups, pots, pouches, and bowls. The volume of plastic allocated to food packaging amounts for around 40 per cent of plastics. Plastic containers and packaging protect against the contamination of food and offer adequate mechanical strength. Plastic box packaging has many advantages in which they can be recycled, and generally they are much more durable than paperboard boxes. Airtight plastic packaging containers can help to preserve the quality of food and eliminate any contamination issues. Plastic packaging also does not break easily and can be stored with food under extreme conditions. Another reason why plastic is a popular choice for packing material is because of its ability to showcase the product at any angle without necessarily opening the packaging. It is also flexible, lightweight and can be applied with films or coating to enhance packaging appearance. They are characterized by a low production cost, lower energy consumption, good mechanical and barrier properties. In addition, they are able to preserve and protect the food for longer, minimizing the use of preservatives. In relation to the consumer, they are easy to handle and open, and offer an effective surface for printing labels or brands. Due to these properties, nowadays they substitute glass, metal and paper, have easy processability and modification for getting required properties *viz.*, light, aesthetic, unbreakable, elastic, transparent or colored, non-degradable in nature; can be recycled or burned. Using plastic as packaging material also offers marketing advantage. Unlike metal and aluminum packaging materials, harnessing the transparency of film packaging for product visibility is now widely practiced, enabling consumers to assess the visual quality of the product before purchase. However, the variable permeability to light, gases and vapours of plastics is a major drawback. Although plastics are recyclable materials, yet they are pollutants. Despite the health and safety concerns regarding residual components from plastic, plastic use continues to grow compared to the aforementioned conventional materials because of its inexpensiveness, thermosealability, microwaveability, and ease of fabrication into countless shapes and sizes.

Biodegradable Polymers

They are biopolymers on the base of hydroxybutyrate or hydroxyvalerate acids, which are produced in nature during biosynthesis; biodegradable (undergo decomposition under enzymatic action of microorganisms *i.e.* bacteria and fungi), non- recyclable. Hence, these are originated from replenish-able stocks from animal sources, marine food processing industry wastes, or microbial sources. In addition to renewable raw ingredients, biodegradable materials break down to produce environmentally friendly products such as carbon dioxide, water, and quality compost. Biodegradable polymers made from cellulose and starches have been in existence for decades. Cellophane is the most common cellulose-based biopolymer. Starch-based polymers, which swell and deform when exposed to moisture, include amylose, hydroxylpropylated starch, and dextrin. Other starch-based polymers are polylactide, polyhydroxybuterate (PHB), polyhydroxyalkanoate (PHA) and a copolymer of PHB and valeric acid (PHB/V). The polyactides made from lactic acid bacterial fermentation of starch derivatives, does not degrade even when exposed to moisture. PHA, PHB, and PHB/V can also be created by microbial fermentation on starches. Another type of biodegradable plastic are blends of synthetic polymers with starch (from potatoes, rise, corn) or cellulose, in which only natural component undergoes decomposition to oxygen dioxide and water (in air conditions) or to water and methane (in oxygen-free circumstance), whereas the synthetic component only breaks up to small pieces and dissipates. They are relatively expensive.

Thin layers of edible materials applied to food as a coating or placed on or between food components, are another form of biodegradable polymer. Edible films are defined as a thin layer of material which can be consumed and provides a barrier to moisture, oxygen and solute movement for the food. The material can be a complete food coating or can be disposed as a continuous layer between food components. Edible films can be formed as food coatings and free-standing films, and have the potential to be used with food as gas aroma barrier. Edible films are derived from plant and animal sources such as zein (corn protein), whey (milk protein), collagen (constituent of skin, tendon, and connective tissue), and gelatine.

6. Flexible Plastic Films

In general, flexible plastic films have relatively low cost and good barrier properties against moisture and gases; they are heat sealable to prevent leakage of contents; they add little weight to the product and they fit closely to the shape of the food, thereby wasting little space during storage and distribution; they have wet and dry strength, and they are easy to handle and convenient for the manufacturer, retailer and consumer. The main disadvantages are that (except cellulose) they are produced from non-renewable oil reserves and are not biodegradable. Concern over the environmental effects of non-biodegradable oil-based plastic packaging materials has increased research into the development of 'bioplastics' that are derived from renewable sources, and are biodegradable. However, these materials are not yet available commercially in developing countries. There is a very wide choice of plastic films made from different types of plastic polymer. Each can have ranges of mechanical, optical, thermal and moisture/gas barrier properties. These are produced by variations in film thickness and the amount and type of additives that are used in their production. Some films (*e.g.* polyester, polyethylene, polypropylene) can be 'oriented' by stretching the material to align the molecules in either one direction (uniaxial orientation) or two (biaxial orientation) to increase their strength, clarity, flexibility and moisture/gas barrier properties. There are thus a very large number of plastic films and small-scale processors should obtain professional advice when selecting a material to ensure that it is suitable for the intended product and shelf life. Typically, the information required includes: type of plastic polymer(s) required; thickness/strength; moisture and gas permeability; heat seal temperature; printability on one or both sides; and suitability for use on the intended filling machinery.

A summary of the main different types of flexible plastic films is as follows:

Cellulose

Plain cellulose is a glossy transparent film that is odourless, tasteless and biodegradable (within approximately 100 days). It is tough and puncture resistant, although it tears easily. It has dead-folding properties that make it suitable for twist-wrapping (*e.g.* sugar confectionery). However, it is not heat sealable and the dimensions and permeability of the film vary with changes in humidity. It is used for foods that do not require a complete moisture or gas barrier, including fresh bread and some types of sugar confectionery. Cellulose acetate is a clear, glossy transparent, sparkling film that is permeable to water vapour, odours and gases and is mainly used as a window material for paperboard cartons.

Polyethylene (or polythene) Low-density Polyethylene (LDPE)

It is heat sealable, inert, odour free and shrinks when heated. It is a good moisture barrier but is relatively permeable to oxygen and is a poor odour barrier. It is less expensive than most films and is therefore, widely used for bags, for coating papers or boards and as a component in laminates. LDPE is also used for shrink- or stretch-wrapping than shrink-wrapping (45 - 75 μm), or alternatively, linear low-density polyethylene is used at thicknesses of 17 - 24 μm. The cling properties of both films are adjusted to increase adhesion between layers of the film and to reduce adhesion between adjacent packages. High-density polyethylene (HDPE) is stronger, thicker, less flexible and more brittle than LDPE and a better barrier to gases and moisture. Sacks made from HDPE have high tear and puncture resistance and have good seal strength. They are waterproof and chemically resistant and are increasingly used instead of paper or sisal sacks.

Polypropylene

Polypropylene is a clear glossy film with a high strength and puncture resistance. It has a moderate barrier to moisture, gases and odours, which is not affected by changes in humidity. It stretches, although less than polyethylene. It is used in similar applications to LDPE. Oriented polypropylene is a clear glossy film with good optical properties and a high tensile strength and puncture resistance. It has moderate permeability to gases and odours and a higher barrier to water vapour, which is not affected by changes in humidity. It is widely used to pack biscuits, snack foods and dried foods.

Other Films

Polyvinylidene chloride is very strong and is therefore, used in thin films. It has a high barrier to gas and water vapour and is heat shrinkable and heat sealable. However, it has a brown tint which limits its use in some applications. Polyamides (or Nylons) are clear, strong films over a wide temperature range (from – 60 to 200°C) that have low permeability to gases and are greaseproof. However, the films are expensive to produce, require high temperatures to heat seal, and the permeability changes at different storage humidities. They are used with other polymers to make them heat sealable at lower temperatures and to improve the barrier properties, and are used to pack meats and cheeses.

Coated Films

Films are coated with other polymers or aluminum to improve their barrier properties or to impart heat sealability. For example, a nitrocellulose coating on both sides of cellulose film improves the barrier to oxygen, moisture and odours, and enables the film to be heat sealed when broad seals are used. Packs made from cellulose that has a coating of vinyl acetate are tough, stretchable and permeable to air, smoke and moisture. They are used for packaging meats before smoking and cooking. A thin coating of aluminum (termed 'metallization') produces a very good barrier to oils, gases, moisture, odours and light. This metallized film is less expensive and more flexible than plastic/aluminum foil laminates. The properties of selected coated films are shown in below table.

Film	*Coating*	*Barrier to*		*Strength*	*Clarity*	*Thickness (μm)*
		Moisture	*Air/Odour*			
Cellulose	PvDC	*	***	*	***	21-40
	Aluminum	***	***	*	***	19-42
	Nitro-cellulose	***	***	*	-	21-42
LDPE	-	**	*	**	*	25-200
HDPE	-	***	**	***	*	350-1000
Polyprylene	-	***	***	***	***	20-40
	PvDC	***	***	***	***	18-34
	Aluminum	***	***	***	-	20-30
Polyester	-	**	**	***	**	12-23
	-	***	***	***	-	20-30

* = Low, ** = Medium, *** = High. PvDC = Polyvinylidene chloride, LPDE = Low density polyethylene, HDPE = High density polyethylene. Thicker films of each type have better barrier properties than thinner films.

Laminated Films

Lamination (bonding together) of two or more films improves the appearance, barrier properties or mechanical strength of a package. Some examples of laminated films are shown in below table.

Laminated Film	*Typical Food Applications*
Polyvinylidene chloride coated poly-propylene (2 layers)	Crisps, snackfoods, confectionery, ice cream, biscuits, chocolate
Polyvinylidene chloride coated poly-propylene polyethylene	Bakery products, cheeses, confectionery, dried fruit, frozen vegetables
Cellulose-polyethylene-cellulose	Pies, crusty bread, bacon, coffee, cooked meats, cheeses
Cellulose acetate-paper-foil-polyethylene	Dried soups
Metallised polyester-polyethylene	Coffee, dried milk
Polyethylene-aluminium-paper	Dried soup, dried vegetables, chocolate

Coextruded Films

Coextrusion is the simultaneous extrusion of two or more layers of different polymers to make a film. Coextruded films have three main advantages over other types of film: they have very high barrier properties, similar to laminates but produced at a lower cost; they are thinner than laminates and are therefore, easier to use on filling equipment; and the layers do not separate.

There are three main groups of polymers that are coextruded:

1. Low-density and high-density polyethylene, and polypropylene.
2. Polystyrene and acrylonitrile-butadiene-styrene.
3. Polyvinyl chloride.

Typically a three-layer coextrusion has an outside layer that has a high gloss and printability, a middle bulk layer which provides stiffness and strength, and an inner layer which is suitable for heat sealing. They are used, for example, for confectionery, snack foods, cereals and dried foods. Thicker coextrusions (75 - 3000 μm) are formed into pots, tubs or trays.

Coextruded Film	*Typical Food Applications*
High impact polystyrene-PET	Margarine, butter tubs
Polystyrene- polystyrene-PvDC- polystyrene	Juice and milk bottles
Polystyrene- polystyrene-PvDC-poly-ethylene Tubs	Butter, cheese, margarine, bottles for coffee, mayonnaise, sauces.

Types of Food Packaging

Packaging depends on the type of food being packaged, and it varies over a wide array of materials and forms. There are various types of food packaging available in modern era. Packaging materials come in different types and shapes with various functions relative to their properties. It is essential for the packaging material to have a balance between its shape and its function. Given the packaging's main purpose of preservation, containment, and protection of food, the packaging material can be rigid, flexible, or semi-flexible.

The different types of packages can be classified into two groups:

Retail Containers

These containers protect food or the content from different damages and at the same time they advertise the product for retail sale. For example, glass bottles, sachets, over wraps, plastic bottles, metal cans *etc.* They can be used for home storage also.

Shipping Containers

These containers contain and protect food and other items during distribution and transport or any other marketing function. For example, sacks, stretch, or shrink wrapped containers, corrugated fire board cartons, drums, barrels, crates, and foil bags.

Industrial packaging materials are different from the traditional packaging materials. In addition to material, attractive design is also preferred. Packing comes in many different forms, based on technical requirements throughout the supply chain, as well as marketing need (like brand identity or consumer information) and other criteria.

Packaging is very essential to every and to any type of industry. Packaging is crucial in all types of industries. It protects the product from any potential damage that will deem the product useless. Thus, a good packaging system is a must.

Having a high-quality taping machine and a case erector will definitely aid in achieving the very essence of packaging that is to secure the product and to keep its original state or how it was after production. Packaging will only vary in every industry depending on how the product will be marketed. Different industries will definitely use different types of packaging but the objective will still remain constant. As most businessmen would always say packaging can be the difference in successfully shipping a product to the market in one piece or in pieces.

Packages can be:

i) **Rigid packages** include bottles, trays, cans, jars, and caps.
ii) **Flexible packages** include bags, cling wraps, bubble wraps, shrink wraps, squeezable tubes, foam trays, stand-up packets, and vacuum bags.
iii) **Semi-flexible packages** include caps and closures, boxes, and tetra packs.

Food packaging types differ in various ways, such as weight, size, durability, and barrier properties. Listed below are different types of packaging being used for various food items:

Boxes

Boxes are used for the most easy form of transportation of a food product. Most common materials found in boxes are metal, corrugated fiberboard, and wood. Frozen pizzas, cereals, and snack crackers are examples of foods packaged in boxes.

Paperboard Boxes

Paperboard is a paper-based material that is lightweight, yet strong. It can be easily cut and manipulated to create custom shapes and structures. These characteristics make it ideal to be used in personalized packaging. It is made by turning fibrous materials that come from wood or from recycled waste

paper into pulp, and then bleaching it. Paperboard packaging comes in various grades, each suitable for different packaging requirements.

SBS (or solid bleached sulfate) paperboard can be used for packing cosmetics, medicines, milk and juice, cosmetics, frozen food and more. Choosing kraft, or CUK (coated unbleached kraft) paperboard packaging are for those who prefer the natural and environmentally-friendly look of recycled paper, which can be used for similar packaging applications. Kraft is often seen to be less resistant to moisture, making it less suitable for food-related products, or frozen-goods packaging. With the right combination of design options, paperboard packaging can look high-end, without high-end pricing.

Chipboard Packaging

Chipboard packaging is used in industries such as electronic, medical, food, cosmetic, and beverage. A chipboard basically is a type of paperboard that is made out of reclaimed paper stock. It can be easily cut, folded, and formed. It is a cost-effective packing option for products.

It comes in various densities and strength is determined by how high the density of the material is. If someone wants images to be directly printed onto the chipboard, one can treat the chipboard with bleach sulfate, and with CCNB (Clay Coated News Back) which makes the material even more durable.

One cannot use chipboard packaging if the business deals with heavy items, since the chipboard sheet is a lightweight material, made for many grocery items such as cereal, crackers, tissue boxes, and so forth. Also, if the storage environment is heavy with moisture, chipboard can easily weaken, and result in discoloration and expansion.

Corrugated Boxes

Corrugated boxes simply refer to what is commonly known as cardboard. Corrugated boxes are the ones many probably consider as 'cardboard' as it produces the large shipping, shoe and storage boxes. What a lot of people do not realize is that corrugated boxes also come in different types depending on the durability and strength of the box. Identifying a certain corrugated material, however, is easy. One can determine its material through its corrugated medium (also known as fluting). It consists of 3 layers of paper, an outside liner, an inside liner and a corrugated medium (also known as fluting). The corrugated medium that givens it strength and rigidity.

The main raw material that is used to construct the corrugated board is most recycled paper, made on large high-precision machinery known as corrugators. These type of boards can reused and recycled again and again as a source of pulp fiber. Corrugated boards are of different types, single faced, double faced (single wall), twin wall, and triple wall. They can be used to make packaging with different characteristics, performances, and strength. The board is cut and folded into different sizes and shapes to become corrugated packaging. Other applications of corrugated board packaging include retail packaging, pizza delivery boxes, small consumer goods packages, and so forth.

Rigid Boxes and Containers

A rigid box is made out of highly condensed paperboard that is four times thicker than the paperboard used in the construction of a standard folding carton. Compared to paperboard and corrugated boxed, rigid boxes are definitely among the most expensive box styles. The rigid boxes usually do not require dies that are expensive or massive machinery and are often hand-made. Their non-collapsible nature also gives them a higher volume during shipping, which easily incurs higher shipping fees. These boxes are commonly used in merchandising cosmetics, jewelry, technology, and high-end luxury couture. It is easy to incorporate features such as platforms, windows, lids, hinges, compartments, domes, and embossing in a rigid box.

During shipping, rigid containers must protect the contents from physical damage during transport and distribution which may be rough and unsympathetic to quality maintenance. They are commonly fabricated from wood, plastic, metal or card-board, and come in the form of crates, barrels, drums and sacks for ultimate transport in shipping containers. They are designed to withstand rough and impacted handling but with the exception of cans rarely suitable for consumer convenience or attractive retail display.

Plastic Crates and Boxes

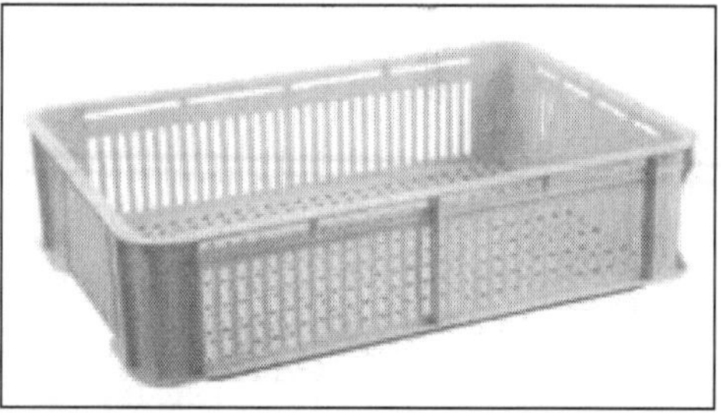

Plastic is used in a wide range of products, from spaceships to paper clips. A number of traditional materials, such as wood, leather, glass, ceramic, and so on, have already been replaced by plastic. Plastic crates are being used increasingly in fields during harvest of such items as root crops, fruits and vegetable. They prevent bruising which will assist in retaining quality because bacteria will attack bruised fruits and vegetable first.

Plastic box packaging has many advantages in which they can be recycled, and generally they are much more durable than paperboard boxes. Airtight plastic packaging containers can help to preserve the quality of food and eliminate any contamination issues. Plastic packaging also does not break easily and can be stored with food under extreme conditions. Another reason why plastic is a popular choice for packing material is because of its ability to showcase the product at any angle without necessarily opening the packaging. It is also flexible, lightweight and can be applied with films or coating to enhance packaging appearance. Contrary to popular belief, plastic is in fact recyclable, in the sense that it takes less energy to produce new plastic, compared to glass, and other materials.

Cartons

Boxes and cartons are often times interchangeable. Like boxes, cartons are also made out of corrugated fiberboard and are used for transporting food. Within the food packaging type of cartons, there are four sub-categories. The most well-known is the egg carton which is molded to the shape of the egg to add protection while the food product is mobile. Aseptic cartons also lie in this category. Examples of this are milk, juice, and soup cartons. Another sub-category of cartons are the folding cartons, which begin as flat pieces of corrugated fiberboard and then assembled by the food manufacturer. Lastly, there are gable tops. These cartons usually hold milk or juice and require the gables at the top to be pinched then pulled in order to be opened.

Laminated Paperboard Cartons

Very complex laminates consisting of LDLPE-paper-LDPE-aluminum-saran-LDPE are enjoying increasing use for fruit juices and UHT milk. Most commonly the cartons are purchased as a roll with the spout preformed and they are filled as the carton is formed. Some headspace is allowed to facilitate mixing and avoid spillage on opening. The advantages of this type of package are:

- They have excellent impact resistance;
- No additional labelling or capping is required;
- They require less energy to produce and are lighter than cans or bottles;
- They can be used for aseptic foods which therefore, suffer less heat damage;
- They stack well on supermarket shelves.

Wooden Containers

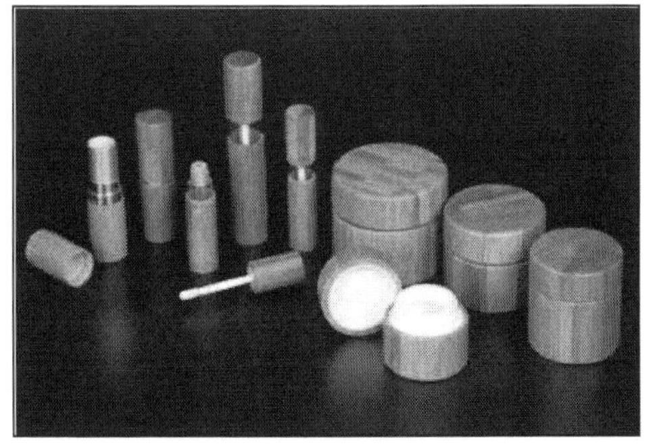

Wooden shipping containers are still used extensively as a protective outer package for bottles and other delicate items and enjoy increasing use for single-use field crates. Some lidded or hinged boxes are produced *e.g.* They remain in common use for wine, cigars, cheese, gifts, tea and beer. High value spirits use wood and a few caps incorporate wood. Wooden chests have high impact resistance and are stackable. They are also relatively inexpensive. High impact-resistant plastic crates are superseding wood as the container of choice because they can be used for many cycles.

The most basic container used in some countries is the woven coconut leaf basket which is still commonly used to transport, display and sell commodities such as root crops, papaya and drinking coconuts and sea urchins. Foods are exposed to the air, to sources of contamination (*e.g.* soil and wind-blown dust, animal droppings) and the weight of the contents is variable.

Trays

This is mostly self-explanatory. Trays are what meats, plant seeds, and drinks can be carried in. They are mostly flat or shallow containers with or without lid with raised edges to keep the product in place, and are made of various materials such as paperboard. There are plastic trays of all kind: those used as a primary food-contact packaging, such as foamed, transparent, high-barrier, peelable and reclosable trays and those used as secondary packages, which are normally thermoformed for containing other food packages.

Cans

Globally, metal cans have been used for two hundred years or so. They offer total protection of the contents and are convenient for ambient storage. They are also an excellent way of preserving and transporting foods. They are tamper-proof but the technology required for canning is expensive. Most cans are made of steel or other thin metal when found in stores, as well as glass jars when foods are canned in the home.

Cans are heavy and bulky and importing transport costs for empty cans is high in the Pacific. Tin is relatively expensive so that the coating used on cans is very thin and susceptible to scratching and ultimately rusting. Cans for pale coloured fruits are traditionally not lacquered on the inside and this allows the fruit acids to attack the tin. The reducing environment so created prevents discolouration of natural colours including anthocyanins (often red) and carotenoids (orange) but it can increase the tin content of the product. Tin-free steel (using a chromium-steel alloy) cans are extremely durable and withstand corrosion more effectively than tin coated steel.

Cans are completely impervious to oxygen, carbon dioxide, moisture and other aroma compounds. Food remains in good condition for several years although the shelf life is commonly set at two years. During storage even canned foods will deteriorate. Most commonly the food attacks the inside of the can stripping the tin from the can. Many cans now have lacquers to prevent this. The effect of interaction of food with the lacquers is not well documented. If cans are stored badly, they can rust and eventually deteriorate to the point that pinholes occur which will allow microorganisms and other contaminants to enter the can. Some modern cans are often fabricated by stamping (more accurately 'deep drawing') and these cans have only two components *i.e.* the body and the lid. There is no seam at the bottom of the can. These two-piece cans are used for fish, soft drinks and beer and often are equipped with a ring-pull opening device. The traditional three piece can with a lid and a base attached to a cylinder are much more common. They are heated to the extent that they are commercially sterile usually in a pressure vessel known as a retort. The coldest part of the coldest can must reach a temperature of 121° C for 2.54 minutes (this is sufficient to reduce the theoretical load of *Clostridium botulinum* spores by a factor of 1012). The disadvantage of such severe heat treatment causes the contents to suffer considerable heat damage and is not ideal for all products.

Furthermore, canning requires significant amounts of energy and large throughput volumes, both factors which can make it not viable in some countries, except for a certain products for which flexible packaging is less appropriate. Cans are also used for soft drinks and beers. Cans will not allow bacteria to enter and the heat process is sufficient to stop all life processes so that canned food is shelf stable and is often given a 'best before,' date of up to three years. Occasionally, some foods attack the interior surface of the can changing the metal content but modern cans have a lacquer that prevents this.

Plastic Can

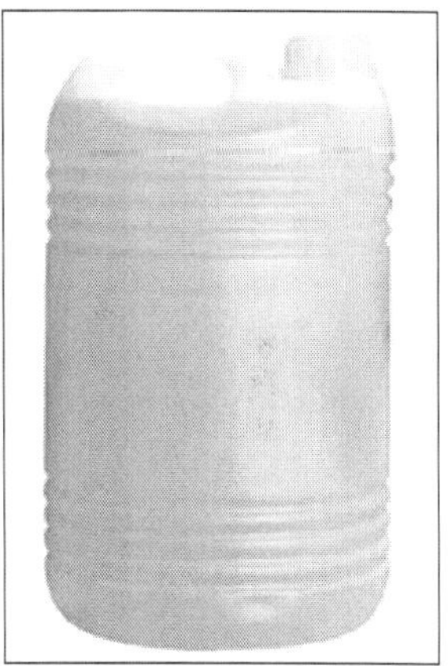

Rigid containers consisting of a round neck of smaller diameter than the body and with an opening where there is a cap to keep the product contained inside. They differ from bottles and jars in the greater capacity, as well as in the applications in which they are used. The body section can be round, oval, square, oblong or a combination of them.

Bottles

Rigid containers consisting of a round neck relatively smaller than the body of the bottle and with an opening, where there is a cap to keep the product contained inside. The body section can be round, oval, square, oblong or a combination of them. They are used to store water, oil, carbonated drinks *etc.*

Bag in Box

They are used for storing and transporting liquids. It is formed by an outer packaging, normally made of cardboard and an inner plastic bag. In many cases, the plastic bag incorporates a tap for dosing the product. It is widely used to package wine.

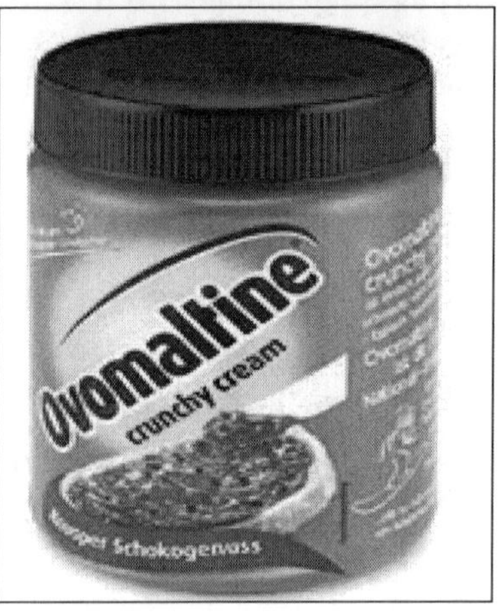

Jars

Rigid containers consisting of a round neck with a diameter similar to that of the body diameter and with a relatively large opening capable of supporting a lid to keep the product inside. The section of the body is usually round or square. Mostly, jams, jellies, pickles, peanut butter *etc.* are packed in them.

Flexible Packaging

Similar to bags, flexible packaging protects the food protect from the environment and create an adequate means of transporting foods. Bagged salad is a common food found in flexible packaging.

Common Packaging Formats Used for different Products

Packaging Formats	*Example of Produce*
Paperboard cartons	Fresh produce (apple, strawberry)
Polyethylene-laminated cartons	Processed produce (orange juice)
Wooden box	Fresh produce (strawberry)
Tetra recart carton	Processed produce (meat)
Tetra wedge package	Processed produce (meat)
Can	Processed food (minimally processed tomato pulp)
Glass bottle	Minimally processed food (tomato sauce, orange juice)
Plastic bottle	Processed food (Citrus juice)

Source: Opara, 2011.

Pallets

Pallets are used for mass transportation of a product. Boxes of the product are placed and stacked on the pallet, then wrapped to secure and decrease food movement.

Wrappers

Used for individual items, wrappers provide protection between the food and the environment or the food and a person's hands. Candy bars are most commonly thought of.

Bags

Like trays, bags are a common form of food packaging. Most commonly known are bagged snacks (chips, pretzels) and fruit (apples, potatoes). "Bagging" separates the food from the environment, namely the air. They can be made of paper, plastic or cloth.

Wooven Jute Sacks

Woven jute sacks are chemically treated to prevent rotting and reduce flammability. They are non-slip and have a high tear resistance, have low extensibility and good durability. They are frequently used a number of times and are used to transport root crops, flour, sugar and salt. Woven polypropylene (PP) sacks are used for coffee and cocoa bean export. These materials occasionally find their way as retail packs but they are usually confined to those used in shipping. This group offers little protection from microbial invasion and normal life processes continue.

Poly Bags

A poly bag, also known as a pouch or a plastic bag, is manufactured out of flexible, thin, plastic film fabric. It is one of the common types of packaging and can carry a wide range of products including food items, waste, chemicals, magazines and so on.

Poly bags are durable yet lightweight, reusable and flexible. Since poly bags are structurally simple to make, it can be fully customized in design, style and sizes but still remain cost-effective. Plastic recycling is also possible with poly bags, depending on the construction. Most of the poly bags are made with security features, tape attachments, hanging holes, and carrying handles to make sure the products are well secured and visually appealing to the customer.

Foil Sealed Bags

Foil sealed bags can be seen typically in most coffee and tea packaging. It keeps the products dense to maintain the flavor, protects it from bacteria coming in and helps increase shelf life. Apart from food, foil sealed bags are also used to package bedding and clothing products. The process involves removing the oxygen from the bag to keep the fabric tight and secure in order to prevent the growth of fungi and other bacteria. Nuts, cereals, smoked fish, cheese and cured meats are also packaged with foil sealed bags to prevent from spoilage.

Plastic Retort Pouches

More recently, plastic pouches have been used and because they are slimmer than cans and heat can penetrate the pack more efficiently and the product does not suffer the same heat damage. Heat processing costs are lower. They have the same functionality as cans but can be imported much more cheaply than empty cans because they are trans-shipped collapsed.

Single Use Bags

Flexible containers, generally closed in all their parts except one, to carry products such as fruits and vegetables. They can be T-shirt type, loop handle type or punched handle type.

Reusable Bag

Similar to single-use bags in shape and designed to be reused normally at least 15 times. Their dimensions vary with respect to single-use bags. They are thicker and have higher capacity and higher mechanical resistance.

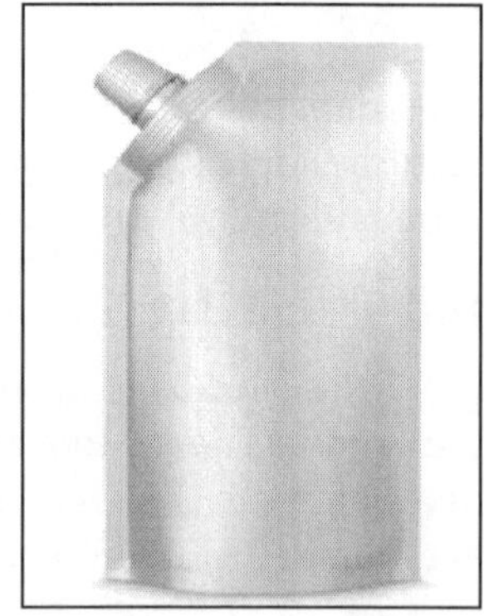

Doy-Pack

Flexible containers designed so that the base allows them to be held in an upright position both filled and empty. They usually incorporate two side longitudinal welds and another one in the upper part. In many cases, they have a mouthpiece to ration the product, closed with a cap.

Nets

Containers formed by crossed plastic threads creating a regular pattern. They are used to contain products like fruits inside, mainly to carry or transport them, since the product is not isolated from the outside environment. They are usually tubular and are closed by two staples.

Film

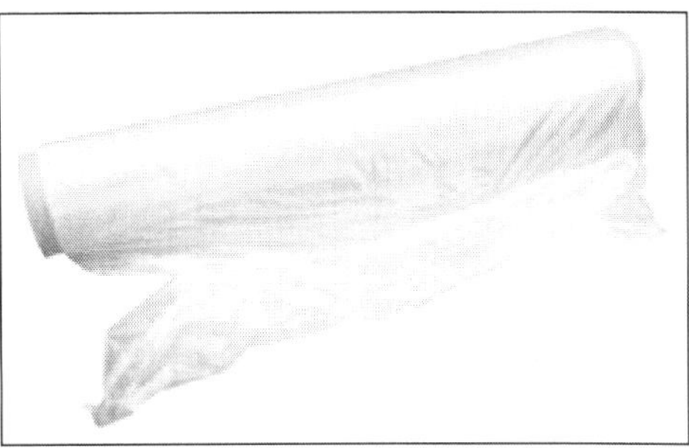

A continuous and thin layer of plastic material with a small thickness. When the thickness is higher than approximately 250 microns, we do not consider it a film, but a sheet. It is used to cover food items.

Stretch Film

Type of film that has a good tensile strength and very high stretching percentages. Due to this property, it is mainly used in packaging for grouping small packaging units. It can also be used in other applications such as combined with trays when packing bakery products and fruit.

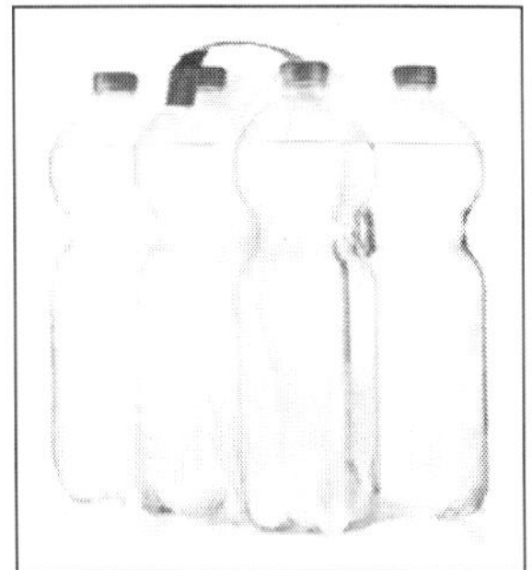

Shrink Film

Type of film that, when subjected to heat, shrinks in one or two directions (longitudinal and transverse). It is used to protect and immobilize the content, thus being in contact with the product. Some examples are packages for cheese wedges and packs of water.

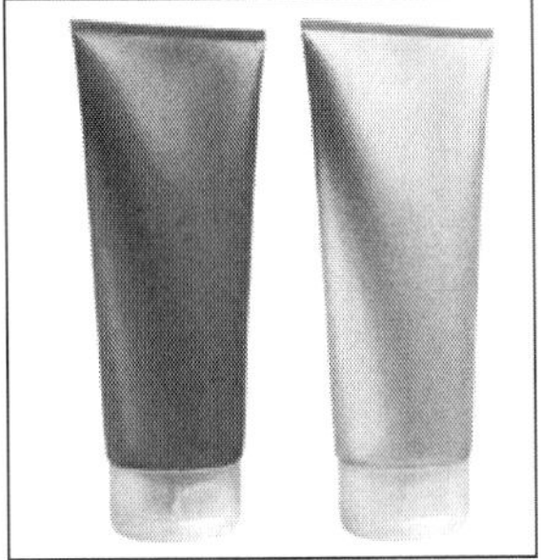

Tubes

Cylindrical containers with a sealed end and a dispenser in the other end. The dispenser is closed with a cap.

Considerations for Use of Different Packaging Materials

The key to successful packaging is to select the package material and design that best satisfy competing needs with regard to product characteristics, marketing considerations (including distribution needs and consumer needs), environmental and waste management issues, and cost. Not only is balancing so many factors difficult, but also it requires a different analysis for each product, considering factors such as the properties of the packaging material, the type of food to be packaged, possible food/package interactions, the intended market for the product, desired product shelf-life, environmental conditions during storage and distribution, product end use, eventual package disposal, and costs related to the package throughout the production and distribution process. Some of these factors are interrelated: for example, the type of food and the properties of the packaging material determine the nature of food *i.e.* package interactions during storage. Other times, the factors are at odds with each other; for example, single-serving packaging meets consumer needs, but bulk packaging is better for environmental reasons. Following considerations provide an overview of the variety of factors at play in package selection:

Product Characteristics

A thorough knowledge of product characteristics, including deterioration mechanisms, distribution needs, and potential interactions with the package, is essential for package design and development. These characteristics concern the physical, chemical, biochemical, and microbiological nature of the product.

Materials

They provide optimum protection of product quality and safety are most preferred. Similarly, distribution systems and conditions help determine the type of packaging material used. In particular, food/package interaction plays an important role in the proper selection of packaging materials for various food applications. Each packaging material has different inherent properties (for example, rigidity and permeability to gases). These properties affect the selection of which material is best for a particular food, given the characteristics of that food (for example, acidity and light sensitivity).

Food/Package Interaction

They involve the transportation of low molecular weight compounds such as gases or vapors and water from (1) the food through the package, (2) the environment through the package, (3) the food into the package, and/or (4) the package into the food. It may also include chemical changes in the food, package, or both. These interactions result in food contamination (a potential health issue), loss of package integrity (a potential safety issue), or decrease in quality. The most common food–package interactions are the migration of low molecular weight substances such as stabilizers, plasticizers, antioxidants, monomers, and oligomers from plastic packaging materials into food. Furthermore, low molecular weight compounds (volatile and nonvolatile) may migrate from food into packaging materials through the sorption mechanism. The volatile substances such as flavors and aromas directly affect food quality while the nonvolatile compounds such as fat and pigments affect the package.

Marketing

Marketing is a prerequisite to successful innovation in the packaging industry; it promotes products in a competitive marketplace and increases consumer choice. Consumers are consistently looking for packages that offer convenience attributes such as resealability, container portability (lightweight materials preferred), ease of opening, convenient preparation features, and product visibility.

Environmental Characteristics

As a comprehensive analysis of the material from production to disposal, life cycle analysis is important in determining the environmental impact of a package. The analysis incorporates a quantitative evaluation of environmental costs, considering issues such as material use, energy consumption, and waste generation. The sustainability goal inherent within the cradle-tocradle concept (imposing zero impact on future generations) builds on life cycle analysis to address material and energy recovery as well. Furthermore, new packaging materials are being developed to facilitate the goal of true sustainability.

Balancing Priorities

Ideally, a food package would consist of materials that maintain the quality and safety of the food indefinitely with no degradation over time; are attractive, convenient, and easy to use while conveying all pertinent information; are made from renewable resources, generating no waste for disposal; and are inexpensive. Rarely, if ever, do today's food packages meet this lofty goal. Creating a food package is as much art as science, trying to achieve the best overall result without falling below acceptable standards in any single category (an exercise in balancing and negotiation).

From a product characteristic perspective, the inertness and absolute barrier properties of glass make it the best choice material for most packaging applications. However, the economic disadvantage of glass boosts the use of alternatives such as plastic. While plastics offer a wide range of properties and are used in various food applications, their permeability is less optimal unlike metal, which is totally impervious to light, moisture, and air. Attempts to balance competing needs can sometimes be addressed by mixing packaging materials such as combining different plastics through co-extrusion or lamination or by laminating plastics with foil or paper. Ultimately, the consumer plays a significant role in package design. Consumer desires drive product sales, and the package is a significant sales tool. Although a bulk glass bottle might be the best material for fruit juice or a sport beverage, sales will be affected if competitors continue to use plastic to meet the consumer desire for shatterproof, portable, single- serving containers.

Levels of Packaging

Packaging systems are categorized into the following four groups:

1. **Primary packaging:** This is the first level of packaging that comes in contact with the product. For example, a tea bag, tomato sachet, beverage tin, canned food *etc.* Their main function is to preserve and for contain of the product. These primary packages should always be made from non-toxic materials and should be compatible with the food been packaged not to cause change in color, flavor, chemical reactions *etc.*
2. **Secondary packaging:** A secondary package contains one or more primary packages and protects it from damage during storage and shipment preventing contaminants from soiling the primary packages. Example of a secondary packages is shrink wrap that holds two or more cans together to ease handling.
3. **Tertiary packaging:** It is made up of number of secondary packages. Example of a tertiary package is the shipping container. Its main function is to protect the product during distribution and to provide for efficient handling. For example, a stretch-wrapped pallet of fibre board boxes of canned food.

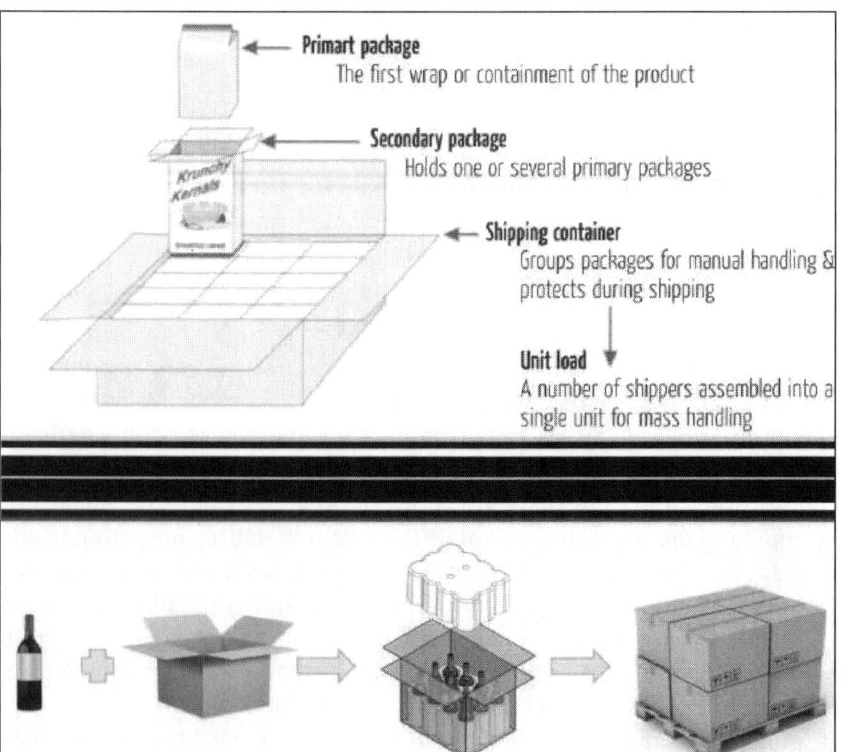

4. **Quaternary packaging or Unit load:** It is used to facilitate handling during transportation and shipping *etc.* For example, a large container containing number of stretch-wrapped pallets of fibre board boxes of canned food.

A unit load means a group of tertiary packages assembled into a single unit. If the corrugated boxes are placed on a pallet and stretch wrapped for mechanical handling, shipping and storage, the single unit is referred to as a "unit load." The objective is to aid in the automated handling of larger amounts of product. A fork-lift truck or similar equipment is used to transport the unit load.

Packaging is defined as a socio-scientific discipline which operates in a society to deliver goods to the ultimate consumer in the best conditions intended for their use. Food packaging performs a number of disparate tasks: it protects the food from contamination and spoilage; it makes it easier to transport and store foods; and it provides uniform measurement of contents. By allowing brands to be created and standardized, it makes advertising meaningful and large scale distribution and mass merchandising possible. Food packages with dispensing caps, sprays, Reclosable openings, and other features make products more usable and convenient. A distinction is usually made between the various level of packaging. A primary package is one that is in direct contact with the contained product. It provides the initial, and usually the major, protective barrier. Examples of primary packages include metal cans, paperboard cartons, glass bottles, and plastic pouches. It is frequently only the primary package that the consumer purchases at retail outlets. A secondary package contains a number of primary packages, for example, a corrugated case. It is the physical distribution carrier and is sometimes designed so that it can be used in retail outlets for the display of primary packages. A tertiary package is made up of a number of secondary packages, the most common example being a stretch-wrapped pallet of corrugated cases. In interstate and international trade, a quaternary package is frequently used to facilitate the handling of tertiary packages. This is generally a metal container up to 40 m in length that can hold many pallets and is intermodal in nature.

Chapter 5

Novel Packaging Technologies: Emerging Approaches in Food Packaging Systems

Need for novel food packaging technologies arose as a result of consumer's desire for convenient, ready to eat, tasty and mild processed food products with extended shelf life and maintained quality. Moreover, modern trend of retail practices and changing modern lifestyle are the incentives for the evolution of novel and innovative packaging techniques without compromising food safety and quality characteristics. Although traditional packaging covers the basic needs of food containment, advances in food packaging are both anticipated and expected. Society is becoming increasingly complex and innovative novel packaging is the result of consumers' demand for packaging that is more advanced and creative than what is currently offered. Rapid growth of novel packaging in food segment is contributed by the enormous use of packaged foods, rising need of prepared foods like use of microwave meals and growing use of smaller size food packages too. Another important reason for innovative food packaging is the rising issues of food borne microbial outbreaks which demand the use of packaging with antimicrobial effects along with retention of food quality.

Innovations in packaging started earlier in the form of electrically driven packaging machinery, metallic cans, aseptic packaging, flexible packaging, aluminium foils and flexographic printing. Additionally, the introduction of various materials, *viz.* polyester, polypropylene, and ethylene vinyl alcohol polymers led to drastic evacuation from metal, paperboard and glass packaging to plastic and flexible packaging. Moreover, in 20th century more advancement in packaging technology appeared as intelligent or smart packaging and active packaging (oxygen scavengers, antimicrobial agents, respiration controllers, and aroma/odour absorbers). These novel technologies besides the basic function of containment increase the margin of food quality and safety. The novel food packaging techniques would thus, help in fulfilling the demands throughout the food supply chain by gearing up towards persons own life style. The emerging changes in packaging industry will strengthen the economy by improving food safety, quality and by minimizing product losses (Vanderroost *et al.,* 2014). Some of the novel food packaging technologies and their way of preservative action, effectiveness and suitability in various types of foods is discussed as below:

Active Packaging

According to the Actipak project, active packaging changes the condition of the packed food to extend shelf-life or to improve safety or sensory properties, while maintaining the quality of the packaged food. Food condition in the definition of active packaging includes various aspects that may play a role in determining the shelf-life of packaged foods, such as physiological processes (*e.g.*, respiration of fresh fruit and vegetables), chemical processes (*e.g.*, lipid oxidation), physical processes (*e.g.*, staling of bread, dehydration), microbiological aspects (*e.g.*, spoilage by micro-organisms) and infestation (*e.g.*, by insects). Through the application of appropriate active packaging systems, these conditions can be regulated in numerous ways and, depending on the requirements of the packaged food, food deterioration can be significantly reduced.

Hence, active packaging is an innovative concept that can be defined as a mode of packaging in which the package, the product and the environment interact to prolong shelf-life or enhance safety or sensory properties, while maintaining the quality of the product Active packaging refers to the incorporation of additives into the package with the aim of maintaining or extending the product quality and shelf life. It can be defined as a type of packaging that changes the condition of the packaging to extend shelf life or to improve safety or sensory properties while maintaining the quality of the packaged food. They are designed to intentionally incorporate components that would release or absorb substances into or from the packaged food or environment surrounding the food (Sivertsvik, 2007). Active packaging is also accurately defined as 'packaging in which subsidiary constituents have been deliberately included in or on either the packaging material or the package headspace to enhance the performance of the package system '(Robertson, 2006). This phrase emphasizes the importance of deliberately including a substance with the intention of enhancing the food product. Active packaging is an extension of the protection function of a package and is commonly used to protect against oxygen and moisture.

Active packaging systems are developed with the goal of extending shelf life for foods and increasing the period of time that the food is of high quality and safe. This goal of active packaging involves application of various strategies like temperature control, oxygen removal, moisture control, addition of chemicals such as salt, sugar, carbon dioxide or natural acids or a combination of these with effective packaging (Restuccia *et al.*, 2010). Hence, active packaging technologies include some physical, chemical or biological action which changes interactions between a package, product, and/or headspace of the package in order to get a desired outcome. In other words, it allows the active preservation of foods, according to their needs, by modification of the environment inside the package by removing undesired gases or by regulating the composition of the gas in the package headspace. Oxygen scavengers, antimicrobial agents, moisture absorbers, ethylene absorbers, ethanol and carbon dioxide emitters can be used as active features in order to address the weak points in the shelf life of packed goods. The internal atmosphere may be regulated by substances that absorb (scavenge) or release (emit) gases or vapors.

Classification of Active Packaging Systems

Active packaging techniques for preservation and improving quality and safety of foods can be classified according to their functionality as scavengers, regulators and emitters, and their action can be specific for several substances (O_2, CO_2, ethylene *etc.*). Active packaging is categorized into three systems:

- **Active scavenging systems (absorbers):** Absorbing (scavenging) systems remove undesired compounds such as oxygen, carbon dioxide, ethylene, excessive water, taints and other specific compounds.
- **Active releasing systems (emitters):** Releasing systems actively add or emit compounds to the packaged food or into the head-space of the package such as carbon dioxide, antioxidants *etc.*
- Other systems

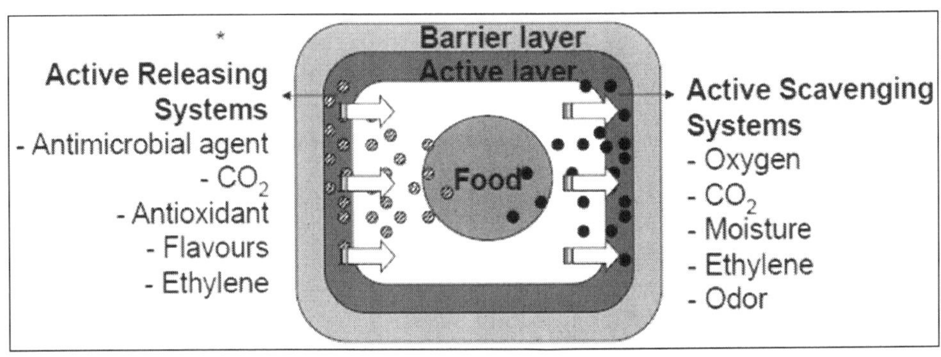

***Source*: Caleb, 2012.**

I. Active Scavenging Systems (Absorbers)

Oxygen Scavengers

Under scavenging packaging system, unwanted compounds such as oxygen, excessive moisture and ethylene which accelerate the spoilage process in foods are removed from the product.

Though oxygen is a key element in supporting life, when it comes to food shelf life, quality and, in some cases, safety, it can become an adversary. The element oxygen is essential to the biological functions in plants and animals involving respiration and photosynthesis. However, in foods it causes the degradative oxidation reactions that result in off-flavors, odors, nutrient loss (through oxidation), colour changes and sometimes-harmful compounds; hence the usage of oxygen scavengers is to maintain quality and extend shelf life of some food products.

i) Oxygen Scavengers through Sachets

Oxygen scavengers are packaging films or devices that chemically react with oxygen to remove it from the packaging environment. This, in turn, protects the product from degradative oxidation reactions. Oxygen scavengers exist in many forms, including sachets, films and enzymes, all of which can be used in food systems and function in a variety of ways. Sachets were developed in the late 1970s in Japan.

Oxygen scavenging sachets are packets of various sizes that are contained within a food package to absorb oxygen. Primarily used for shelf-life extension, sachets are highly effective and can reduce oxygen levels in a package to as low as 100 ppm. However, when dealing with food, oxygen scavenging sachets must be safe, easily handled, compact in size and must not produce toxic substances or offensive odors/gases. They also must absorb a large amount of oxygen, have an appropriate oxygen absorption speed and be economically priced.

The two most common substances used in sachets are iron powder and ascorbic acid. For oxygen scavenging, the sachets essentially utilize the process of rusting, or the oxidation of iron compounds in the presence of oxygen and water. Oxygen absorbers are usually made of powdered iron or ascorbic acid. Iron based scavengers typically do not pass the metal detector inspections on most packaging lines, and in these incidences ascorbic acid is advantageous. Oxygen absorbers in sachets are commonly found in meat and poultry products, coffee, pizzas, baked goods and dried foods.

There are many commercially available oxygen scavenger sachets. Different sachets are used for different food formulations depending on aw and other intrinsic factors. The choice of sachet to use in a given application depends on the intrinsic factors (*a*w, pH, *etc.*) of the food involved. For example, a moist food that could cause bacterial or mold growth would typically require a very fast oxygen-absorbing sachet, such as Mitsubishi's Ageless FX, while a dry food could use a less expensive, slower oxygen absorber like Ageless Z-PK. Oxygen sachets are also used in conjunction with other sachets to improve overall product quality. These sachets include moisture absorbers and carbon dioxide absorbers and are mainly used in roasted or ground coffee packages.

Principles on which scavengers are based:

Oxygen scavenging sachets: Principle is that in general, O_2 scavenging technologies are based on one of the following concepts: iron powder oxidation, ascorbic acid oxidation, catechol oxidation, photosensitive dye oxidation, enzymatic oxidation (*e.g.* glucose oxidase and alcohol oxidase), unsaturated fatty acids (*e.g.* oleic acid or linolenic acid) or immobilised yeast on a solid material (Floros *et al.,* 1997). The majority of presently available oxygen scavengers are based on the principle of iron oxidation.

☆ **Iron Powder Oxidation:** Mechanism:

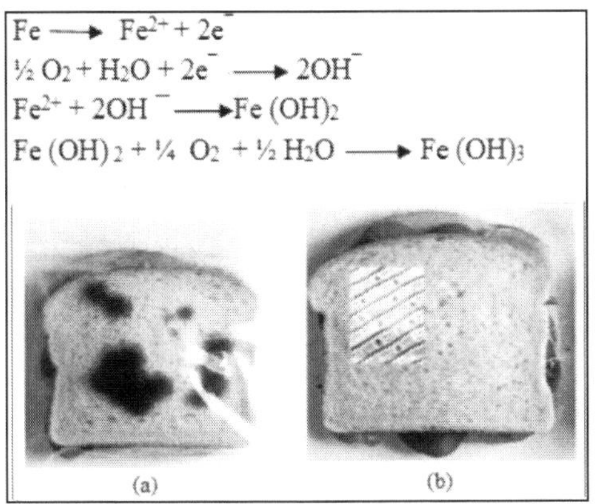

Bread Packaged without (a) and with (b) Oxygen Scavenger.

☆ **Enzymatic Oxidation:** A combination of two enzymes *i.e.* glucose oxidase and catalase, has been applied for oxygen removal. The reaction is:

$$2\ \text{Glucose} + 2\ O_2 + 2\ H_2O \longrightarrow 2\ \text{Gluconic Acid} + 2\ H_2O_2$$

Since H_2O_2 is an objectionable end product, catalase is introduced to break down the peroxide (Rooney, 1995; Vermeiren *et al.,* 1999).

$$2\ H_2O_2 + \text{Catalase} \longrightarrow 2\ H_2O + O_2 + \text{Catalase}$$

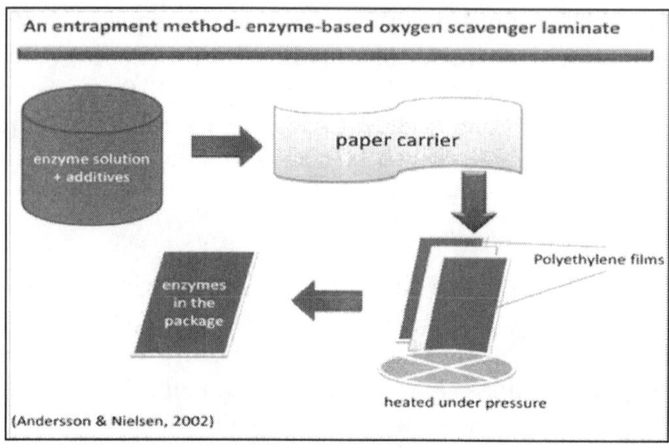

Enzyme-based Oxygen Scavengers.

Benefits of Oxygen Scavenging Through Sachets

Sachets were one of the first promising technologies for oxygen scavenging. Since the sachet is an applied packet, there is very little disruption to the package production line. Oxygen scavenger packaging material technology currently being utilized in beer, fresh pastas, coffee, cured meats, beverages, baked goods *etc.* offers many benefits to food packaging, including the following:

- Product shelf life is extended considerably.
- There is prevention of aerobic pathogens and spoilage organisms.
- The risk of vitamin oxidation is reduced (*i.e.* Vitamin A, C, and E).
- The growth of insects and hatching of insect eggs is prevented (*i.e.* weevils, beetles, *etc.*).
- The color, flavor and overall freshness of products are maintained.
- There are no additives to the food.

Examples of Active Packaging Systems

Absorbers

1. Oxygen Absorbers (Sachets, labels, films, corks)

Purpose

- Reduction/preventing of mould, yeast and aerobic bacteria growth
- Prevention of oxidation of fats, oils, vitamins, colours
- Prevention of damage by worms, insects and insect eggs

- Applications in: Cheese, meat products, ready-to-eat products, bakery products, coffee, tea, nuts, milk powder
- Examples: Ferro-compounds, ascorbic acid, metal salts, glucose oxidases

2. *Carbon Dioxide Absorbers*

Purpose

- Removing of carbon dioxide formed during storage in order to prevent bursting of a package
- Applications in: Roasted coffee, Beef jerkey, Dehydrated poultry products
- Examples: Calcium hydroxide and sodium hydroxide or potassium hydroxide, Calcium oxide and silica gel

3. *Ethylene Absorbers*

Purpose

- Prevention of too fast ripening and softening
- Applications in: Fruits like apples, apricots, mango, tomatoes, avocados and vegetables like carrots, potatoes and Brussels sprouts
- Examples: Aluminium oxide and potassium permanganate (sachets), Activated carbon + metal catalyst, Clay

4. *Humidity Absorbers (Drip-absorbent sheets, films, sachets)*

Purpose

- Control of excess moisture in packed food
- Reduction of water activity on the surface of food in order to prevent the growth of moulds, yeast and spoilage bacteria
- Applications in: Meat, fish, poultry, bakery products, cuts of fruits and vegetables
- Examples: Polyacrylates, Propylene glycol, Silica gel, Clay

5. *Lactose Remover*

Purpose

- Serving milk products to the people suffering lactose intolerance
- Applications in: Milk and other dairy products
- Example: Immobilised lactase in the packaging material

6. *UV - light Absorbers*

Purpose

- Restricting light-induced oxidation

- Applications in: Light-sensitive foods such as ham, Drinks
- Examples: Polyolefins like polyethylene and polypropylene doped with a UV-absorbent agent

7. *Cholesterol Remover*

Purpose

- Improving the healthiness of milk products
- Applications in: Milk and other dairy products
- Example: Immobilised cholesterol reductase in the packaging material

- Markets can be extended for global distribution due to prolonged shelf life.
- Sachets are also fairly inexpensive in relation to the added value that they provide for the quality of the product. Cost savings are realized through reduced waste caused by expired shelf life and less frequently having to replace dated stock.
- Because sachets are made in a variety in sizes, shapes, absorbency levels and forms, many different foods can be catered towards the specific bio-chemical needs of that food.

Limitations

- Sachets cannot be used in liquid foods.
- They may not be used in a package made of flexible film, as the film will cling to the sachet and prevent it from performing its function.
- Sachets have the risk of accidental ingestion by consumers

ii) Oxygen Scavengers Directly in Packaging Materials

In addition to sachets, oxygen-scavenging compounds can be incorporated directly into the packaging material itself. These materials include flexible films, rigid plastics (blow-molded or injection-molded polymers) and liners in closures. Oxygen scavenging components added to the film would include an oxidizable metal, oxidation promoters and fillers. All of these components are usually food-contact approved compounds. They would be directly added to the resin or be a separate layer laminated or co-extruded with the other layers. Typically, this oxygen-scavenging mechanism is moisture-activated. For dry food products, a different approach must be taken due to their low moisture content (aw). Initiating the reaction can be done by incorporating photon dyes into the film. The photosensitive dyes will increase the reaction of the oxygen with iron compounds. This was a result of the discovery that iron will not react quickly enough with ground state oxygen to successfully extend shelf life. This exposure will catalyze the reaction to scavenge the free oxygen in the headspace of the package. The film is illuminated with UV light that will excite the dye molecules, which sensitize oxygen molecules that are diffused into the film, changing them to the singlet state. The singlet oxygen molecules react with acceptor molecules and are consumed.

Benefits

- This method has potential for use in polyethylene terephthalate (PETE) bottles and can be included in many plastic containers and closures.
- Adding scavengers to the plastic rather than a sachet can save many problems. For example, in a packaging film that is tight fitting such as a cheese pack, a sachet to absorb oxygen cannot be used because the tight-fitting film would stifle its functionality.
- Incorporating oxygen absorbing materials into the plastic components of the packaging material could be more efficient.

Potential Risks of Oxygen Scavengers

Although there are many benefits to the safety and quality of food products using an oxygen scavenger, there are facts that should be considered when determining its application. One precaution for oxygen scavenger packaging is the prevention of anaerobic pathogen proliferation. Product developers must thoroughly investigate the food system with which they are intending to use oxygen scavengers. Studies have been done with oxygen scavengers that show pathogens such as *Clostridium botulinium* Type E surviving in an oxygen scavenger environment. The real risk is that the normal aerobic spoilage organisms that act as indicators for the consumer are not present because of the lack of oxygen in the package. Consumers could potentially eat a product with a pathogen unknowingly if other hurdle technology is not applied to the product. Another risk in using this method is that some sachets need a certain amount of moisture to activate the oxygen absorption reaction. Lack of moisture present will result in no activation of the sachet, thus potentially causing quality issues in the product. There are also some commercial sachets that are very sensitive and could therefore, activate prematurely. If this occurs the sachet must be disposed. Since the sachet is only exposed to the main area of the package, it is important that the package does not block the path of the absorber. For this reason, sachets are not capable of being used with any liquid foods or any packaging that may have compartments that may entrap oxygen. Another risk is that the sachet has the potential hazard of being consumed. Also, there is the chance that the sachet may leak. Although the reactants within the sachet are not harmful, there may be a problem that could result as product adulteration or product identity.

Ethylene Scavengers

Ethylene (C_2H_4) is a plant hormone that accelerates the respiration rate and subsequent senescence of horticultural products such as fruit, vegetables and flowers. Many of the effects of ethylene are necessary, *e.g.* induction of flowering in pineapples and colour development in citrus fruits, bananas and tomatoes, but in most horticultural situations, it is desirable to remove ethylene or to suppress its effects. Effective systems utilize potassium permanganate ($KMnO_4$) immobilized on an inert mineral substrate such as alumina or silica gel. $KMnO_4$ oxidizes ethylene to acetate and ethanol and in the process changes colour from purple to brown and hence indicates its remaining ethylene-scavenging capacity. $KMnO_4$-based ethylene scavengers are available in sachets to be placed inside produce packages or inside blankets or tubes that can be placed in produce storage warehouses.

II. Active Releasing Systems (Emitters)

The use of a carbon dioxide emitting package should not induce anaerobic metabolism when carbon dioxide presents at excess amounts. Therefore, the film permeability and the respiration rate should be taken into consideration. Most carbon dioxide emission processes are activated by moisture that usually comes from packaged foods.

Carbon Dioxide Scavenger/Emitter

Another type of active packaging is releasing systems (emitters). There are many commercial sachet and label devices that can be used to either scavenge or emit carbon dioxide. The high level of carbon dioxide in the headspace of packaged fruits and vegetables retards the growth of aerobic microorganisms and reduces the respiration and senescence processes. The use of carbon dioxide scavengers is particularly applicable for fresh roasted or ground coffees that produce significant volumes of carbon dioxide. Fresh roasted or ground coffees cannot be left unpackaged since they absorb moisture and oxygen and lose desirable volatile aromas and flavours. If coffee is hermetically sealed in packs directly after roasting, the carbon dioxide released builds up within the packs and eventually causes them to burst. To circumvent this problem, two solutions are currently used. The first is to use packaging with patented one-way valves that allow excess carbon dioxide to escape. The second solution is to use a carbon dioxide scavenger or a dual action oxygen and carbon dioxide scavenger system. These dual-action sachets and labels typically contain iron powder for scavenging oxygen and calcium hydroxide which scavenges carbon dioxide when it is converted to calcium carbonate under sufficiently high humidity conditions.

Principle

The reactant commonly used to scavenge CO_2 is calcium hydroxide, which, at a high enough water activity, reacts with CO_2 to form calcium carbonate:

$$Ca(OH)_2 + CO_2 \longrightarrow CaCO_3 + H_2O$$

Examples of Sachet and Film Type Releasing Active Packaging Systems

1. *Carbon Dioxide Emitters (Sachets)*

Purpose

- ☆ Growth inhibition of gram-negative bacteria and moulds

Applications in: Vegetables and fruits, fish, meat, poultry

Examples: Ascorbic acid, Sodium hydrogen carbonate

2. *Ethanol Emitters (Sachets)*

Purpose

- ☆ Growth inhibition of moulds and yeast

Applications in: Bakery products, Dry fish

Examples: Ethanol/water mixture absorbed onto silicon dioxide powder generating ethanol vapour

3. *Antimicrobial Preservative Releasers (films)*

Purpose

- ☆ Growth inhibition of spoilage and pathogenic bacteria

Applications in: Meat, poultry, fish, bread, cheese, fruit and vegetables

Examples: Organic acids, *e.g.* sorbic acid, Spice and herb extracts, Allylisothiocyanate

4. *Sulphur Dioxide Emitters (sachets)*

Purpose

- ☆ Inhibition of mould growth

Examples: Sodium - metabisulfite incorporated in microporous material

Applications in: Fruits

5. *Antioxidant Releasers (films)*

Purpose

- ☆ Inhibition of oxidation of fat and oil

Examples: BHA, BHT, Tocopherol

Applications in: Dried foodstuffs, Fat-containing foodstuffs

6. *Flavouring Emitters (films)*

Purpose

- ✰ Minimisation of flavour scalping
- ✰ Masking off-odours
- ✰ Improving the flavour of food

Examples: Various flavours in polymers

7. *Pesticide Emitters (the outer or inner layer of packaging material)*

Purpose

- ✰ Prevention of growth of spoilage bacteria
- ✰ Fungicidal or pest control

Examples: Imazalil, Pyrethrins

Applications in: Dried, sacked foodstuffs, *e.g.*, flour, rice, grains

III. Other Examples of Active Packaging Systems

1. Self-heating Cans

A Self-heating can is an enhancement of the common food can. Self-heating cans have dual chambers, one surrounding the other, making a self-heating food package. The inner chamber holds the food or drink, and the outer chamber houses chemicals that undergo an exothermic reaction when combined. When the user wants to heat the contents of the can, a ring on the can when pulled breaks the barrier that keeps apart the chemicals in the outer chamber from the water. In another type, the chemicals are in the inner chamber and the beverage surrounds it in the outer chamber. To heat the contents of the can, the user pushes on the bottom of the can to break the barrier separating the chemical from the water. This design has the advantages of being more efficient (less heat is lost to the surrounding air) as well as reducing excessive heating of the product's exterior, causing possible discomfort to the user. These are usually made up of aluminium or steel. The heating agent and responsible reaction vary from product to product. However, calcium oxide is mostly used.

2. Self-cooling Cans

These are also made up of aluminium or steel. These are based on liquid nitrogen or other cooling gas. The beverage can have a built- in Heat Exchange Unit (HEU) which contains the technology necessary to chill the drink in under a minute. Upon activation (pressing the button at the base of the can), activates the environmentally safe reclaimed CO_2, or other cooling gas being used, in the HEU that leads to the chilling of the beverage.

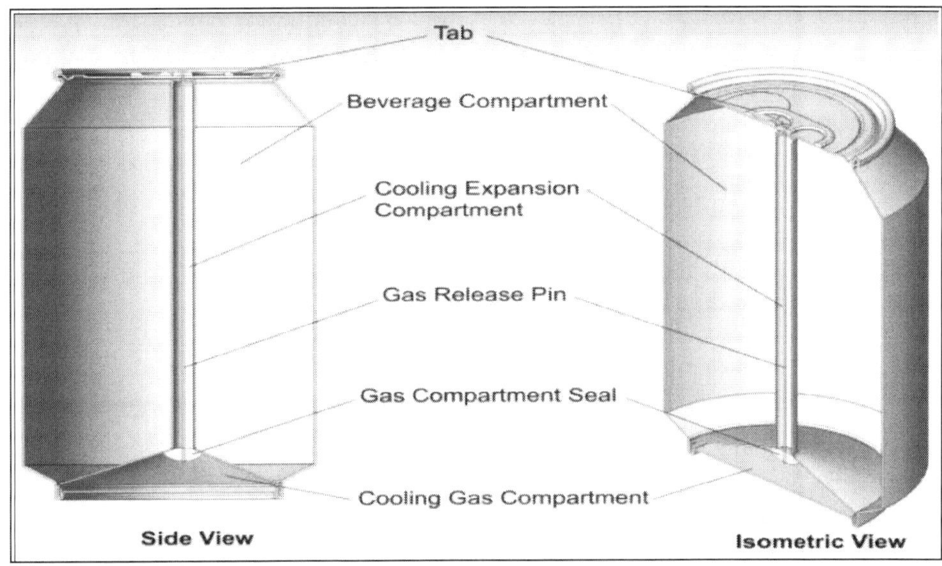

Self-cooling Can

Antimicrobial Food Packaging

Antimicrobial packaging is a form of active packaging. Active packaging interacts with the food product directly or the space between the package and the food to preserve product quality and elongate shelf life. Antimicrobial food packaging acts to decrease, inhibit or hinder the growth of microorganisms that maybe present in the packed food or packaging material. Antimicrobial agent in food packaging is incorporated either directly into food particle or to the packaging material where it is released over a period of time to maintain the products quality, in that case leading to its extended shelf life. Antimicrobial packaging can effectively inhibit targeted bacteria when appropriate amount of antimicrobial agents are incorporated into polymer film. The efficacy of antimicrobial packaging is greater compared to direct addition of preservative agents into food due to two important reasons.

Benefits of Antimicrobial Food Packaging as Active Packaging

The usage of antimicrobial packaging has more advantages compared to direct adding of antimicrobial agents onto foods because antimicrobial agents added on food surfaces by sprays or drips are not effective enough to inhibit microorganisms. This is due to rapid diffusion of antimicrobial agent into foods and denaturation of the active substances by food constituents which reduce the reactivity of antimicrobial agent. Whereas, antimicrobial packaging offers slow and continuous migration of antimicrobial agent from packaging material to food surfaces which enable antimicrobial agent to maintain a high concentration over a long period. The attachment of antimicrobial agents with polymer film enables slow release of antimicrobial agents function over longer period. Moreover, antimicrobial activities of preservative agents may experience inactivation (such as neutralization, hydrolysis, dilution and *etc.*) by food matrixes and components when added directly into the food.

Besides, direct addition of preservative agents into food can reduce the food quality as resulted by changing the organoleptic and textural qualities of the food. Thus, antimicrobial packaging is playing important roles to inhibit the growth of targeted bacteria on foods while improving foods safety and prolong shelf-life without scarifying the foods quality.

Antimicrobial Food Packaging

Antimicrobial Substance	*Packaging Materials*	*Food*	*Microorganisms*
Benzoic acids	PE	Culture media	Total bacteria
Parabens	LDPE	Simulants	*S. cerevisiae, Asp. niger, Pen. spp.*
Benzoic and sorbic acids	PE	Culture media	*S. cerevisiae*
Acetic, propionic acid	Chitosan	Water	*Pen. spp.*
Lysozyme, nisin, EDTA	SPI, zein	Culture media	*E. coli, Lb. plantarum*
Imazalil	LDPE	Cheese	Moulds
Clove extract	LDPE	Culture media	Microorganisms
Eugenol, cinnamaldehyde	Chitosan	Ham	Enterobac., lactic acid bacteria
Allyl isothiocyanate	PE film	Chicken, meat	*E. coli, L. monocytogenes*

Effectiveness of Antimicrobial Agents

Antimicrobial functions which are achieved by adding antimicrobial agents in the packaging system or using antimicrobial polymeric materials show generally three types of mode; (i) release; (ii) absorption; and (iii) immobilization.

Antimicrobial agents have different activities on different pathogenic microorganisms due to their diverse compositions. Classification of microorganisms can be very helpful in the type of antimicrobial agent to be used. This can consist of cell wall composition:

- ✰ Gram-Negative and Gram-Positive,
- ✰ Oxygen Requirements: Aerobes and Anaerobes,
- ✰ Growth Stage: Spores and Vegetative Cells,
- ✰ Acid/Osmosis Resistance,
- ✰ Optimal Growth Temperatures- Mesophilic, thermophilic, psychotropic.

Some antimicrobial agents inhibit the essential metabolic pathways of microorganisms and some cell wall/membrane structure. Example, lysozyme without the inhibition of metabolic pathways causes the destruction of cell wall while EDTA and lactoferin behave as coupling agents of charged polymers and cationic ions. Antimicrobial agents, therefore, perform two important functions;

- Microbial cidal
- Microbial static effects

For microbial static effects, the antimicrobial substance has to possess the active function of maintaining the concentration above the minimal inhibitory concentration during the entire storage period or shelf life in order to prevent the growth of target microorganisms. Various antimicrobial substances that have been incorporated into food packaging systems. These commonly include antioxidants, antimicrobials, biotechnology products, natural antimicrobials, antimicrobial polymers, essential oils *etc.*

Considerations for Designing Antimicrobial Packaging System

Several conditions should be taken into consideration when designing an antimicrobial packaging system. First, the regulatory status of the antimicrobial agent is important. Some antimicrobial systems rely on diffusion or release of the antimicrobial agent in order for it to be effective. It is really important to know the variation in antibacterial activity of the agents when incorporated into the packaging film from its original activity in order to establish the levels that need to be incorporated for effective bacterial inhibition. This would place the packaging material under the classification of an indirect food additive and the material would require review by the US Food and Drug Administration (FDA) along with similar examination in other countries. Generally, approval may not be difficult if the additive has already been effectively used as a direct food additive as long as migratory concentrations and conditions of use are addressed. Another issue is the cost-to-benefit ratio. Some antimicrobial systems can be effective, but if produced on a large scale, they might require expenses beyond the benefits obtained by an extended shelf life or improvement in quality. Lastly, there are numerous technical challenges related to coating methods, the rate of curing, the ease of heat sealing, the effects on physical and mechanical properties of film, the effects on colour, the texture or flavour of the food, and the ability of the antimicrobial agent to provide effectiveness throughout the product life cycle.

Types of Antimicrobial Agents that can be Incorporated into Packaging Materials

i) Antimicrobial Packaging with Volatile Gases

The most popular commercialized products available in the market are those packaging that use volatile gas-form antimicrobial agents such as chlorine dioxide, ethanol and sulfur dioxide. These AM agents are often enclosed separately in sachets/pads that are attached to the internal part of the package. The AM agents will release in vapor form to the headspace of packaging to contact with food products. Commercial product includes Microgarde as marketed by BarrierSafe Solutions International Inc., USA exists in stickers or sheets form containing sodium chlorite and acid precursors. When humidity or moisture in the air is high, it enters the Microgarde™ sheet and initiates the solid-state dry reaction subsequently producing chlorine dioxide that diffuses throughout the package to inhibit microbial contamination and control odor. Microgarde™ is found effective to inhibit aerobic total viable count on iceberg lettuce. Another commercial product called Ethicap™ from Japan, developed by Freund Corp., is ethanol vapor emitter. Ethicap is a paper wafer contains ethanol-silicon dioxide in acetic acid that sandwiched between layers of plastic films that permeable to ethanol. The ethanol vapor released into packaging headspace is able to retard mold growth. This technology is often applied for bakery and dried fish products. Besides, silver compound is also employed as the AM agent in food packaging film. Silver is generally recognized by public as hygiene-improving and antitoxic material. In fact, silver substituted Zeolite products have been widely accepted by public when applied in plastic films for food packaging. Examples of commercialized products included Zeomic™, AgIon™, Novaron™ and Cleanaid™. Sodium ions of Zeolites are substituted by silver ions to inhibit a wide range of microorganisms such as bacteria and mold by disrupting the microbial enzymes activities. For example, Zeomic, product of Sinanen Zeomic Co. is able to control the growth of gram-positive bacteria (methicillin-resistant *Staphylococcus aureus*), gram-negative bacteria (*Escherichia coli* and *Pseudomonas aeruginosa*), and fungi (*Aspergillus niger* and *Penicillium nigricans*). Zeomic™ has been widely applied in chopping board, food packaging films, glove and lunch box.

Sanitizer and fungicide that are considered as non-food grade antimicrobial agents are allowed to be added into food packaging material. These antimicrobial agents are usually immobilized through covalently bonding to polymer materials for limiting the migration of AM agents into food products. Example of commercialized product that incorporates sanitizer to food packaging system is Microban which is developed by Microban International Ltd in USA. Microban utilizes triclosan as an antimicrobial agent and this triclosan-incorporated food packaging was approved by European Union who regulates that migration of triclosan to foods should not exceed 5 mg/kg. There were a wide range of microorganisms claimed sensitive to this packaging. By using plate overlay assay method, it was found that this product inhibited the growth of *Salmonella typhimurium* (ATCC 14028), *Staphylococcus aureus* (ATCC 12598), *Brochothrix thermosphacta* (ATCC 11509), *Bacillus subtilis* (ATCC 6051), *E. coli* (ATCC25922), *Shigella flexneri* (ATCC 12022) and several strains of *E. coli* O157:H7. However, when tested on refrigerated, vacuum-packaged meat surfaces, that product did not effectively reduce the number of bacteria strains. Among the plant extract antimicrobial agents, only wasabi extract antimicrobial products have been commercialized. The antimicrobial packaging used wasabi extract called Wasapower is well known developed by Sekisui Plastics Co. in Japan. Wasabi is a traditional food that often served by Japanese together with sushi. They believe wasabi has the ability to kill microorganisms on raw meat. Volatile allyl isothiocyanate (AIT) is the main antimicrobial component in wasabi extract that inhibits bacteria such *E. coli* and *S. aureus* as well as fungi *A. niger* and *Penicillium italicum.*

ii) Antimicrobial Packaging with Essential Oils and Plant Extracts

The antimicrobial activities of essential oils and plant extracts are well known and numerous of research outcomes have been published on the antimicrobial activities of plant essential oils against food-borne pathogens. Since essential oils are rich in volatile terpenoids and phenolic particles, they are highly potential to inhibit a wide spectrum of microorganisms. Generally, the active components of plant essential oils inhibit microorganisms through disruption of the cytoplasmic membrane, disrupting the proton motive force, electron flow, active transport and inhibition of protein synthesis. Examples of plant extracts and essential oils that most widely incorporated into food packaging are linalool, thymol, carvacrol, clove oil, cinnamaldehyde and basil essential oils.

iii) Antimicrobial Packaging with Enzyme

Lysozyme, obtained from hen egg white is an enzyme classified as natural antimicrobial agent and commonly incorporated into packaging materials. Similar with other antimicrobial agents, lysozyme is less effective towards gram-negative bacteria due to the existing of protective lipopolysaccharide (LPS) layer on the bacteria cell wall. Gram-positive bacteria are very susceptible to lysozyme, because their membrane is made up of 90 per cent peptidoglycan where hydrolyzation of β-1-4 glycosidic linkage between the N-acetyl muramic acid and N-acetyl glucosamine take place. In order to further improve the antimicrobial effectiveness of enzyme, other agents such as detergents and chelators were usually added. Lactoferin can also be added to improve the activity of lysozyme towards gram-negative bacteria. Lactoferin is a natural component of milk with large cationic patches that disrupting the outer membrane permeability resulting the release of LPS. The combination usage of lysozyme and disodium ethylenediaminetetraacetic acid (Na_2EDTA) as the antimicrobial agent. EDTA is a chelating agent that has the ability to destabilize the LPS layer of gram-negative bacteria. Zein films incorporated with lysozyme showed antimicrobial effect on *B. subtilis* (ATCC 6633) and *L. plantarum* (DSM 1954). By the addition of Na_2EDTA, the films also became effective on gram-negative bacteria, *E. coli* (ATCC 53868).

iv) Antimicrobial Packaging with Bacteriocin

Antimicrobial agents generated by bacteria called bacteriocin were gradually gaining popularity attribute to their ability to withstand elevated temperatures and acidic environment. Bacteriocin is metabolic byproduct (antimicrobial peptide) produced by bacteria defense system from almost all types of bacteria. This naturally occurred activity allowed bacteria of one strain to inhibit the growth of other strains in adjacent. Bacteriocin produced by lactic acid bacteria (LAB) has highly acceptance from public, because LAB has been the important bacteria for food fermentation for centuries.

There are different methods of incorporating these antimicrobial agents into food packaging materials, they include:

- **Addition of sachets/pads containing antimicrobial agents to packages:** One of the most successful commercial application of antimicrobial packaging has been sachets that are attached to the internal part of the package. Examples of antimicrobials attached to the interior of packaging materials are oxygen absorbers, moisture absorbers and ethanol vapor generators. Oxygen and moisture absorbers are used primarily in bakery, pasta produce and meat packaging to prevent oxidation and water condensation. Oxygen absorbers should ordinarily not be classified as an antimicrobial but a decrease in oxygen inhibits the growth of aerobes, particularly molds. Moisture absorbers can reduce the moisture content also indirectly affecting microbial growth. Ethanol vapor generators consist of ethanol absorbed or encapsulated in packaging materials and enclosed in polymer packets. The ethanol invades the selective barrier and is released into the headspace within the package. Organic acids and surfactants have been incorporated into these pads to prevent microbial growth in the exudates, which are rich in nutrients.
- **Coating or adsorbing antimicrobials to polymer surfaces:** Antimicrobials that cannot tolerate the temperatures used in polymer processing are often coated onto the material after forming or are added to cast films. Cast edible films, for example, have been used as carriers for antimicrobials and applied as coatings onto packaging materials or foods. Examples include nisin/methylcellulose coatings for polyethylene films *etc.* Proteins have an increased capacity for adsorption due to their amphiphilic structure. Glucose oxidase has been coated onto moisture proof fabric sheets by using polyvinyl alcohol, starch and casein as adhesives.
- **Immobilization of antimicrobials by ionic or covalent linkages to polymers:** This type of immobilization requires the presence of functional groups on both the antimicrobial and the polymer. Examples of antimicrobials functional groups are peptides, enzymes, polyamines and organic acids. In addition to functional antimicrobials and polymer supports, immobilization may require the use of 'spacer' molecules that link the polymer surface to the bioactive agent. These spacers allow sufficient freedom of motion so the active portion of the agent can contact microorganisms on the food surface. Spacers that could potentially be used for food.

- **Incorporation of antimicrobial agents directly into Polymers:** Antimicrobials have been incorporated into paper, thermoplastics and thermosets and have been tested against a variety of microorganisms including *Listeria monocytogenes*, pathogenic *E. coli*, and spoilage organisms including molds. Silver substituted zeolites are the most widely used as polymer additives for food applications, especially in Japan. Sodium ions present in zeolites are substituted by silver ions, which are antimicrobial against a wide range of bacteria and molds. These substituted zeolites are incorporated into polymers like polyethylene, polypropylene, nylon and butadiene styrene at levels of 13 per cent. Silver ions are taken up by microbial cells disrupting the cells' enzymatic activity. Antimicrobial enzymes such as lactoperoxidase and lactoferin, antimicrobial peptides such as magainins, cecropins, defensins, natural phenols like hydroquinones and catechins, fatty acid esters, antioxidant phenolics, antibiotics have also been added. The rationale for incorporating antimicrobials into the packaging is to prevent surface growth in foods were a large portion of spoilage and contamination occurs. This approach can reduce the addition of larger quantities of antimicrobials that are usually incorporated into the bulk of the food. The gradual release of an antimicrobial from a packaging film to the food surface may have an advantage over dipping and spraying. In the latter processes, antimicrobial activity may be rapidly lost due to inactivation of the antimicrobials by food components or dilution below active concentration due to migration into the bulk food matrix. Emulsifiers and fatty acids, for example, are known to interact with nisin reducing the bacteriocin activity.

Antimicrobial Packaging with Inherently Bioactive Synthetic Polymers

Non-Migratory Bioactive Polymers (NMBP) in Food Packaging

Non-migratory bioactive polymers (NMBP) are a class of polymers that possess biological activity without the active components migrating from the polymer to the substrate. NMBPs are inherently bioactive polymers and polymers with covalently immobilized bioactive agents. For inherently bioactive polymers, the structural polymer itself is bioactive. For example, polymers containing free amines have been shown to be antimicrobial. Included in this definition are structural polymers with modified backbones. These polymers differ from those with immobilized bioactive compounds in that no previously synthesized bioactive compound is attached to the polymer chain. Most examples of inherently bioactive polymers involve antimicrobial activity.

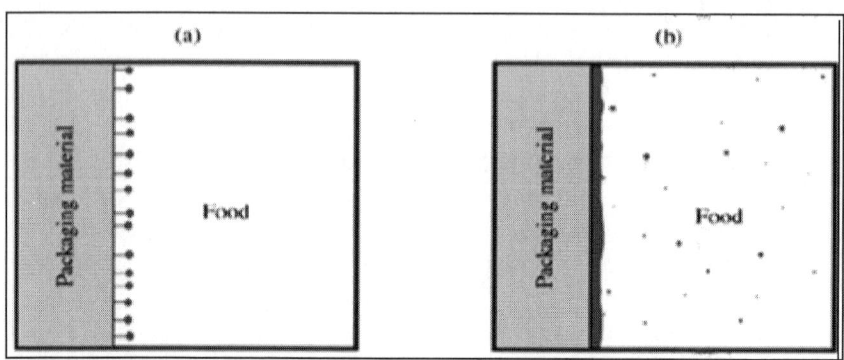

Simplified Visual Comparison of (a) Non-migratory and (b) Migratory Bioactive Packing (Adapted Form Han, 2000)

This concept has existed for some time and has been applied primarily to immobilized enzyme processing. It is only now becoming of interest in packaging applications. The goal is to use bioactive materials for which the response is desirable from the standpoint of the package or the product, for example inhibition of microbial growth or flavour improvement. Enzymes are classic examples of bioactive substances, as are many peptides, proteins, and other organic compounds. The definition, from the perspective of packaging, is based on function *i.e.* the way the substance interacts with living systems. Bioactive polymers can be formed by attachment of bioactive molecules to synthetic polymers, as in the case of enzyme immobilization, or may result from an inherent bioactive effect of the polymer structure, as with chitosan. They have potential applications in the packaging of food and other biological materials, in food processing equipment, on biomedical devices and in textiles.

Types and Applications of Inherently Bioactive Synthetic Polymers

Non-migratory bioactive polymers (NMBP) are a class of polymers that possess biological activity without the active components migrating from the polymer to the substrate. This concept has existed for some time and has been applied primarily to immobilized enzyme processing. It is only now becoming of interest in packaging applications. Bioactive materials are based on molecules that elicit a response from living systems. The goal is to use bioactive materials for which the response is desirable from the standpoint of the package or the product, for example inhibition of microbial growth or flavour improvement. Enzymes are classic examples of bioactive substances, as are many peptides, proteins, and other organic compounds. The definition, from the perspective of packaging, is based on function: the way the substance interacts with living systems. Purely physical processes, for example adsorption or diffusion, are excluded from this definition.

Bioactive polymers can be formed by attachment of bioactive molecules to synthetic polymers, as in the case of enzyme immobilization, or may result from an inherent bioactive effect of the polymer structure, as with chitosan. They have potential applications in the packaging of food and other biological materials, in food processing equipment, on biomedical devices and in textiles. Non-migratory bio polymers are defined to be those for which the bioactive component does not migrate out of the polymer system into the surrounding medium. Typically, this is achieved through covalent attachment of the active component to the polymer backbone, inherently bioactive polymer backbones, or entrapment of the active component within the polymer matrix.

There are two main types of NMBP *i.e.* inherently bioactive polymers and polymers with covalently immobilized bioactive agents. For inherently bioactive polymers, the structural polymer itself is bioactive. For example, polymers containing free amines have been shown to be antimicrobial. Included in this definition are structural polymers with modified backbones. These polymers differ from those with immobilized bioactive compounds in that no previously synthesised bioactive compound is attached to the polymer chain. Most examples of inherently bioactive polymers involve antimicrobial activity and are discussed as under:

i) Chitosan

Chitosan is the deacetylated form of chitin (poly-β-(1,4)-*N*-acetyl-D-glucosamine), a common natural biopolymer extracted from the shells of crustaceans. Production of chitosan from chitin involves demineralization, deproteinization, and deacetylation. The key feature of the antimicrobial effect of chitosan is probably the positive charge that exists on the amino group at C-2 below pH 6.3. The properties of chitosan films, including antimicrobial efficacy, mechanical and barrier properties, are significantly affected by the degree of deacetylation.

Chitosan is probably the most studied inherently bioactive NMBP to date. Chitosan has some good properties such as it is natural polymer, non-toxic, film-forming ability and biodegradable. Besides, it has inherent antimicrobial properties and able to inhibit the growth of wide range of food pathogens. It possesses broad spectrum antimicrobial activity in simple media and is available commercially as an antifungal coating for shelf-life extension of fresh fruit.

However, owing to the poor solubility of chitosan at high pH value, its application is only effective in acid medium. The only weakness of chitosan film is the poor mechanical properties compared to synthetic polymer. Thus, chitosan can be coated on plastic film to compromise mechanical properties and enhance antimicrobial activities. Most researchers found that chitosan is highly effective towards gram-negative bacteria. This is because the positively charged chitosan molecules can interact better with negatively charged teichoic acid backbone in cell wall of gram-negative bacteria. In order to further enhance the antimicrobial properties of chitosan, other antimicrobial agents were also added into chitosan-coated films.

These synergistic combinations can inhibit wider range of microorganisms due to the ability of chitosan to function as carrier for other additives.

Applications of chitosan are used in:

- ✰ Food packaging
- ✰ In microbial growth broths, chitosan has been found effective against gram-positive and gram-negative bacteria, along with some moulds and yeasts
- ✰ As an edible antimicrobial film for fish fillets

ii) UV Irradiated Nylon

A recent development is surface modification of polymers leading to antimicrobial activity, for example treatment of nylon with an excimer laser at UV frequencies (193 nm). This has been described as a physical modification, although the actual change which leads to the induced antimicrobial activity is a chemical change: amides on the nylon surface are converted to amines, which remain bound to the polymer chains, as observed with X-ray photoemission spectroscopy (XPS). Antimicrobial nylon-6,6 is prepared by irradiating with an UV excimer laser at 193 nm for a total exposure of 1-3 J/cm^2. This results in conversion of approximately 10 per cent of the surface amides and some *etc.*hing of the film surface. The antimicrobial effect is strongly dependent on the wavelength of the laser used, with films treated at 193 nm showing a 5-log reduction in *K. pneumoniae* in one hour, while film treated at 248 nm has no antimicrobial effect. XPS analysis of the surface of film treated at 193 nm indicated that surface amide groups were converted to amines, while film treated at 248 nm showed no such change. The mechanism of antimicrobial activity is presumably similar to that of chitosan, poly-L-lysine and other cationic polymers, involving interaction with negatively charged microbial membranes leading to membrane disruption and leakage of cellular constituents.

Intelligent or Smart Packaging

The term smart packaging was coined in the mid-1980s to describe package structures that sensed changes in the internal or surrounding environment and altered some of their relevant properties in response. Intelligent, or smart packaging is basically designed to monitor and communicate information about food quality. Intelligent packaging refers to the use of packaging as an intelligent messenger to monitor the condition and provide quality information of packed foods to the consumers. In other words, intelligent packaging is an extension of the communication function of traditional packaging, and communicates sense, detect, or record external or internal changes in the product's environment (Biji *et al.*, 2015). Intelligent packaging could be defined as a packaging system that is capable of carrying about intelligent functions (such as sensing, detecting, tracing, recording and communicating) to facilitate decision making to extend shelf life, improve quality, enhance safety, provide information, and warn about possible problems. It contains an external or internal indicator to provide information about aspects of the history of the package and/or the quality of the food. Intelligent packaging systems have components that sense the environment and process the information and then allow action to protect the product by conducting communication functions.

The purpose of the intelligent systems is to improve the quality or value of a product, to provide more convenience, or to provide tamper or theft resistance. It is a great tool for monitoring possible abuse that has taken place during the food supply chain. Intelligent packaging can report the conditions on the outside of the package, or measure the quality of the food product inside the package. There is currently work being developed with labels or seals that are transparent until a package is opened. Once the package is tampered with, the label or seal will undergo a permanent colour change and may even spell out 'opened' or 'stop'. In order to measure product quality within the package, there must be direct contact between the food product or headspace and the quantity marker. Examples of intelligent packaging are those that give the time-temperature history (TTIs), indicate microbial growth, ripeness indicators, shock, leakage, spoilage, product authenticity, pack integrity *etc.* The intelligent packaging monitors the condition of packaged foods and gives information about the quality of the packaged food during transport and storage. At the moment, freshness and leakage indicators are commercially available for monitoring food quality. High temperatures are often correlated with food deterioration as result of irreversible biochemical reactions combined with microbial growth. It is essentially an integrating method that deals with mechanical, chemical, electrical and/or electronically-driven functions that enhance the usability or effectiveness of the food products in a proven way (Kit *et al.*, 2017). Some of the examples of intelligent packaging are explained below:

Indicator	Principle/reagents	Gives information about	Application
Time-temperature indicators (external)	Mechanical Chemical Enzymatic	Storage conditions	Foods stored under chilled and frozen conditions
Oxygen indicators (internal)	Redox dyes pH dyes Enzymes	Storage conditions Package leak	Foods stored in packages with reduced oxygen concentration
Carbon dioxide indicator (internal)	Chemical	Storage conditions Package leak	Modified or controlled atmosphere food packaging
Microbial growth indicators (internal/external) i.e. freshness indicators	pH dyes All dyes reacting with certain metabolites (volatiles or non-volatiles)	Microbial quality of food (i.e. spoilage)	Perishable foods such as meat, fish and poultry
Pathogen indicators (internal)	Various chemical and immunochemical methods reacting with toxins	Specific pathogenic bacteria such as *Escherichia coli* O157	Perishable foods such as meat, fish and poultry

Temperature Controlling Packaging

These could be a good application to prevent temperature changes in the produce. A time–temperature indicator (TTI), being used as an indicator of product safety and quality is leading the way in packaging technology. It is a small measuring device attached to the package surface that exploits a change in a physical or physiochemical property to produce irreversible evidence of exceeding a predetermined temperature threshold or record the cumulative time-temperature history. The TTI reaction is highly accurate and precise, and is based on simple irreversible mechanical, chemical, enzymatic, or microbial changes. A TTI can be placed on shipping containers or individual packages as a small self-adhesive label, and an irreversible change, like a colour change, will result when the TTI experiences abusive conditions. TTIs are particularly useful with chilled or frozen foods, where the cold storage during transportation and distribution are important for food quality and safety.

TTIs are provided in different versions, with different activation temperatures and flow times. Depending on their type, they should be appropriate for specific product categories with defined temperature sensitivity. Systematic kinetic models for both the TTI reaction and quality deterioration reaction of each particular produce have been developed and used for the prediction of the shelf life in the market

- ✰ **Thermochromic Inks:** Inks are available that are temperature sensitive and can change colours based on temperature. These inks can be printed onto packages or labels such that a message can be conveyed to the consumer based on the colour of the ink they are seeing. These inks are available with colour change activation at temperatures from -10°C up to 70°C and are reversible, changing either way as the ink warms or cools. Thermochromic inks can let a consumer know whether a package is too hot to touch, or cold enough drink. If appropriate colors are chosen, then hidden messages such as "DRINK NOW" or "TOO HOT" become visible. They are becoming popular technology for beverages (Robertson, 2006). The inks used can be adversely affected by UV light and temperature over 121°C, so consumers should not fully rely on the inks message when it comes to deciding the proper time to consume a food.
- ✰ **Freshness indicators:** They detect the presence of microbiological metabolites. They are attached inside the packaging and since they are dependent on any inference from temperature history, they signal directly product quality of the packed food. They detect volatile or non-volatile compounds or changes in the product itself. Used methods are visible indicator tags in contact with package headspace, labels and optical detectors. The indicator systems for metabolites include colour change of a dye or liquid crystals, formation of colour compounds, changes in optical properties or a miniaturized electronic nose. The sole use of freshness indicators can be enhanced, and most important of all, brought

to a universally interpretable level by extending the indicator with interpretation instructions which are accepted and understood by all parties involved in the handling chain of the perishable goods in question.

- **Gas Indicators:** Food is a complicated material to package because it is capable of respiration and therefore, may change its own atmosphere when inside a package. The gas composition within a package can easily change due to the interaction of food with its environment. Gas indicators are a helpful means of monitoring the composition of gases inside a package by introducing a change in the colour of the indicator through a chemical or enzymatic reaction. The indicators must be in direct contact with the gaseous environment directly surrounding the food in a package. Indicators are capable of signaling whether there is a gas leakage in the package, or they may be used to verify the efficiency of an oxygen scavenger. They typically signal the presence or absence of oxygen and/or carbon dioxide. Oxygen in the air can cause oxidative rancidity, unwanted colour changes in foods, and allow aerobic microbes to grow on foods. Oxygen indicators typically result in a colour change when oxygen is present, and the presence of oxygen can indicate that the package has a leak or has been tampered with. Oxygen indicators can also indicate improper sealing of a package.

  Gas indicators are also being developed to detect water vapours, ethanol, and hydrogen sulfide.

- **Microwave Doneness Indicators (MDIs):** MDIs detect and indicate visually the state of readiness of foods heated in a microwave/oven. These indicators are able to detect the readiness of foods that are heating in microwave ovens and signal to consumers when foods are safe to eat. The biggest challenge in this field at the moment is the ability to evenly heat foods in the microwave so that there is a defined stage in which a food could be called safe to eat.

Currently, food heat non-uniformly, and hot spots occur throughout the food. These hot spots would trigger a doneness indicator while cooler regions would not have reached acceptable cooking temperatures. An ideal MDI would be located on the lid or dome of the microwave container so that consumer could easily see the visual indicator for doneness (Robertson, 2006). This would be functional as a food is heated in the microwave and the space above the food would heat and transfer to the lid. The relationship between the temperatures of the food and the temperature of the lid would be the basis of the indicator system.

It would be important that the indicator not give false reading because the device itself heats in the microwave. The indicators must be viewable by the consumer without having to open the microwave. As of yet, MDIs do not exist commercially, but their arrival is much anticipated. The issue with this system would be that different ovens heat foods differently, different foods have different dielectric and thermal properties, different packages come in different shapes and designs, and cooking from one item to the next will never be perfectly replicated.

Radio Frequency Identification: Another intelligent packaging is radio Frequency Identification (RFID). RFID is the use of radio frequencies to read information on a small device known as tag. RFID is the term used for any device that can be sensed at a distance by radio frequencies with few problems from disorientation. This technology uses radio waves for product identification and traceability. It includes incorporating an RFID tag which is an advanced form of data information carrier that can identify and trace a product. RFID tag is incorporated into the package, from which a proper sensor is used to collect data about the item's status. The data stored in tags are activated by the sensor, which is then transmitted to a reader for decoding and processing by a computer system. The data are about the product identification (*e.g.*, description of the label content) and its history (*e.g.*, how long the product took to move though the supply chain, its temperature, pressure, humidity, and gas leakage) and can be collected at any point during processing and distribution. The information obtained from data analysis can be used for judgment of the produce status such as traceability in the case of outbreaks of food borne infections.

The RFID tag could offer much more than a conventional barcode. In contrast to a barcode, RFID does not need to be in direct line of sight to be recognized by a scanner. This could revolutionize the way check-out works at a grocery store. Many RFID tags can be read simultaneously at a rapid rate. RFID tags could also store information such as temperature and relative humidity data, nutritional information and cooking instructions. They could be integrated with time - temperature indicator or a biosensor to carry time-temperature information or microbiological data. These are used to increase convenience/efficiency in supply chain management, and are being normally applied to secondary or tertiary packaging. If costs can be reduced significantly, then they could find application on individual consumer packages.

RFID technology in the food system is still in the early stages. If the quality of foods is limited by expiry, it becomes important to meet correct decisions upon their disposal during transportation, storage and sale, otherwise, losses may occur due to premature expiry before reaching the point of sale, or food safety rules could be violated. The key decisions preventing premature spoilage can be made in following operational areas:

i) Food processing
ii) Packing process
iii) Logistic and shelf management at the point of sale.

It is evident that such decisions can only be met properly if information about the status of goods is provided *i.e.*, the packed good can be tracked over the delivery and storage processes.

At present, simple applications like tracking and identification are the focus of most food science matters, and these must be perfected before more complex applications can come to light.

Modified Atmosphere Packaging (MAP)

The examination of gas atmosphere on the effect on food shelf life has a long history, and the first pioneers started experiments more than 100 years ago. Since the early development of MA, it has been shown that MA can, on its own, inhibit the growth of microorganisms. Based on MA technology, Modified atmosphere (MA) packaging has been used for decades and the effects on different raw materials are well understood. MAP technology emerged in 1960s as a new technology that extends storage life of perishable agricultural produce and reduces its spoilage and decay. MA packaging has now become a widely commercially used technology for the storage and distribution of foods. This trend is mainly driven by the modern consumer's demand for preprocessed products that have a fresh appearance and are convenient and easy to prepare, combined with the industry's desire for longer shelf life. Product safety is a major concern associated with the use of MA packaging. The desired suppression of spoilage microorganisms extends the shelf life, compared to food products stored in a normal air environment, which may create opportunities for slower-growing pathogenic bacteria.

MAP, one of the food preservation methods maintains the natural quality of food products in addition to extending the storage life. The fact that gas modifications only are bacteriostatic has encouraged producers to use additional preservation technology to enhance or support safety and retard spoilage. By combining different preservation factors or techniques, reliable and mild multi-targeted preservation can be achieved that facilitates improvements in food safety, quality, and economical aspects. In other words, MAP is a technique used for prolonging the shelf-life of fresh or minimally processed foods by changing the composition of the air surrounding the food in the package. The use of MAP reduces the respiration rate and activity of insects or microorganisms and provides control of fruit and vegetable ripening, retardation of senescence, or browning in cut produce, and ultimately prolong the shelf life. Modified Atmosphere enables fresh and minimally processed packaged food products to maintain visual, textural and nutritional appeal.

MAP is a procedure which involves replacing air inside a package with a predetermined mixture of gases prior to sealing it. The gases involved in modified atmosphere packaging, as applied commercially today, are optimal blend of carbon dioxide, nitrogen and oxygen within a high barrier or permeable package. Replacing the normal atmosphere with a modified atmosphere, *i.e.*, altered concentrations of O_2, CO_2, N_2, or even other gases, will add additional stress to microorganisms and change the composition of the initial microbial flora.

i) High levels of CO_2 have a bacteriostatic effect on microorganisms, and properly designed MA packaging can double a product's shelf life. Carbon dioxide reacts with water in the product to form carbonic acid which lowers the pH of the food. It also inhibits the growth of certain microorganisms, mainly moulds and some aerobic bacteria.

ii) Nitrogen has no direct effect on microorganisms or foods, other than to replace oxygen, which can inhibit the oxidation of fats.

iii) Oxygen is included in MAP packages of red meat to maintain the red colour, which is due to the oxygenation of the myoglobin pigments.

A finely adjusted and carefully controlled gas blend is developed to meet the specific respiration needs for each packaged food product.

MAP is an active or passive dynamic process of altering gaseous composition within a package. It relies on the interaction between the respiration rate of the produce, and the transfer of gases through the packaging material, with no further control exerted over the initial gas composition. Passive MAP can be generated inside a package by relying on the natural process of produce respiration and film permeability to attain the desired gas composition over time. While, active MAP is a rapid process of gas replacement or displacement, or the use of gas scavengers or absorbers to establish a desired gas mixture within a package. This involves the addition of active agents into packaged food product, such as O_2, CO_2 and ethylene scavengers. For example, CO_2 absorbers can prevent a build-up of CO_2 gas to deleterious levels.

The physiological processes of produce (mainly respiration and transpiration) play significant roles in the postharvest quality of MA-packaged fresh and fresh-cut fruit and vegetables. Respiration is a metabolic activity that provides the energy needed for other plant biochemical reactions. Aerobic respiration involves the oxidative breakdown of complex organic compounds such as carbohydrates, lipids, and organic acids into simpler molecules, including CO_2 and water with the release of energy. Decreasing respiration rate via MA and lowering temperature delays enzymatic degradation of complex substrates and reduces sensitivity to ethylene synthesis, thereby extending the shelf life and avoiding senescence of the produce.

The most common way for increasing the OTR of plastic films, used in a wide range of fresh produce, is the addition of micro-perforations by electrostatic discharge, laser technology, or mechanically by small needles. The diameter of the micro-perforation generally ranges from 40 to 250 μm. It is therefore, possible to tailor a specific micro-perforated polymeric film (number and size of the holes) in accordance with the rate of respiration of the produce, permeability of the film, film surface area as related to produce weight, void volume of the packaging, storage temperature, and anticipated fluctuation in temperature during shipment. This technology is also suitable for high-respiring fresh produce such as strawberry, asparagus, broccoli, leek, fresh herbs, sweet corn, mushrooms, and others, that are packed in either consumer or bulk packaging and can tolerate medium–high levels of CO_2. MAP in micro-perforated film is suitable for those products that benefit from medium–high levels of CO_2 but not for those requiring low O_2 and low CO_2.

The effectiveness of Xtend films in maintaining quality of MA-packed fresh produce has been reported to be greater than that of other commercially available films such as polyethylene, polypropylene, and polyvinyl chloride. The micro-perforated Xtend packaging allowed the formation of a desirable modified atmosphere, retarding ripening and senescence of the produce. Additional beneficial effects of Xtend films included reduction of decay, chilling injury, leaf elongation, leaf sprouting, tissue discoloration, peel blemishes, and formation of off-odors, and inhibition of bacterial growth on the produce surface. Most of the technologies based on the Xtend packaging reviewed in have been successfully implemented in postharvest practice in Israel, the U.S., and other countries. Another way of increasing gas exchange in plastic packaging is by mixing the polymeric film with inert inorganic material such as $CaCO_3$ and SiO_2, thereby generating microporous films.

Advantages of MAP

- MAP maintains freshness and extends shelf-life from several days to several weeks, compared to conventional storage.
- Reduction of O_2 and increment of CO_2 environment suppresses respiration rate of the commodity thereby slows vital processes and prolongs the maintenance of post-harvest quality.
- Reduction of respiration rate, loss of moisture, production of metabolic heat, yellowing, browning decay and C2H4 sensitivity and production occurs.
- Delay of ripening takes place.
- Reduction of weight loss, desiccation/water loss and shriveling take place in MAP.
- In MAP reduction of physiological injury, disorder and pathological deteriorations takes place.
- Quality advantage such as colour, moisture, flavours and maturity retention occurs.
- Reduction of fungal growth and diseases is common.
- MAP retards softening and compositional changes.
- Alleviation of chilling injury is common.
- Increased shelf-life allowing less frequent loading of retail display in shelves occurs.
- Improved presentation and clear visibility of product all around the package occurs.
- In MAP, little or no preservative is used.
- Centralized/semi-centralized packaging option is possible.
- Expanded distribution are and reduced transport costs due to less frequent deliveries is possible.

- Reduction of labour and waste at the retail level occurs.
- Excellent branding options.
- Reduction of handling and distribution of unwanted or low-grade produce.
- Quality advantages transferred to the consumer.
- Reduction in production and storage cost due to better utilization of labour, space and equipments.
- MAP has great advantage in developing countries because it can economically be done by hand saving the high cost of new machinery. Additionally, the need therefore such a technique is much greater because of the dearth of refrigerated storage.

Disadvantages and Limitations

- Requirement of additional investment in machinery and labour in the packaging line.
- Risks of spoilage of produce may occur due to improper packaging or temperature abuse.
- Possible occurrence of new risks of microbiological safety due to possible development of anaerobic pathogenic flora.
- Plastic films may be environmentally undesirable unless effective recycling is installed.
- MAP technology is still unavailable for most produce.
- Although the principles of modified atmosphere packaging (MAP) of fresh produce are well known, the technology is still applied mainly in retail packages characterized by a high ratio between the film surface and produce weight. The method is still limited in bulk packaging, especially with highly respiring products, because of the difficulties in controlling in-package atmosphere and humidity during commercial shipments. These problems very often occur when the produce is exposed to temperature fluctuations during storage or shipment.
- In many cases, the commercialization of modified atmosphere packaging (MAP) is limited due to the accumulation of condensed water in the package causing increased pathological and physiological disorders.
- No particular standard is available for MA packaging, because the intrinsic properties of the commodity varies greatly with cultivar, place of cultivation, maturity stages *etc.* and the permeability of the films varies with the manufacturing company and process, *etc.*

Edible Packaging

An edible film or coating is simply defined as a thin continuous layer of edible material formed on, placed on, or between the foods or food components.

The package is an integral part of the food, which can be eaten as a part of the whole food product. Selection of material for use in edible packaging is based on its properties to act as barrier to moisture and gases, mechanical strength, physical properties, and resistance to microbial growth. Natural polymers have been studied extensively for the development of edible packaging. A variety of polysaccharides (starch, cellulose, alginates, pectin and hydrocolloids), proteins (whey proteins, gluten, zein, soybean proteins and fish proteins) and lipids (waxes) have been used, either individually or in mixtures, to produce edible films. Many lipid compounds, such as animal and vegetable fats, acetoglycerides have been used in the formulation of edible packaging for fresh produces because of their excellent moisture barrier properties. Edible packaging materials are not normally meant to entirely replace conventional packaging. Rather, the efficiency of food preservation can be improved by using primary edible packaging together with non-edible packaging as secondary packaging to add additional protection from the atmosphere and prevent contamination from microorganisms or foreign particles. Edible films and coatings have some advantages such as edibility, biocompatibility, barrier properties, absence of toxicity, the fact that they are nonpolluting, and low cost. Moreover, biofilms and coatings, by themselves or acting as carriers of food additives (*i.e.,* antioxidants and antimicrobials), have been considered particularly for food preservation because of their ability to extend the shelf – life.

Nanotechnology in Food Packaging (Nano-packaging)

The concept of nanotechnology was first introduced by Noble Laureate Physicist Richard Feynman in 1959. The word 'nano' comes from the Latin word 'dwarf'. Eric Drexler popularized the term 'nano-tech' in 1980's during his works on these incredible devices. Nanotechnology involves manipulating with particles, called nano-particles that are in the scale of nanometers (a billionth of a meter). In other words, nanotechnology is the manufacture of structures, materials and machines using nanoparticles with programmed precision. Hence, nanotechnology is an emerging technology which enable us to create materials at molecular level into functional structures with nanometer dimensions. This will make products cheaper, production more efficient and more sustainable by using less water and chemicals.

Nanotechnology can enable designers to alter the structure of packaging materials on the molecular scale, in order to give the material the desired properties. With a different nanostructure, plastics can have improved gas/water vapor permeability to fit the requirements for the preservation of fruits, vegetables, beverages, and other foods. By adding nanoparticles, it is possible to produce bottles and packages with lighter resistance, stronger mechanical and thermal properties, and less gas absorption. These properties can significantly increase the shelf life, efficiently preserve the flavor and color, and facilitate the transport of food. Use of polymer nano-composites (bionanocomposites), the hybrid nanostructured materials with improved mechanical, thermal and gas barrier properties for food packaging is more environmentally friendly solution since it reduces the requirement to use plastics as packaging material. Furthermore, nanostructured film can effectively prevent the invasion of bacteria and microorganisms in food and thus, ensure food safety. These properties can significantly increase the shelf - life and sensory characteristics of food products, and facilitate transportation and usage. The addition of nanosensors to food packages could be used to detect chemicals, pathogens and toxins in foods. Hence, nanoscale innovation could potentially introduce many amazing improvements to food packaging in the form of barrier and mechanical properties, detection of pathogens, and smart and active packaging with food safety and quality benefits. Nanotechnology is expected to change the entire packaging industry.

Overall, actually to summarize, the main applications of nanomaterial materials in packaging are as under:

(i) Improved packaging, such as barrier performance pertaining to gases such as oxygen, carbon dioxide, and ultraviolet rays, as well as to add strength, stiffness, dimensional stability, and heat resistance;

(ii) Active packaging, such as antimicrobial or other properties (*e.g.*, antioxidant, UV absorbance) with intentional release into and consequent effect on the packaged food in term of taste, freshness and self-life; Antibodies attached to fluorescent nanoparticles to detect chemicals or food borne pathogens; Antimicrobial and antifungal surface coatings with nanoparticles (Ag, Mg, Zn) smart/intelligent packaging, such as oxygen indicators, freshness indicators and pathogen;

(iii) Biodegradable nano sensors for temperature, moisture and time monitoring

(iv) Nanoclay and nanofilms as barrier materials to prevent spoilage and oxygen absorption; Modified permeation behavior of foils

(v) Electrochemical nanosensors to detect ethylene and other obnoxious gases

(vi) Lighter, stronger and more heat resistant films with silicate nanoparticles

Examples of nano-food packaging include:

- ☆ Biocidal nanocompounds: Biocidal nanocompounds (*e.g.* nanosilver) are main components of antibacterial "active" protection. They are used against pathogenic organisms. Biocidal agents are released on food surface through food packages in controlled way with time. Fluorescent nanoparticles can be attached to antibodies which could detect food spoilage and development of pathogenic bacteria.
- ☆ Clay nanocomposites: Clay nanocomposites in food packaging act as a non-permeable membrane for gasses.
- ☆ Adhesive nanoagents: Adhesive nanoagents are used for establishing better contact between colour and food packaging.

In this way, new food packaging materials with improved mechanical, barrier and antimicrobial properties enable preserving of taste, color, flavor, texture, consistency and nutrients of foodstuffs. The importance of food traceability and monitoring during the transport and storage lead to incorporation of new technologies with embedded nano-sensors into food packages to track environmental conditions throughout the supply chain. Self-assembly will reduce fabrication costs and infrastructure, and more flexible packaging methods will provide consumers with fresher and more customized products.

Whilst most nanotechnology applications for food and beverages are currently at research and development phase or near market stages, applications for food packaging are already becoming a commercial reality. However, the potential risks of nanomaterials to human health and to the environment are unknown. There may be a need for a pragmatic regulatory oversight to ensure a case-by-case pre-market safety evaluation of the nanotechnology-derived products to safeguard the consumer from any potential risks.

Aseptic Packaging

- Process in which a food product, such as ultra high temperature (UHT) milk and its package is sterilized separately and then combined and sealed under sterilized atmosphere.
- It increases the shelf-life.

Aseptic packaging is the process by which a sterile (aseptic) product (typically food or pharmaceutical) is packaged in a sterile container in a way that maintains sterility. Sterility is achieved with a flash-heating process (temperature between 195 and 295 °F (91 to 146 °C)), which retains more nutrients and uses less energy than conventional sterilization techniques such as retort or hot-fill canning. Aseptic food preservation methods allow processed food to keep for long periods of time without preservatives, as long as they are not opened. The aseptic packages are typically a mix of paper (70 per cent), polyethylene (LDPE) (24 per cent), and aluminum (6 per cent), with a tight polyethylene inside layer. (Aseptic packaging council, 2011).

Vacuum Packaging

Vacuum packaging is another way to increase the shelf life of food products. Here the product is placed in an air-tight pack, the air sucked out and the package sealed. By removing air from around the product, the levels of oxygen in the packaging are reduced, impeding the ability of oxygen-breathing microorganisms to grow and spoil the product. The lack of oxygen also reduces the amount of spoilage due to oxidation, limiting the growth of aerobic bacteria or fungi, and preventing the evaporation of volatile components. It is also commonly used to store dry foods over a long period of time, such as cereals, nuts, cured meats, cheese, smoked fish, coffee, and potato chips (crisps). On a more short term basis, vacuum packing can also be used to store fresh foods, such as vegetables, meats, and liquids, because it inhibits bacterial growth. In vacuum packaging the product to be packed is put in a vacuum bag (made of special, hermetic fills) that is then evacuated in a vacuum chamber and then sealed hermetically in order to provide a total barrier against air and moisture. If some of the product cannot bear the atmosphere pressure due to vacuum inside the package then the packages are flushed with inert gases like Nitrogen and CO_2 after evacuation. In an oxygen-depleted environment, anaerobic bacteria can proliferate, potentially causing food-safety issues. Vacuum packing is often used in combination with other packaging and food processing techniques to stop the reproduction of anaerobic organism.

- It is a procedure in which air is drawn out of the package prior to sealing but no other gases are introduced.
- This technique has been used for many years for products such as cured meats and cheese.

Vacuum packaging is removing air from the product pouch and hermetically sealing it. This increases storage or shelf life by inhibiting the growth of microorganisms and improves hygiene by reducing the danger of cross contamination. Vacuum packing also preserves flavour and protects against dehydration and weight loss.6 Advantages of vacuum packaging i. A simple solution to packaging goods requiring protection from oxygen ii. Positive control of the moisture content of the produce iii. Inhibits the growth of aerobic spoilage bacteria iv. Lower costs than those of rigid containers v. Longer shelf life for goods Disadvantages of vacuum packaging i. It is virtually impossible to remove all the oxygen, since small quantities will be trapped within food cells. ii. Microorganisms which are not affected by oxygen are not advantageously affected by vacuum packing.

Emerging Approaches in Food Packaging Systems

Green Plastics for Food Packaging

1. Polylactic Acid (PLA)

Polylactic acid (PLA) is a polymer of lactic acid. It behaves quite similarly to polyolefins. PLA, a fermentation product of low cost polysaccharides, is a product which is produced from a combination of biotechnology and chemical technology, and can be converted into plastic products by standard processing methods such as injection moulding and extrusion. It has potential for use in the packaging industry as well as hygiene applications. The current global market for lactic acid demand is 100,000 tons per annum, of which more than 75 per cent is used in the food industry. Perhaps the biggest opportunities for PLA lie in fibres and films. An important market niche for PLA can be found in the agricultural industry such as crop covers and compostable bags. Biodegradations of PLA have been a subject of interest and so far proteinase K (EC 3.4.21.64) is the only reported enzyme that will degrade

PLA amorphous regions of low MW (Reeve, *et al.*, 1994) [21]. Microbial degradation studies of PLA have been inclusive (Torres, *et al.*, 1996) [24]. Although most microorganisms studied can utilise lactic acid and its dimer, microbial degradation of oligomers and polymers of PLA have not yet been observed at appreciable rates. A microbial degradation study on PLA/CL only showed the degradation of the PCL segments (Lostocco, *et al.*, 1998). Compost, field and environmental degradations of PLA are primarily due to hydrolysis (Ho, *et al.*, 1999)

2. Native Starch

Starch is nature's primary means of storing energy and is found in granule form in seeds, roots and tubers as well as in stems, leaves and fruits of plants. Starch is totally biodegradable in a wide variety of environments and allows the development of totally degradable products for specific market needs. The two main components of starch are polymers of glucose: amylose (MW 10^5 -10^6), an essentially linear molecule and amylopectin (MW 10^7-10^9), a highly branched molecule. Amylopectin is the major component of starch and may be considered as one of the largest naturally occurring macromolecules. Starch granules are semi-crystalline, with crystallinity varying from 15 to 45 per cent depending on the source. The term 'native starch' is mostly used for industrially extracted starch. It is an inexpensive (< 0.5 Euro/kg) and abundant product, available from potato, maize, wheat and tapioca.

3. Thermoplastic Starch

Thermoplastic starch (TPS) or destructurised starch (DS) is a homogeneous thermoplastic substance made from native starch by swelling in a solvent (plasticiser) and a consecutive 'extrusion' treatment consisting of a combined kneading and heating process. Due to the destructurisation treatment, the starch undergoes a thermo- mechanical transformation from the semi-crystalline starch granules into a homogeneous amorphous polymeric material. Water and glycerol are mainly used as plasticisers, with glycerol having a less plasticising effect in TPS compared to water, which plays a dominant role with respect to the properties of thermoplastic starch.

4. Chitin and Chitosan

Chitin is the second most abundant naturally occurring biopolymer (after cellulose) and is found in the exoskeleton of crustaceans, in fungal cell walls and other biological materials (Andrady and Xu, 1997) [1]. It is mainly poly (β-(1-4)-2-acetamide-D-glucose), which is structurally identical to cellulose. Chitosan is derived from chitin by deacetylation in the presence of alkali. Chitosan can form semi-permeable coatings, which can modify the internal atmosphere, thereby delaying ripening and decreasing transpiration rates in fruits and vegetables. Films from aqueous chitosan are clear, tough, flexible and good oxygen barriers (Sandford, 1989; Kalplan *et al.*, 1993) [7, 23, 15]. Carbon dioxide permeability could be improved by methylation of polymers. Butler *et al.* (1996) observed that films from chitosan were rather stable and their mechanical and barrier properties changed only slightly during storage. Chitosan coatings are usually used on fruit and vegetable products such as strawberries, cucumbers, bell peppers as antimicrobial coating (El Ghaouth *et al.,* 1991a, 1991b) [8, 9], and on apples, pears, peaches and plums as gas barrier (Elson and Hayes, 1985; Davies *et al.,* 1989) [7].

5. Cellulose

Cellulose is the most abundant occurring natural polymer on earth, being the predominant constituent in cell walls of all plants. Cellulose is composed of a unique monomer: glucose under its β-d-glucopyranose form (Credo and Berthelot, 2014) [5]. Due to its regular structure and array of hydroxyl groups, it tends to form strong hydrogen bonded crystalline microfibrils and fibers and is most familiar in the form of paper, paperboard and corrugated paperboard in the packaging context (Cruz-Romero and Kerry, 2008; Babu, *et al.,* 2013) [6, 3].

6. Shellac Resins

Shellac is a resin secreted by the female lac bug (*Laccifer lacca*), on trees in the forests of India and Thailand. It is processed and sold as dry flakes and dissolved in ethanol to make liquid shellac, which is used as a brush -on colorant, food glaze and wood finish. Shellac functions as a tough natural primer, sanding sealant, tannin -blocker, odour-blocker, stain, and high-gloss varnish. Shellac is a natural bio-adhesive polymer and is chemically similar to synthetic polymers, and thus can be considered a natural form of plastic. It can be turned into a moulding compound when mixed with wood flour and moulded under heat and pressure methods, so it can also be classified as thermoplastic. This resin is soluble in alcohols and in alkaline solutions. Shellac is not a GRAS substance; it is only permitted as an indirect food additive in food coatings and adhesives. It is mostly used in coatings for the pharmaceutical industry (Hernandez, 1994) [13]. Shellac is also used as a coating on citrus fruits as it has good gas barrier properties. Shellac- and wood resin-based coatings also tend to increase prevalence of post-harvest pitting (Petracek *et al.,* 1997, 1998) [19, 20].

Conclusion

The food industry has been under growing pressure to feed an exponentially increasing world population and challenged to meet rigorous food safety laws and regulations. The plethora of media consumption has provoked consumer demand for safe, sustainable, organic, and wholesome products with "clean" labels. The application of active and intelligent packaging has been commercially adopted by food and pharmaceutical industries as a solution for the future for extending shelf life and simplifying production processes; facilitating complex distribution logistics; reducing, if not eliminating the need for preservatives. Similarly antimicrobial packaging can play an important role in reducing the risk of pathogen contamination, as well as extending the shelf-life of foods; it should never substitute for good quality raw materials, properly processed foods and good manufacturing practices. It should be considered as a hurdle technology that in addition with other non-thermal processes such as pulsed light, high pressure and irradiation could reduce the risk of pathogen contamination and extend the shelf-life of perishable food products. Participation and collaboration of research institutions, industry and government regulatory agencies will be key on the success of antimicrobial packaging technologies for food applications.

Biodegradable Packaging

The present global concern about petrochemical - based plastic materials has generated much interest in biodegradable, or " green" packaging materials. — According to the ASTM,2003 guidelines, a "biodegradable plastic" is defined as a degradable plastic in which the degradation results from the action of naturally occurring microorganisms such as bacteria, fungi and algae.

Biodegradable or green packaging must satisfy some basic requirements to be an ideal candidate for food packaging. These requirements include barrier properties (to water vapor, gases, light and aromas), optical properties (transparency), strength, welding and molding properties, printing properties, migration resistance, chemical and temperature resistance, the ability to satisfy disposal requirements, antistatic properties, and the ability to retain sensory properties. Bio-based polymers, or biopolymers, are obtained from renewable resources (Weber *et al.,* 2002) } These renewable resources consist of proteins (whey protein, soy protein, collagen, gelatin, wheat protein *etc.*), polysaccharides (starch, alginates, pectin, carrageenans and chitosan/chitin) and lipids (fats, waxes and oils) (Comstock *et al.,* 2004).

Polymers such as polylactide (PLA) have gained growing attention in the last decade as food packaging materials because they can easily be obtained from renewable resources, their production consumes quantities of carbon dioxide, they can be recycled and composted, and their physical and mechanical properties can be tailored through the polymer architecture (Sinclair, 1996; Siracusa *et al.,* 2008).

Active and Intelligent Packaging of the Future As exciting as the prospect of active and intelligent packaging is, all packaging material must be approved for use, and the legislation that applies to traditional packaging also applies to active and intelligent packaging (Hurme *et al.,* 2002). To date, no specific methods exist to determine whether or not active or intelligent packaging is suitable when in direct contact with foods (Robertson, 2006). A major issue is that most active and intelligent packaging systems require that food be in direct contact with a sensor of some kind, and substances from the sensor may migrate into foods. Whether these migrations are intentional or unintentional, the substance, amount of the substance, and possible health affects of the substance must be determined in order for the substances to be allowed and regulated. In addition, the cost of active and intelligent packaging limits its commercial use. Most active or intelligent systems add cost to the package, so innovations in packaging must have a final beneficial outcome that outweighs the extra expenses of adding the technology. In addition, systems must be reliable and effective. This requires that the system be validated to assure the information being 9 conveyed is true and the consumer is not let down when they trust these news technologies over the old ways. It is also important the food producer, retailer, and consumer be in tuned with the active or intelligent system on a large scale. Attitudes must be willing to accept new technologies and those involved in each step of the food chain must be sure that the new system is safe and true for the user. Despite these hurdles, many developments are still on their way. The attitude toward active and intelligent packaging is positive and there is still much potential for exciting innovation to come. The following are some active or intelligent packaging idea that are in the works.

Biosensors for Pathogen or Toxin Identification

Foodborne pathogens are of great concern to the food industry and many consumers have become increasingly aware of this problem. The need to rapidly and accurately detect small amounts of pathogen or toxins in food is an essential step in keeping the consumer safe. A biosensor is an analytical device used to detect a substance, in this case a pathogen, and then transmit this information into some sort of signal that is quantifiable. An intelligent system in the works aims attach antibodies to a plastic packaging surface to detect pathogens or toxins (Yam *et al.,* 2005). If the antibodies come into contact with the target pathogen, the packaging material would display a visual cue to alert the consumer. This intelligent system would only be useful when foods were contaminated with very high concentrations of pathogen or toxin. In reality, a consumer could get ill from just small concentrations of pathogen or toxin and this intelligent system could give the consumer a false sense of security. Also, this system would work only to detect pathogens or toxin on the surface of a food product, and would not alert consumers of the dangerous substances that could potentially be distributed throughout the product. This system has a long way to go before it becomes commercially available.

Although this system is highly technical, it is not an impossibility. There are also several ways in which this system could be altered or simplified based on consumer demand. Although the idea has been planted, this is truly a vision of the future. Conclusions Active and intelligent systems are a branch of packaging that is truly innovative and offers exciting opportunities for food safety, quality and convenience. Many active and intelligent packaging concepts are commercially available in the United States.

Impact of Packaging on Food Quality, Shelf Life and Safety

Food Quality and Food Safety

Interactions between foods and packaging can be detrimental to quality and safety. Changes in product flavour due to aroma sorption and the transfer of undesirable flavours from packaging to foods are important mechanisms of deterioration when foods are packaged in polymer-based materials. Careful consideration must be given to those factors affecting such interactions when selecting packaging materials in order to maximize product quality, safety, and shelf-life while minimizing undesirable changes. Product considerations include sensitivity to flavour and related deteriorations, colour changes, vitamin loss, microbial activity, and amount of flavour available. Storage considerations include temperature, time, and processing method (Hayk, 2014).

Food safety is defined as cooperation with necessary requirements during production, processing, storage, transport and distribution of food and taking essential measurements to ensure healthy food production. It is ensuring the safety of materials in contact with food from field to table as the final link of food safety chain, determination of its effect on food quality and safety and identification of health risks resulting from these materials are compulsory for the protection of consumer rights (Aysel *et al.,* 2016). Food safety is an important aspect to maintain our health and well-being. One of the common ways we can be affected is through our consumption of food. The carelessness towards food handling and preparation safety procedures can expose hundreds, thousands or millions of people to health issues. In case of an outbreak of disease because of a particular food been consumed, correctly packaged and labelling food products will make it easier for the food to be recalled.

Shelf Life

Shelf life is an established time under a defined set of conditions during which a product is acceptable for use. The three largest variables affecting shelf life are product characteristics, packaging materials and storage conditions. The correct packaging will improve the food's shelf life, one of the biggest issues that leads to food health problems is because of the shelf life of the food. Many people believe that just by placing food inside a cold area, the food will be preserved for a longer shelf life, it is not totally true. The issue is food will still decompose if exposed to high oxygen levels. By removing the food's exposure to the air, you can preserve the food safely for longer within the packaging.

Conclusion and Recommendation

Packaging is an essential component of the food system and plays a critical role in containing, protecting and preservation food from field to the end user. Researchers have shown that the use of appropriate packaging can contribute to reducing food losses and waste, and maintenance of product quality and safety. Food packing has to fulfill a broad spectrum of requirements all being related to the preservation of food quality and the proper presentation and handling of the food item. With respect to preservation of food quality an effective barrier function is mandatory preventing the transfer from/into the food related to its surrounding/ environment. A number of food packaging materials are used which includes; glass, metals, and plastic polymers and paper/cardboard. These materials most times when used as the only component, cannot fulfill this task in a sufficient way, other substances are added to ensure quality packaging of processed foods.

Antimicrobial food packaging play an important role in reducing the risk of post processing pathogen contamination, as well as extending the shelf life of food. It is gaining global interest across the food industry and research community as it provides quality and safe food. Proper care should be taken while selecting and incorporating the antimicrobial into the food packaging film by evaluating all the implications of food safety and harsh reality of microbial resistance.

Different Types of Packaging Methods

Different types of packaging methods are:

- ✰ Anti-corrosive Packaging
- ✰ Pharma Packaging
- ✰ Plastics Packaging
- ✰ Flexible Packaging

1. Anti-corrosive Packaging

Anti-corrosion refers to the protection of metal surfaces from corroding in high-risk (corrosive) environments.

Products can be protected against corrosion and avoid the tedious sequences using anti-corrosive packaging. Materials such as VCI bubble film, paper, oil, chips, and bags that are used as a part of anti-corrosive packaging helps to prevent goods from exposing to various climate zones.

Moreover, this packaging can be used in the scenarios such as export of goods over long distances, long-term storage of metal goods, and metals that are sensitive for corrosion. Several companies including VCORR Packsys Pvt. Ltd with product range VCI Oils, Stretch Films, and Bio-Degradable VCI Film *etc.* provide anti-corrosive packaging.

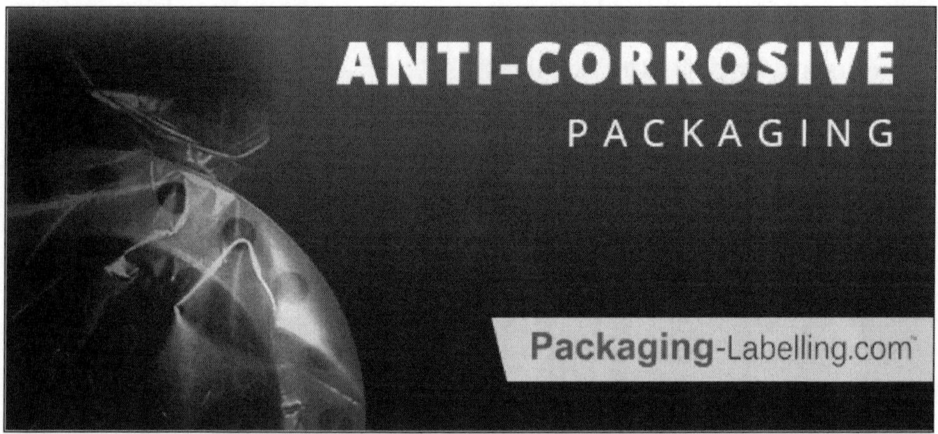

Anti-corrosive Packaging.

"The "International Measures of Prevention, Application and Economics of Corrosion Technology (IMPACT)" study estimates that the global cost of corrosion to be US$2.5 trillion, a year, which is equivalent to roughly 3.4 percent of the global Gross Domestic Product (GDP)."

Methods Used for Anti-corrosive Packaging are:

- ✰ Volatile Corrosion Inhibitors (VCI) Papers and Films
- ✰ Aluminum barrier foil in combination with desiccants
- ✰ Coating methods (oils or liquids)

Anti-corrosive Packaging can be used in the following cases

- ✰ Exporting/shipping goods by sea or moisture prone environments
- ✰ Storing metallic goods for extended periods of time
- ✰ Manufacturing metal goods sensitive to corrosion
- ✰ Shipping any unassembled parts
- ✰ CKD shipment *e.g.* of cars
- ✰ Temporarily laying of machines, planes

2. Plastics Packaging

Plastics Packaging is used for packaging a variety of items *i.e.* fragile or non-perishable products. Plastics packaging materials are used to cover the plastic related materials or products. Most Plastics Packaging companies recycle waste or scrap plastic for re-use in their own facilities, and even offer alternatives.

Without packing, these sophisticated goods with low shelf life will be easily affected by climatic conditions and eventually result in their malfunction or destruction. This packaging is done in such a way that no air or dust can sneak in and affect the goods.

According to Plastic Packaging Market, "The plastics packaging market was valued at USD 318.92 billion in 2017 and is expected to reach a value of USD 399.44 billion by 2023 at a CAGR of 3.44% over the forecast period of 2018–2023."

Trends in Plastics Packaging are:

- ✰ Changing consumer habits
- ✰ Increased urbanization and growing young population
- ✰ Increased use of home and personal care products
- ✰ Developments in the field of polymer science
- ✰ Increased investment in food processing and production in China
- ✰ Increased preference of plastic pouches, by consumers and retailers

Types of Plastics Packaging

The most common types of plastics used and their applications are illustrated in the following table:

Polymer Types	Examples of applications
Polyethylene Terephthalate	Fizzy drink and water bottles. Salad trays
High-Density Polyethylene	Milk bottles, bleach, cleaners and most shampoo bottles.
Polyvinyl Chloride	Pipes, fittings, window and door frames (rigid PVC). Thermal insulation (PVC foam) and automotive parts.
Low-Density Polyethylene	Carrier bags, bin liners, and packaging films.
Polypropylene	Margarine tubs, microwaveable meal trays, also produced as fibres and filaments for carpets, wall coverings, and vehicle Upholstery.
Polystyrene	Yoghurt pots, foam hamburger boxes and egg cartons, plastic cutlery, protective packaging for electronic goods and toys. Insulating material in the building and construction industry.
Unallocated References	Any other plastics that do not fall into any of the above categories - for example, polycarbonate which is often used in glazing for the aircraft industry

Benefits of Plastics Packaging

Here are the benefits of choosing plastics packaging:

- ☆ Lightweight
- ☆ Durable
- ☆ Sustainable
- ☆ Cost-effective
- ☆ Versatile

3. Flexible Packaging

Flexible packaging can be made with flexible materials that can be packed, sealed and can retain the freshness of its content. Be it moulding the packages into a variety of shapes or preserving the flavor and safety of food or logistical aspects, all are covered under the flexible packaging.

This is the major benefit that makes flexible packaging the first choice for food packaging. The package can last for a very long time; if it is stored in the right conditions. Flexible packaging is environmentally friendly as it uses less material and less wastages to make it.

Flexible Packaging

> "The global flexible packaging market will grow at a rate of 3.5% PA until the year 2018, according to a major new study by Smithers Pira."

Why Flexible Packaging?

Flexible Packaging can be defined as any package whose shape can be readily changed. Choosing the flexible packaging,

- Adds value and marketability to food and non-food products alike from ensuring food safety and extended shelf life by providing heat seal, barrier protection, ease of use, press-to-seal and high-quality print finishes.
- This type of packaging produces a fraction of waste during the printing process, overall reducing landfill discards.
- Improvements in productions processes have reduced water and energy consumption, greenhouse gas and emissions and volatile organic compounds.
- Finally, lighter-weight flexible packages result in less transportation-related energy and fossil fuel consumption, and environmental pollution.

Choosing these methods will provide the ideal results for your products. Keeping abreast of these packaging methods will give your products a better shelf life and a competitive edge in the market. Many companies are using these methods to create awareness among their customers and help them in increasing their sales. Adopting these packaging methods can draw attention to latest trends to attract the consumers. Innovative packaging designs can also maximize the profitability of the product.

Methods

- Field Packing – products are placed in their shipping materials during the harvesting process. The filled containers are then taken to a precooling facility where field heat is removed. Field packing is a common practice for strawberries as this method ensures optimal freshness.
- Shed Packing – products are processed and packed indoors at a central location. Produce is brought in from the field to the packing shed in field crates, bins or trucks. Products are precooled at the shed.
- Repacking – products are taken out of one container, regraded and placed in another. This might often occur at store or distribution warehouse level, when smaller containers are required for consumers.

Types

- Volume fill – products are placed into the container until the desired capacity, weight, or count is reached. Common for oranges and avocados.
- Tray/Cell Pack – products are placed into molded trays; this provides separation and can work to reduce bruising.
- Place Pack – products are wrapped and carefully placed into containers; often used to reduce bruising and gives a nice look.
- Film/Shrink Wrap – each piece of produce are individually wrapped and sealed in film to reduce moisture and prevent loss and decay. The film is often treated with a chemical solution that will aid in preservation.

☆ Modified Atmosphere – products are placed in individual consumer packs or containers and sealed with plastic film or bags. This reduces product respiration and slows the ripening process. Tectrol is a well-known brand of modified atmosphere package that is used for berries.

Table 3: Selected Commercial Active and Intelligent Systems that are Used or can Potentially be Used for Fresh-Cut Fruits and Vegetables

Packaging System	*Commercial Name*	*Mechanism of Action*	*Manufacturer*	*Packaging Form*
Oxygen scavenger	Ageless	All iron based	Mitsubishi gas chemical LTD.	Sachets and labels
	Freshilizer Oxyguard Zero$_2$	Dye/organic compound	Toppon printing Co. LTD.	Sachets Plastic trays Plastic films and container
Ethylene scavenger	N/A	Potassium permanganate	Air repair products LTD.	Sackets/ blankets
	Neuphalon Evert fresh Bio-fresh	Activated carbon Activated zeolites Activated clays		Sachet Plastic film
Moisture absorber	Toppan sheet	Not known	Toppan printing	Absorber sheet
	Thermarite	Not known	Thermarite Pyt. LTD	Absorber sheet
Time-temperature integraters	Checkpoint TT sensor	Enzymatic activity Microbial growth Diffusion of polar compound	VITS A.B CRYLOGO Avery dennission Corp.	Adhesive label
Gas and volatiles indicator	It's fresh Freshness guard	Chemical reaction Enzymatic detection	It's fresh Inc. UPM raflatac	Indicator tag Test strips or biochemical biosensars
Microbial growth	Toxin guard	Immunochemical reaction	Toxin alert Inc.	Indicator tag

Source, Day (2008), L′opez-Rubio *et al.* (2008), and Smolander (2008); Farber *et al.*, 2003; Mahajan *et al.*, 2007; Caleb *et al.*, 2012); (Charles *et al.*, 2003; Farber *et al.*, 2003) (Ares *et al.*, 2007); (Fränkel, 1889; Kolbe, 1882) (Kader and Watkins 2000; Charles *et al.*, 2003; Farber *et al.*, 2003) (Fonseca *et al.*, 2002a, b); (Mathooko 1996; Fonseca *et al.*, 2002a, b)(Muriel-Galet *et al.*, 2012) (Lucia L *et al.*,2017). (Bhanu *et al.*, 2015). (Manuel R *et al.*, 2016); (Quintavalla and Vicini, 2002) (Sung S.Y *et al.*, 2013).

Active Packaging

Flavor absorbers are also being used in active packaging. It has been known for decades that packaging materials can scalp or absorb flavors from foods such as fruit juices. Scalping is now being used in a positive way to absorb unwanted flavors and odors. An exciting innovation in active packaging is the potential for the controlled release of antimicrobials from packaging materials. Antimicrobials incorporated in packaging materials could extend shelf live by preventing bacterial growth and spoilage. In one system, known as "BioSwitch" (de Jong *et al.,* 2005), an antimicrobial is released on command when bacterial growth occurs. The basic concept is that a change in the environment such as pH, temperature, or UV light occurs and the antimicrobial responds accordingly. The external stimulus results in a release of the antimicrobial component of the package. In this system, the antimicrobial is released on command, and the system is active only at specific conditions. This system could potentially increase the stability and specificity of preservation and reduce the amount of chemicals needed in foods. A common example of release on command antimicrobials in food packaging is the inclusion of polysaccharide particles that encapsulate antimicrobial compounds. Many bacteria will digest polysaccharide when they grow, and so if a bacterial contamination occurs, the growth of bacteria will release the antimicrobial compounds and should inhibit subsequent microbial growth.

Chapter 6

Mechanical and Chemical Pulps

6.1 Introduction

Pulp is the raw material for the production of paper, paperboard, corrugated board and similar manufactured products. It is obtained from plant fiber and it is a renewable resource.

The origin of papermaking, which is the formation of a cohesive sheet from the rebonding of separated fibers, has been attributed to Ts'ai-Lun of China in 105 AD, who used bamboo, mulberry bark and rags.

Since then many fibers have been used for the manufacture of paper including those from flax, bamboo and other grasses, various leaves, cottonseed hair and the woody fibers of trees. Today about 97 per cent of the world's paper and board is made from woodpulp, and about 85 per cent of the woodpulp used is from spruces, firs and pines-coniferous trees that predominate in the forests of the north temperate zone (brit). The influence of the raw material can be largely assigned to the length and wall thickness of the fibers rather than to their chemical composition. The main difference between low and high density woods is less paper-bonding of the latter at equal treatment.

Now let us study about the main constituents of the wood cell wall.

Cellulose

A French chemist by name Anselme Payen was the first to discover cellulose in the year 1838. Cellulose is one of the most abundant biomaterials with high strength and other superior mechanical properties on the earth. It is generally synthesized by plants, but it is also produced by some bacteria. It is the main polysaccharide used for structural function in plants. It is a complex carbohydrate consisting of

oxygen, carbon, and hydrogen. This is a long chain, linear polymer built up of a large number of glucose molecules (weight average degree of polymerization 8000–10,000 for native wood cellulose) put together via β-1,4 linkages.

Many properties of cellulose depend upon the degree of polymerization or chain length and the number of glucose molecules constituting the polymer molecule. The fiber-forming properties of cellulose depend on the fact that it consists of long, relatively straight chains that tend to lie parallel to one another. Cellulose is a tough, fibrous, odourless, tasteless and is insoluble in water and most organic solvents. It is biodegradable and chiral. It is moderately resistant to the action of chlorine and dilute sodium hydroxide under mild conditions, but is modified or dissolved under more severe conditions. It is relatively resistant to oxidation (*e.g.*, with bleaching agents) and therefore, bleaching operations can be used to remove small amounts of impurities such as lignin without appreciable damage to the strength of the pulp. Amount of cellulose present in cotton is 90 per cent. The amount of cellulose present in the wood is 40-45 per cent and dried hemp is 57 per cent respectively. Cellulose is used mainly in paperboard and paper production. Smaller amounts are converted into a wide variety of derivatives, such as cellophane and rayon.

Since it is a non-toxic, bio-degradable polymer with high tensile and compressive strength, it has widespread use in various fields such as nanotechnology, pharmaceutical industry, food industry, cosmetics, textile and paper industry, drug-delivery systems in treating cancer and other diseases. Micro-crystalline cellulose in particular is among the most frequently used cellulose derivatives in the food, cosmetics, pharma industry, *etc.* and is an important excipient due to its binding and tableting properties, characterized by its plasticity and cohesiveness when wet.

Hemicelluloses

Hemicellulose is a natural polymer like cellulose, consisting of carbohydrate monomers. They are polysaccharides in plant cell walls that have β-(1→4)-linked backbones with an equatorial configuration. Like cellulose most hemicellulose function as supporting material in the cell wall. However, hemicelluloses are lower molecular weight (weight average degree of polymerization 100–200) mixed-sugar polysaccharides consisting of one or more of the following molecules: xylose, mannose, arabinose, galactose and uronic acids, the composition differing from timber to timber. The degree of polymerization of hemicellulose in wood is about 100–200, which is very low when compared to that of cellulose, which is 10,000. The difference between wood and non-wood fibers does not appear in the content of cellulose, however, it does appear in the contents of hemicellulose and lignin. Non-wood fibers such as grasses (wheat, corn, rice) contain up to 40 per cent of the major hemicellulose found in grasses, whereas wood fibers are composed of 25 per cent –35 per cent of hemicellulose by dry weight. Since hemicellulose has considerable side-branch groups, its structure is more open and non-crystalline. It is more hygroscopic than cellulose and attracts more water molecules due to its open structure. Hemicelluloses are usually soluble in dilute alkalis. The quantity rather than the chemical nature of the hemicelluloses appears to determine the paper properties. Hemi-celluloses are largely responsible for hydration and development of bonding during beating of chemical pulps.

Lignin

Lignin, present in nearly all plant materials, is a highly branched, alkylaromatic, thermoplastic polymer of uncertain size, built up largely from substituted phenyl-propane units. The lignin macromolecules contain a variety of functional groups, such as phenolic and aliphatic hydroxyl, carbonyl, and methoxy groups. It is a major chemical constituent of wood, the most widely used raw material for the production of pulp and paper. After all, while plants, and even some bacteria, can produce cellulose and hemicelluloses, only lignified plants can be described as "woody." Most of the unique chemical and physical properties of wood are determined by lignin. Indeed, the very word "lignin" is derived from the Latin "Lignum" meaning wood. Lignin has been described as a random, three-dimensional network polymer comprised of variously linked phenylpropane units. Lignin is the second most abundant biological material on the planet, exceeded only by cellulose and hemicellulose, and comprises 15-25 per cent of the dry weight of woody plants. This macromolecule plays a vital role in providing mechanical support to bind plant fibers together. Although lignin is necessary to trees, it is undesirable in most chemical papermaking fibers and is removed by pulping and bleaching processes.

When papermakers refer to "fibers" they include all of the long cells useful for papermaking. In this context, "fibers" refer not only to the fibers of angiosperm woods but also to gymnosperm tracheids. In contrast, the "fiber" that is common to materials science is generally thought of as a thin, thread-like substance of rather uniform chemical composition.

Plant lignins can be broadly divided into three classes: softwood (gymnosperm), hardwood (angiosperm) and grass or annual plant (graminaceous) lignin. Softwoods are generally preferred for paper manufacturing because of their long fibers, but hardwoods and agricultural residues (*e.g.*, bagasse) may also be pulped. Wood can be treated mechanically and/or chemically to separate the cells into pulps for paper and fiberboard production. Pulping necessarily requires that the lignin be cleaved into smaller fragments which may be washed away from the fiber. The cell wall of softwoods, which are preferred for most pulp products, typically contain 40–45 per cent cellulose, 15–25 per cent hemicellulose and 26–30 per cent lignin by weight (brit). Compared to hardwoods, softwoods have fibers which are generally up to 2.5 times longer. As a result, hardwoods produce a finer, smoother but less strong sheet. The purpose of pulping is to separate the fibers without damaging them so that they can then be reformed into a paper sheet in the papermaking process. Commercial pulping methods take advantage of the differences between the properties of cellulose and lignin in order to separate fibers, but breaking and weakening of the fibers does occur to a greater or lesser degree at various stages during the pulping process.

Just as the properties of wood are influenced by lignin, so are the properties of products made from wood fibers. Pulps which retain most of the wood lignin consist of stiff fibers that do not produce strong papers; they deteriorate in color and strength quite rapidly. These properties can be improved by removing most or all of the lignin by cooking wood with solutions of various chemicals, the pulps so produced being known as *chemical pulps.*

Chemical pulping is a means of removing lignin from wood (delignification) by chemically altering it to produce soluble fragments of the polymer. There are several chemical pulping methods, each based either directly or indirectly on the use of sodium hydroxide. Sufficient lignin is dissolved from the middle lamella to allow the fibers to separate with little, if any mechanical action. The nature of the pulping chemicals influences the properties of the residual lignin and the residual carbohydrates. For production of chemical pulps, the bark is removed and the logs passed through a chipper. The chipped wood is charged into a digester with the cooking chemicals, and the digestion carried out under pressure at the required temperature.

During this process, as lignin has no fiber-forming properties and is insoluble in water, it must be converted into water-soluble derivatives. The intercellular substance (primarily lignin) must be softened or dissolved to free individual fibers. Advanced organosolv pulping uses an organic solvent in which the lignin is soluble. It is attacked by chlorine and sodium hydroxide with formation of soluble, dark brown derivatives. It softens at about 160°C. Pulping is unable to remove all lignin, so high-quality pulp is subsequently bleached with oxidants to remove residual lignin and color. The two major chemical pulping technologies are the kraft and sulfite processes, with kraft dominating about 96 per cent of the market. The removal of lignin allows individual fibers to be freed from the wood matrix with mild mechanical treatment. Pulping must be able to remove lignin from fibers through chemical degradation, while minimizing damage to the cellulosic portion of the fibers to maintain strength. The kraft pulping process is the most commonly used chemical pulping process, accounting for 98 per cent of chemical pulp production in the United States and 92 per cent of chemical pulp production in the world.

In contrast to *chemical pulps, mechanical pulps* are produced by pressing logs on to a grindstone, when the heat generated by friction softens the lignin so that the fibers separate with very little damage. Mechanical pulps can also be formed by grinding wood chips between two rotating refiner plates. In addition, there are some processes which are categorized as *semi-chemical* and *chemi-mechanical.* In the 1950s the *refiner mechanical pulping* (RMP) process was developed which produced a stronger pulp and utilized various supplies of wood chips, sawmill residues and sawdust. However, the energy requirements of RMP are higher, and the pulp does not have the opacity of groundwood fibers.

Groundwood pulp is produced by forcing wood against a rapidly revolving grindstone. Practically all the wood fiber (both cellulose and lignin) is utilized compared to the several chemical processes where the lignin is dissolved to varying degrees. As a result, the yield of chemical pulp is about one half that of the mechanical process. The fibers vary in length and composition since they are in effect torn from the pulpwood. Groundwood pulp contains a considerable proportion (70–80 per cent) of fiber bundles, broken fibers, and fines in addition to the individual fibers. The fibers are essentially wood with the original cell-wall lignin intact. Therefore, they are very stiff and bulky, and do not collapse like the chemical pulp fibers. Most groundwood pulp is used in the manufacture of newsprint and magazine papers because of its low cost and quick ink absorbing properties (a consequence of the frayed and broken fibers). It is also used as board for folding and molded cartons, wallpapers, tissues and similar products. The paper has high bulk and excellent opacity, but relatively low mechanical strength.

To illustrate the extremes of the effect of lignin on paper products, the comparison between a bleached chemical pulp and a mechanical pulp often is made. Mechanical pulps often are prepared by grinding the wood to produce the papermaking fibers. Sometimes a relatively mild chemical treatment is used to reduce the grinding energy and improve pulp properties (referred to as chemo-mechanical pulps). These pulps have high lignin contents, and papers made from them reflect this contribution by lignin. An example of a high lignin paper is newsprint. Newsprint is not very strong, will rapidly discolor on exposure to light, and becomes brittle on aging. These properties do not impair seriously the utility of newsprint because of the short time that the product is used. The most common chemical pulp, kraft, is produced by treating wood at high temperature (150°C to 170°C) with strong sodium hydroxide in the presence of sulfide ions. Most of the lignin is depolymerized and dissolved. The resulting brown-colored fibers are very strong (hence the German word "kraft"); paper grocery bags are one example of a kraft 18 paper product. Kraft paper can be bleached to remove the color of the residual lignin to give white papers for writing and printing uses. Very high-yield pulps have not been delignified significantly and the lignin distribution has not changed appreciably from the untreated wood. However, depending on the pulping method, the surfaces of the papermaking fibers can be quite different. Mechanical pulps generally contain not only individual fibers, but also a large number of fiber bundles and broken fibers. Because the wood cells will be split mechanically at various locations, the exposed surfaces of mechanical pulp fibers are mixes of the various layers (middle lamella, primary wall, and secondary wall). Thus, the papermaking fibers will have both lignin-rich and cellulose-rich surfaces.

Thermomechanical pulping (TMP) pre-steams chips to 110–150°C so that they become malleable and do not fracture readily under the impact of the refiner bars. This material is highly flexible and gives good bonding and surface smoothing to the paper. The production of TMP pulps increased dramatically after its introduction in the early 1970s because they could be substituted for conventional groundwood pulps in newsprint blends to give a stronger paper. In this process, thermomechanical pulp (TMP) is prepared by pretreating wood chips with steam followed by fiberization in a disk refiner. At high temperatures (165°C to 185°C), the middle lamella is softened, and fiberization occurs due to separation at this layer (Asplund process). Although the lignin distribution has not changed, the exposed surface of the fiber is now mainly the middle lamella. This type of fiber therefore has a lignin-rich surface and is generally used in fiberboard production. If a TMP is produced at lower temperature (115°C to 130°C), the secondary wall is ruptured. Many of these fibers have surfaces composed mainly of cellulose. Also, because these surfaces become fibrillated, this kind of TMP fiber bonds better than some mechanical pulps.

There are a number of other high- and moderate-yield pulp types. In general, if the pulping process does not remove much lignin, the lignin distribution in the pulp fibers will be determined by the mechanical degradation that occurs on fiberization. For pulps that have undergone significant delignification, the distribution of lignin in the pulp fibers will be determined by the topochemistry of delignification and any mechanical degradation that occurs in processing.

In chemical pulping, delignification of the wood cells may not be uniform. For example, upon kraft or acid sulfite pulping, the secondary wall delignifies faster than the middle lamella in the initial part of the cook. Oxygen delignification of soda pulp fibers gives the opposite result: the middle lamella is delignified before the secondary wall. Little topochemical effect was seen on neutral sulfite pulping and there was no effect reported for acid chlorite delignification. Thus, the lignin distribution of the pulp fiber in chemical pulps will be determined to a large degree by the topochemistry of the pulping procedure used and by the extent of delignification.

The topochemistry of alkaline delignification and the distribution of lignin type in birch have led to an interesting quandary. If the secondary wall delignifies preferentially, and if the secondary wall contains a syringyl-rich lignin, then the prediction is made that the lignin removed in the initial part of a kraft cook of birch should be syringyl rich. Not only is this prediction wrong, the opposite occurs: guaiacyl-rich lignin is dissolved in the early stages of delignification.

One explanation offered for this failed prediction is that vessels, which may contain guaiacyl-type lignin, delignify first due to better accessibility. However, recent results in a study (which included using vessel-enriched and fiber-enriched fractions of black oak in one series of experiments; madrone, which has syringyl-rich vessel lignin, in another series; and vessel-less angiosperm woods in a third series), indicated that vessels were not the sole source of the guaiacyl-rich dissolved lignin. These results were consistent, however, with another suggestion that a low molecular weight, reactive, guaiacyl-rich lignin may be distributed throughout the wood.

6.2 The Pulping Process

After pulping process, the pulps vary considerably in their color and nature. Now let us study about sulfite processes.

The invention of this process is generally credited to the American chemist Benjamin Chew Tilghman who found in 1866 that cellulose fibers could be obtained by treating wood with a solution of bisulfite and sulfurous acid. A Swedish chemist, one C.D. Ekman, treated wood with magnesium bisulfite in 1870 and constructed the first sulfite paper mill in 1874 in Bergvik, Sweden. A German modification of the American sulfite process was developed by the German chemist Mitscherlich and involved cooking (using indirect steam heating) at lower temperatures and pressures and for longer times than previously. The Austrians Eugen Ritter and Carl Kellner first practiced the direct injection of steam into a digester about 1878, reducing the time required for heating up the digester and thus, the total cooking time. Their procedure subsequently became known as the Ritter-Kellner process.

Several pulping processes are based on the use of sulfur dioxide as the essential component of the pulping liquor. Sulfur dioxide dissolves in water to form sulfurous acid, and a part of the acid is neutralized by a base in preparation of the pulping liquor. The various sulfite processes differ in the kind of base used and in the amount of base added, which governs the resulting acidity or pH of the liquor. These processes depend on the ability of sulfite solutions to render lignin partially soluble.

Acid Sulfite Process

By 1900 sulfite pulping had become the dominant means of producing wood pulp, surpassing mechanical pulping methods. The production of acid sulfite pulp proceeds similarly to kraft pulping, except that different chemicals are used in the cooking liquor. In place of the caustic solution used to dissolve the lignin in the wood, sulfurous acid is employed. This process (no longer of great commercial importance) uses an acidic liquor composed of calcium bisulfite and a considerable excess of sulfur dioxide. The main cation used in acid sulfite pulping is the calcium base and it operates in the highly acidic range of pH 1-2. The other soluble cations like sodium, magnesium and ammonium were introduced in sulfite pulping which helped to operate the sulfite cooking range above pH 2. The use of calcium as the base in the sulfite process has the disadvantage in the low calcium solubility which restricts calcium sulfite pulping to low pH (<2 pH) cooking only. It is also important to have a high free SO_2 concentration in the liquor phase. Finally, it is not possible to recycle and reuse the cooking chemicals after the cook which leads to high pollution of the water recipients.

Wood cooked in acid liquor gives a light-colored pulp which is easily bleached, most of the lignin and hemicelluloses dissolving in the liquor. The effectiveness of delignification during acid sulfite pulping has been attributed to both sulfonation and hydrolysis reactions. The former makes the lignin more hydrophilic by introducing sulfonic groups, and the latter breaks ether bonds and creates new phenolic groups, thereby increasing the hydrophilicity of lignin as well as lowering its molecular weight. The sulfite process is characterized by its high flexibility compared to the kraft process, which is a very uniform method, which can be carried out only with highly alkaline cooking liquor. One of the main advantages that the sulfite processes had over the kraft process before the 1950's was the lower capital investment and lower operating cost for small and medium scale industries which, in many cases, had no chemical recovery systems. Another was that they produced unbleached pulp of high brightness, unlike the dark unbleached pulp produced by the kraft process. Moreover, the bleaching demand was very low and, in the case of the softwood sulfite process, the pulp yield was higher. In principle, the entire pH range can be used for sulfite pulping by changing the dosage and composition of the chemicals. Thus, the use of sulfite pulping permits the production of many different types and qualities of pulps for a broad range of applications. The sulfite process can be distinguished according to the pH adjusted into different types of pulping. The main sulfite pulping processes are acid (bi)sulfite, bisulfite (Magnefite), neutral sulfite (NSSC), and alkaline sulfite. Sulfite pulps now account for less than 10 per cent of the total chemical pulp production and the number of sulfite mills continues to decrease.

Bisulfite Process

The bisulfite cooking process is suitable for producing paper pulp for printing and writing paper, but not dissolving pulps. The dominant cooking reagent in the cooking liquor is the bisulfite (HSO_3^-) ion; the pH range of the process is 3- 5. The bisulfite process became commercially successful during the 1950's when the sodium bisulfite (Arbiso) and magnesium bisulfite (Magnefite) processes were developed (Casey, 1980). Resinous wood raw materials, *e.g.*, pine and Douglas fir, can be easily cooked with the bisulfite process: wood that is difficult to cook in the acid sulfite process because of problems associated with the lignin condensation reactions generated by resins are starting to be used. If such resinous raw materials are to be used for producing dissolving grade pulp, the sulfite cooking process will have to be carried out as a two-stage process: bisulfite cooking followed by acid sulfite cooking. The common base used in the bisulfite cooking stage is either sodium or magnesium.

This process uses a liquor composed of magnesium or ammonium bisulfite containing a slight excess of sulfur dioxide, resulting in a liquor which is not so strongly acidic. The delignification of wood chips takes place by cooking in bisulfite liquor at pH 4, the lignin reacting with the sulfurous acid to form soluble ligno-sulfonates which are later removed. The hemicelluloses are easily dissolved and removed accordingly.

The semichemical pulping idea was apparently first expressed by Mitscherlich in 1874 in a process involving softening of wood chips with sulfurous acid or bisulfite followed by rubbing or grinding to produce a pulp. It was started commercially through the development of a process to utilize certain by-product hardwood chips and then extended to other hardwoods which were used to make corrugated board.

The object of this process is to produce as high a yield as possible commensurate with the best possible strength and cleanliness. The first stage involves a mild chemical digestion using conventional type chemicals such as sodium hydroxide, sodium carbonate or sodium sulfate; the second stage involves mechanical disintegration. The hemicelluloses, mostly lost in conventional chemical digestion processes, are retained to a greater degree and result in an improvement in potential strength development. Semichemical pulps, although less flexible, resemble chemical pulps more than mechanical pulps.

The *neutral-sulfite semichemical* (NSSC) process makes use of sodium sulfite and a small amount of sodium hydroxide or sodium carbonate to give a slightly alkaline liquor. In this method, wood chips are cooked in a neutral solution of sodium sulfite and sodium carbonate. Sulfite ions react with the lignin in wood, and the sodium bicarbonate acts as a buffer to maintain a neutral solution. The major difference between all semichemical techniques and those of kraft and acid sulfite processes is that only a portion of the lignin is removed during the cook, after which the pulp is further reduced by mechanical disintegration. This method achieves yields as high as 60 to 80 per cent, as opposed to 50 to 55 per cent for other chemical processes. The NSSC process varies from mill to mill. Some mills dispose of their spent liquor, some mills recover the cooking chemicals, and some, when operated in conjunction with kraft mills, mix their spent liquor with the kraft liquor as a source of makeup chemicals. When recovery is practiced, the involved steps parallel those of the sulfite process.

The NSSC pulp is obtained in higher yield but with a higher lignin content than in the other sulfite processes. It was developed specifically as a semichemical process for the production of corrugating medium, using unbleached pulp, and lends itself to small mills with minimal capital investment.

The digestion process consists essentially of the treatment of wood in chip form in a pressurized vessel under controlled conditions of time, liquor concentration and pressure/temperature. The main objects of digestion are:

1. To produce a well-cooked pulp, free from the non-cellulosic portions of the wood, *i.e.*, lignin and to a certain extent hemicelluloses.
2. To achieve a maximum yield of raw material, *i.e.*, wood pulp from pulp wood, commensurate with pulp quality.
3. To ensure a constant supply of pulp of the correct quality.

Today most pulping processes are continuous, and to give an indication of the processing conditions encountered, the widely used Kamyr kraft pulping process will be briefly described. After steaming at low pressure during which time turpentine and gases are vented to the condenser, the chips are brought to the digester pressure of 1000 kPa. They are picked up in a stream of pulping solution and their temperature raised to 170°C over 1.5 hours. After holding at this temperature for a further 1.5 hours, the digestion process is essentially complete.

After digestion, the liquor containing the soluble residue from the cook is washed out of the pulp which is then screened to remove knots and fiber bundles that have not fully disintegrated. The pulp is then sent to the bleach plant or paper mill.

Pulps vary considerably in their color after pulping, depending on the wood species, method of processing and extraneous components. The whiteness of pulp is measured by its ability to reflect monochromatic light in comparison to a known standard (usually magnesium oxide). Cellulose and hemicellulose are inherently white and do not contribute to color; it is the chromophoric groups on the lignin that are largely responsible for the color of the pulp.

Table 6.1: Approximate Brightness Ranges of Various Unbleached Pulps

Brightness Range	*Type of Pulp*
15–30	Kraft
40–50	NSSC, Bisulfite, Soda
50–60	Groundwood, Sulfite

Basically there are two types of bleaching operations: those that chemically modify the chromophoric groups by oxidation or reduction but remove very little lignin or other substances from the fibers, and those that complete the delignification process and remove some carbohydrate material.

Chemical methods must be used to improve the color and appearance of the pulp; these are bleaching treatments and involve both the oxidation of colored bodies and the removal of residual encrusting materials (the principal one being lignin) remaining from the digestion and washing stages. Since bleaching reduces the strength of the pulp, it is necessary to reach a compromise between the brightness of the finished sheet and its tensile properties.

The most effective bleaching agent for most groundwoods is hydrogen peroxide, and since the bleaching is performed in alkaline solutions, sodium peroxide is also used. The reaction requires 3 hours at 40°C and is followed by neutralization and destruction of excess peroxide with SO_2. These pulps may be improved in color to only a limited extent since they contain virtually all the lignin from the original wood. Peroxide bleaching allows brightness to be increased by nearly 20 per cent.

Chemical pulps vary a little when while compared to mechanical pulps.

Here the reagents for full bleaching are mostly oxidative, and since the carbohydrates are also susceptible to oxidation, bleaching is accomplished under the mildest conditions. Bleaching of chemical pulps is basically stepwise purification of colloidal cellulose, and thus, bleaching can be regarded as a continuation of the cooking process. The bleaching of pulp is done through chemical reactions of bleaching agents with lignin and coloring matter of the pulp. The bleaching is performed in a number of stages utilizing chlorine, hypochlorite, chlorine dioxide, oxygen and peroxide. Between these stages, the pulp is treated with alkali to dissolve degradation products.

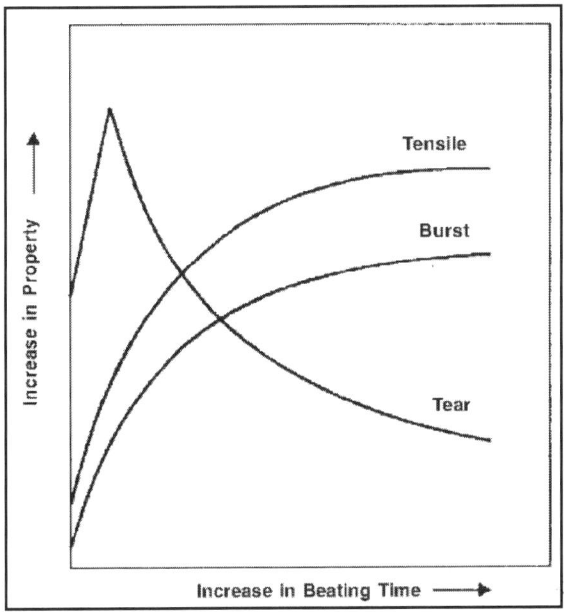

Figure 6.1: Typical Beating Curves for a Softwood Kraft Pulp.

Chlorine gas and chlorine dioxide are the major bleaching chemicals used, and concerns have been expressed about the formation of trace levels of dioxins as a result of the bleaching process. A relatively new development which is now fairly common in the industry is the replacement of conventional bleaching agents such as chlorine, hypochlorites and chlorine dioxide with oxygen, ozone and peroxides; details of these processes have been given elsewhere. These changes have been introduced to enable pulp and paper mills to meet tough new anti-pollution laws and regulations and to conserve wood, chemicals and energy.

Stock preparation is conducted to convert raw stock into finished stock (furnish) for the paper machine. It is defined as that part of the pulp and papermaking process in which pulp is treated mechanically and, in some instances, chemically by the use of additives, and is thus, made ready for forming into a sheet or board on the paper machine. During the stock preparation steps, the pulps are most conveniently handled as aqueous slurries. However, in the papermaking process, utilizing purchased pulps and waste paper which are received as dry sheets, the first step is the separation of all the fibers from one another, and their dispersion in water with a minimum of mechanical work to avoid altering the fiber properties. This process is known as slushing and is carried out on machines such as the Hydrapulper or the Sydrapulper. When the pulping and papermaking operations are adjacent to one another, pulps are usually delivered to the paper mill in slush form directly from the pulping operation.

Now, let us study about stock preparation operations. The following operations are in practice in the papermills *i.e.,*

1. Dispersion
2. Beating
3. Metering and blending of fiber and additive

Dispersion: Pulpers are used to disperse dry pulp into water to form a slush or a slurry. The stock in the pulper is accelerated and decelerated repeatedly, and hydrodynamic shear forces are produced by the severe velocity gradients. The resulting forces serve to loosen fibers and reduce any flakes into individual fibers.

Beating/Refining: Pulp produced in a mill without mechanical treatment is unsuitable for most paper grades. Paper made from unbeaten virgin pulp has a low strength, is bulky, and has a rough surface. In good-quality paper, the fibers must be matted into a uniform sheet and must develop strong bonds at the points of contact. Beating and refining are the processes by which the undesirable characteristics are changed. Hence, the object of beating is to increase the surface area of the fibers by assisting them to imbibe water. As a result, additional bonding opportunities are provided for between cellulose molecules of neighbouring fibers. The beating also makes the fibers more flexible, causing them to become relatively mobile and to deform plastically on the paper machine.

The mixture of pulp (known as the furnish) is passed into the beater and brought to a consistency of 5–7 per cent. The fibers are then beaten while suspended in the water in order to impart to them many of the properties that will determine the character of the final product.

The quality and characteristics of the finished paper depend to a great extent on the treatment in the beater. Because papermaking fibers are stiff and springy, the resulting paper would be flabby and weak if made into a sheet of paper without beating. There would be little adhesion between the fibers and they could not be consolidated under the presses of the papermachine. A sheet formed from an unbeaten pulp has a low density, and is rather soft and weak, whereas if the same pulp is beaten, the resultant paper is much more dense, hard and strong. If taken to the extreme, beating produces very dense, translucent and glassine-type sheets. Thus, beating can be controlled to produce papers as widely different as blotting and greaseproof.

A small amount of beating will produce a highly absorbent sheet with high tear strength but low burst and tensile strength. As beating progresses, burst and tensile strength values rise to a maximum, followed by a fairly rapid fall as the fibers become shortened and lose their structure. The bulk decreases and the density increases, opacity is lowered and transparency increased. The tensile strength, bursting strength and tearing resistance are shown as functions of beating time for a softwood kraft pulp.

The term "beating" is applied to the batch treatment of stock in a Hollander beater or one of its modifications. Since its invention in Holland around 1690, the principle of the batch-operated hollander beater has remained substantially the same. It consists of a cylindrical roll containing knives which revolves over a stationary bedplate which also contains a set of knives. Circulating stock passes between the roll and the bedplate, the severity of beating being controlled by adjusting the load of one against the other. Circulation is continued until the pulp is considered ready to be made into the desired paper.

In many papermills, beaters have been replaced by continuous refiners, of which the disk refiner (rotary disks rotate against a working surface) is the most recent development. However, the batch beater is a convenient vessel for adding chemicals and mixing them intimately with the pulp in order to give special properties to the final papers.

The term "refining" is used when the pulps are passed continuously through one or more refiners, whether in series or in parallel. Refining develops different fiber properties in different ways for specific grades of paper. Usually, it aims to develop the bonding ability of the fibers without reducing their individual strength by damaging them too much, while minimizing the development of drainage resistance. So, the refining process is based on the properties required in the final paper. Different types of fiber react differently because of differences in their morphological. The refining process must take into account the type of fibers.

Most of the strength properties of paper increase with pulp refining because they rely on fiber–fiber bonding. However, the tear strength, which depends highly on the strength of the individual fibers, decreases with refining. After a certain point, the factor limiting the strength is not the fiber–fiber bonding, but the strength of the individual fibers. Refining beyond this causes a decrease in other strength properties besides tear.

Refining a pulp increases the fibers' flexibility and results in denser paper, which means that bulk, opacity, and porosity values decrease during the process. Mechanical and hydraulic forces are employed to alter the fiber characteristics. Shear stresses are imposed by the rolling, twisting, and tensional actions, occurring between the bars and in the grooves and channels of the refiner. Normal stresses (either tensional or compressive) are imposed by the bending, crushing, and pulling/pushing actions on the fiber clumps caught between the bar-to-bar surfaces. During beating and refining, fibers randomly and repeatedly undergo tensile, compressive, shear, and bending forces. They respond in three ways

1. Fibers develop new surfaces externally through fibrillation and internally through fiber wall delamination.

2. Fibers deform, resulting in changes in their geometric shape and the fibrillar alignment along their length. Overall, the fibers flatten or collapse. Fiber curl changes and kinks are induced or straightened. On the small scale, dislocations, crimps, and microcompressions are induced or diminished.
3. Fibers break, resulting in changes in length distribution and a decrease in mean fiber length. A small amount of fiber wall material also dissolves. All these changes occur simultaneously and are primarily irreversible. The extent of the changes depends on the morphology of the fibers, the temperature, the chemical environment, and the treatment conditions.

As different types of fibers have different lengths and cell wall thicknesses and also vary in the width of their central lumen (canal), some fibers are fibrillated more by one kind of beating than another. Fibers that have thick cell walls and narrow central canals, such as linen, are less prone to cutting but readily fibrillate. So choosing the right type of fiber to produce the refining effect appropriate to the desired paper properties is important. The problem in producing paper with the required properties is that cutting and fibrillation are not independent of one another. Cutting is necessary to produce smaller fibers that pack to give a good, smooth paper, while fibrillation is required for strength. The degree to which each of these is achieved depends on the characteristics of refining.

Blending/Chemical Addition

Accurate proportioning of pulps and additives into a blend is the major task of this stage. Many chemical additives are added to the stock at different points before the paper machine. In papermaking, chemicals such as strength additives, adhesives, mineral fillers and sizing agents may be added at the beater stage prior to sheet formation (*i.e.*, internal addition), or to the resulting sheet after complete or partial drying, depending primarily on the desired effects. Strength additives usually are added internally if uniform strength throughout the sheet is wanted, but they are applied to the surface if increased surface strength is needed. Fillers can improve brightness, opacity, softness, smoothness and ink receptivity, and are essentially insoluble in water under the conditions of use. The main drawback is that the materials added may be lost through the wire of the paper machine in the large amount of water used. Therefore, if an additive cannot be retained efficiently from a dilute pulp slurry, it is better to apply it to the surface of the sheet.

Sizing

Sizing is the process of adding materials to the paper in order to render the sheet more resistant to penetration by liquids, particularly water. Adequate sizing maintains consistent runnability through the size press and helps to provide uniform printability in uncoated papers. For coated paper base stock, adequate sizing helps to control binder migration. The process of sizing can be accomplished in two principal ways: (1) internal sizing and (2) surface sizing. The purpose of internal sizing is to inhibit penetration of liquids into the internal structure of paper. Internal sizes are introduced at the wet end of the papermaking system, usually as colloidal suspensions, which are retained in the fiber network during sheet formation. Adequate sizing maintains consistent runnability through the size press and helps to provide uniform printability in uncoated papers.

On the basis of the pH, the internal sizing can be categorized into three types: (1) acid sizing, (2) neutral sizing, and (3) alkaline sizing. The three most common internal sizing agents currently being used are rosin systems, alkene ketene dimers (AKD), and alkenyl succinic anhydrides (ASA). For paper being made with acidic conditions in the wet end, a papermaker would choose to use a rosin size as an internal sizing agent. In the past, mills used to prepare their own rosin size. But nowadays, ready-made fortified rosins either in powder or in liquid form are available for use. Rosin is the most widely used sizing agent, but starches, glues, caseins, synthetic resins and cellulose derivatives are also used. For neutral/alkaline paper machines, a papermaker would prefer to use synthetic sizing agents such as AKD and ASA. AKD is used as a sizing agent in pH range of 6–9 that is both in neutral and alkaline media. Internal sizing at neutral/alkaline pH gives the possibility to use calcium carbonate as a filler. ASA is another type of cellulose-reactive sizing agent that has found wide domestic acceptance in recent years due to its advantages. The ASAs are quite reactive molecules and can promote sizing without heating. Therefore, alkaline papermaking with an ASA size offers the papermaker an opportunity for enormous saving in raw materials and production cost. AKD use is more common in Europe, while in the United States the application of ASA is growing and today surpasses that of AKD. The sizing agents may be added directly to the stock as beater additives to produce internal or engine sizing, or the dry sheet may be passed through a size solution to produce a surface size.

6.3 Paper Conversion and Surface Treatments

Almost all paper is converted by undergoing further treatment after manufacture, such as embossing, impregnating, saturating, laminating and the forming of special shapes and sizes such as bags and boxes. Further surface treatment involving the application of adhesives, functional products and pigments are common, depending on the end use of the paper. Surface treatment applies similar materials on the web.

Because of the widespread use of paper and paperboard in direct contact with foods, most mills require paper chemicals that have been cleared for use with food by regulatory authorities such as the FDA. It is recommended that the regulatory status of any paper-chemical additive be determined with the supplier prior to use.

Two general approaches for the treatment of the paper surface are usually considered *i.e.*, surface sizing and coating. Surface treatments such as sizing and coating are extensively applied to improve the appearance of products. Surface sizing is usually performed by the application of an aqueous suspension of cationic starch alone or combined with a synthetic polymer, such as poly(styrene-co-acrylate) or poly(styrene-co-maleic anhydride), among others. Surface-sizing agents prevent excess water penetration and improve the strength of the paper. Sizing imparts resistance to liquids on the paper surface, a property necessary for paper used for writing or printing. Without external sizing, ink would bleed and feather. The sizing agent penetrates far enough into the paper to increase the fiber bonding and the dependent properties such as bursting, tensile and folding strengths. An additional effect is an improvement in the scuffing resistance of the paper surface.

The rate of flow of a liquid through a very thin tube or capillary can be represented by the Washburn equation which combines the equation representing the natural driving force for fluid movement in a capillary tube with the Poiseuille equation for laminar flow through a tube. Two of the five parameters that govern the rate of flow, *viz.*, surface tension and viscosity of the penetrating liquid, depend on the use to which the sized paper product is put. Two more, *viz.*, the radius and length of the pores, are governed by the papermaker's products, *i.e.*, the basis weight, bulk density and porosity of the sheet. It is obviously important that the appropriate chemical is selected to produce the sizing that is required by the paper producer and the user.

The most commonly used materials for surface sizing are starches, usually chemically modified *e.g.*, oxidized starches, cationic starches and hydroxyethylated derivatives. Other film-forming materials which can be used for surface sizing include carboxymethyl cellulose and poly (vinyl alcohol) which provide oil- and grease-repellent coatings and improve paper strength. Other polymeric sizing agents such as polyurethanes are used as surface-sizing and strength-enhancing agents. Their cost is relatively high compared with other sizing agents.

Fluorochemical emulsion sizing agents can be applied to the surface of paper or paperboard to provide good oil and grease repellancy. They find application for pet-food bag papers, meat, fish and poultry wrap, cookie bags and candy wrappers.

Coating paper may be desirable or necessary to improve optical, printing/writing, and/or functional properties. Functional properties can be for protection from liquids, oils, gases, and chemicals, improve adhesion characteristics, improve wear, or some other property. Coatings can be classified as aqueous, solvent, high solids, or extrusion. Aqueous coatings, used for commodity papers, contain water-soluble binders and are applied as a liquid. Common aqueous binders are casein, starch, protein, acrylics, and polyvinyl acetates. Solvent coatings are used in situations where the binders are not soluble in water and are used with specialty papers. High-solid and extrusion coatings are used for specialized papers, where chemical, gas, or liquid resistance is necessary. The coating processes use a different type of chemical formulation which includes a pigment (with kaolin, calcium carbonate, titanium dioxide, aluminum oxide, and silica being the most common ones used), a binder (mostly polyvinyl alcohol or latexes such as styrene butadiene), a co-binder (*e.g.*, carboxymethylcellulose), a thickener (*e.g.*, carboxymethyl cellulose) and a dispersant (a cationic polymer additive). Recently, sericite, a mineral composed mainly of silicon dioxide and aluminum oxide has also been proposed as a potential coating pigment.

Paper may be coated either on equipment that is an integral part of the paper machine (*i.e.*, on-machine coating), or on separate converting equipment. The most common method for the application of chemicals to the surface of a paper web is by a size press. In the size press, dry paper is passed through a flooded nip and a solution or dispersion of the functional chemical contacts both sides of the paper. Excess liquid is squeezed out in the press and the paper is re-dried.

6.4 Pigments and Adhesives

Pigments comprise 70–90 per cent of the dry solids in paper coatings and are generally designed to mask or change the appearance of the base stock, improve opacity, impart a smooth and receptive surface for printing, or developing a practical paper machine, eventually becoming bankrupt and dying in poverty. Despite their misfortune, their name has been familiar to generations of papermakers for the development of a machine, the essential principles of which are still in use today.

Paper is made by depositing a very dilute suspension of fibers from a very low consistency aqueous suspension (greater than 99 per cent water) on to a relatively fine woven screen, over 95 per cent of the water being removed by drainage through the wire. The fibers interlace in a generally random manner as they are deposited on the wire and become part of the filter medium.

Although paper is the general term for a wide range of matted or felted webs of vegetable fiber that have been formed on a screen from a water suspension, it is usually subdivided into paper and paperboard. However, there is no rigid line of demarcation between the two, with board often being defined as stiff and thick paper. ISO standards define paperboard as paper with a basis weight (grammage) generally above 250 grams per square meter, but there are exceptions.

The modern fourdrinier paper machine consists essentially of an endless woven wire gauze or forming fabric stretched over rollers. The suspension of pulp in water is forced on to the wire in a regular stream, the relative speeds of the stock and wire affecting the degree to which the fibers are aligned along the direction of travel. As it progresses down the wire, the fiber slurry loses water, first by free drainage and then under vacuum.

Fourdriniers are standard in the industry and are used to produce all grades of paper and paperboard. Heavy paperboards require a long drying time and machine speeds are 10–250 meters per minute. Very dense papers (*e.g.*, glassine and greaseproof) are difficult to dewater in the forming and press sections and speeds range from 20 to 300 meters per minute, depending on the product. Brown grades (*e.g.*, paper bags and linerboard) are produced at 200–800 meters per minute depending on basis weight. The majority of newsprint machines operate at 600–900 meters per minute.

We have studied about pigments. Now let us have a bird view on types and functions of adhesives and chemicals.

Paper industry is one of the largest users of adhesives and chemicals. Overall world production of paper and paperboard is more than 100 million tons. The generic term "paper" includes a very wide range of products with very different coatings. Many of these products are produced with the help of adhesives and chemicals. Different products have led to increase in number of adhesives, each optimized for a specific purpose. An adhesive can be defined as a material used for holding two surfaces together. An adhesive must wet the surfaces, adhere to the surfaces, develop strength after it has been applied, and remain stable. In a single statement it is a material used for bonding that exhibits flow at the time of application.

In paper industry, the primary function of the adhesive in pigment coating is to bind the pigment particles together and to the raw stock. The type and proportion of the adhesive controls many of the characteristics of the finished paper such as surface strength, gloss, brightness, opacity, smoothness, ink receptivity and firmness of the surface. The strength must be sufficient to prevent the coating from being picked up by tacky printing inks.

Adhesives are available in various types:

Natural

- **Starch and dextrin** - These materials are derived from cereals or roots *etc.* The basic constituent is polysaccharide which on hydrolysis yields long chain glucose units. Different versions can occur with the use of different food materials and with changes in hydrolysis methods. The main use for these adhesives is the paper industry where they are used in multiwall bags, corrugated paper *etc.* Starches are used in many coated papers, and are the principal adhesives if resistance to moisture is not required. Dextrin which is dry roasted starch is used in remoistenable adhesives.
- **Gelatine (animal, fish, vegetable glues):** These are proteins which are derived from the hydrolysis of either collagen or soya flour, or by separating casein from skim milk. Animal glue was the first material used for bonding paper and as an adhesive in paper coating. Animal glues from bones and hides are used in the paper industry such as book-binding and case making. But animal glue has been replaced by casein and soy protein now. Caseirs from skim milk are used mainly in wood-to-wood bonds. Soybean glues are used in paper backs.
- **Asphalt and Bitumin:** These high fractions of crude oil are more used as sealers rather than adhesives except in the bonding of coarse grade papers to produce waterproof building papers.

Semi-synthetic

- **Cellulosic:** Cellulose is in the structural elements of plants and is a polymer made up of glucosidic rings joined by oxygen. Ethers and esters can be formed to give either water soluble or solvent soluble polymers. Cellulose acetate butyrate is mainly used as paper to paper and plastic adhesives.

Synthetic

- **Vinyls:** A range of vinyl products are being used worldwide in paper industry. The number of vinyl derivatives is very large and all have some specific properties. Some members are produced either as an emulsion or as a solvent soluble type.

  Polyvinyl acetate: Used in paper convertion, wood.

  Polyvinyl alcohol: Used in paper applications.
- **Acrylics:** As same as vinyls but with few different compositions are used in pressure sensitive adhesives, and for flooring, paper lamination, flocking adhesives *etc.* Acrylic-based emulsions are used mostly on paperboard, and their freedom from odour makes them ideally suited for use in food packaging. Various rubber latexes and other emulsions are also employed as adhesives.

In many packaging applications, a barrier may be needed against water vapor or gases such as oxygen. A water barrier can be formed by changing the wettability of the paper surface with sizing agents. Coating the paper with a continuous film of a suitable material will confer gas or vapor barrier properties. Paraffin wax applied in a molten form was commonly used to give a water vapor barrier, but polyethylene applied by extrusion gives a more durable and flexible coating. Highly sophisticated coatings based on a wide range of polymers and modifiers can be formulated using solvent systems. The polymers include cellulose derivatives, rubber derivatives, vinyl copolymers, polyamides, polyesters and butadiene-styrene copolymers.

Apart from Adhesives there are various chemicals used in manufacturing paper industry.

- **Optical brightening agent:** Optical brightener is used to make paper appear whiter.
- **Pigment:** Pigments that absorb in the yellow and red part of the visible spectrum can be added. As the dye absorbs light, the brightness of the paper will decrease. To increase whiteness, a combination of pigments and optical brightening agent are often used. The most commonly used pigments are blue and violet dyes.
- **Retention agent:** It is added to bind fillers to the paper. This retention agent is a polymer with high cationic, positively charged groups. An additional feature of retention means to accelerate the dewatering in

the wire section of the paper. Polyethyleneimine and polyacrylamide are some of the chemicals used.

- **Caustic soda:** Caustic soda is added to increase the pH in the pulping process of fibers. The higher pH of the paper fiber solution causes the fibers to smoothen and swell, which helps in the grinding process of the fibers.
- **Sizing agent:** Internal sizing of paper, most paper types must have some water-resistance in order to have a specific writing quality and/ or printability. Today mainly AKD (alkyl ketene dimer) and ASA (alkenyl succinic anhydride) are used.
- **Wet strength additive:** These additives ensure that when the paper becomes wet, it retains its strength. This is especially important in a tissue paper. Typical chemicals used are as epichlorohydrin, melamine, urea formaldehyde and polyimines. These substances polymerize in the paper and result in construction of a strengthening bond.
- **Dry strength additives**: Also known as dry strengthening agents are chemicals that improve paper strength of normal or not wet condition. Those strength including compression strength, bursting strength, tensile breaking strength, delamination resistance *etc.* Typical chemicals used are as cationic starch and polyacrylamide (PAM) derivatives. These Dry strength additives act as binder of fibers, often under aid of aluminum ion in paper sheet.
- **Cationic starch:** It is used to enhance the paper strength; cationic starch is added to wet pulp in the manufacturing process. Quaternary salts that are used include 2.3-epoxy propyl trimethyl ammonium chloride (EPTAC, also known as or Glytac Quab) and (3-chloro-2-hydroxypropyl) trimethyl ammonium chloride (CHPTAC, also known as Quat 188).
- **Mineral Fillers:** Usually found as China Clay, Calcium carbonate
- **Coating binders:** Available as Styrene butadiene latex, Styrene acrylic, dextrin, oxidized starch.
- **Pulping chemicals:** Are used for the production of chemical pulp from wood chips, following chemicals can be used: Caustic soda, Sodium sulfide for the Kraft process, Sulfurous acid for the Sulfite process, Caustic soda, Anthraquinone for the Soda pulping.
- **Bleaching chemicals:** Sodium dithionite, Chlorine dioxide, Hydrogen peroxide, Ozone

Today the production of paper, cardboard and cartons without usage of waste paper is unimaginable. Usage of adhesives and chemicals is a vital part in manufacturing any kind of paper. When compared, adhesives with other substances brought into the paper system during paper production (*e.g.*, coating binders) and paper processing (*e.g.*, ink binders) and that have properties similar to adhesives, one can see that adhesives play a little role with respect to quantity. We can negate the fact that how important may be, adhesives and chemical are harmful, careful usage of it can create better effect on manufacturing paper and its products.

6.5 Fine Papers and Coarse Papers

Paper is mainly divided into two main types:

1. Fine Papers
2. Coarse Papers

Fine papers, generally made of bleached pulp, and typically used for writing paper, bond, ledger, book and cover papers, and coarse papers, generally made of unbleached kraft softwood pulps and used for packaging.

Kraft Paper

Kraft paper is the most used material in packaging. It is made by the sulfate pulping process. It usually has basis weight in the ranges of 18–200 lbs (approximately 8.2–90.7 kg), with the common weight between 25 and 80 lbs approximately 11.3–36.3 kg. This is typically a coarse paper with exceptional strength, often made on a fourdrinier machine and then either machine-glazed on a Yankee dryer or machine-finished on a calender. It is sometimes made with no calendering so that when it is converted into bags, the rough surface will prevent them from sliding over one another when stacked on pallets.

Natural or unbleached kraft paper has light brown color, while bleached kraft paper is white. Natural kraft paper is the strongest of the common packaging papers and is used when maximum strength is required, such as in industrial bags, grocery bags, inner plies of multiwall sacks, or plain wrapping paper. Kraft fibers are also used to produce liner of corrugated board. Bleached kraft paper is the strongest white paper. It is used where strength, printability, and appearance are important, such as flour and sugar bags, labels, and envelopes. Bleached kraft fibers are also used in production of solid bleached sulfate (SBS), a high quality paperboard.

Kraft paper can be coated or laminated to improve its barrier properties and provide additional strength. Polyamide or polyamine resin can be added into kraft fiber and cured under heat to produce wet strength kraft paper. When wet, this type of paper can retain a third of its dry strength and can be used in high humidity conditions.

Bleached Paper

These are manufactured from pulps which are relatively white, bright and soft and receptive to the special chemicals necessary to develop many functional properties. They are generally more expensive and weaker than unbleached papers. Their aesthetic appeal is frequently augmented by clay coating on one or both sides.

Greaseproof Paper

Greaseproof paper was first developed as a replacement for parchment by the engineer Otto Munthe Tobiesen in 1894 at the Granfoss mill in Norway (Mathiesen and Danielsen, 1994). The cooking was performed at a low temperature and with high amount of calcium in the white liquor. This resulted in a sulphite pulp which after a long beating time gave a grease-resistant paper. The special properties of greaseproof paper are mainly due to the high degree of beating which creates many bonding sites on each fibre, which in turn create a paper of high density. The production of greaseproof paper is a costly and an energy demanding process both through beating energy and drying energy.

This is a translucent, machine-finished paper which has been hydrated to give oil and grease resistance. Prolonged beating or mechanical refining is used to fibrillate and break the cellulose fibers which absorb so much water that they become superficially gelatinized and sticky. This physical phenomenon is called hydration and results in consolidation of the web in the paper machine with many of the interstitial spaces filled in. Greaseproof paper has a closed surface structure with a small number of large surface pores. This suggests that greaseproof paper with its closed surface structure can be an excellent substrate for a barrier coating.

The satisfactory performance of greaseproof papers depends on the extent to which the pores have been closed. Provided that there are few interconnecting pores between the fibers, the passage of liquids is difficult. However, they are not strictly "greaseproof" since oils and fats will penetrate them after a sufficient interval of time. Despite this, they are often used for packaging butter and similar fatty foods since they resist the penetration of fat for a reasonable period.

Waxed Paper

Waxed papers provide a barrier against penetration of liquids and vapors. A great many base papers are suitable for waxing, including greaseproof and glassine papers. The major types are wet-waxed, dry-waxed and wax-laminated. Wax-sized papers, in which the wax is added at the beater during the papermaking process, have the least amount of wax and therefore, give the least amount of protection.

Wet-waxed papers have a continuous surface film on one or both sides, achieved by shock-chilling the waxed web immediately after application of the wax. This also imparts a high degree of gloss on the coated surface. Dry-waxed papers are produced using heated rolls and do not have a continuous film on the surfaces. Consequently, exposed fibers act as wicks and transport moisture into the paper. Wax-laminated papers are bonded with a continuous film of wax which acts as an adhesive. The primary purpose of the wax is to provide a moisture barrier and a heat sealable laminant.

Often special resins or plastic polymers are added to the wax to improve adhesion and low temperature performance, and to prevent cracking as a result of folding and bending of the paper.

Table 6.2: Main Types of Packaging Papers

Basic Material	*Source*	*Weight Range kg/1000 m²*	*Tensile Strength kg/m*	*Properties and Uses*
Kraft papers	Sulfate pulp from softwoods	70–300	250–1150	Heavy-duty paper, bleached, natural or colored; may be wet-strengthened or made water-repellent. Used for bags, multi-wall sacks and liners for corrugated board. Bleached varieties are used for food packaging where strength is required.
Sulfite papers	Usually bleached generally made from mixture of softwood and hardwood	35–300	Very variable	Clean, bright paper of excellent printing nature and used for smaller bags, pouches, envelopes, waxed papers, labels and for foil laminating, *etc.*
Greaseproof papers	From heavily beaten pulp	70–150	180–450	Grease-resistant for baked goods and fatty foods
Glassine papers	Similar to grease-proof but super-calendered	40–150	140–535	Oil- and grease-resistant, used as an odor barrier for lining bags, boxes, *etc.* and for greasy foods
Vegetable parchment	Treatment of unsized paper with concentrated sulfuric acid	12–75	215–1450	Non-toxic, high wet strength, grease- and oil-resistant for wet and greasy food
Tissue papers	Lightweight paper from most pulps	20–50	Low	Lightweight, soft wrapping paper

Glassine Paper

Glassine paper derives its name from its glassy, smooth surface, high density and transparency. It is produced by further treating greaseproof paper in a supercalender where it is carefully dampened with water and run through a battery of steam-heated rollers. This results in such intimate interfiber hydrogen bonding that the refractive index of the glassine paper approaches the 1.02 value of amorphous cellulose, indicating that very few pores or other fiber/air interfaces exist for scattering light or allowing liquid penetration. The transparency can vary widely depending on the degree of hydration of the pulp and the basis weight of the paper. The addition of titanium dioxide makes the paper opaque, and it is frequently plasticized to increase its toughness.

Vegetable Parchment

Vegetable parchment takes its name from its physical similarity to animal parchment (vellum) which is made from animal skins. The process for producing parchment paper was developed in the 1850s, and involves passing a web of high-quality, unsized chemical pulp through a bath of concentrated sulfuric acid. The cellulosic fibers swell and partially dissolve, filling the interstices between the fibers and resulting in extensive hydrogen bonding. Thorough washing in water, followed by drying on conventional papermaking dryers, causes reprecipitation and consolidation of the network, resulting in a paper that is stronger wet than dry (it has excellent wet strength, even in boiling water), free of lint, odor and taste, and resistant to grease and oils. Unless specially coated or of a heavy weight, it is not a good barrier for gases.

Because of its grease resistance and wet strength, it strips away easily from food material without defibering, thus finding use as an interleaver between slices of food such as meat or pastry. Labels and inserts in products with high oil or grease content are frequently made from parchment. It can be treated with mold inhibitors and used to wrap foods such as cheese.

Parchment paper with great shock-absorbing capability can be produced by wet creping, resulting in extensibility combined with natural tensile toughness. Special finishing processes provide qualities ranging from rough to smooth, brittle to soft, and sticky to releasable. It was first used for wrapping fatty foods such as butter, and still finds such an application today.

Glazed imitation parchment (GIP) is made from strong sulfite pulp which is heavily engine-sized and glazed to give the necessary degree of protection.

Tissue Paper

Tissue papers range from semi-transparent to totally opaque, and can be waxed. They are generally either machine-finished (MF) or machine-glazed (MG), the former process involving calendering between highly-polished rollers which squeeze and polish both sides of the paper, and the latter process producing a smooth glazed side and a rough back side. MG papers may also be machine finished to improve the smoothness on both sides.

6.6 The Uses of Papers

A variety of papers are used in the manufacture of paper bags depending on the end use of the bags. Both bleached and unbleached papers have been used, but the most common are the brown paper bags made from kraft paper. Bags can differ in shape, style and the number of piles, *e.g.*, single, double and multiwall. Multiply (or multiwall) paper bags are also referred to as paper sacks.

Single-wall bags can be simple flat bags with a lengthwise seam and a base where the paper is folded under the glued; satchel-bottom bags which provide a flat base when filled; and square-bottom bags which are similar to flat-bottom bags except that they contain bellows folds at the sides to reduce the width of the closed bag without reducing capacity.

Multiwall bags are used primarily for the packaging of materials such as powders or granules that need no protection against compressive forces, since their principal function is to absorb energy without rupturing, contain the contents and protect them from contamination. It is important that the paper has strength when under load, and that it should stretch under these conditions. Therefore, tensile strength and stretch characteristics of the paper are important. The strength of the paper (and therefore, the performance of the bags made from it) varies with the moisture content of the paper. Dry conditions tend to lower the strength of the bag, while the strength increases with moisture content up to a certain level.

Multiwall bags contain from two to six plies of high quality paper, sometimes with special coatings such as polyethylene to provide a liquid water and water vapor barrier. Each ply is fabricated as a tube and arranged one within the other so that each layer bears its share of any applied or induced stress. Generally better performance is obtained against shock or impact by the use of several plies of relatively lightweight papers rather than by the use of fewer plies of heavier-weight papers, although the latter is considered to be more effective against externally applied point stresses such as protruding nails or sharp corners on pallets.

The average heavy-duty multiwall bag is constructed of a number of plies of paper with basis weights of 65–114 gsm, the most frequently used basis weights being 65, 81 and 98 gsm.

Paper is generally termed board when its substance exceeds 250 gsm, except in the United Kingdom where the dividing line is arbitrarily taken as 220 gsm.

A variety of types of paperboards are manufactured. The simplest types are single-ply thick papers made from 100 per cent bleached chemical wood pulp; they are used in food packaging where purity and clean appearance is required together with a degree of strength.

The methods of fiber treatment are essentially the same as those used in manufacturing paper grades. After the sheet has been formed, the methods used for removing excess water and finishing the web are also essentially the same as those used in the manufacture of paper. The main difference is found at the forming section of the machine where the web is formed.

Rigid paper containers are generally constructed of paperboard or a combination of paperboard and paper.

Folding Cartons

Folding cartons are containers made from sheets of paperboard (typically with thicknesses between 300 μm and 1100 μm) which have been cut and scored for bending into desired shapes; they are delivered in a collapsed state for erection at the packaging point.

The boards used for cartons have a ply structure and many different structures are possible, ranging from recycled fibers from a variety of sources (as in chipboard), through fibers where the outer ply is replaced with better quality pulps to give white-lined chipboards, to duplex boards without any waste pulp and solid white boards made entirely from bleached chemical pulp. The latter are used for frozen food packaging and other applications where the board must be waxed.

A number of steps are involved in converting paperboard into cartons. Where special barrier properties are required, coating and laminating is carried out; wax lamination provides a moisture barrier, lining with glassine provides grease resistance, and laminating or extrusion coating with plastic materials confers special properties including heat sealing. The use of barrier materials in cartonboard is restricted by the inability of the normal types of carton closure to prevent the ingress of moisture directly. Consequently, considerable effort has been directed towards the development of one-piece coated cartons which have improved barrier properties. Liquid-tight, hermetically-sealed cartons are now widely used for the packaging of a wide range of liquid foods, such packaging being able to be carried out under aseptic conditions to give packs which will retain the product in a commercially sterile state. Incorporation of a liner which can be sealed into a bag can give a carton able to contain a liquid, or give gas and water vapor protection to the contents, and such systems are available. Other developments have included the incorporation of various types of tear-off, easy-open and dispensing devices into cartons.

Coating of the outer board greatly enhances the external appearance and printing quality, and clay and other minerals are used for such purposes. The coating can be applied either during the board-making operation or subsequently. Foil-lined boards are also used for various types of carton, not only to provide barrier properties (as in aseptic cartons) but also (in certain applications) to improve reheatability of the contents, especially in microwave ovens.

The conventional methods of carton manufacture involve printing of the board, followed by creasing and cutting to permit the subsequent folding to shape, the stripping of any waste material which is not required in the final construction, and the finishing operation of joining appropriate parts of the board, either by gluing or (occasionally) stitching. During creasing and folding, cartonboard is subjected to complex stresses, and the ability of a board to make a good carton depends on its rigidity, ease of ply delamination, and the stretch properties of the printed liner. It is important that the surface layer on the top of the board is of an elastic nature and relatively high strength compared with the properties of the underlying layers since they will be in compression.

6.7 The Construction of Corrugated Board

Corrugated cardboard is a stiff, strong, and light-weight material made up of three layers of brown kraft paper. In 1884, Swedish chemist, Carl F. Dahl, developed a process for pulping wood chips into a strong paper that resists tearing, splitting, and bursting. He named it the kraft process because it produces a strong paper that resists tearing, splitting, and bursting. Corrugated board is widely used in the packing industry. The main advantages are lightness, recyclability and low cost. This makes the material the best choice to produce containers devoted to the shipping of goods. The corrugated box was initially used for packaging glass and pottery containers. Later, the case enabled fruit and produce to be brought from the farm to the retailer without bruising, improving the return to the producers and opening up export markets.

Corrugated board is essentially a paper sandwich consisting of corrugated medium layered between inside and outside linerboard. On the production side, corrugated is a sub-category of the paperboard industry, which is a sub-category of the paper industry, which is a sub-category of the forest products industry. On the marketing side, it is a part of the packaging industry. Throughout the distribution cycle of moving goods from producer to consumer, corrugated is the most widely used kind of shipping container. Traditionally, corrugated is best known for its structural strength that offers protection to packaged products throughout the transportation cycle. However, it has evolved over the course of time and today it is a much more versatile product.

Containerboard, also referred to as CCM (corrugated case material), is a type of paperboard specially manufactured for the production of corrugated board. The term encompasses both linerboard and corrugating medium (or fluting), the two types of paper that make up corrugated board. Since containerboard is made mainly out of natural unbleached wood fibers, it is generally brown, although its shade can vary depending on the type of wood, pulping process, recycling rate and impurities content. For certain boxes that demand good presentation, white bleached pulp or coating is used on the top ply of the linerboard that goes outside the box.

Corrugated board is manufactured on a very large and specialized machine (really a series of machines) called a corrugator which can be divided into three sections. In the first (known as the single facer), the medium (a relatively thin web of paperboard) is conditioned with heat and steam and then passed between two fluted metal rolls to form the corrugation. Starch adhesive is applied to the tips of the flutes, and the conditioned liner is brought into contact with the flute tips to form the adhesive bond.

In the second section the single-face web is preheated, adhesive applied to the flutes and the double-backer liner joined to the single-face web. After cooling, the board enters the third section where it is trimmed using a slitter and then scored to define the flaps and depth of the box, a process which involves crushing the flutes in a narrow groove in the machine direction so that the box can be folded along these lines later when it is assembled. The board is then cut to predetermined lengths.

Double-wall corrugated board is made by combining two fluted corrugating mediums with a central liner and then adding two outer liner facings. These have extra padding and strength and are great for stacking heavy items. Sometimes three corrugated mediums with two inner liners and two outer facings are combined to form triple-wall corrugated board. This 7-ply construction is used where large container sizes are involved, such as pallet packs. These are very strong and crush resistant. These are excellent for storage and transit.

After leaving the corrugator, the blanks pass to a printer-slotter. Here they are printed using soft rubber dies, and body scores and slots are made. The body scores are similar to the flap scores made in the corrugator except that they are parallel to the flutes, thereby forming the vertical edges of the box when it is assembled. The slots permit the side flaps to fold over the end flaps, the flaps forming the top and bottom of the box. When the printing and slotting operations have been completed, the blanks are formed into a flat tube by a variety of methods: gluing the two ends of the box; stitching with metal staples; or taping with paper or cloth tape. The flat tubes are then palletized for shipment.

A number of variations to the above procedures are possible: different weight facings and different flutes can be used. As well, innovative designs have been developed to produce boxes which can be set up by the packager without the need for additional gluing or stapling to form special purpose containers.

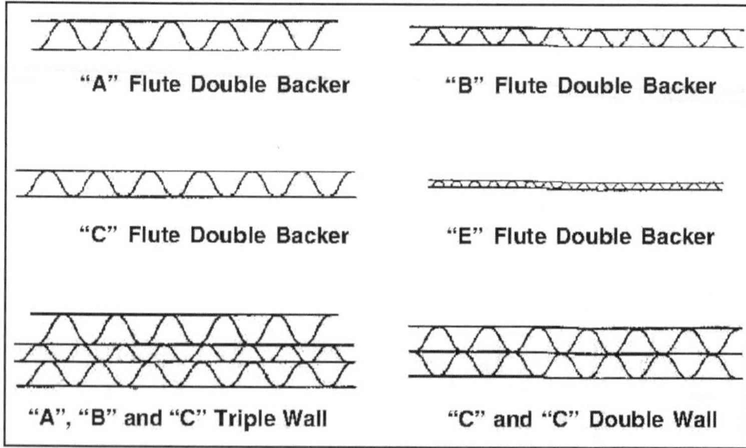

Figure 2.2: Various Types of Corrugated Board Construction (Actual Size)

An international code (known as the FEFCO/ASSCO Fiberboard Case Code) has been agreed on between the various fiberboard case-making organizations of many countries, and it is widely used. Although there are a wide variety of styles of corrugated fiberboard containers, the most popular is the regular slotted container (RSC), due mainly to its simple construction and economical board usage in terms of board area to volume ratio of the fabricated container. However, its poor stacking strength is its major disadvantage, *i.e.,* it rates poorly in terms of volume over strength of fabricated containers. Therefore, when the product can be relied upon to carry most of the static and dynamic stacking load forces (as is the case for canned foods), or when internal partitions can carry a significant proportion, the economy of the RSC is unbeatable. However, when the container has to provide protection in addition to simple containment, other less economical container styles must be resorted to.

Because there are a number of products where it is undesirable for them to carry a significant part of the stacking loads (*e.g.,* fresh produce), there is a continual program of research and development to devise new container styles which will use less board and provide greater stacking strength. Since most of the stacking load is carried by the corners of the container, increasing the wall thickness at the corners makes most efficient use of the material.

The use of mathematical models to aid the design of corrugated containers has become increasing prevalent over the last 20 years. Two basic approaches are possible in constructing a mathematical model to predict corrugated container stacking strength: the properties of the material (*e.g.,* paperboard facings and corrugating medium) are correlated to the stacking strength of the fabricated container; or the container is viewed as a basic structure, *e.g.,* single degree of freedom system with inherent viscoelastic properties.

The former is the traditional approach and was used by McKee who adapted a well-known semiempirical formula commonly used in prediction of failure in shell-type structures for prediction of maximal top-to-bottom strength of corrugated shipping containers. The ultimate compressive strength of a RSC can be related to the board caliper, container perimeter and edgewise compressive strength of the corrugated paperboard by a simplified version of this formula. This enables a direct link to be established between the corrugated paperboard manufacturing process and the ultimate strength of the finished container, providing an invaluable tool for paper manufacturers and corrugated paperboard converters, many of whom have developed in-house computer-aided design systems to assist in container development. However, the end user of the container is primarily interested in the performance characteristics of the container in the field, irrespective of the type of paper it is made from, and ultimate static compressive strength in ideal environmental conditions is only one of several indices of container performance as Peleg has shown.

Advantages

The use of corrugated paper packaging made of corrugated cardboard containers, it has many advantages: light weight, structure, good performance. Its inherent structure is similar to corrugated arch structure, and can play a role in anti-red shock, has good mechanical properties. Of packaged goods have many good protections. For example, moisture, heat, easy to transport and so on. Transportation costs are low and easy to implement packaging and transport mechanization and automation. Changes in specifications and size are easy to implement and can quickly adapt to all kinds of items packaging. Sealing, bundling are both convenient and have easy automation. They can adapt to all types of cartons decor printing, can be a good solution to the protection and promotion of goods issue. Waste recycling boxes are easy to meet environmental requirements. There is a great scope to its use through with a variety of cover material or combination of moisture.

Disadvantages

Due to its being readily available, inexpensive and strong relative to its weight, corrugated cardboard is a popular packing material, as well as a popular material for a variety of large-structure arts and crafts products. However, there are times when corrugated cardboard is not the best choice for a project or packing need. Some of its following parameters may not be suitable for a few packaging needs of the project:

Appearance: Corrugated cardboard has an appearance that's generally less than desirable for craft projects where beauty is a main goal for the finished product. Corrugated cardboard, due to its internal structure, has a rough, ridged appearance over its surface that's prone to wrinkling and bending. Any structures built from it will have an untidy appearance, whereas non-corrugated cardboard holds a smooth, neat shape.

Strength: Corrugated cardboard is relatively strong for being so lightweight, but for projects and purposes that need a bit more heft, one may prefer a denser material. Much of the mass of corrugated cardboard is air, due to the shape of the corrugation inside. If placed under much sideways pressure, corrugated cardboard will bend, pucker and crumple.

Water resistance: As a packing material, corrugated cardboard provides very little resistance to moisture. When made wet, either from ambient moisture in the environment or from direct contact, corrugated cardboard will soften, eventually becoming pulpy. Corrugated cardboard is also highly absorbent, and will thus readily transfer moisture to its contents.

Edges: Due to the layered construction of corrugated cardboard, cutting any kind of intricate or curved shape from a sheet of this material is likely to yield untidy-looking results. The corrugated inner layer will often show on the edge of a cut.

Folding: It is difficult to create neat folds in corrugated cardboard. The inner layers will both dictate the shape of the folds (not necessarily in the direction you want) and cause the material to stretch and crumple in unintended (and undesirable) ways.

Bulk: While the curvy, air-filled corrugation inside the layers of the cardboard provide excellent cushioning, their bulk makes them inconvenient for many craft projects or structures that might require more detail and finesse in shaping. The thickness of cardboard, relative to the structural strength it provides, is sometimes a disappointment for such designs.

Cutting: When cutting a piece of corrugated cardboard into a certain shape, it can be difficult to get the angle just right so as to make a clean cut through both the two outer layers and the inner corrugated layer. Without an even cut with a blade positioned at a consistent 90-degree angle to the cardboard, the cut will angle, making the two sides different sizes.

6.8 Solid Fiberboard

Solid fiberboard consists of numerous bonded plies (typically two to five, with three- and four-ply being the most common) of container board lined on one or both faces with kraft or similar paper between 0.13 and 0.30 mm thick to form a solid board of high strength. The total caliper of the lined board ranges from 0.80 to 2.8 mm. Being solid, it is consequently much heavier in weight for a given thickness than corrugated board, the combined weight of the component plies ranging from 556–1758 gsm.

Solid fiberboard containers are generally two to three times the cost of comparable corrugated containers and are therefore, used almost exclusively for applications in which container return and reuse are possible. Solid fiberboard containers can be re-used satisfactorily from 10 to 15 times.

Solid fiberboard can be made by passing two or more webs or plies of paperboard between a number of sets of press rolls, adhesive being applied to each ply before it passes through the press nips. In multiple structures it is usual to use a poor grade of paperboard (*e.g.*, chipboard) in the central plies and a strong linerboard as the outside facings or liners [baum]. After formation of the solid fiberboard, the subsequent operations are similar to those described for corrugated boxes. Many variations of style, quality and properties are possible in both materials.

Now let us study about the common tests used in the manufacture of corrugated board.

The more common tests performed on linerboard used in the manufacture of corrugated board are: bursting strength, basis weight, puncture, stiffness, tear, tensile, moisture, caliper, water absorbency, pH, compression resistance and water drop. There are appropriate ASTM, BS and TAPPI standards for these tests, but the more important ones will be briefly mentioned here.

Bursting Strength

The bursting strength of paper, paperboard and corrugated paperboard is a composite strength property that is affected by various other properties of the sheet, principally tensile strength and stretch. Generally, bursting strength depends upon the kind, proportion, and amount of fibers present in the sheet, their method of preparation, their degree of beating and refining, upon sheet formation, and the use of additives.

Bursting strength is measured by means of a Mullen tester. The test specimen, held between annular clamps, is subjected to an increasing pressure by a rubber diaphragm, which is expanded by hydraulic pressure at a controlled rate, until the test specimen ruptures. The pressure reading at the instant of rupture on a pressure gauge at this point gives the bursting strength value in force per unit area. This is an important control test in the paper mill or corrugating plant because the bursting strength of the linerboard I has an important influence over the finished container properties. The units of expression are pounds per square inch or "points". The measurement is variously termed the "Mullen", "pop test", or simply the "burst". The burst test for paperboard differs from that of lighter papers in that a heavier Mullen tester is used with a larger diaphragm opening.

Two quite different Mullen testers are available commercially and are both used in the Paper Testing Laboratory:

- The Model C Mullen tester; this is a comparatively light instrument for the testing of paper.
- The Model A Mullen tester; this is a much heavier instrument for the testing of paperboard.

The light Model C instrument for testing paper has two gages, one ranging from 0 to 30 psi, the other from 0 to 120 psi. Never exceed the range of a gage, because serious damage to the gage may result. Obviously, this instrument is intended for testing papers of bursting strength less than 120 psi. The circular specimen area tested has diameter 30.50 ± 0.05 mm.

The heavy Model A tester for paperboard also has two gages, with ranges 0 - 200 psi and 0 - 800 psi, respectively. The specimen area tested has diameter 31.50 ± 0.03 mm, different from the light instrument. The Model A tester also has a much heavier molded rubber diaphragm, whereas the Model C tester has a fairly thin and pliable rubber diaphragm.

Because of these differences between the two instruments, they do not have a common scale. This means that a paper sample would have a totally different value of bursting strength when tested on the Model A tester. It is therefore, a serious mistake to measure the bursting strength of paper on the Model A tester, as the test value obtained is meaningless.

In both instruments, the specimen is firmly clamped between two concentric platens, each having a circular opening in the center. The specimen exposed between the openings is the test area. Pressure is applied at a uniform rate to the underside of the specimen by the rubber diaphragm which is expanded by hydraulic pressure. A maximum reading gauge is included in the hydraulic system. The bursting test value is the maximum gauge pressure required to produce rupture of the specimen.

The observed gauge value includes the hydraulic pressure required to distend the rubber diaphragm in addition to that required to rupture the specimen.

Compressive Strength

However, of more importance than bursting strength in determining the end-use performance of the finished corrugated container is compressive strength. A measure of the edgewise resistance of corrugating medium to compression is given by the ring crush or stiffness test which is used to indicate edgewise rigidity of the board and the probable crushing resistance of the finished container. The test essentially involves crushing a ring specimen rolled up from a precisely cut strip of paper 12.7 mm wide and 152 mm long which is inserted on edge in a circular groove in a specially designed machine, several different makes of instrument being available. The crushing force is applied to the long edges of the sample until it is collapsed, the maximum load being recorded as the ring crush strength. Ten specimens are tested from both the machine and transverse directions of the sample.

A variety of tests are performed on corrugated board before it is made into the finished container, including bursting strength, edge compression, flat crush, pin adhesion and puncture.

The bursting strength of corrugated and solid fiberboard is described in TAPPI T 810 and is primarily an indication of the character of the materials used, giving no direct information regarding the ultimate performance of the finished container.

The edgewise compression strength of corrugated board is the major contributor to the box stacking strength. Edgewise compression strength of corrugated board gives information about the compressive strength, parallel to the flutes, of single-, double-, and triple-wall corrugated board. A specimen 31.75 × 50.8 mm (long edges perpendicular to the flute axis) is dipped into molten paraffin to a depth of 6.35 mm and then placed in a compression-testing instrument. The load required to failure is then determined, and this value can be related to the top-to-bottom compressive strength of the finished corrugated container. If the load is applied at an angle to the flute direction, the maximum load that the board can withstand decreases.

The flat crush test (performed only on single-wall board due to the difficulty with lateral movement of the liners in double- and triple-wall corrugated board) measures the resistance of the flutes in corrugated board to a crushing force applied perpendicular to the surface. From this test some indication of the durability of the finished corrugated board container is obtained. A circular specimen about 3225 mm^2 in size is placed between two parallel platens in a compression-testing instrument, and the force required to crush the sample is measured. Low flat crush values indicate poor flute formation or bad corrugating medium.

The pin adhesion test evaluates the strength of the adhesive bond and is probably the most commonly test performed on corrugated board. It measures the force required to separate the bond between the flutes and the linerboard, using pins (attached to a special fixture) which are inserted edgewise into a small rectangular sample of corrugated board along the open spaces in the flutes. The entire fixture is then placed in the jaws of a compression tester and a separating force applied.

Tests performed on the finished container include the puncture, compression resistance, impact resistance, stiffness, and drop tests. Of these, the compression test is the most common.

The compression test measures the ability of the container (usually empty but occasionally with inner packing or fully loaded) to resist external compressive forces and is used by many corrugated paperboard converters and box manufacturers as a routine quality control test. The container can be positioned in various ways to measure the resistance: top-to- bottom, end-to-end, diagonally opposite edges or any other desired direction. The sealed container is placed in the center of the bottom platen of the compression instrument and force applied at a constant rate; readings are taken until failure of the container occurs.

As Peleg has pointed out, specification of container strength by compression test rather than board bursting strength would be more valuable since the main concern of container users is the stacking strength of the container rather than the material it is made of. A main obstacle to the use of compression tests is the lack of performance standards specifying compression test tolerances for optimal performance of the various containers.

Chapter 7

The Kinds of Deteriorative Reactions

7.1 Introduction

Knowledge of the kinds of deteriorative reactions that influence food quality is the first step in developing food packaging that will minimize undesirable changes in sensory properties and maximize the development and maintenance of desirable properties.

Once the nature of the reaction is understood, a knowledge of the factors that control the rates of these reactions is necessary in order to fully control the changes occurring in foods during storage, *i.e.,* while packaged. The nature of the deteriorative reactions in foods is reviewed in this section, and the factors which control the rates of these reactions are discussed in the following section.

Let us try to understand few more details about enzymes.

Enzymes are complex, globular proteins that act as biocatalysts in a variety of reactions. A wide variety of enzymes are therefore used as supplements for various food processing operations. There are, however, certain enzymes whose activities result in a rapid deterioration of food quality to be controlled in order to maintain quality and to extend product shelflife.

Enzymes have three important characteristics. They are extremely efficient and possess great catalytic power, transforming about 100 to 10,000 substrate molecules into product per second and proceeding from 103 to 108 times faster than the uncatalyzed reaction. They can act as catalysts, accelerating the rate of chemical reactions by factors of 1012–1020 over that of uncatalyzed reactions. Biochemical catalysis of enzyme also depends on the pH of the solution. An enzyme exerts its full potential at an optimum pH ranging between 5-7 pH values.

The catalytic activity of enzymes can be inhibited by competitive inhibitors, noncompetitive inhibitors or irreversible inhibitors. Competitive inhibitors reversibly bind to the active site of the enzyme, thus blocking the substrate from binding to the enzyme. Noncompetitive inhibitors bind to any site of the enzyme other than the active site, enabling the enzyme less active or inactive. Irreversible inhibitors form bonds with enzymes to make them inactive.

A second characteristic is their specificity, enabling the food processor to selectively modify individual food components and not affect others. The active site of an enzyme shows a strong affinity for a specific substrate, and further slightly changes the conformation of substrate. It is known as an "induced fit" to accommodate the reacting substrate, and this induced fit further leads to a stabilized transition state to lower the activation energy of the reaction. The same catalyst cannot be applied in more than one reaction. Stereo specificity means that enzymes can detect the different optical isomers and react to only one type of isomer. Group specific enzymes refer to those catalyzing a group of substances with specific groups. As well, they can have a regulatory role, controlling the biochemical processes associated with various membranous components of a tissue, thus having a profound effect on regulating life processes and defining the quality of foods during postharvest or post mortem processing and storage.

During cheese-making (*e.g.*, rennin) and those produced by added and adventitious microorganisms, combine to produce the desired characteristic flavors and textures of the cheese. Enzymes supplied by microorganisms (adventitious or intentionally added) can have profound roles in altering flavors, colors and textures in food during intentional fermentation (desirable), and during food spoilage (undesirable).

Enzymes endogenous to plant or animal tissues or biological fluids can have undesirable or desirable consequences in foods. Examples involving endogenous enzymes include autolysis of fish postharvest, the senescence and spoilage of fruits and vegetables postharvest, glycolysis in postmortem, prerigor muscle and suberization (wound healing) of plant tissues postharvest.

Examples involving endogenous enzymes include oxidation of lipids by lipases and lipoxygenase (affecting the color, flavor and texture), oxidation of phenolic substances in plant tissues by phenolase (leading to browning), hydrolysis of phospholipids in fish by phospholipases (leading to changes in texture), sugar-starch conversions in plant tissues by amylases, invertase and invertase inhibitor, and postharvest demethylation of pectic substances in plant tissues (leading to softening of plant tissues during ripening, and firming of plant tissues during processing).

The importance of enzymes to the food processor is often determined by the conditions prevailing within and outside the food. Control of these conditions is necessary to control enzymic activity during food processing and storage. The major factors useful in controlling enzyme activity are temperature, water activity, pH, chemicals which can inhibit enzyme action, alteration of substrates, alteration of products, and preprocessing control.

Three of these factors are particularly relevant in a packaging context. The first is temperature: the ability of a package to maintain a low product temperature and thus retard enzyme action will often increase product shelf life. The second important factor is water activity, since the rate of enzyme activity is dependent on the amount of water available; low levels of water can severely restrict enzymic activities and even alter their pattern of activity. Finally, alteration of substrate (in particular, the ingress of oxygen into a package) is important in many oxygen-dependent reactions which are catalyzed by enzymes.

Table 7.1: Chemical Reactions that can Lead to Deterioration of Food Quality or Impairment of Safety

Nonenzymic browning
Lipid hydrolysis
Lipid oxidation
Protein denaturation
Protein cross-linking
Oligo- and polysaccharide hydrolysis
Protein hydrolysis
Polysaccharide synthesis
Degradation of specific natural pigments
Glycolytic changes

Many of the chemical reactions occurring in foods can lead to a deterioration in food quality (both nutritional and sensory) or the impairment of food safety. The more important classes of these reactions are discussed fully in standard textbooks on food chemistry. Worthy of note in the present context is the fact that such reaction classes can involve different reactants or substrates depending on the specific food and the particular conditions for processing or storage.

The rates of these chemical reactions are dependent on a variety of factors amenable to control by packaging including light, oxygen concentration, temperature and water activity. Therefore, the package can in certain circumstances play a major role in controlling these factors, and thus indirectly the rate of the deteriorative chemical reactions.

7.2 Major Changes in Food Quality

The two major chemical changes which occur during the processing and storage of foods and lead to a deterioration in sensory quality are lipid oxidation and non-enzymic browning. Chemical reactions are also responsible for changes in the color and flavor of foods during processing and storage.

Lipid Oxidation

Autoxidation is the reaction of molecular oxygen by a free radical mechanism with hydrocarbons and other compounds. The reaction of free radicals with oxygen is extremely rapid, and many mechanisms for initiation of free radical reactions have been described. Lipids which contain multiple double-bonded systems such as polyunsaturated fatty acids and phospholipids are particularly susceptible to autoxidation. The crucial role which autoxidation plays in the development of undesirable flavors and aromas in foods is well documented, and is a major cause of food deterioration.

The mechanism of lipid oxidation is characterized by four major steps initiation, propagation, branching and termination. Initiation takes place by loss of a hydrogen radical due to heat, light or trace metals. In propagation, the lipid-free radical reacts with oxygen to form peroxy free radicals, which in turn react with more lipid molecules to form hydro-peroxides. During the branching process, there is a geometrical increase in free radicals from decomposition of hydroperoxide. Termination involves the elimination of free radicals by addition of two free radicals or transfer of the radical to a compound to form a stable radical. The four processes are summarized in Table 7.2. RH refers to any unsaturated fatty acid in which the hydrogen is labile by reason of its position on a carbon adjacent to a double bond. Termination occurs through recombination of radicals to produce relatively unreactive products.

Table 7.2: Four Major Steps Characterizing Autoxidation of Lipids

Initiation	RH → R● + H●
Propagation	R● + O_2 → ROO●
	ROO● + RH → ROOH + R●
Branching	ROOH → RO● + OH → ●2RH + ROH + H_2O (monomolecular decomposition)
	2ROOH → ROO● + RO● + H_2O (bimolecular decomposition)
Termination	ROO● +ROO● → ROOR + O_2
	R● + R● → R–R
	R● + ROO● → ROOR

As well as being responsible for the development of off-flavors in foods, the products of lipid oxidation may also react with other food constituents such as proteins, resulting in extensive cross-linking of the protein chains through either protein-protein or protein-lipid crosslinks.

Factors which influence the rate and course of oxidation of lipids are well known and include light, local oxygen concentration, high temperature, the presence of catalysts (generally transition metals such as iron and copper, but also heme pigments in muscle foods) and water activity. Control of these factors can significantly reduce the extent of lipid oxidation in foods.

Non-enzymatic Browning

Nonenzymic browning is one of the major deteriorative chemical reactions which occurs during storage of dried and concentrated foods. The Maillard reaction, first reported by Louis-Camille Maillard is a type of non-enzymic browning which involves the reaction of carbonyl compounds, especially reducing sugars, with compounds which possess a free amino group, such as amino acids, amines and proteins. In most foods, the ε-amino groups of the lysine residues of proteins are the most important source of free amino groups, and the ease with which they take part in the reaction explains why the Maillard reaction is the most important route to nutritional damage of food proteins. The Maillard reaction in fact comprises a complex network of intertwining reactions and takes place during food processing, especially when heat treatment is involved, and also on storage. Apart from resulting in nutritional damage, the Maillard reaction is also primarily responsible for the development of aroma and colour, which may be desirable or undesirable, in heated foods. It also results in the formation of potentially toxic compounds and in the development of components with antioxidant properties.

Although the reactions involved are complex and, as yet, not completely elucidated, the non-enzymic browning or Maillard reaction can be divided into three stages:

1. Early Maillard reactions which are chemically well-defined steps without browning;
2. Advanced Maillard reactions which lead to the formation of volatile or soluble substances; and
3. Final Maillard reactions consists of the condensation of amino compounds and sugar fragments into polymerized protein and brown pigments, called melanoidins (brown compound).

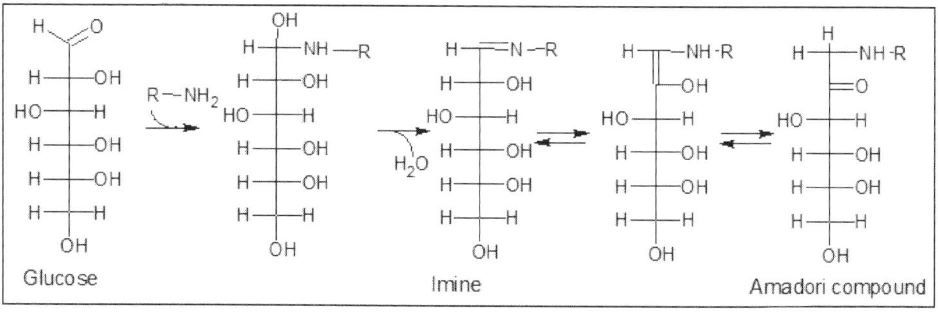

Glucose Imine Amadori compound

The initial reaction involves a simple condensation between an aldehyde (usually a reducing sugar) and an amine (usually a protein or amino acid) to give a glycosylamine via a Schiff's base. The glycosylamine then undergoes an Amadori rearrangement to form an Amadori 1-amino-1-deoxy-2-ketose derivative.

The formation of Amadori compounds accounts for the observed loss of both reducing sugar and amine during the Maillard reaction. Although the early Maillard reactions forming. Amadori compounds do not cause browning, they do reduce nutritive value.

There are five pathways in the advanced Maillard reactions. The first two pathways start from the 1–2 enol or 2–3 enol forms of the Amadori product and yield various flavor compounds. The third pathway is the Strecker degradation, which involves oxidative degradation of amino acids by the dicarbonyls produced in the first two pathways. The fourth pathway involves transamination of the Schiff base, and the final pathway starts with a second substitution of the amino-deoxy-ketose.

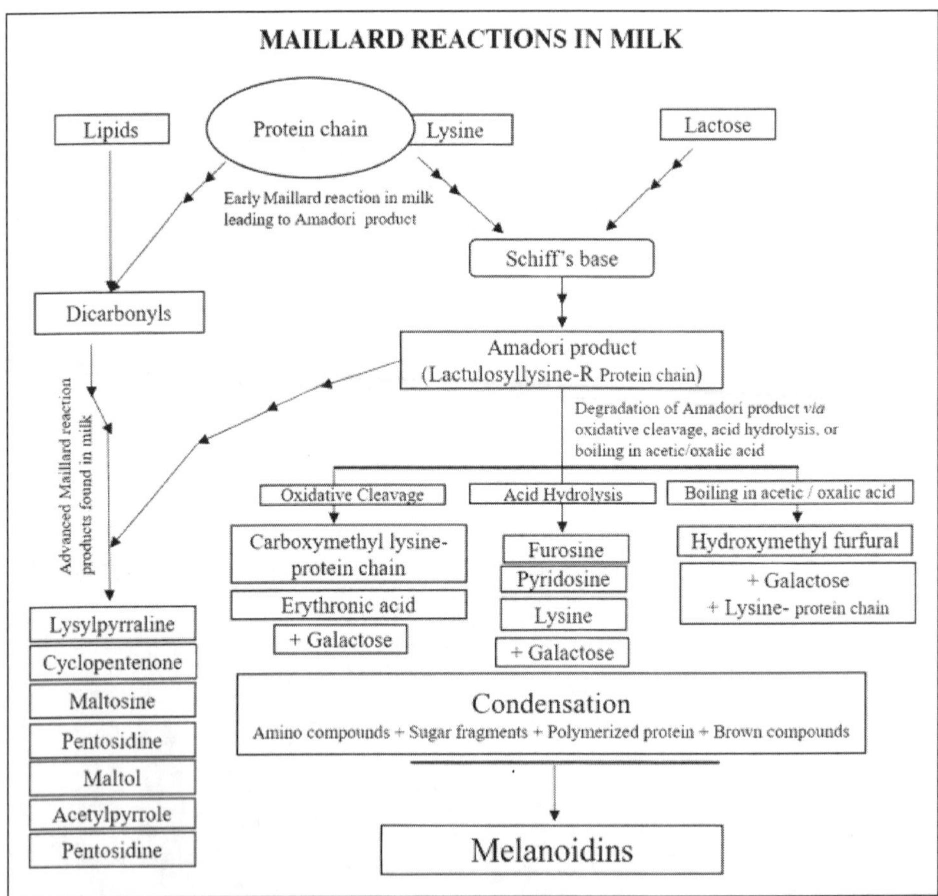

The final step of the advanced Maillard reaction is the formation of many heterocyclic compounds such as pyrazines and pyrroles, as well as brown melanoldin pigments. These pigments are formed by polymerization of the reactive compounds produced during the advanced Maillard reactions, such as unsaturated carbonyl compounds and furfural. The polymers are relatively inert and have a molecular weight greater than 1000.

Acceptability of color in a given food is influenced by many diverse factors, including cultural, geographical and sociological aspects of the population. However, regardless of these many factors, certain food groups are only acceptable if they fall within a certain color range. The color of many foods is due to the presence of natural pigments. The nature and major changes which these can undergo are briefly described; a more detailed discussion can be found elsewhere.

Chlorophylls

The name chlorophyll describes those green pigments involved in the photosynthesis of higher plants. In general terms, chlorophylls are magnesium-chelated tetrapyrrole structures with various substitutions (most commonly methyl) at different positions. The major change which chlorophylls can undergo is pheophytinization, the replacement of the central magnesium by hydrogen and the consequent formation of a dull olive-brown pheophytin. Because this reaction is accelerated by heat and is acid catalyzed, it is therefore unlikely to be influenced by the choice of packaging.

Almost any type of food processing or storage causes some deterioration of the chlorophyll pigments. Although pheophytinization is the major change, other reactions are possible. For example, dehydrated products such as green peas and beans packed in clear glass containers undergo photooxidation and loss of desirable color.

Heme Pigments

Meat is an important part of many diets, and the color of red meat is due to the presence of the heme pigment myoglobin. Myoglobin is a complex muscle protein contained within the cells of the tissues where it acts as a temporary storehouse for the oxygen brought by the hemoglobin in the blood.

7.3 Biological Changes in the Food Quality

The biological changes which occur in food are due to microorganisms.

Microorganisms can make both desirable and undesirable changes to the quality of foods, depending on whether or not they are introduced as an essential part of the food preservation process (*e.g.*, as inocula in food fermentations) or arise adventitiously and subsequently grow to produce food spoilage. In the latter case, they only reach readily observable proportions when they are present in the food in large numbers. Since the initial population or microbial load is usually small, observable levels are only reached after extensive multiplication of the microorganism(s) in the food.

The two major groups of microorganisms found in foods are bacteria and fungi, the latter consisting of yeasts and molds. Bacteria are generally the fastest growing, so that in conditions favourable to both, bacteria will usually outgrow fungi. The phases through which the two groups pass are broadly similar: a period of adjustment or adaptation (known as the lag phase) is followed by accelerating growth until a steady, rapid rate (known as the logarithmic phase since growth is exponential) is achieved. After a time the growth rate slows until growth and death are balanced and the population remains constant (known as the stationary phase). Eventually, death exceeds growth and the organisms enter the phase of decline.

Foods are frequently classified on the basis of their stability as non-perishable, semi perishable and perishable. An example of the first classification is sugar; provided it is kept dry, at ambient temperature and free from contamination, it should have a very long shelf life. However, few foods are truly non-perishable, and an important factor in determining their perishability is their packaging.

For example, hermetically sealed and heat processed (*e.g.*, canned) foods are generally regarded as non-perishable. However, they may become perishable under certain circumstances when an opportunity for recontamination is afforded following processing. Such an opportunity may arise if the can seams are faulty, or if there is excessive corrosion resulting in internal gas formation and eventual bursting of the can. Spoilage may also take place when the canned food is stored at unusually high temperatures: thermophilic spore-forming bacteria may multiply, causing undesirable changes such as flat sour spoilage.

Low moisture content foods such as flour, dried fruits and vegetables, and baked goods are classified as semi perishable. Frozen foods, though basically perishable, may be classified as semi perishable provided that they are properly stored at freezer temperatures.

Table 7.3: Intrinsic and Extrinsic Parameters Influencing Microbial Growth in Foods

Intrinsic Factors	*Extrinsic Factors*
pH	Storage temperature
a_w	Relative humidity of environment
E_h	
Nutrient content	Presence and concentration of gases in the environment
Antimicrobial constituents	
Biological structures	

The majority of foods (*e.g.*, flesh foods such as meat and fish; milk, eggs and most fruits and vegetables) are classified as perishable unless they have been processed in some way. Often, the only form of processing which such food receive is to be packaged and kept under controlled temperature conditions.

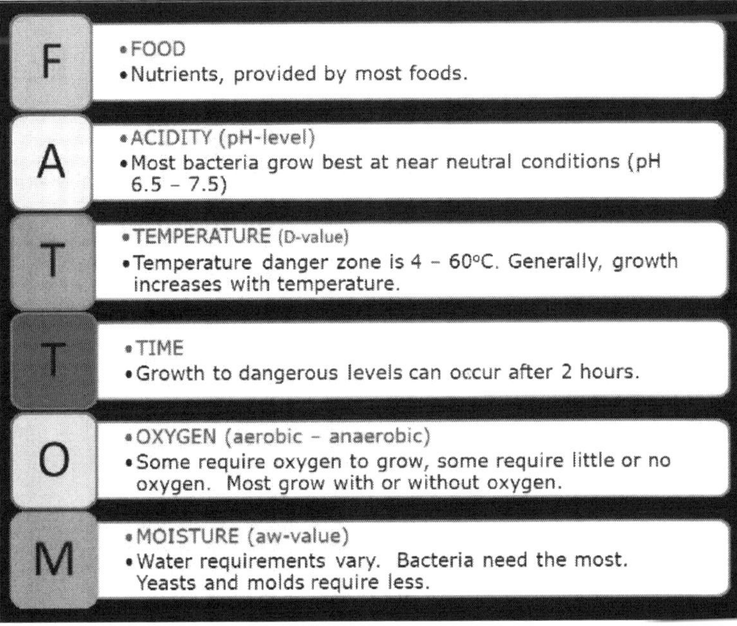

Factors affecting growth of microbes

- Intrinsic
  - ✓ Nutrients
  - ✓ a_w
  - ✓ pH
  - ✓ Redox potential
  - ✓ Inhibitors
- Extrinsic
  - ✓ Temperature
  - ✓ Humidity
  - ✓ Atmosphere
- Implicit
  - ✓ Interactions of microorganisms

The species of microorganisms which cause the spoilage of particular foods are influenced by two factors: the nature of the foods and their surroundings. These factors are referred to as intrinsic and extrinsic parameters.

Most microorganisms grow best at pH values around 7.0, while few grow below pH 4.0. Bacteria tend to be more fastidious in their relationships to pH than molds and yeasts, with the pathogenic bacteria being the most fastidious.

The minimum water activity (a_w) values reported for growth of some microorganisms in foods are presented in Table7.4. Worthy of note is the fact that yeasts and molds grow over a wide a_w range than bacteria.

Table 7.4: Approximate Minimum *a*w Values for the Growth of Microorganisms of Importance in Foods

Organism	*Minimum aw*
Most spoilage bacteria	0.91
Most spoilage yeasts	0.88
Most spoilage molds	0.80
Halophilic bacteria	0.75
Xerophilic molds	0.65
Osmophilic yeasts	0.60

Certain relationships have been shown to exist between a_w, temperature and nutrition. Firstly, at any temperature, the ability of microorganism to grow is reduced as the a_w is lowered. Secondly, the range of a_w over which growth occurs is the greatest at the optimum temperature for growth; thirdly, the presence of nutrients increases the range of a_w over which the organisms can survive.

The oxidation-reduction potential (E_h) of a substrate may be defined as the ease with which the substrate loses or gains electrons. The more highly oxidized a substance, the more positive will be its electrical potential. Aerobic microorganisms require positive E_h values (oxidized) for growth while anaerobes require negative E_h values. Among the substances in foods that help to maintain reducing conditions are the sulfhydryl groups on proteins, and ascorbic acid and reducing sugars in fruits and vegetables. Deteriorative chemical reactions can alter the E_h value of foods during storage.

In order to grow and function normally, microorganisms require several nutrients: water, a source of energy, a source of nitrogen, vitamins and related growth factors, and minerals. The availability of water is related directly to the water activity of the food. The primary source of nitrogen utilized by microorganisms is amino acids. Growth factor and vitamin requirements tend to be specific to individual groups of microorganisms. Some foods contain certain naturally occurring substances which have been shown to have antimicrobial activities, thus preventing or retarding the growth of specific microorganisms in those foods.

Microbial Nutrition

Nutrients: Substances in the environment used by organisms for catabolism and anabolism.

1. Macronutrients: required in large amounts, including: carbon, oxygen, hydrogen, nitrogen, sulfur, phosphorus (Components of carbonhydrates, lipids, proteins, and mucleic acids); potassium, calcium, magnesium and iron (cations and part of enzymes and cofactors).
2. Micronutrients: **Microbes require very small amounts of other mineral elements, such as iron, copper, molybdenum, and zinc; these are referred to as trace elements. Most are essential for activity of certain enzymes, usually as cofactors.**

   contaminants in water, glassware, and regular media components often are adequate for growth.

The extrinsic parameters of foods are those properties of the storage environment that affect both the foods and their microorganisms. The growth rate of the microorganisms responsible for spoilage primarily depends on these extrinsic parameters: storage temperature, relative humidity and gas composition of the surrounding atmosphere.

The temperature of storage is particularly important, and several food preservation techniques (*e.g.*, chilling) rely on reducing the temperature of the food to extend its shelf life. Although there is a very wide range of temperatures over which the growth of microorganisms has been reported, specific microorganisms have relatively narrow temperature ranges over which growth is possible. Those which grow well below 20°C and have their optimum between 20° and 30°C are referred to as *psychrophiles* or *psychrotrophs.* Those that grow well between 2° and 45°C with optima between 30° and 40°C are referred to as *mesophiles,* and those that grow well at and above 45°C with optima between 55° and 65°C are referred to as *thermophiles.* Molds are able to grow over a wider range of temperature than bacteria, many being capable of growth at refrigerator temperature. Yeasts grow over the psychrophilic and mesophilic temperature ranges but generally not within the thermophilic range.

The relative humidity (RH) of environment is important and can influence the water activity of the food unless the package provides a barrier. Many flexible plastic packaging materials provide good moisture barriers but none are completely impermeable, thus limiting the shelf life of low a_w foods.

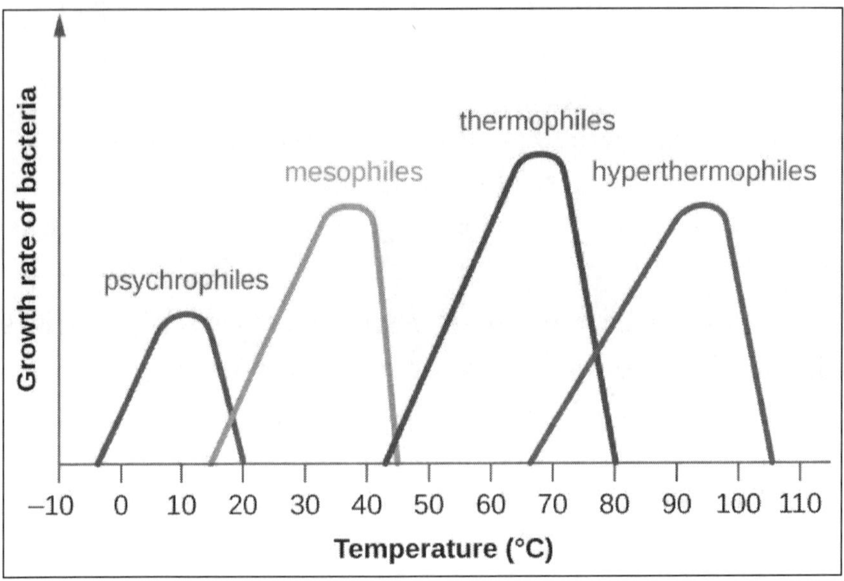

The presence and concentration of gases in the environment has a considerable influence on the growth of microorganisms. Increased concentrations of gases such as CO_2 are used to retard microbial growth and thus extend the shelf life of foods. As well, vacuum packaging (*i.e.*, removal of air (and thus oxygen) from a package prior to sealing) can also have a beneficial effect by preventing the growth of aerobic microorganisms. This type of packaging (known as modified atmosphere packaging or MAP) raises certain safety issues. Most food pathogens do not grow at refrigerator temperatures, and carbon dioxide is not highly effective at non-refrigeration temperatures. Therefore, most MA packaged food is usually held under refrigeration. Temperature abuse of the product (*i.e.*, holding at non-refrigerated temperatures) could allow the growth of organisms (including pathogens) which had been inhibited by carbon dioxide during storage at lower temperatures. For these reasons, it is difficult to evaluate MAP safety solely on the growth of certain pathogens at abusive temperatures.

7.4 The Common Insect Pests

The common insect pests of fresh food are flies (from the order Diptera), and cockroaches. They are attracted by food odours regardless of whether the food is fresh or beginning to decay. Any insect in food is a pest since not only does the food become contaminated with their bodies and excreta, but they are also capable of transmitting pathogens, including food poisoning organisms.

In contrast, the main insect species important as pests of stored foods are entirely from the orders *Lepidoptera* (moths) and *Coleoptera* (beetles). They regularly damage and destroy large quantities of stored foods around the world every year. The number of species involved is not large, there being no more than about twelve really important ones. They include weevils, various other beetles and the larvae of several moths. Most stored-product insects are cosmopolitan in that any given species is for the most part found worldwide in areas with similar climatic conditions. Warm humid environments promote insect growth, although most insects will not breed if the temperature exceeds about 35°C or falls below 10°C. Also many insects cannot reproduce satisfactorily unless the moisture content of their food is greater than about 11 per cent.

Moths and beetles are generally found in dry storage areas. They are able to survive on very small amounts of food, and thus can persist on food residues in improperly cleaned premises or equipment. Good ventilation, the use of cool storage areas and rotation of stock assists in keeping these pests at bay. Basic to effective control of insect pests is an understanding of their life cycles and feeding habits.

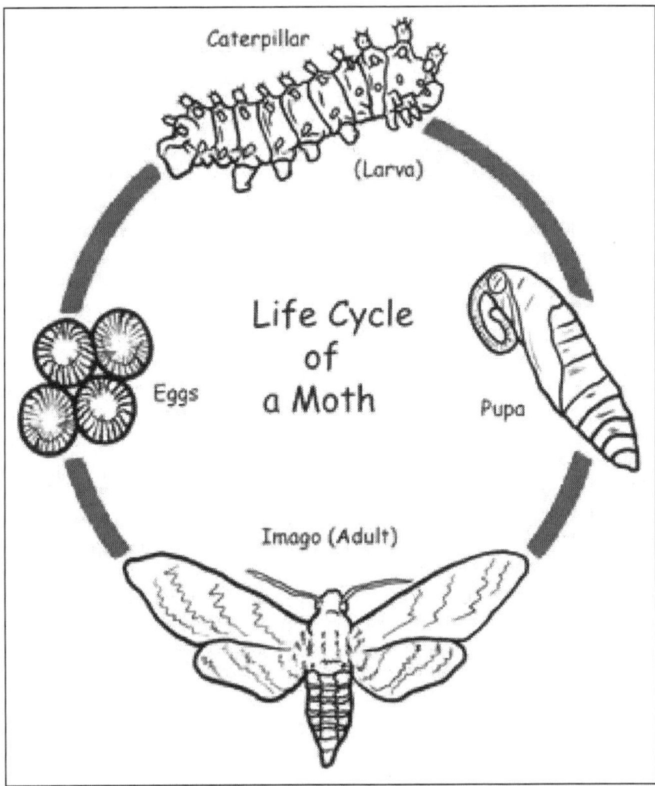

In common with many insects, moths pass through four stages during their development: the egg, the larva (caterpillar), the pupa (chrysalis) and the adult moth. The larval stage is the only one which consumes food; their presence can be recognized by a characteristic mixture of silken threads and frass (droppings) which they produce. Beetles have the same four life stages as moths, but they differ in that the adult beetle is a considerably harder-bodied insect than the moth, and may live and feed for months or even years. Cockroaches are larger, more robust insects which are highly mobile. Their young ones are like small versions of the adult and unlike beetles and moths, they do not inhabit packaged foods. Ants are a highly specialized group of insects which form nests, normally outside buildings. Worker ants may travel considerable distances and collect almost any type of food (especially sweet or high-protein foods) and take it back to the nest.

Table 7.5: Resistance of Various Materials to Insect Penetration

	Excellent	*Good*	*Fair*	*Poor*
Polycarbonate	×			
Poly (ethylene terephthalate)	×			
Cellulose acetate		×		
Polyamide		×		
Polyethylene (0.254 mm)		×		
Polypropylene (blaxially oriented)		×		
Poly(vinyl chloride) (unplasticized)		×		
Acrylonitrile			×	
Poly(tetrafluoroethylene)			×	
Polyethylene (0.123 mm)			×	
Regenerated cellulose				×
Corrugated paperboard				×
Ethylene vinyl acetate copolymer				×
Ionomer				×
Kraft paper				×
Paper/foil/polyethylene laminate pouch				×
Polyethylene (0.0254–0.100 mm)				×
Poly(vinyl chloride) (plasticized)				×
Vinyl chloride/vinylidene chloride copolymer			×	

Mites which sometimes occur in stored foods are not insects but are closely related to spiders, having eight legs. They are minute in size, requiring a lens to see them, and are primarily pests of cereals and other foods with a moisture content of at least 12 per cent. Mites are so small that the presence of a few would pass unnoticed. They produce a sour odour in the food.

The main categories of foods subject to pest attack are cereal grains and products derived from cereal grains, other seeds used as food (especially legumes), dairy products such as cheese and milk powders, dried fruits, dried and smoked meats, and nuts. As well as their possible health significance, the presence of insects and insect excreta in packaged foods may render products unsalable, causing considerable economic loss, as well as reduction in nutritional quality, production of off-flavors, and acceleration of decay processes due to the creation of higher temperatures and moisture levels.

Early stages of infestations are often difficult to detect due to the small size of the insects and the fact that some species are only active at night. However, infestations can generally be recognized not only by the presence of the insects themselves but also by the products of their activities such as webbing, clumped-together food particles and holes in packaging materials. Each pest species tends to have a particular preference as to the kind of material attacked, so that the species that occurs in any particular situation depends very largely on the nature of the food involved.

Unlike microorganisms, some insects species (penetrators) have the ability to bore through one or more of the flexible packaging materials in use today and take up residence inside. Other species (invaders) usually do not enter packages unless there is an existing opening. However, such openings need not be very large. For example, the adult sawtoothed grain beetle can enter an opening less than 1 mm in diameter. Newly hatched larvae can enter much smaller openings; holes only 0.1 mm in diameter are sufficient to admit the larvae of some insects. Thus, package seals are critical in the design and use of packages that are expected to protect foods from insect infestations.

Unless plastic films are laminated with foil or paper, insects are able to penetrate most of them quite easily, the rate of penetration usually being directly related to film thickness. In general, thicker films are more resistant than thinner films, and oriented films tend to be more effective than cast films. The looseness of the film has also been reported to be an important factor, loose films being more easily penetrated than tightly fitted films. Regenerated cellulose, polyester and polyethylene films are inadequate barriers to insects if used singly; however, if laminated with biaxially oriented polypropylene, increased resistance can be obtained.

Generally, the penetration varies depending on the basic resin from which the film is made, on the combination of materials, on the package structure, and on the species and stage of insects involved. Absolute values are difficult to determine because resistance to penetration is influenced by factors such as package configuration and the presence or absence of folds, tucks and other harborage sites. Therefore, after appropriate packaging materials have been selected they must be evaluated *in situ* for insect resistance.

Aluminum foil (0.10 mm thick) has been shown to be penetrated by cadelle larvae and by *Trogodenna inclusum,* a dermestid. Lamination of foil to either paper or plastic films usually improves its resistance to insect penetration. More recently it has been demonstrated that vacuumized polyethylene bags of shelled peanuts were penetrated by the flour beetle *Tribolium castaneum,* whereas unvacuumized bags of the same material, and both vacuumized and unvacuumized bags made from poly(ethylene terephthalate) film were not. The author had no ready explanation for the differences in behavior of the various packs.

In an evaluation of films for their intrinsic capacity to exclude insects, it was noted that impregnation of these materials with pesticides conferred a greater degree of resistance. However, only pyrethrin synergized with piperonyl butoxide is registered in the USA for use as an insecticide for package treatment. Packaging applications that are currently permitted are large cotton bags, the outer ply of large multiwall paper bags, and two-ply regenerated cellulose/polyolefin sheets, in which case the insecticide combination is added to the adhesive.

Insects readily infest paper and can find ideal conditions for growth and reproduction within this material. Synergized pyrethrins have been added to paper at levels of 16.4 mg m^{-2} in multiwall paper bags. Corrugated paper shipping containers, although not directly in contact with food, may greatly influence the potential for contamination of foods in individual packages as it provides ideal shelter for insects to hide away in.

Table 7.6: Methods for Obtaining Insect Resistance of Packaging Materials

1.	Select a film and film thickness that are inherently resistant to insect penetration.
2.	Use a film adhesive containing an approved insecticide.
3.	Add an insecticide directly to the packaging material.
4.	Use shrink film overwraps to provide an additional barrier.
5.	Seal carton flaps completely.

The rodents - rats and mice are among humanity's most cunning and capable enemies. They have highly developed senses of touch, smell and hearing, and can identify new or unfamiliar objects in their environment. Rats can wriggle through openings the size of a quarter; a mouse needs a hole only as large as a nickel to gain access. Rats and mice carry disease-producing organisms on their feet and/or in their intestinal tracts and are known to harbor salmonellae of serotypes frequently associated with foodborne infections in humans. In addition to the public health consequences of rodent populations in close proximity to humans, these animals also compete intensively with humans for food.

Rats and mice gnaw to reach sources of food and drink and to keep their teeth short. Their incisor teeth are so strong that rats have been known to gnaw through lead pipes and unhardened concrete, as well as sacks, wood, and flexible packaging materials. Obviously proper sanitation in food processing and storage areas is the most effective weapon in the fight against rodents, since all packaging materials apart from metal and glass containers can be attacked by rats and mice.

7.5 The Influence of Water on Chemical Reactivity

Water may influence chemical reactivity in different ways. It may act as a reactant or as a solvent, where it may exert a dilution effect on the substrates, thus decreasing the reaction rate.

Water may also change the mobility of the reactants by affecting the viscosity of the food systems, and form hydrogen bonds or complexes with the reacting species. Thus, a very important practical aspect of a_w is to control undesirable chemical and enzymic reactions that reduce the shelf life of foods. It is a well-known generality that rates of changes in food properties can be minimized or accelerated over widely different values of a_w. Small changes in a_w can result in large changes in reaction rates.

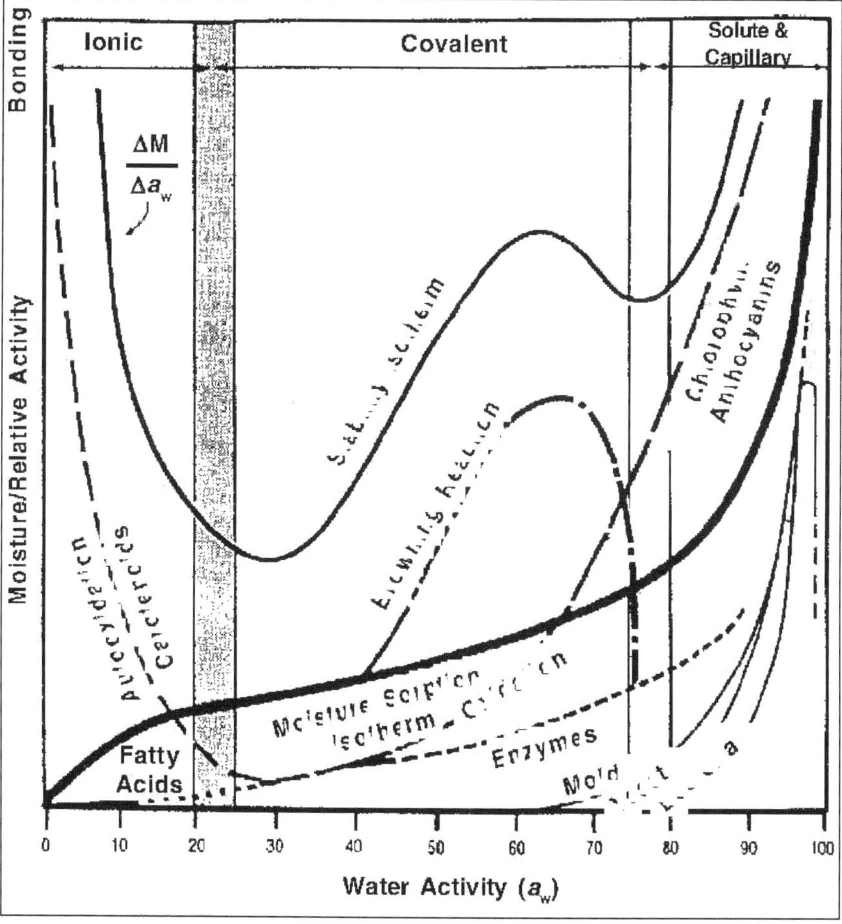

Figure 7.1: Rates of Reactions as a Function of Water Activity.

The influence of a_w on lipid oxidation has been studied extensively, mainly with the use of model systems. The general effect of a_w on lipid oxidation is shown in Figure 7.1. At very low a_w levels, foods containing unsaturated fats and exposed to atmospheric oxygen are highly susceptible to the development of oxidative rancidity. This high oxidative activity occurs at a_w levels below the monolayer level, and as a_w increases, both the rate and the extent of autoxidation increase until an a_w in the range of 0.3–0.5 is reached. Above this point, the rate of oxidation increases until a steady state is reached, normally at a_w levels in excess of 0.75. At a_w below the monolayer value, the oxidation rate decreases with increasing a_w. The rate reaches a minimum around the monolayer value and increases with further increase in a_w. Water may influence lipid oxidation by influencing the concentrations of initiating radicals present, the degree of contact and mobility of reactants, and the relative importance of radical transfer versus recombination reactions.

Browning

Water may accelerate browning by imparting mobility to the substrates, or it may decrease the browning rate by diluting the reactive species. In the low a_w range, the mobility factor predominates, whereas the dilution factor predominates in the high a_w range. As a consequence, browning rate generally increases with increasing a_w at low moisture content, reaches a maximum at a_w's of 0.4–0.8, and decreases with further increase in a_w.

Activation energies for the formation of Amadori compounds decrease with increasing a_w and level off at a_w of about 0.50. Therefore, the reaction rate is a_w highly temperature dependent at low a_w and becomes much less dependent on temperature at a_w greater than 0.50. This is the reason for the use of low air temperatures during the final stage of drying.

The effect of a_w on vitamin degradation in food systems has been reviewed. The rate of degradation of vitamins A, B_1, B_2 and C increased as a_w increased over the range 0.24–0.65. In contrast, the degradation of α-totocopherol in a fat-free model system increased with increasing a_w over the range 0.10–0.65. Generally, the rate of ascorbic acid degradation increases exponentially with increase in a_w. The photodegradation of riboflavin has been shown to increase with increasing a_w.

Near or below the monolayer a_w value, enzyme activities are generally minimized or cease. Above the monolayer value, enzyme activity increases with increasing a_w or increased substrate mobility, as illustrated in Figure 7.2. Substrates of high molecular weight such as protein and starch are less mobile than low molecular weight substrates such as glucose, and generally the latter have a lower a_w threshold for enzyme activity. At subfreezing temperatures, the reaction rate generally decreases with decreasing temperature and a_w, due partly to the lower temperature, partly to the increase in viscosity of the partially frozen system, and partly to enzyme denaturation.

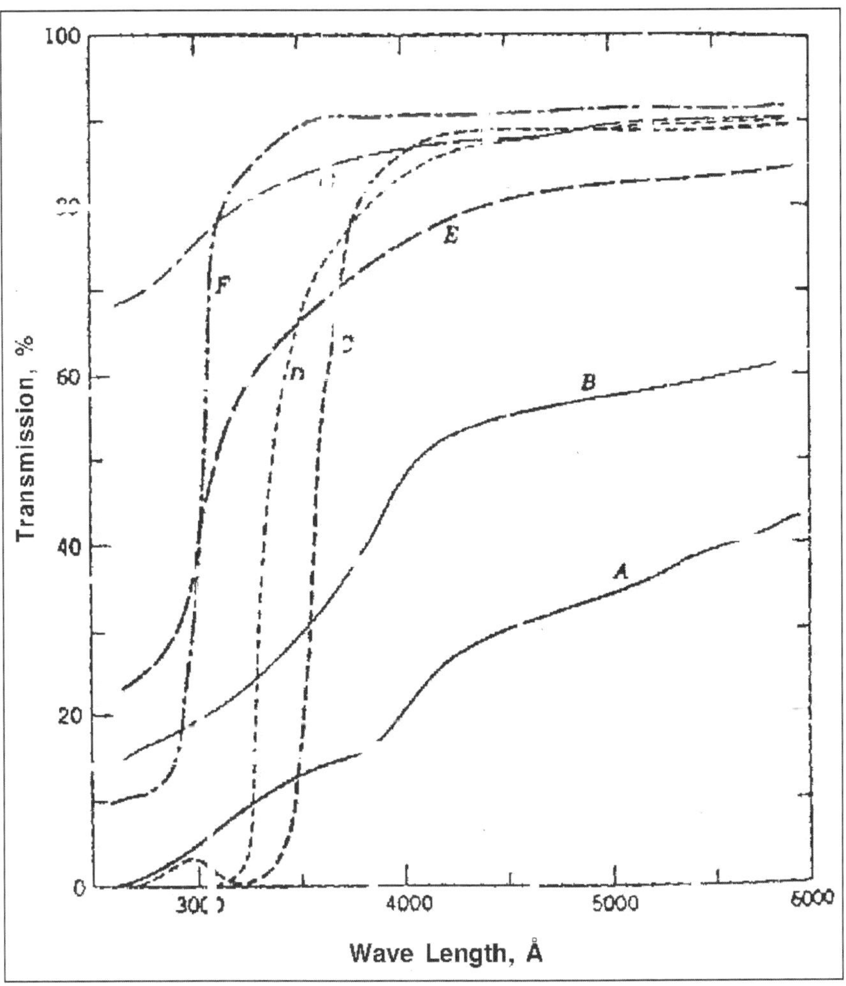

Figure 7.2: Light Transmission Characteristics of Various Flexible Packaging Materials.

Every microorganism has a limiting a_w value below which it will not grow, form spores or produce toxic metabolites. Table 7.7 lists the range of water activity permitting the growth of various common microorganisms, together with common foods categorized according to their a_w.

Water activity can influence each of the four main growth cycle phases by its effect on the germination time, the length of the lag phase, the growth rate phase, the size of the stationary population, and the subsequent death rate. Generally, reducing the a_w of a given food increases the lag period and decreases the growth rate during the logarithmic phase, the maximum of which becomes lower.

Table 7.7: Water Activity and Growth of Microorganisms in Food

Range	*Microorganisms Generally Inhibited by Lowest a_w in this Range*	*Foods Generally within this Range*
1.00–0.95	*Pseudomonas, Escherichia, Proteus, Shigella, Klebsiella, Bacillus, Clostrium perfringens,* some Yeasts	Highly perishable (fresh) foods and canned fruits, vegetables, meat, fish and milk; cooked sausages and breads; foods containing up to approximately 40 per cent (w/w) sucrose or 7 per cent NaCl
0.95–0.91	*Salmonella, Vibrio parahaemolyticus, C. botulinum, Serratia, Lactobacillus, Pediococcus,* some molds, yeasts	Some cheeses (cheddar, Swiss, Muenster, provolone), cured meat (ham), some fruit juice concentrates; foods containing 55 (w/w) sucrose or 12 NaCl
0.91–0.87	Many yeasts (*Candida, Torulopsis, Hansenula*), *Micrococcus*	Fermented sausage (salami), sponge cakes, dry cheese, margarine; foods containing 65 per cent (w/w) *sucrose* (saturated) or 15 per cent NaCl
0.87–0.80	Most molds (mycotoxigenic penicillia), *Staphylococcus aureus,* most *Saccharomyces* (*bailii*) spp., *Debaryomyces*	Most fruit juice concentrates, sweetened condensed milk, chocolate, syrup, maple and fruit syrups; flour, rice, pulses containing 15–17 per cent moisture; fruit cake, country-style ham, fondants, high-ratio cakes
0.80–0.75	Most halophilic bacteria, mycotoxigenic aspergilli	Jam, marmalade, marzipan, glaced fruits, some marsh-mallows
0.75–0.65	Xerophilic molds (*Aspergillus chevalieri, A. Candidus, Wallemia sebid*), *Saccharomyces bisporus*	Rolled oats containing approximately 10 per cent moisture, grained nougats, fudge, marshmallows, jelly, molasses, raw cane sugar, some dried fruits, nuts
0.65–0.60	Osmophilic yeasts (*Saccharomyces rouxii*), few molds (*Aspergillus echinulatus, Monascus bisporus*)	Dried fruits containing 15–20 per cent moisture, some toffees and caramels; honey
0.50	No microbial proliferation	Pasta containing approximately 12 per cent moisture; spices containing approximately 10 per cent moisture
0.40	No microbial proliferation	Whole-egg powder containing approximately 5 per cent moisture
0.30	No microbial proliferation	Cookies, crackers, bread crusts, and so on containing 3–5 per cent moisture
0.20	No microbial proliferation	Whole milk powder containing 2–3 per cent moisture, dried vegetables and corn flakes containing approximately 5 per cent moisture, country-style cookies, crackers

Sporulation may occur at or slightly below the minimum a_w for growth. In contrast, germination of spores of some microorganisms may occur at a_w values below that required for growth. The minimum a_w for growth of microorganisms is, without exception, less than or equal to the minimum a_w for toxin production. Optimal conditions of temperature, pH, oxygen tension and nutrient availability are necessary to permit sporulation, germination and toxin production at reduced a_w.

Whether a microorganism survives or dies in a low a_w environment is influenced by intrinsic factors that are also responsible for its growth at a higher a_w. These factors include water-binding properties, nutritive potential, pH, E_h and the presence of antimicrobial compounds. The influences exerted by these factors interact with a_w both singularly and in combination. Microbial growth and survival are not entirely ascribed to reduced a_w but also to the nature of the solute. However, the exact nature of the role that water plays in the mechanism of cell survival is not clearly understood.

Key extrinsic factors relative to a_w that influence microbial deterioration in foods include temperature, oxygen and chemical treatments. These factors can all combine in a complex way to either encourage or discourage microbial growth. Detailed discussion of these inter-relationships is beyond the scope of this book, and the interested reader is referred to standard textbooks on food microbiology.

Although the water vapor permeability of the packaging is a decisive factor in controlling changes in moisture content and thus water activity of packaged foods, the a_w of food in impermeable packaging can alter as a consequence of microbial growth. For example, experiments with slices of cake contaminated with pure strains of xerotolerant (*i.e.*, tolerant of dry conditions) molds stored in water-impermeable packaging showed two phases in the a_w evolution when molds grew. During the first stage, a_w decreased significantly, probably due to water uptake by spores and water binding in the young cytoplasm. In the second phase, the respiration of the growing mycelium produced metabolic water in excess and the a_w rose, progressively allowing the growth of less xerotolerant species which were present.

Chapter 8

Food Packaging Metals and their Corrosion

8.1 Introduction

The chemical structure of metals which gives them their valuable practical properties is also responsible for their main weakness and their susceptibility to corrosion.

Corrosion is the term used to describe the chemical reaction between a metal and its environment to form compounds; it is a universal process affecting all metals to a greater or lesser extent. This reaction causes damage and disintegration of the metal starting from the portion of the metal exposed to the environment and spreading to the entire bulk of the metal. During corrosion, a refined metal is naturally converted to a more stable form such as its oxide, hydroxide or sulphide state that leads to deterioration of the material. Because the reaction takes place at the metal surface, the rate of attack can be reduced and controlled by modifying the conditions at the surface.

Metals are chemically reactive and can be readily oxidized by oxygen and other agents to form largely useless corrosion products. This vulnerability to oxidation accounts for the fact that with few exceptions (copper, silver and gold), metals do not occur naturally in the metallic state but are found combined with oxygen or sulfur in their ores. A considerable amount of energy is required to extract metals from their ores, and the reverse process (which releases energy) is strongly favoured as the metal reverts back to its natural state. As a very broad generalization it can be said that the more difficult it has been to win the metal from its natural form, the greater will be its tendency to return to that form by corroding, but the rate of return will of course depend on the environment.

Metals placed higher in the reactivity series such as iron, zinc, *etc.* get corroded very easily and metals placed lower in the reactivity series like gold, platinum and palladium do not corrode. The explanation lies in the fact that corrosion involves oxidation of the metals. As we go down the reactivity series tendency to get oxidised is very low (oxidation potentials is very low). Interestingly, aluminium doesn't corrode unlike other metals even though it is reactive. This is because aluminium is covered by a layer of aluminium oxide already. This layer of aluminium oxide protects it from further corrosion.

The reaction of metals in aqueous solutions or under moist conditions (known as wet corrosion) is electrochemical in nature, involving the transfer of electric charge across the boundary formed by the metal surface and its environment. An electrolyte is a medium which conducts electricity by movement of ions, the cations (*e.g.,* Fe_{2+}) and anions (*e.g.,* Cl^-) moving in opposite directions. When an electrode reaction takes places at a metal surface, the electron flow in the metal corresponds to an ion flow in the electrolyte.

When a metal corrodes, atoms of the metal are lost from the surface as cations, leaving behind the requisite number of electrons in the body of the metal. This dissolution of the metal is called an anodic reaction and takes place at a surface termed an anode; an anodic reaction involves the release of electrons or electrochemical oxidation. Thus, for the case of a metal M:

$$\underset{\text{(reduced)}}{M} \longrightarrow \underset{\text{(oxidized)}}{M_{n+}} + n \text{ electrons}$$

Simultaneously, reagents in the electrolyte solution react with the metal surface to remove electrons left behind by the departing metal ions. This removal of electrons is termed a cathodic reaction and takes place at a surface called a cathode. The cathodic reaction always involves consumption of electrons or electrochemical reduction.

Since practically all metals are covered with an oxide film, this must be removed before the metal can be exposed to an electrolyte. A metal covered with an oxide has different properties in a solution from a bare metal, but in studying electrochemical corrosion, it is the simplest to begin with the ideal case of a pure, bare metal electrode on which the only reaction occurring is metal dissolution.

8.2 Factors Affecting the Rate of Corrosion

The factors that affect the rate of corrosion can be studied under the following headings:

1. Polarization of the electrodes
2. Supply of oxygen

3. Temperature
4. Passivity

Polarization of the Electrodes

Polarization is the change of potential from a stabilized state, *e.g.* from the open-circuit electrode potential as the result of the passage of current. When a current flows, there is a change in the potential of an electrode; this is known as polarization. As the current begins to flow, the potential of the cathode becomes increasingly negative and the anode increasingly positive. Hence, polarization also refers to the change in the potential of an electrode during electrolysis, such that the potential of an anode becomes more noble, and that of a cathode more active, than their respective reversible potentials. Often accomplished by formation of a film on the electrode surface.

Consequently, the potential difference between the anode and the cathode decreases until a steady state is reached when corrosion proceeds at a constant rate. Thus, the corrosion current and therefore, the corrosion rate will be affected by anything which affects the polarization of the electrodes.

The potential at which the reaction takes place changes by an amount called the overpotential, η, which is defined as:

$$\eta = E_{corr} - E_{I}$$

where E_i is the polarized potential. The anodic overpotential η_a drives the metal dissolution process, and the cathodic overpotential η_c drives the cathodic deposition process.

Overpotential increases with current density, i, in accord with the Tafel equation:

$$\eta = \beta \log \frac{i}{i_0}$$

where β and i_0 are constants for a given metal and environment; both depend on temperature. The dominant polarization term controlling the corrosion rate of many metals in deaerated water is the hydrogen overpotential at cathodic areas of the metal. The hydrogen overpotential for iron at 16°C in 1 N hydrochloric acid is 0.45 V, and for tin at 20°C in 1 N hydrochloric acid is 0.75 V.

Supply of Oxygen

Metal + Oxygen $\rightarrow$ Metal oxide (corrosion product). When oxidation starts, a thin layer of oxide is formed on the metal surface and the nature of this film decides the further action. The mechanism of reactions of oxidation corrosion are as follows:

1. Oxidation takes place at the surface of the metal forming metal ions

   M2+ M $\rightarrow$ M2+ + 2e

2. Oxygen is converted to oxide ion (O_2^-) due to the transfer of electrons from metal.

   $n/2\ O_2 + 2n\ e- \rightarrow nOO_2^-$

3. The overall reaction is of oxide ion reacts with the metal ions to form metal oxide film.

   $2\ M + n/2\ O_2 \rightarrow 2\ Mn+ + nO_2^-$

The Nature of the Oxide formed plays an important part in oxidation corrosion process.

The rate of which oxygen is supplied largely governs the rate of corrosion, since corrosion by oxygen reduction requires the presence of oxygen for the cathodic reaction to proceed. The rate of supply is proportional to the rate at which oxygen diffuses to the metal surface, and this depends on the concentration of dissolved oxygen in solution. This is further justification for the practice of attempting to remove all the oxygen from canned foods prior to seaming on the can end.

Temperature

The rate of corrosion generally increases with increase in temperature as more reactant molecules or ions are activated and are able to cross over the energy barrier. As well, increasing the temperature tends to increase the rate of diffusion of molecules or ions in a solution, although the solubility of oxygen in water decreases with increasing temperature.

Passivity

The dissolution process a metal was described as an oxidation process of the general form $M \rightarrow M_n$ + n electrons. From the Pourbaix diagrams, it is evident that if the metal can be oxidized to an oxide that is stable in the electrolyte, then the metal is rendered passive (*i.e.*, passivated). Passivation usually requires strong oxidizing conditions.

Thus, corrosion resistant metals and alloys can withstand an aggressive environment because of the presence of thin films of adherent oxides on their surfaces. The oxide layer will completely stop the anodic reaction which is the direct cause of corrosion, and if the film is insoluble in the electrolyte solution, it will form an insulation barrier which will reduce the rate of the cathodic reaction.

For example, iron is readily attacked by dilute nitric acid, but is inert in concentrated nitric acid because a thin, protective film is formed. As a result, iron behaves in concentrated nitric acid like a much more noble metal than it actually is. Iron can also be passivated by chromate solutions, as can tinplate, the latter being a very important step in the manufacture of tinplate. Passivation of tinplate can be achieved using an aqueous solution of chromic acid, although an electrolytic treatment in a sodium dichromate electrolyte has gained widespread favour. The resultant film is composed of chromium and chromium oxides and tin oxide, its properties varying depending on the quantity and form of these basic components.

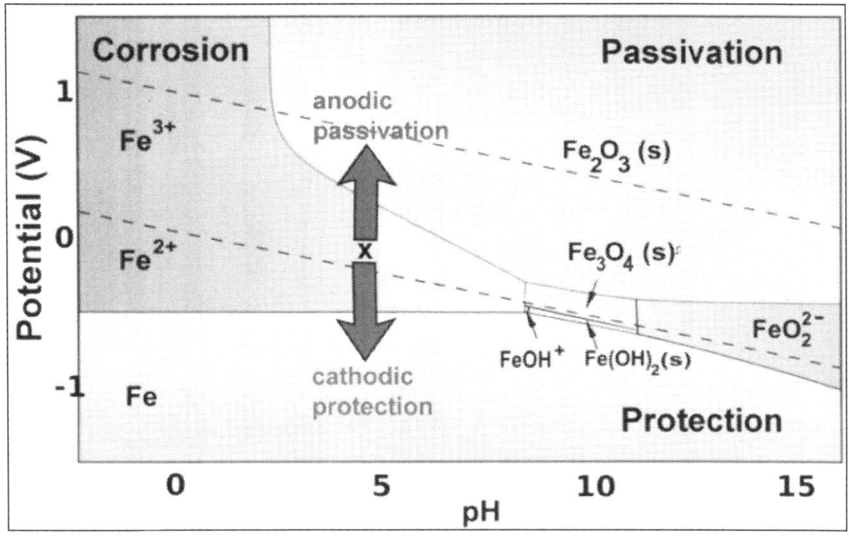

The tinplate surface consists of a large area of tin and tiny areas of exposed tin-iron alloy ($FeSn_2$) and steel as a result of pores and scratches in the tin coating. Although the now obsolete hot dipped tinplate had a substantial tin-iron alloy layer, electrolytic tinplate has a much thinner layer which is electropositive to the base and also to the tin, thus acting as a chemically inert-barrier to attack on the steel base. The effect of this barrier is to prevent a significant increase in the steel cathode area. Thus, the density or degree of continuity of the alloy layer has a material effect on the rate of corrosion. The alloy-tin couple test gives a good indication of the continuity of the alloy layer in tinplate.

In the case of tinplate exposed to an aerated aqueous environment, tin is noble (*i.e.,* cathodic) to iron according to the electrochemical series. Therefore, all the anodic corrosion is concentrated on the minute areas of steel and the iron dissolves, *i.e.,* rusts. In extreme cases, perforation of the sheet may occur. This is the process which occurs on the external surface of tinplate containers.

However, inside a tinplate can, the tin may be either the anode or the cathode depending on the nature of the food. In a dilute aerated acid medium, the iron is the anode and it dissolves, liberating hydrogen. In deaerated acidic food, iron is the anode initially, but later reversal of polarity occurs and the tin becomes the anode, thus protecting the steel; tin has been described in this situation as a sacrificial anode. This reversal occurs because certain constituents of foods can combine chemically with Sn_{2+} ions to form soluble tin complexes. As a consequence, the activity of Sn_{2+} ions with which the tin is in equilibrium is greatly lowered, and the tin becomes less noble (*i.e.,* more electropositive) than iron.

By using the Nernst equation, it is possible to calculate the ratio of Sn_{2+} to Fe_{2+} within the can for the reversal of polarity to occur. For the system 0.1 M citric acid at pH 3.8:

$Fe_{2+} + Sn \longrightarrow Sn_{2+} + Fe$

$$E = [0.14 - 0.44] - \frac{0.0592}{2} \log \frac{(Sn_{2+})}{(Fe_{2+})}$$

The cell reverses polarity when E = 0. Hence:

$$\log \frac{(Sn_{2+})}{(Fe_{2+})} = -0.30 \times \frac{2}{0.0592} = -10.30$$

and

$Sn_{2+} = 5 \times 10^{-1}{}_{1} Fe_{2+}$

This very low concentration of Sn_{2+} relative to Fe_{2+} can only occur through the formation of tin complexes and these are discussed later. Generally, complexing agents tend to increase the corrosion rates of many metals by reducing the metal ion activity, thereby shifting metal potentials markedly in the active direction.

The reversal has also been related to the high hydrogen overpotential of tin compared with that of iron, and to the low solubility constant of stannous hydroxide ($5 \times 10_{26}$) which forms and precipitates on the cathode and reduces its size, compared with that of iron ($1.6 \times 10_{15}$).

Now let us study about corrosion of enameled cans.

Food cans with enamel (lacquer) coatings are used to protect against excessive dissolution of tin, sulfide staining, local *etc.*hing and change in colour of pigmented products such as berry fruits. However, the use of enamels will not guarantee the prevention of corrosion, and in some cases may actually accelerate it. Therefore, careful consideration must be given before selecting an enamel system for a particular canned food.

The general pattern of corrosion in enameled cans is very different from that in plain cans, and is generally more complex. It depends not only on the quality of the base steel plate, the tin-iron alloy layer and the tin coating, but also on the passivation layers and the nature of the enamel coating. The only exposure of metal in an enameled can is at pores and scratches in the enamel coating and at cracks along the side seam. Some of these discontinuities in the enamel coating may coincide with pores in the tin coating, thus resulting in exposure of the steel. Even if defects in the enamel film expose only the tin coating, the availability of all the corrosion promoters in the can for attack on the limited areas of tin ensures that steel is soon exposed at them. Because these areas of exposed steel are almost unprotected either by the electromechanical action of a tin coating or by dissolved tin, corrosion may proceed at a rapid rate, resulting in hydrogen swelling or perforation of the can. Thus, it is easily possible for the use of an enamel to actually reduce the shelf life of a canned product.

The effectiveness of an enamel coating is related directly to its ability to act as an impermeable barrier to gases, liquids and ions, thereby preventing corrosive action on the protected surface. The transport of ions through the enamel is governed by the electrochemical characteristics of the film, in contrast to the transport of gases and liquids which involves dissolution in the enamel film and diffusion through it under a concentration gradient.

Because ions are electrically charged, their transfer through the enamel coating (actually the flow of an electric current) is complex and depends not only on their electric charge but also of the concentration of the electrolyte. If the transport rate of cations and anions through the coating differs, the coating itself may become charged. Thus, the protection offered by the enamel coating depends on its resistance to ion transfer which may take place even in the absence of pores, scratches or blisters.

The performance of enameled food cans is greatly affected by the thickness of the enamel coating. A thickness of 4 to 6 μm is sufficient for nonaggressive products such as apricots and beans, but aggressive products such as tomato paste require thicknesses of 8 to 12 μm, the heavier coatings having much lower porosities.

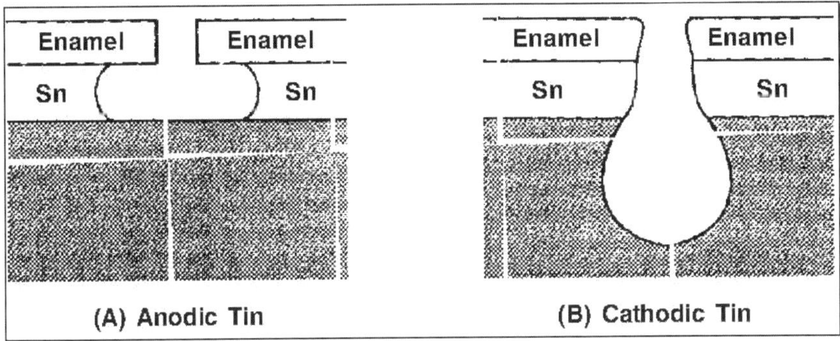

Figure 8.1: Schematic Diagram Illustrating two possible Corrosion Mechanisms of Enameled Tinplate.

A possible electrochemical corrosion mechanism in enameled cans has been proposed and a schematic diagram is presented as Figure 8.1. In a plain can, the product would attack the tin layer, causing it to dissolve. The presence of an enamel coating protects the surface, and tin dissolution occurs only from both sides of a scratch under the enamel or through a pore, causing anodic undermining. Because only a very small area is in contact with the product in the can, dissolution is slow. With time the exposed area increases, and detachment of the enamel may be observed, resulting in the appearance of enlarged pores in the enamel coating. The tin-iron alloy layer may also be visible as a grayish color. Since (in theory) the alloy layer is more passive than tin and iron, it should dissolve only after all the tin and iron. However, because of its extreme thinness and its coupling effect with the tin, the potential of the can may be more noble than that of iron, which may also go into solution. In this situation, the enamel coating may serve as the cathode due to diffusion of protons through it.

Mannheim and Passy have used tomato concentrate as an example of the above scenario: while the overall rate of tin dissolution is reduced in the presence of an enamel coating, protection of the steel by the tin is also impaired due to reduction in the ratio of exposed tin to steel areas. Corrosion is concentrated in small areas and strong local currents may occur. The presence of corrosion accelerators such as nitrates aggravates the situation.

However, since no tin is dissolved, the alloy layer is not laid bare and thus, has no influence on the corrosion process. Failure occurs eventually due to pinhole formation (also known as pitting corrosion) but there is no undermining of the enamel, corrosion usually starting at a point of discontinuity such as a scratch or pore in the coating. Aggressive products such as beets in acetic acid and berries are responsible for this type of corrosion. The cans do not appear to be corroded, and it is only on closer inspection (usually with a, hands lens) that spots of corrosion (often with deep penetration into the steel) are visible.

Failure of enameled cans is often due to a reduction in the bond between the enamel and the metal surface, resulting in eventual lifting of the enamel coating. Thus, good adhesion is required to prevent anodic reactions, to counteract forces developed under the coating due to physical or chemical factors, and to ensure an aesthetic appearance.

Loss of adhesion may commence prior to corrosion due to stresses induced during fabrication, but typically detachment of the enamel coating is a result of breakdown processes taking place through or under the coating, or to underfilm corrosion spreading from exposed metal. The quality of the surface of the tinplate prior to enameling has a major effect on adhesion and enamel performance. The presence of salt residues can cause failure due to their ability to attract water and therefore, establish a conducting film beneath the coating, and to provide ions for carrying the corrosion current.

8.4 The Corrosion Accelerators

Food products and beverages are extremely complex chemical systems covering a wide range of pH and buffering properties, as well as a variable content of corrosion inhibitors or accelerators, Factors which influence the corrosiveness of food products and beverages can be divided into two groups hartwell: intensity and type of corrosive attack inherent in the food itself, and corrosiveness due to the processing and storage conditions, All these factors are interrelated and may combine in a synergistic manner to accelerate corrosion.

The most important corrosion accelerators in foods include oxygen, anthocyanins, nitrates, sulfur compounds and trimethylamines.

From a corrosiveness point of view, it is convenient to divide foods into five classes: those that are highly corrosive such as apple and grape juices, berries, cherries, prunes, pickles and sauerkraut; those that are moderately corrosive such as apples, peaches, pears, citrus fruits and tomato juice; those that are mildly corrosive such as peas, corn, meat and fish; and strong detinners such as green beans, spinach, asparagus and tomato products; beverages are conveniently considered as a fifth class.

While the above classification offers a good, broad guide, it is important to note that different lots or varieties of the same food can exhibit as much variation in their corrosiveness as may exist between different types of foods, Thus, for example, the same variety of fruit from different growing regions may vary several folds in corrosiveness. According to Hartwell, finding a twofold or greater variation between different packs of the same food in identical containers is the usual experience of investigators in this field.

The various factors that can influence the corrosiveness of food products and beverages will now be considered in more detail.

Acidity

No direct proportionality exists between the acidity of a product and the degree of corrosion of tinplate. In other words, two products of the same acidity will not necessarily be equally corrosive. It also appears that pure solutions of organic acids are less corrosive than the fruit juices containing them, suggesting that fruit juices contain unidentified depolarizers which enhance the corrosive action of organic acids. It has been well established that the tendency of an acid to form a complex with dissolved tin has an important bearing on the relative polarity of tin and steel, and hence the degree of corrosion.

In a study comparing six organic acids in a model system, acetic and lactic acids were the most corrosive based on subjective severity of can pitting; the type of organic acid was more important than its concentration, and the addition of nitrate increased pitting severity in all acid solutions.

pH

As was the case with acidity, no direct proportionality exists between pH and the degree of corrosion of tinplate; this is not surprising when it is considered that the reaction product when a metal is dissolved is not always an ionic species but often a solid oxide or hydroxide. Pourbaix diagrams indicate the relationship between pH and the equilibrium potential, and show regions of immunity, corrosion and passivity. The pH of the system also determines the relative cathodic protection given to steel. In some cases, tin is cathodic to steel over a certain pH range (in the case of acetic acid the range is pH 2.0 to 4.5), while in others it offers protection up to pH 4; above that level it may accelerate corrosion.

Sulfur Compounds

Sulfur and sulfur compounds may be introduced into the can in a number of ways: as spray residues from agricultural chemicals; as residues from sulfur-containing preservatives, and as components in sulfur-containing compounds such as proteins in meat, fish and certain vegetables: the proteins are degraded during heat processing, releasing free sulfide or hydrosulfide ions and evolving hydrogen sulfide gas into the headspace.

Trace amounts of sulfur compounds from agricultural chemicals (for example, derivatives of thio- and dithiocarbamic acid fungicides) can lead to accelerated corrosion and failure of plain tinplate cans containing acid foods such as apricots and peaches. As well, pitting corrosion can occur, and this has been attributed to inactivation of the tin coating by a protective film of sulfide having a more cathodic potential. As a consequence, there is a significant reduction in the tin dissolution rate, and no electrochemical protection of the steel by the tin.

Sulfur dioxide may be directly reduced on the tin surface to sulfide or even to sulfur, with tin passing into solution and the development of unpleasant odors and flavors. Residual SO_2 accelerates corrosion through its action as a depolarizer, inducing a negative charge in the double layer: this repels the electrons from the electrode, thus shifting the potential in the positive direction. Trace amounts of SO_2 as low as 1 mg kg^{-1} are sufficient to accelerate corrosion; such corrosion problems may be overcome by the use of enameled cans.

Table 8.1: Some Corrosion Promoting Agents and their Mode of Reaction

Corrosion Accelerator	*Reduction Product*	*Equivalent in Weight*
Proton (H^+)	H_2	1 mL H_2 = 5.3 mg Sn^{2+}
Oxygen (O_2)	H_2O	1 mL O_2 = 10.6 mg Sn^{2+}
Sulfur dioxide (SO_2)	H_2S	1 mL SO_2 = 5.5 mg Sn^{2+}
Sulfur (S)	H_2S	1 mL S = 3.7 mg Sn^{2+}
Nitrate (NO_3)	NH_3	1 mg NO_3 = 7.65 mg Sn^{2+}
Trimethylamine oxide (TMAO)	TMA	1 mg TMAO = 1.57 mg Sn^{2+}

There are two types of sulfide staining: one involves iron sulfide and is sometimes called sulfide black, and the other involves tin sulfides. Neither type constitutes a health hazard nor leads to failure of the can, but both types may cause adverse reactions from the consumer.

Iron sulfide stains are characteristically black and usually occur at isolated points on the can (mainly in the headspace region) during or immediately after heat processing. Iron sulfide is not formed at pH values below about 6 so that it is uncommon to find it in the portion of the can in contact with the contents. However, the pH of condensed volatile matter in the headspace may be above 6. The problem may be overcome by using enameled cans or plain cans with enameled ends.

Tin sulfide staining is usually widespread throughout the can and is blue-black or sometimes brown. Two stages are believed to be involved: an oxidation of the tin, followed by deposition of an insoluble tin sulfide precipitate on the surface.

It occurs during or soon after heat processing and shows little or no increase in intensity during storage. It may be reduced by the use of CDC plate or prevented by using sulfur-resistant enamels into which quantities of zinc oxide or carbonate are added before being applied to the plate surface. These react with sulfur-bearing gases to form almost invisible white zinc sulfide.

Nitrates

Nitrates are found in fruits and vegetables grown in heavily fertilized soils, and may also occur in water supplies as a result of pollution by fertilizers. Vegetables such as green beans, spinach, turnips, lettuce, beets and radishes have been shown to contain on occasion several thousand mg kg^{-1} of nitrates. Nitrates are very efficient cathode depolarizers since they are capable of being reduced all the way to ammonia. They have been responsible for serious economic and toxicological problems in some canned foods, notably tomato products. Although nitrates and nitrites are also present as intentional additives in processed meats, they present no problem; this is because meat products are above the critical pH range for detinning to occur via the nitrate-tin reduction system.

According to Britton, the reduction of 60 mg of nitrate could account for the dissolution of 470 mg of tin; a tin coating of 11.2 gsm may be dissolved completely within 6 months by tomatoes containing 100 mg kg^{-1} of nitrate.

Nitrates act as electron acceptors, replacing the hydrogen evolution reaction with the electron-nitrate two-step reduction reaction and shifting the reaction towards increased tin dissolution. The reduction reactions involved are believed to be as follows:

$$4Sn \longrightarrow 4Sn^{2+} + 8e^-$$

$$NO^-_3 + 2e^- + 2H^+ \longrightarrow NO^-_2 + H_2O$$

$$NO^-_2 + 6e^- + 8H^+ \longrightarrow NH_4^+ + 2H_2O$$

The first reduction reaction equation is probably rate determining. The equations indicate that the rate of detinning depends on the nitrate concentration and pH. Nitrate does not immediately affect the corrosion rate, but begins to act after tin and iron ions have passed into solution. Although the ammonium ion is the major conversion product at pH 5 and less, above this pH other products are formed, including nitrous oxide (N_2O), nitric oxide (NO) and hydroxylamine. Oxygen present in the can at the time of processing triggers the nitrate-detinning reaction by increasing the initial rate of formation of Sn_{2+}.

Overcoming the problem by restricting the use of nitrate fertilizers has proved difficult, and instead efforts have been directed towards finding cultivars which do not accumulate high concentrations of nitrate. It is also possible to avoid the use of waters with a high nitrate content for canning operations; a suggested maximum is 5 mg L^{-1} nitrate. While raising the pH of the product above 6.0 will effectively inhibit nitrate detinning and can be used with some vegetables, such a procedure cannot be used with acid foods such as tomato or citrus products where the acidity determines the organoleptic properties and the heat treatment required. The best solution to the problem at present is the use of enameled cans, although the development of corrosion inhibitors (the search for such substances is continuing) would offer an alternative solution.

Phosphates

Phosphates are naturally present in meat and are often intentionally added as polyphosphates to processed meat products such as hams to reduce the loss of water during processing. The presence of phosphates leads to increased discoloration due to iron phosphate and sulfide formation. One countermeasure used against the effect of polyphosphate in ham cans has been the introduction of a small area of aluminum which acts as a sacrificial anode, protecting the tin surface. This is not always wholly successful, and parchment liners may be used in addition or as an alternative.

Plant Pigments

The anthocyanins and related pigments are among the most important potential corrosion accelerators (cathodic depolarizers) since they are easily reduced. Anthocyanin pigments can also act as anodic depolarizers through their ability to form complexes with cations, particularly those of iron and tin salts. Analysis of samples of canned fruit after a period of storage usually shows a greater amount of tin in the drained fruit than in the syrup, indicating that at least part of the tin is combined in an insoluble form with some constituent within the fruit.

The nature of the anthocyanin pigment is also important. For example, raspberries contain cyanidin glucosides which have *ortho*-dihydroxy groups in their structures; it is these groups which are involved in the formation of blue-tinted complexes with metals such as tin. The major pigment of strawberries is pelargonidin-3-g1ucoside which does not possess the necessary *ortho*-dihydroxy groups for complex formation. Therefore, strawberries do not show the same shift to a blue color in the presence of tin salts as do raspberries.

Combination of metal ions with tannins has been observed in other fruits. For example, discoloration in canned cranberry has been attributed to the formation of a complex between tannins present in the fruit and tin salts. Darkening in canned Maraschino cherries has been observed to be more severe when fruit of a high tannin content is used, a high tannin content being found in unripe fruit and fruit that has been stored in wooden barrels.

It should be noted that not all reactions between plant pigments and metal ions will produce undesirable colors, although this is usually the case with anthocyanin pigments.

Synthetic Colorings

The canned products which most commonly contain synthetic colorings are soft drinks, which consist basically of sugar-based syrups and carbonated water containing flavors, acidulants and colors. The behavior of soft drinks depends to a considerable extent on the presence of azo dyes (*e.g.*, amaranth) and the amount of residual oxygen in the filled can. Both of these components are capable of acting as corrosion accelerators and are potentially active corrosive agents. Tin dissolution may adversely affect the color of some products, and iron dissolution may lead to perforation and flavor defects. Thus fully enameled cans are essential, it being important to obtain near- perfect coverage by the enamel.

Soft drinks have been categorized as follows:

1. Products containing azo dyes or other oxidants which may promote dissolution of tin in the absence of oxygen.
2. Products based on citric acid where the tin is anodic and has a high dissolution rate.
3. Products based on phosphoric acid with fruit flavors where tin is strongly cathodic to iron and there is rapid corrosion of exposed steel because of the depolarizing action of the dyes.
4. Cola beverages containing phosphoric acid where the tin-iron couple action is insignificant; the tin serves as an inert barrier, resulting in relatively low iron pick-up.
5. Products based on citric acid without active dyes where tin is anodic to iron and reaction rates are generally slower. Protection of the exposed steel can be obtained by the deliberate addition of stannous chloride which acts as an effective corrosion inhibitor, but such addition is banned in many countries.

Copper

Foods containing dissolved copper will deposit it when put in a metallic container, and in acid products, this can lead to accelerated corrosion, either by tending to strip the tin or by producing local attack on the steel. The basic reason why copper accelerates the corrosion of steel is because copper catalyzes the reduction of oxygen.

In the past, copper all too frequently entered products by the solution of copper oxide from copper-bearing metal food contact equipment. However, with the replacement of much of this equipment with stainless steel equipment, the problem of copper in canned foods has been dramatically reduced. Another possible source of copper is from certain fungicides, and occasionally some cannery water supplies may have unacceptably high copper levels.

8.5 Effect of Processing and Storage

Oxygen

Oxygen may be dissolved in food products: it is naturally present in food tissues such as fruit and vegetable cells, and is inevitably entrained or absorbed by particulate products prior to their being filled into cans. Removal of as much of this oxygen as possible is an essential part of good cannery practice and a variety of methods are used: hot-filling, vacuum filling, exhausting, closure under vacuum, steam flow closure and vacuum syruping. As well, positive control of the headspace volume is essential.

A larger headspace is likely to contain a higher residual oxygen concentration than a smaller one. However, the larger the headspace, the more room there is for accumulation of hydrogen resulting from corrosion and thus, the greater the time required to form a hydrogen swell. In practice, the cannery technologist has little room to maneuver the headspace if the declared weight on the can label is to be met.

Despite the above procedures, there will always be some oxygen present in the headspace of newly filled cans. The rate of oxygen consumption at this stage is quite rapid but decreases with time, the rate being a function of initial concentration, headspace volume, can vacuum, nature of the product and type of container.

Oxygen acts as a depolarizer, accelerating corrosion by reacting with the hydrogen formed in the can through a cathodic reaction as shown in earlier equation. Calculations suggest that if the residual pressure in the headspace of a filled can of a common size is 0.5 atmospheres, then the oxygen present would be capable of removing about 0.25 gsm of tin from the coating, producing 15 mg kg^{-1} of dissolved tin in the can contents. Although this is a relatively small amount in comparison with the total coating weight, the action of the oxygen may be localized on a narrow zone of the gas/liquid interface in the headspace. This may result in the production of an unsightly etched line and even exposure of an area of steel much larger than that exposed at pores in the initial coating.

Thermal Processing

Little is known about the effect of heat sterilization process on corrosion rates, except that the quantity of metal dissolved during the process is very small. This is hardly surprising given the comparatively short processing times (typically within the range of 30 to 120 minutes) relative to the total shelf life of the canned product (typically from 1 to 2 years).

Storage Temperature

The rate of a chemical reaction increases as the temperature is raised, and for many reactions, the rate doubles for each 10°C increase in temperature, *i.e.,* the temperature quotient Q_{10} equals 2. Thus to minimize undesirable reactions such as nonenzymic browning in canned foods, it is preferable that storage temperatures be kept as low as practicable.

The situation with respect to the rate of corrosion of tinplate cans is less clear; data on the relationship between temperature and the corrosion current measured between tin and iron for some test solutions and food products showed a wide variation, and in all cases studied, Q_{10} was less than 2. For 3 per cent acetic acid, a negative Q_{10} was observed, since it is a tin dissolver at 20°C but an iron dissolver between 50 and 60°C. Similar reversal of the corrosion effect has been observed in fish products, but as a critical review commented "it is wrong to conclude from this that increasing the temperature improves storage stability, since corrosion is only one of the processes affecting shelf life."

Chapter 9

Packaging of Foods in Metal Containers

9.1 Introduction

Metal packaging plays an important role in the process of food preservation. The commercial packaging of foods in metal containers began in the early 19_{th} century, following on from the discovery in the 1790s by the French confectioner Nicolas Appert of a method of conserving all kinds of food substances in containers a method to which the term canning is now applied indiscriminately, whether the container is made from, aluminium, glass or plastics.

Metal packaging has a double function as a protection against any external influence on the foodstuff during heat treatment and storage and as a sales and information pack. The basic requirement for such a package is the hermetic tightness of the container. The food, which is sterilised by the heat process, ought to be protected against any re-infection with microorganisms or any other kind of influence from the outside.

Metal packages for food products must perform the following basic functions if the contents are to be delivered to the ultimate consumer in a safe and wholesome manner:

- Preserve and protect the product
- Resist chemical actions of product
- Withstand the handling and processing conditions
- Withstand the external environment conditions

- ✰ Have the correct dimensions and the ability to be practically interchangeable with similar products from other supply sources (when necessary)
- ✰ Have the required shelf display properties at the point of sale
- ✰ Provide easy opening and simple/safe product removal
- ✰ Should be constructed from recyclable raw materials.

In addition, these functions must continue to be performed satisfactorily until well after the end of the stated shelf-life period.

Four metals have become commonly used materials for the packaging of foods: steel, aluminum, tin and chromium. Tin and steel, and chromium and steel, are used as composite materials in the form of tinplate and electrolytic chrome-coated steel (ECCS), the latter being somewhat unhelpfully referred to for many years as tin-free steel (TFS). Aluminium for light metal packaging is used in a relatively pure form, with manganese and magnesium added to improve the strength properties. This material cannot be welded by can making systems and can only be used for seamless (two-piece) containers. The internal surfaces of aluminium containers are always coated with an organic lacquer because of the products normally packed. Aluminium is inherently less strong than steel, so this limits where it may be economically used for the construction of metal packaging. Aluminium cannot be used for making three-piece welded side seam containers. Aluminium is, however, used for pressurised containers, where wall strength is not so critical, for products, such as beer, carbonated drinks in DWI cans and also for seamless impact extruded aerosols and collapsible tubes. Two other metals are used during the soldering or welding of three-piece tinplate and ECCS containers: lead and copper. However, since they are not used for the fabrication of containers in their own right, they will not be discussed in detail in this chapter. The health aspects of these different metals, together with can coatings, are discussed elsewhere.

The common expression used to describe metal packaging is "canning". Canned food has become an important part of the human diet in developed countries during the past century. It is of particular value in those parts of the world where no or limited refrigeration exists for storing food. It is a means of safely preserving foodstuffs without microbiological deterioration. After almost two centuries of history, there is still controversy as to who introduced the tin can as a package. The 15_{th} edition of the Encyclopaedia Britannica (1987) states in the section on food processing that "Peter Durand, an Englishman, conceived and patented the idea of using tin cans instead of bottles....", in conflict with the British Dictionary of National biography's (1882–1900) assertion that Bryan Donkin "devised the method of preserving meat and vegetables in air-tight cases."

The first commercial manufacture of tinplate commenced in England in 1699 and in France in 1720 where it was used for a variety of purposes including household utensils such as plates. Sometime about the middle of the eighteenth century, the Dutch navy began to use foods preserved by packing them in fat in tinned iron canisters. After cooking and while still hot, the material to be preserved was placed into the canister, covered with hot fat and the lid immediately soldered on, a forerunner of what today could be described as a form of aseptic canning! Records show that from 1772 to 1777 the Dutch Government supplied their navy (which had been sent out to Suriname (formerly Dutch Guiana in South America) to quell a revolt) with roast beef packed in this way.

Before the end of the eighteenth century the Dutch had also established a small industry to preserve salmon in a similar manner. Freshly caught salmon were cleaned, cooked in boiling brine, smoked over a wood fire for two days and then placed in a tinplated iron box. The spaces were filled up with hot salted butter or olive oil and a lid soldered onto the box. A famous London firm of snuff merchants supplied 13 tins of Dutch salmon to one of their clients in 1797. Thus, a canning industry of sorts had been established in Holland independently of Appert's work.

Durand's patent granted in 1810 speaks of "an invention communicated to him by a certain foreigner residing abroad" and covered the use of "glass, pottery, tin or other metals or fit materials." It is open to doubt whether the patent contributed anything other than the "tin" among a list of metals and materials for preserving vessels. Appert deliberately avoided tinplate in his early work because of the poor quality of the French product, according to the 4_{th} (1831) and 5_{th} (1858) editions of his book. However, the quality of tinplatc in England was good and it was freely available.

The Durand patent was acquired by the Englishman John Gamble who was involved with Bryan Donkin in developing the Fourdrinier papermaking machine. Donkin had become interested in the tinning of iron as early as 1808 and in 1811, in association with John Hall, founder of the famous Dartford Iron Works, developed a process based on that of Appert. It is not clear how this occurred. The same year, using the name Durand (it is thought to avoid the legal problem of explaining the assignment of the patent) Donkin applied to the British admiralty for a test of his product. Although the kind of container used is not known, the first substantial orders were placed in 1814 with the London firm of Donkin, Hall and Gamble for meat preserved in tinplate canisters. By the 1820s canned foods were a recognized article of commerce in Britain and France.

As the review by Bishop demonstrates, flat assertions as to the origin of inventions are generally dangerous. To complete the historical record, William Underwood left London and arrived in New Orleans in 1817; he travelled up to Boston where he started a business preserving food in glass jars by Appert's method. In 1819 Thomas Kennsett, also from England, started a similar business in preserved foods in New York in partnership with his father-in-law Ezra Daggett. The first offering of preserved provisions in tin cans in America is assumed to be the announcement by Daggett and Kensett in the New York Evening Post of July 18, 1822, although it was not until 1825 that they took out a patent in which "vessels of tin" were mentioned.

The American Civil War afforded the opportunity for canning to become a great industry, and by the end of the war in 1865 the canners had increased their output sixfold. For many years the cans were made slowly and laboriously by hand. Both ends were soldered to the can with a hole about 25 mm diameter left in the top. After the can was filled through this hole, a metal disk was soldered into place. The mechanical roll crimping (commonly known as double seaming) of the can ends into a body with a soldered sideseam was patented in 1896 by Max Ams of New York, making it possible to develop high-speed equipment for the making, filling and closing of these cans.

Today, materials like tinplate and aluminum have become universally adopted for the manufacture of containers and closures for foods and beverages, due to largely several important qualities of these metals. These include their mechanical strength and resistance to working, low toxicity, superior barrier properties to gases, moisture and light, their ability to withstand wide extremes of temperature and provide ideal surfaces for decoration and lacquering.

9.2 The Use of Aluminum

Aluminum is the earth's most abundant metallic constituent, comprising 8.8 per cent of the earth's crust; only the nonmetals oxygen and silicon are more abundant. Alumina or aluminum oxide (Al_2O_3) is the only oxide formed by aluminum and is found in nature as the minerals corundum (Al_2O_3); diaspore ($Al_2O_3.H_2O$); gibbsite ($Al_2O_3.3H_2O$); and most commonly bauxite, an impure form of gibbsite.

Hans Christian Oersted, a Danish chemist, first isolated aluminum in 1825 using a chemical process involving potassium amalgam. Between 1827 and 1845 Friedrich Wohler, a German chemist, improved Oersted's process by using metallic potassium. In 1854 Henri Sainte-Claire Deville in France obtained the metal by reducing aluminum chloride with sodium. Aided by the financial backing of Napoleon III, Deville established a large-scale experimental plant and displayed pure aluminum at the Paris Exposition of 1885. In 1886 Charles Martin Hall in the United States and Paul Héroult in France independently and almost simultaneously discovered that alumina would dissolve in fused cryolite (Na_3AlF_6) and could then be decomposed electrolytically to a crude molten metal. A low-cost technique (the Hall-Héroult process) is still the only method used for the commercial production of aluminum, although new methods are under study.

Due to the chemical stability of its oxides, the energy requirements for smelting are extremely high. This has led to the production of aluminum in areas where cheap electrical power is available, but current trends suggest that the days of cheap power may be coming to an end. If aluminum smelters are required to pay the true market price for power, then the use of aluminum as a packaging material is almost certainly likely to decline quite sharply.

Today a typical aluminum process would be as follows: alumina is dissolved in cryolite in carbon-lined steel boxes called pots. A carbon electrode or anode is lowered into the solution and an electric current of 50–150 MA is passed through the mixture to the carbon-cathode lining of the pot. The current reduces the alumina into aluminum and oxygen, the latter combining with the anode's carbon to form carbon dioxide, while the aluminum (denser than cryolite) settles to the bottom of the pot.

Aluminium is widely used in food contact materials. Most commercial uses of aluminum require special properties that the pure metal cannot provide. Therefore, alloying agents are added to impart strength, improve formability characteristics and influence corrosion characteristics. A wide range of aluminum alloys are available commercially for packaging applications, depending on the container design and fabrication method being used. Aluminium alloys used for food contact may contain elements such as magnesium, silicon, iron, manganese, copper and zinc. The composition of aluminium alloys for rigid containers varies according to the intended use, with up to 5 per cent magnesium, 1.5 per cent manganese and traces of iron, silicon, zinc, copper and titanium. Aluminium alloys or tempers commonly used are: a fully hard material such as 3004 H19 and a softer one such as 5052 H-34. Where 3004 or 5052 denote the aluminium alloy sheet and H-19, H-34 denotes the tempering. The hard tempered alloy H-19 allows the very thin gauges which make the container bodies economical. Shallow-drawn parts, such as can ends, use alloys with less ductility and medium temper of H-34 for average conditions. Aluminium and its various alloys are highly resistant to corrosion. When exposed to air, the metal develops a thin film of aluminium oxide. The film is colourless, tough, and non flaking and few chemicals are able to dissolve it. Pure aluminium is attacked by most dilute acids. At neutral pH, aluminium hydroxide has limited solubility. However, the solubility increases markedly at pH below 4.5 and above 8.5.

Table 9.1: Chemical Composition of Some Commonly Used Aluminum Alloys

Alloy Type	*Typical Usage*	*Si*	*Fe*	*Cu*	*Mn*	*Mg*	*Cr*	*Zn*	*Ti*
1050	Foils and flexible tubes	0.25	0.4	0.05	0.05	–	–	0.03	–
1100		1.0	0.20	0.05	–	–	0.10	–	–
3003		0.6	0.7	0.70	1.5	–	–	0.10	–
3004	Beverage closures and D and I can bodies	0.30	0.7	0.25	1.5	1.3	–	0.25	–
5050		0.40	0.7	0.20	0.1	1.8	0.10	0.25	–
5182	Easy open beverages	0.20	0.35	0.15	0.5	5.0	0.10	0.25	0.10
8079		0.30	1.3	0.05	–	–	–	0.10	–

Commercially pure aluminum (types 1100 and 1050) is used for the manufacture of foil and extruded containers since it is the least susceptible to work hardening.

The general effect of several alloying elements on the corrosion behaviour of aluminum has been reported as follows:

1. Copper reduces the corrosion resistance of aluminum more than any other alloying element and leads to a higher rate of general corrosion;
2. Manganese slightly increases corrosion resistance;
3. Magnesium has a beneficial influence and Al-Mg alloys have good corrosion resistance;
4. Zinc has only a small influence on corrosion resistance in most environments, tending to reduce the resistance of alloys to acid media and increase their resistance to alkalis;
5. Silicon slightly decreases corrosion resistance, depending on its form and location in the alloy microstructure;
6. Chromium increases corrosion resistance in the usual amounts added to alloys;
7. Iron reduces corrosion resistance and is probably the most common cause of pitting in aluminum alloys; a high iron content increases the bursting strength but reduces the corrosion resistance.
8. Titanium has little influence on corrosion resistance of aluminum alloys.

Compared with tinplate and ECCS, aluminum is a lighter, weaker but more ductile material that cannot be soldered.

9.3 The Use of Tin

The tin canning process was allegedly created by the Frenchman Philippe de Girard, but the idea eventually passed to British merchant Peter Durand who was used as an agent to patent Girard's idea in 1810. The canning concept was based on experimental food preservation work in glass containers the year before by the French inventor Nicholas Appert. Durand did not pursue food canning, but, in 1812, sold his patent to two Englishmen, Bryan Donkin and John Hall, who refined the process and product, and set up the world's first commercial canning factory on Southwark Park Road, London. By 1813 they were producing their first tin canned goods for the Royal Navy. By 1820, tin canisters or cans were being used for gunpowder, seeds, and turpentine. In 1901 in the United States, the American Can Company was founded, at the time producing 90 per cent of United States tin cans.

Tin can be applied as thin layer on steel used for metal packaging. Generally speaking, the traditional way of producing cans are through tin-plating. Nowadays, you will see a lot of products packaged in tin-plated cans. These cans usually consist of iron or steel materials coated with tin in a process called tin plating. Because of this, tin-plated cans actually consist of about 1 to 2 per cent tin. The tin coating on the metal serves as protection from other elements to avoid damages to the can. It is applied electrolytically during the manufacture of ETP (electrolytic tinplate). The tin layer provides corrosion resistance and in some cases is not coated, as tin can act as an efficient oxygen scavenger. However, using uncoated tin is limited by the various possible interactions between the metal surface and the foodstuff and is therefore, mainly used for *e.g.* fruits in brine or tomato based products.

Packagers prefer tin cans over regular metals because regular metals react with acidic products which cause the material to corrode. When this happens, the corroded substances can contaminate the packaged product while damaging the can itself. Contrastingly, tin cans are totally safe for packaging food products, medicines, and cosmetics for longer time periods. This is because they can resist acidic contents. The only drawback of traditional cans is that they are much heavier compared to aluminum tins (modern cans). This makes tin-plated cans more expensive and less convenient to transport.

9.4 The Use of Electrolytic Chromium Coated Steel

The most common alternative to electrolytic tinplate for food contact applications is ECCS (electrolytic chromium/chromium oxide coated steel), which has equal coating weights on both surfaces of the coil. This material was developed in Japan during the 1960s and was originally known as tin free steel (TFS-CT). ECCS consists of a steel plate like the one used for tinplating, coated with a metal chromium film (50 to 90 mg/m2) and a passivation coating of metal chromium and chromium (III) oxide (10 to 20 mg/m2). The function of a chromium coating is to prevent atmospheric oxidation or sulphur staining of the steel by foodstuffs and to improve lacquer adhesion. ECCS is always used with an additional organic coating (*i.e.*, can coating). Hence, compared to tinplate, electrolytic chromium-coated steel is lower priced, has better resistance to atmospheric oxidation, and has excellent lacquer adhesion. It is normally used for the manufacture of drawn cans, can ends.

Adversely, it shows poorer resistance to corrosion by acid solutions, especially by foods. For this reason, ECCS is always lacquered. The usual welding procedure utilized in the manufacture of three-piece cans cannot be applied to this material.

9.5 Types of Metal Packaging

Metal packaging used for foodstuffs can be arbitrarily divided into cans, pails and drums, aerosol containers, tubes, trays, closures and lids.

Cans

Globally, metal cans have been used for two hundred years or so. Metal cans are used to contain a wide variety of products, including beverages, foods, aerosol products, paints, medicines, and many other products. Metal cans and can parts are made from aluminum or steel. Although most cans are cylindrical in shape, cans may be manufactured in other shapes, including rectangular cans such those used to contain gasoline or paint thinner and oblong cans used for packing hams and other meats. The sizes of cans range from very small (a few gms) to catering pack sizes (typically for contents of 2-10 kgs)

One of the main environmental benefits arising from fresh food packed into metal cans, which are then heat processed, is that the pack is then stored at ambient temperature for the remainder of its life. Metal packaging has excellent barrier properties *viz.*, 100 per cent protection from light, air and water; ensuring maximum product quality. Products packed in metal do not require additives to preserve them. Canned food may be stored for years and no further energy being required purely to preserve the integrity of the product. In many underdeveloped countries, vast quantities of agricultural products go to waste because they have become inedible before they can be consumed. Preserving these products by packing into heat-processed metal containers at source can provide a satisfactory solution to this issue. Furthermore, it does not matter how far away the final customer is located from the original source of the fresh product or how long it takes to arrive, as the contents once opened will still be fresh with no additional energy required to keep it so.

They offer total protection of the contents and are convenient for ambient storage. They are tamper-proof but the technology required for canning is expensive. Cans are heavy and bulky and importing transport costs for empty cans is high. Tin is relatively expensive so that the coating used on cans is very thin and susceptible to scratching and ultimately rusting. Cans for pale coloured fruits are traditionally not lacquered on the inside and this allows the fruit acids to attack the tin. The reducing environment so created prevents discolouration of natural colours including anthocyanins (often red) and carotenoids (orange) but it can increase the tin content of the product. Tin-free steel (using a chromium-steel alloy) cans are extremely durable and withstand corrosion more effectively than tin coated steel. Cans are completely impervious to oxygen, carbon dioxide, moisture and other aroma compounds. Food remains in good condition for several years although the shelf life is commonly set at two years.

If cans are stored badly, they can rust and eventually deteriorate to the point that pinholes occur which will allow microorganisms and other contaminants to enter the can. Some modern cans are often fabricated by stamping (more accurately 'deep drawing') and these cans have only two components – the body and the lid. There is no seam at the bottom of the can. These two-piece cans are used for fish, soft drinks and beer and often are equipped with a ring-pull opening device. The traditional three piece can with a lid and a base attached to a cylinder are much more common. They are heated to the extent that they are commercially sterile usually in a pressure vessel known as a retort. The coldest part of the coldest can must reach a temperature of 121°C for 2.54 minutes (this is sufficient to reduce the theoretical load of *Clostridium botulinum* spores by a factor of 1012).

Table 9.2: Types of Cans and their Uses

Process	*Material Used*	*Products*
3-piece	Steel	Food, juices, spices, aspirin, and other non-food items such as paints and glues (includes decorative tins); includes aerosols
Draw and iron	Aluminum (1- and 2-piece)	2-piece: Primarily beer, carbonated beverages, juices 1-piece: aerosol and pump products (perfume, air freshener, saline solution); fuel additives
	Steel (2-piece only)	Food, other non-food items
2-piece draw/redraw	Steel, aluminum	Food, shoe polish, sterno fuel, car wax, other non-food items
Ends	Steel, aluminum	All (except 1-piece draw and iron)
Crowns and closures	Steel	Food and non-food items

9.6 Can Making Process

There are in essence distinct types of cans and ends (or lids). Packers and fillers purchase empty cans and lids and seam the lids onto the cans. Cans consist of either two or three separate components (two-piece cans and three-piece cans). The three basic types of can configurations are three-piece cans, two-piece drawn and ironed cans and two-piece drawn and redrawn cans.

Three-piece Cans

Three-piece cans are mainly used for food, but may be used for some non-carbonated beverages, particularly fruit juices. They were the original cans, consisting of a bottom (end), walls (body) and lid.

Although the conventional open-top metallic can was introduced about 200 years ago, it is still the most widely used in food packaging. Besides the old cylindrical can, three-piece cans are classified according to their transversal sections: circular, rectangular, oval, or trapezoidal. Despite the fact the basic design has hardly been altered, manufacturing technology has significantly changed with electrical weld, robotization, and computer control of the manufacturing lines. Also, the progressive thickness reduction of materials has led to the introduction of beading profiles to maintain the required rigidity during container sterilization and handling, especially in 500 g or larger formats. Manufacturing is accomplished via a 10-step integrated process.

Three-piece can bodies are made from flat sheets cut from coils of tin-plated or tin-free steel, depending on the end use. The tin plating is applied to prevent rust. Tin-free steel is electrocoated with a layer of metallic chromium covered by a layer of chromium oxide. Before the bodies are formed, coatings are usually applied to the interior and exterior surfaces with a roller onto the flat sheet. Interior coatings are applied to protect the can from corrosion by the contents and/or to protect the contents from being contaminated by dissolved metal from the can. Occasionally, however, pigmented interior coatings are applied to enhance the visual appearance of the inside of the can. After the can is fabricated, some facilities spray the interior with additional coating to cover any defects in the roller-applied coating. End seal compounds and interior side seam striping) are also interior coatings. Exterior coatings are applied for decoration, to protect the can from corrosion, to protect the printed designs from marring or abrasion, or to reduce friction on the bottom of the can to facilitate handling. Typical exterior coating operations are base coating, size coating, decorative ink and overvarnish application, bottom coating, side seam stripe application, and repair coating. Exterior coatings are usually applied with direct-roll coaters except for side seam striping, which is performed with a spray; and repair coatings, which are sprayed. Steel sheets are fed to a conveyor by a sheet feeder.

The sheets are transferred to the coater, where the coating is transferred from a tray to the roll coater by a series of rollers, then is applied to the sheets. After passing horizontally through a short flashoff area, the sheets are picked up by wickets and conveyed through a wicket oven. The curing oven operates at temperatures of up to 425°F. Typically, multiple heating zones are required to achieve the temperature profile that results in proper curing of coatings. The heating zone is followed by a cooling zone that cools the sheet using ambient air from inside or outside the plant. Line speeds range from 60 to 110 sheets per 9 minute depending on the design and age of equipment and the type of coating used. The 60 sheet-per-minute line speed is for lithography printing and varnish, which run more slowly than other sheet coating operations. Oven exhaust rates usually vary between 2,000 and 14,000 standard cubic feet per minute. Most roller-applied coatings (except for printing inks and overvarnishes) can be applied using the same coating equipment, and many facilities use the same equipment to apply a variety of coatings to can bodies, ends, crowns and closures, and decorative tins. Decorations on 3-piece cans may be printed on the can body or on paper labels that are then glued onto the can. Decorative inks are usually applied to an exterior base coat but may be applied directly to the metal. The transfer of inks is influenced by environmental factors such as temperature, draft, and humidity because the inks can become emulsified in the presence of water. An overvarnish is applied on top of the decoration by a direct roll coater while the inks are still wet. The inks and overvarnish are cured in a wicket oven similar to, but usually smaller than, the base coat oven. Exhaust rates range from 1,500 to 8,000 scfm. If the required design has more than two colors, the first set of inks is dried in an oven. Another set of inks is then applied followed by an overvarnish and baking in an oven. At least 100 existing 3-piece printing lines are known to use ultraviolet-radiation-cured (UV-cured) printing inks and more than 30 lines also use UV overvarnishes. These coatings are applied in the same manner as solvent- or waterborne coatings, but are cured by exposure to ultraviolet radiation rather than heat. Consequently, these coatings do not need to pass through a drying oven.

The components of the three-piece can are cut from steel after coating (if used) and the sheets are transported to the fabrication process. The sheets are unloaded from a stacker to a conveyor and transported to the slitter, which cuts the sheet into body blanks. The body blanks enter the body maker where each blank is rolled into a cylinder and the seam is welded or cemented, then sprayed with a protective lacquer (liquid or powder) coating called a "side seam stripe" to protect exposed metal along the seam. A bottom (either a classic or easy open end) is then attached. The side seam stripe is cured in an electric or gas-fired oven, or by exposure to a direct-flame burner. The cylinders are flanged in preparation for the attachment of ends, and are sometimes necked down to reduce the size of the ends, which reduces the amount of material used to make the ends.

After filling, a lid (classic end) is attached to close the can. Both the bottom end and the lid have a gasket applied to ensure a hermetic seal. Earlier technology soldered instead of welded the side seams of the can bodies, which resulted in lead contamination of the contents. This practice started to cease in Europe in the early 1980s and has been replaced by welding all over the world today. Soldered cans are not traded internationally anymore. The use of a pure tin solder is retained for a few very specific applications. In some types of three-piece cans, the side seam is crimped rather than welded or soldered. Non-welded seams are either folded or interlocked folded. A gasket could be used for sealing. This does not apply for the standard food can that is sterilised. It should be noted that some three-piece food can bodies, particularly those with ETP bodies, are not internally lacquered, but their TFS ends are. However, three-piece cans to be filled with dry food, *e.g.*, milk powder, instant baby food, sweets or roasted nuts, are usually not coated at all on the internal surface. Dry foodstuffs such as biscuits or tea are available in tin boxes. These are in essence a cylindrical or rectangular three-piece can. The lid is not seamed on but is formed as a "slip on" or "plug" lid. The food is not processed in such boxes. The finished cans are checked for leaks, and then are stacked on pallets for storage. Line speeds for three-piece can manufacturing range from 350 to 800 cans per minute.

Two-piece Can Bodies

Manufacture of the two-piece can in which the body and the bottom end are formed from a drawing on a flat plate is an old technology used in the production of formats with low height/diameter ratios (<0.6). Two-piece cans are made from a disc of metal that is reformed into a cylinder with an integral end to become a seamless container. To this is seamed a loose end to finally close the can. Two-piece cans are manufactured by new technologies *i.e.* either by the draw-redraw process (DRD) or the draw-and- wall ironing process (DWI). After the fabrication process, various coatings are applied and cured.

Drawn and re-drawn containers are often referred to as DRD cans. Drawing is the operation of reforming sheet metal without changing its thickness. Re-drawing is the operation of reforming a two-piece can into one of smaller diameter, and therefore greater height, also without changing its thickness. Both steel and aluminium substrates can be used. The metal substrate is usually coated on both sides. A punch is used to deform a disc of metal. The discs are "cupped" to form the shape to a short metal beaker, normally by the use of a high-pressure press. This is the first stage of the deformation to make a can. These cans are differentiated by the method used to form them. For shallow cans, a single drawing operation is required. Typical shallow drawn cans are club, sardine, tuna and ready-meal cans. The cans of processed ham are at the limit of a single draw can. For taller cans, such as large salmon cans, two or three operations *i.e.*, draw and redraw (DRD) or multiple redraw are used. Some drawn cans have a larger diameter at the top than at the bottom. The tapered shape enables them to slide into each other for transport, thereby minimizing the costs of transporting large quantities of air for large distances. A typical example is the large North American salmon can. DRD cans maintain the original thickness of the flat metal sheet after the drawing process, and therefore provide enough mechanical resistance against deformations during sterilization. Most of the drawn cans find usage in the beverage as well as the food industry. Drawn cans are also widely used for packing sweets, and these are usually closed with a slip lid.

Drawn and Wall-Ironed (DWI) Cans

Ironing is the operation of thinning the wall of a two-piece can by passing it through hardened circular dies. They are made from either aluminium or steel. To produce a DWI can, a drawn cup is further processed to make a cylinder by deforming the sides of the cup by stretching it through dyes of decreasing diameter ("wall ironing"). The wall thickness of the can is reduced by "ironing" the metal and consequently lengthening the can. The process of making a DWI can is similar for aluminium and steel. However, a major difference comes at the stage when the cans are necked (the diameter is reduced at the top of the can). Steel beverage cans utilize more coating and processing options than aluminium ones. The cans are washed (four to six times) in a multi-stage washer. At this stage, various treatment chemicals can be used in the washers; these may differ for aluminium and steel cans. The washing serves to remove any lubricant from the wall ironing process. Aluminium cans may also undergo a pre-treatment during the washing process to improve the adhesion of the coatings. The last wash is normally with deionised water to create a contaminant-free surface for coating. After drying, the cans are externally decorated. The final stage is the application of the internal lacquer, which is spray-applied and cured in an oven.

The draw and wall ironing process (DWI) is very economical for making cans where the height is greater than the diameter and is particularly suited for making large numbers of cans of the same basic specification. DWI cans are lighter (by 40 to 50 per cent), since the body wall thickness is significantly reduced to 0.09 mm. Thus, the DWI process implies a greater metal reduction compared to DRD technology. Due to their specific characteristics, DWI cans are appropriate for pressurized liquid products (carbonated beverages, beers, *etc.*), as the internal pressure compensates for the low mechanical resistance of very thin walls. Soft drinks are normally cold-filled and carbonated thereby generating a positive pressure, which serves the multiple purpose of displacing headspace oxygen, producing the characteristic "fizz" and, importantly, providing abuse resistance and strength to the thin-walled DWI can. Post-processing is minimal. Beers are normally pasteurized after filling. This involves heating to about 65°C for 20–30 min and then cooling. They can also be used for non-carbonated beverages by introducing liquid nitrogen into the can to generate internal pressure.

Drawn and Ironed Cans (DI)

This is a variation on DWI beer and beverage cans, but these cans are for food, either human or pet, and are not normally externally decorated.

9.6 The Profile of Can End

The lids are always attached after filling the can with foodstuffs, thus the packers and fillers purchase empty cans and lids and seam the lids onto the can. The ends on a can are seamed onto the body to give a hermetic seal. The lid is seamed on when the can is filled. The exceptions are open top cans (for example sweet or biscuit tins. Many of the dimensions are critical to ensure that any potential microbiological ingress is minimised (can integrity).

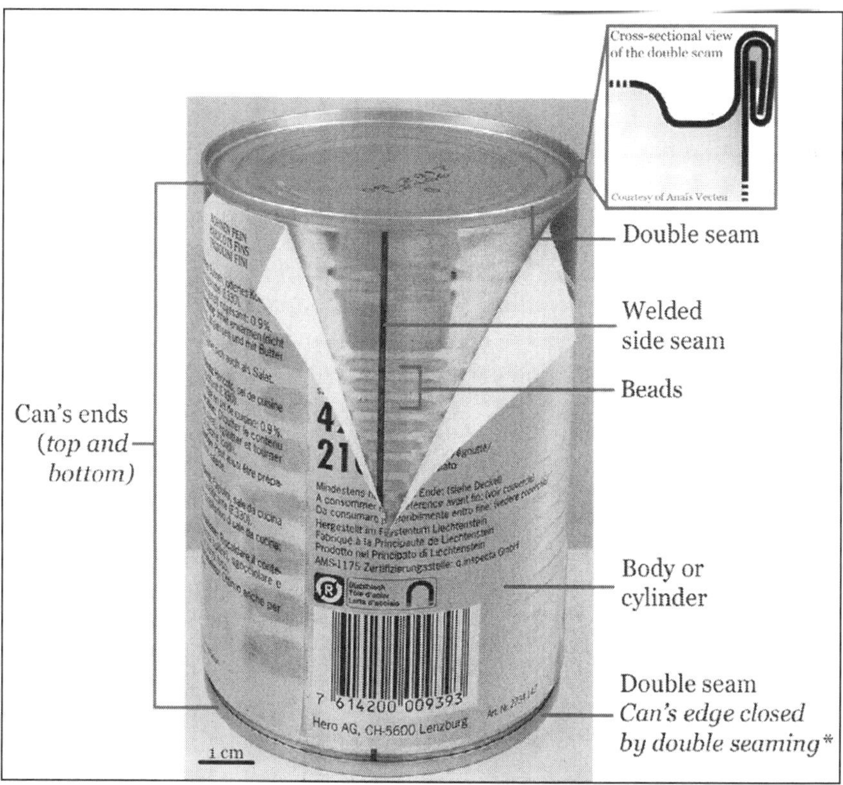

The can end is of complex design developed for optimum deformation behaviour, the latter being dependent on plate thickness, the precise contour of the expansion rings and the countersink depth. It is important that the ends are able to deform under internal and external pressure without becoming permanently distorted. In effect they must act like diaphragms, expanding during thermal processing and returning to a concave profile when vacuum develops inside the can on cooling.

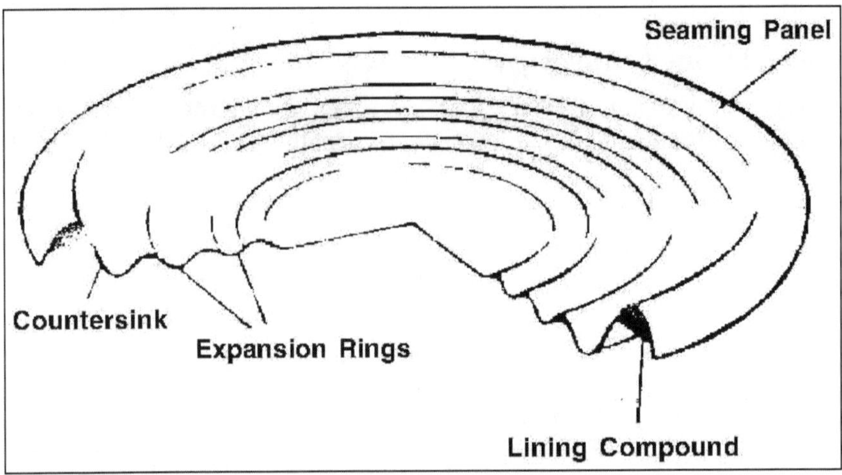

Figure 9.1: Profile of Can End.

The ends are stamped on power presses from tinplate sheet, generally of high temper grade, which has usually been previously lacquered. After stamping the ends fall through the press into the curler to form the outside curl and diameter.

A lining or sealing compound is then applied into the seaming panel; the sealant used is based on natural or synthetic rubber and is dispersed in water or solvent. Its constituents are subject to stringent food regulations. The purpose of the sealant is to assist the formation of a hermetic seal by providing a gasket between adjacent layers of metal.

Several types of easy-opening devices such as the key opening scored strip found in solid meat or shallow fish cans have been available for many years.

Easy-open ends (EOEs) are becoming ever more popular. Not only do they avoid the inconvenience of having to use a can opener but also, due to their design, they are much safer with fewer lacerations caused by the edges of the jagged end produced by a can opener. An increased demand for convenience features has seen the development of easy-open ends of two broad types: those which provide a pouring aperture for dispensing liquid products, and those which give a near full aperture opening for removing more solid products. The first easy-open end for canned beverages was developed in 1963.

The full aperture easy-open end (FAEOE) is gaining widespread acceptance. In this type of end, all of the end of the can lid is effectively removed during the opening process, unlike a beverage end where only part is opened and stays on the end or is separated from the end, to enable the contents to be poured from the can. Ends for food cans can be made from aluminium, ETP or TFS. Whilst designs and metals may differ, the principles of manufacture are similar: The metal is coated. A shell is punched and then the EOE shape is fabricated, including the score line that weakens the metal sufficiently to allow its fracture and easy opening. A tab is then attached to the end by a rivet. This rivet is drawn from the can end (to ensure microbiological integrity), a most demanding deformation process, to ensure the integrity of the end. A repair coat can be used to repair the damage caused by scoring and forming the rivet. Easy-peel ends are also beginning to affect the market for easily removable ends.

Most designs incorporate a ring-pull end consisting of a scored portion in the end panel and a levering tab (formed separately) which is riveted onto a bubble-like structure fabricated during pressing. Modern designs (especially for beverage cans) have attempted to overcome litter problems by incorporating nondetachable tabs. Most but not all of the aperture circumference is scored to leave sufficient unscored portion to function as a hinge when the tab is pressed in.

Because of the greater ease of fabrication, integrated rivet ends have usually been made from aluminum but this presents problems when such ends are used with steel cans. For example, the corrosion of carbonated soft drinks in such cans is accelerated because of the bimetallic container, although the use of the same container for beer is beneficial because of the sacrificial nature of the aluminum and the more tolerable aluminum-beer reaction products than the iron counterpart. These problems have led to the development of steel easy-opening units, but they are much more difficult to fabricate.

Close control of scoring conditions is vital to ensure adequate resistance to bursting without requiring an unduly high tearing load to open. Particular attention must be paid to metal exposure resulting from internal enamel (lacquer) fracture at the score, and post-lacquer spraying is sometimes used to ovecome potential problems.

9.7 Double Seaming and Welded Side Seams

Welding has been used in the manufacture of steel pails for many years, but because of inconsistencies in weld integrity, the process was considered unacceptable for the production of high quality, leak-free food cans. Two processes were developed in about 1960 for welding side seams instead of soldering: the wire weld system developed by Soudronic A.G. of Switzerland, and the Conoweld roll-welding system developed by Continental Can Company in the United States. Both have since been improved and gamed widespread commercial acceptance.

Welding as an alternative to soldering spread widely in the 1980s due largely to the imposition by more and more countries of lower limits of lead in canned foods. Since the 1970s most countries have insisted that only pure tin solder be used on cans intended for baby foods; this added significantly to the cost of such cans and is one of the reasons that glass containers for baby food are so popular. However, in addition to the health aspects, welding offers savings in material, since the overlap needed to produce a weld uses less metal than an interlocked soldered seam. As well, the side seam is stronger, it is easier to seam on the ends, and a greater surface area is available for external decorating.

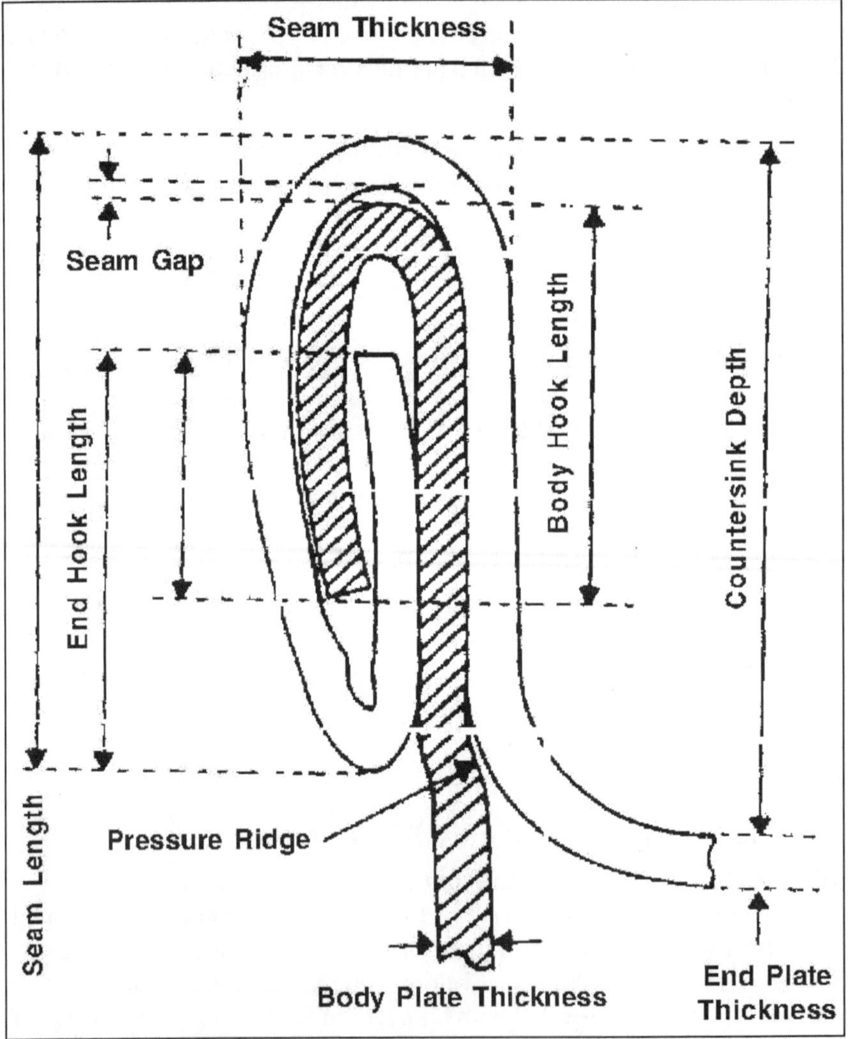

Figure 9.2: Main Components of a Double Seam.

The Conoweld technique was developed initially for the production of beverage cans made from ECCS; it has since been expanded to tinplate aerosol containers and some food cans. To ensure satisfactory welds, thorough removal of the chrome/chrome oxide coating to a width of about 2 mm along each of the blank edges which will form the lap seam is required just prior to welding. A square wave form of alternating current is used to give solid-state welds which are forged together by a wheel.

The Soudronic Wima and Superwima (Wima is an abbreviation for wire mash) processes were developed in Switzerland for the high speed welding of tinplate and ECCS containers using a sine wave alternating current power supply and a continuous copper wire electrode to produce a weld with an extremely low metal overlap (less than 0.4 mm in the case of the Superwima seam). High electrical resistance causes the interface temperature to rise rapidly to at least 900°C and melt, resulting in solid phase bonding at all locations along the seam. The tensile strength of a good weld is equal to that of the base plate.

Although the welded seam is free from the danger of lead pick-up, the weld has to be effectively coated to prevent traces of iron being picked up by some types of beverages and acidic foods; side striping of the internal surface of the seam is required in these situations.

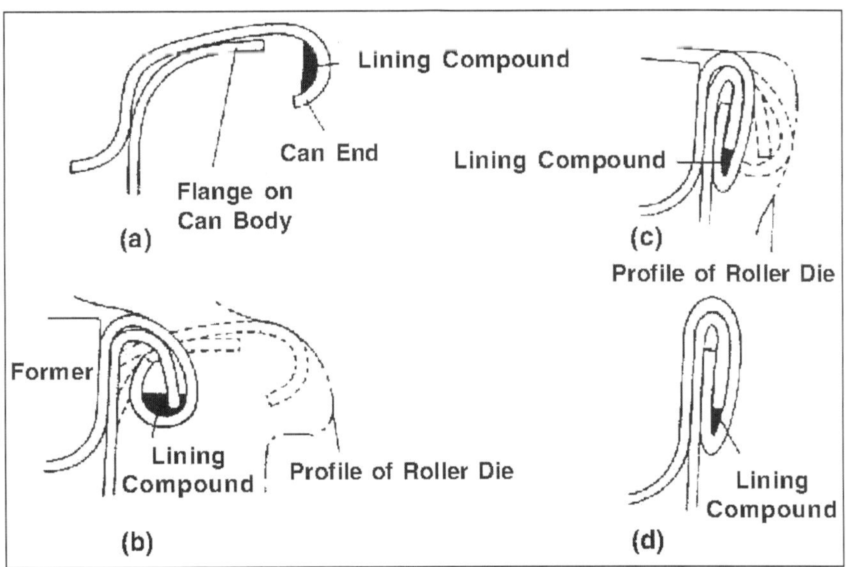

Figure 9.3: Double Seaming of Metal Ends on to Metal Containers.

(a) End and body are brought together; (b) First seaming operation; (c) Second seaming operation; (d) Section through final seam.

A system of chemical bonding of side seams has been developed by the American Can Company for ECCS can bodies. Known as the Miraseam process, it utilizes a thermoplastic polyamide adhesive which is applied to one edge of the pre-heated body blank before it is rolled into a cylinder, giving complete protection of the raw edges of the blank. A strong bonded lap seam is produced that is able to withstand the high in-can pressures generated by beers and carbonated soft drinks during can warming or pasteurisation. This method can only be used with ECCS since the melting point of tin is close to the fusion temperature of the plastic.

The bodies are next transferred to a flanger for the final metal-forming operation: necking and flanging for beverage cans, and beading and flanging for food cans. The can rim is flanged outwards to enable ends to be seamed on. The top of beverage cans is necked to reduce the overall diameter across the seamed end to below that of the can body wall, yielding savings in the cost of metal through the use of smaller ends, and allowing more effective packing methods to be adopted. Simultaneous creation of the neck and flange using a spin process is now increasingly used. Double-, triple- and quadruple-necking is now quite common, the latter reducing the end diameter by over 8 mm (from 68.26 mm to 60.33 mm).

For food products where the can may be subjected to external pressure during retorting or remain under high internal vacuum during storage, the surface of the body may be beaded for strength. There are many bead designs and arrangements, all of which are attempts to meet certain performance criteria. Basically, beading produces shorter can segments that are more resistant to paneling, but such beads reduce the axial load resistance by acting as failure rings.

The end is then mechanically joined to the cylinder by a double seaming operation. This involves mechanically interlocking the two flanges or hooks of the body cylinder and end. It is carried out in two stages. In the first operation, the end curl is gradually rolled inwards radially so that its flange is well tucked up underneath the body hook, the final contour being governed by the shape of the seaming roll. In the second operation, the seam is tightened (closed up) by a shallower seaming roll. The final quality of the double seam is defined by its length, thickness and the extent of the overlap of the end hook with the body hook. Rigid standards are laid down for an acceptable degree of overlap and seam tightness.

Finally the cans are tested for leakage under pressure in large wheel-type testers; leaking cans are automatically rejected.

9.8 Drums and Pails

Drums and pails are in essence large three-piece steel cans. They are supplied empty, with bungs in the "lid" for the customer to fill. They are not subjected to any processing when filled. Drums tend to refer to larger volume containers of typically 100–220 l whilst pails normally refer to 5–25 l containers. There are different grades of drums depending upon the intended contents and method of transportation.

9.10 Tubes

Metallic tubes are extruded from a slug of metal (mostly aluminium). Not all tubes are internally lacquered, particularly toothpaste tubes. Many tubes are no longer based on metal. Only those tubes with contents necessitating minimal interaction with oxygen are metal based.

9.11 Trays and Foils

Rigid and semi-rigid aluminium trays for food application are based on rolled aluminium (with different alloys) of a thickness in the range 70–300 μm. In some cases, containers with polyolefin laminate structures together with polyurethane adhesives are used for the food contact side of the tray to provide retort resistance.

9.12 Aerosols and Aerosol Products

The word aerosol is used in colloid chemistry to describe a suspension in air or gas of small particles with a radius of less than 50 μm.

However, in the present context, an aerosol is defined as a self-contained sprayable product in which the expelling force is supplied by liquefied or compressed gases so that the product is self-dispensing.

The present aerosol industry is generally considered to have received its stimulus from the development of the aerosol insecticides used in the 2_{nd} World War, although a few aerosols had been marketed prior to this time. Considerable information has been published about aerosols and patents granted over the period 1862 to 1942, and the reader is referred to standard texts for a detailed account of their development.

The history of aerosol food products can be traced back to 1925 when a method for fluffing and dispensing ice cream under pressure was patented by Ashley. However, long before that date, a patent was granted to Weyde and Matthews covering the use of nitrous oxide in artificially aerated drinks, and in 1882 the use of compressed gases for charging of milk products was patented. It was not until the mid-1950s that a range of food aerosol products appeared, including a drink mix, barbecue sauce, chocolate and table syrups, a vitamin syrup, a meat tenderizer and a salad dressing. Unfortunately, several of these products were disappointing and development work on new aerosol products dwindled almost to a stop.

In terms of the total worldwide aerosol business, the food aerosol market is very small. However, it is an area that does have significant growth potential. The major applications by volume at the present time appear to be whipped cream, cheese spreads and pan release sprays (*i.e.*, edible cooking lubricants).

Most aerosol products contain three major components: propellants, solvents and active ingredients. Aerosols can be divided into two broad classes (homogeneous or heterogeneous) depending on the manner in which these components are combined in the product. In homogeneous products, all the components are mutually soluble and two phases are present; a liquid phase in equilibrium with a vapor phase. Such products do not have to be shaken before use. Heterogeneous products are where the components are not mutually soluble and several types exist. Suspension systems consist of a solid phase and liquid phase in equilibrium with a vapor phase, and emulsion systems contain two liquid phases in equilibrium with a vapor phase. Virtually all food aerosols are heterogeneous, emulsion systems.

There are two main types of food aerosol systems on the market today: those where the propellant gas is mixed directly with the food system, and those where the propellant system is kept separate from the food product by means of an inert plastic bag or a moving piston device.

Propellant

Propellants provide the pressure that forces the product out of the container when the valve is opened. They also influence the form in which the product is discharged: foam, stream or spray. Variations in the type and concentration of a propellant can change a coarse, wet spray into a fine, dry spray, or a wet, sloppy foam into a dry, coherent foam.

There are three broad classes of propellants: hydrocarbons, fluorocarbons and compressed gases. For use in food aerosols, a propellant should be inert, non-toxic, non-flammable, odorless, tasteless and cheap. Several propellants are available for use in foods, including nitrous oxide (N_2O), carbon dioxide, nitrogen, and two fluorocarbons: Freon C-318 (octafluorocyclobutane) and Freon 115 (chloropentafluoroethane); these are the only two fluorocarbons that have been approved by the FDA for food use.

Freon 115 is very stable and has a high vapor pressure; it is used as a compressed gas or as a liquefied gas in combination with C-318 which acts as a vapor pressure depressant. Generally, mixtures containing about 75/25 wt per cent of nitrous oxide-fluorocarbon have been found most satisfactory, greater foam stability resulting as higher percentages of fluorocarbon are included. However, actual commercial use of fluorocarbons in food aerosols is negligible; this is partly due to their much higher cost compared to the compressed gases, and partly to the role of chlorofluorocarbons in the destruction of the ozone layer. The pressure of the compressed gases changes relatively little with temperature in comparison to the liquefied gases.

For food aerosols, only the compressed gases are employed in any quantity, with nitrous oxide, either by itself or in combination with carbon dioxide, being the most common, the sweet taste of nitrous oxide offsetting the acidic taste of the carbon dioxide. Nitrogen is seldom used, for although it is inexpensive, stable and low in toxicity, it is almost completely insoluble in the liquid phase of aerosols, thus presenting a number of problems. Because it is practically insoluble in liquids, it concentrates in the vapor phase. As the product is discharged, the volume of the vapor phase increases with a consequent decrease in the pressure exerted by the nitrogen and a noticeable change in spray characteristics. Also, because nearly all of the nitrogen is in the vapor phase, the initial volume of this phase has to be considerably larger than is required for conventional liquefied gas propellants in order to have sufficient nitrogen to discharge the product completely. The same disadvantages apply to a lesser extent to carbon dioxide and nitrous oxide. Of course, an advantage of nitrogen is that because it is insoluble in the liquid phase, it does not alter the characteristics of the product. It is considered a good propellant for such products as pastes and syrups as it produces a solid stream.

Since the other compressed gases are soluble to only a limited extent in the liquid phase of aerosols, coarse sprays are produced and mechanical breakup actuators are necessary. The spray characteristics also change noticeably as the pressure drops during use, partly because the gas dissolved in the liquid phase is not released fast enough to maintain equilibrium between the liquid and vapor phases.

Misuse by the consumer of compressed gas aerosols by discharging the container when the end of the dip tube is in the vapor phase has been a major problem. If the valve is actuated when the container is in the inverted (and on occasions horizontal) position, the vapor phase is discharged instead of the liquid phase. This results in loss of the propellant, since most of the propellant is concentrated in the vapor phase, and the product cannot be obtained from the container. This problem is not as great with nitrous oxide and carbon dioxide as it is with nitrogen, because their increased solubility in the liquid phase provides a reservoir of propellant.

A modification of this system can produce solid-stream dispensing; a soluble compressed gas is used and the aerosol (without a dip tube) is used in the inverted position. When the valve is depressed, the material is emitted as a stream. This system was developed to be used with pressurized beverage products.

When selecting a propellant, the nature of the dispensed product must be considered since this is influenced by the type of propellants. The more soluble propellants (nitrous oxide and carbon dioxide) tend to form a product that foams, since when the product is dispensed, the dissolved gas expands and escapes into the atmosphere, resulting in dispersion of the liquid into fine particles. Nitrogen delivers the product in a nonaerated form.

For many food products, use of the typical aerosol systems may be unsuitable because of the viscosity of the product, incompatibility of the product and the propellant, or the desired dispensing characteristics of the finished product. In these situations, barrier containers have been designed. They have a specialized construction designed to keep the product completely separated from the propellant at all times from the initial filling to complete discharge. Two different types of containers are used for this purpose: one is a bag-in-can container and the other uses a plastic piston.

The main advantages of barrier packages are:

1. Products can be packaged without alteration of their composition or reformulation;
2. Very viscous products can be packaged;
3. The ratio of product to propellant is high (a consumer advantage);
4. Sterility is easier to achieve with barrier containers than with conventional aerosol systems;
5. The package can be discharged in any position.

The disadvantages of barrier packages is the higher cost, and they are not considered to be competitive with conventional aerosols.

Although all aerosol propellants are gases at room temperature and pressure, only the compressed gases are loaded into aerosol containers as gases. All other propellants are liquefied prior to loading, either by cooling them below their boiling point or increasing their pressure at room temperature. In the cold fill process, the cold active ingredient and cold propellant are both loaded into the container before the valve is added and crimped on. In the pressure fill process, the active ingredient is added to the container at room temperature and the valve crimped on; the propellant is then loaded through the valve under pressure.

When the compressed gas is added through the valve, the active ingredients are loaded in the container first, air is removed either by evacuation or purging with nitrogen, and the valve is added and crimped on. If carbon dioxide and nitrous oxide are employed as propellants, it is necessary to add a shaking step during the loading of the gas in order to establish equilibrium as rapidly as possible between the gas dissolved in the liquid phase and that in the vapor phase. Since nitrogen is almost insoluble in the active ingredients, it is forced through the valve without the need for shaking.

The most common material used for aerosol containers is tinplate, but ECCS, aluminum, glass, stainless steel and plastics are also used to a lesser extent. Tinplate three-piece aerosol containers were traditionally soldered at the side seam but were one of the first groups of soldered metal containers to benefit from the conversion to welding. For food use, protective coatings as described previously for metal containers must be used. Aerosol cans are either three-piece or two-piece. In essence, the principles of two- or three-piece can manufacture apply, whilst the ends differ and are fitted as a unit. All three-piece aerosols are made from steel. The main difference between aerosol and can manufacturing is with the two-piece aerosol, known as a monobloc. The aerosol container is manufactured of aluminium by impact extrusion. The diameter of the open end is reduced (swaging) to receive the spray nozzle and an internal lacquer is spray applied. Typical rates of production are about one fifth of those of a DWI B and B can process, partly due to the parameters surrounding the application of the internal spray.

9.13 Coatings for Protection and Decorations

The primary function of interior can coatings is to prevent interaction between the can and its contents, although some enamels have special properties which allow products such as meat loaf to be easily removed from the cans, while others are used merely to improve the appearance of the pack. Exterior can coatings may be used to provide protection against the environment (*e.g.*, when the cans will be marketed in particularly humid or salt-laden climates), or as decoration to give product identity as well as protection.

There are several essential requirements needed in an interior coating: it has to act as an inert barrier, separating the container from its contents and not imparting any flavor to the contents; it must resist physical deformation during fabrication of the container and still provide the required chemical resistance; and the enamel must be flexible, spread evenly, completely cover the substrate and adhere to the metal surface. Adhesion failure may occur during, or as a result of, mechanical deformation during heat processing or undermining by corrosion.

For most containers the enamel is applied to the metal in the flat before fabrication, typical film masses being in the range 3–9 gsm (4–1.2 micron). However, because of the considerable amount of metal deformation with substantial disruption of the surface which takes place in the D and I operation, such containers must be coated internally after fabrication. Control of the dry film weight is essential if the enamel is to function correctly. A dry film which is too thin may not cover the surface completely, while an overthick film leads to brittleness and impaired protection, as well as being uneconomical. As the enamel film has imperfections and is damaged during container manufacture, it does not give complete protection to the can. Where it is essential to minimize product-container interactions, *e.g.*, for canned beer and soft drinks where metal pick-up can affect flavor and clarity, the cans are given a post-fabrication repair lacquering.

Many types of internal enamel are available for food containers including oleoresinous, vinyl, acrylic, phenolic and epoxy-phenolic. The original can lacquers were based on oleoresinous products which includes all those coating materials which are made by fusing natural gums and resins and blending them with drying oils such as linseed or tung. Although oleoresinous coatings are still used today, their open micellar structure means that they are prone to corrosion/staining problems with sulfur bearing products unless they are pigmented with zinc oxide. It was for this reason that a move has been made to synthetic phenolic resins dissolved in a blend of solvents.

Phenolic resins produced by the action of formaldehyde on phenol or other substituted phenols, have limited flexibility and high bake requirements, but are still used on three piece cans where flexibility is not required.

Vinyl coatings are based on copolymers of vinyl chloride and vinyl acetate of low molecular weight dissolved in strong ketonic solvents. The long carbon-carbon chains make them thermoplastic and they can be blended with alkyd, expoxy and phenolic resins to enhance their performance. Their flexibility allows them to be used for caps and closures as well as drawn cans. Their main disadvantage is their high sensitivity to heat and retorting processes, restricting their application to cans which are hot filled, and to beer and beverage products.

Epoxy phenolic coatings are made either by straight blending of a solid epoxy resin with a phenolic resin or are the products of the precondensation of a mixture of two resins in appropriate solvents. A three dimensional structure is formed during curing and baking which combines the good adhesion properties of the epoxy resin with the high chemical resistance properties of the phenolic resin. The balanced properties of epoxy phenolic coatings have made them almost universal in their application on food cans with the exception of deep multistage DRD cans.

Vinyl organosol coatings incorporate a high molecular weight PVC organosol dispersion resin which is thermoplastic in nature and extremely flexible. Soluble thermosetting resins (including epoxy, phenolic and polyesters) are added in order to enhance the film's product resistance and adhesion. Plasticizers are also added to aid film formation. These coatings are the most common on DRD cans in the USA and are typically white or buff colored due to the addition of titanium dioxide.

The types of resins most commonly used for internal coating of containers, their properties and general applications are summarized in Table 9.2. Generally the lacquers may be divided into three broad classes:

GP (General Purpose)

These are based on epoxy phenolic resins and are often used for the more acidic products. General purpose enamels are applied as either a single coat (they are then designated GP1), or as two coats (designated GP2). GP2 systems are used for products such as acidified beetroot and colored berry fruits which are especially corrosives. GP enamels also have some sulfur-resisting properties. GP enamels may be pigmented or have a clear appearance.

SR (Sulfur Resistant)

These enamels are used to prevent staining of tinplate surfaces by sulfur compounds released from foods such as meat, fish and vegetables which have sulfur-containing amino acids that breakdown during heat processing and storage to release sulfides. These react with tin to form black tin sulfide, or accumulate in the headspace and give out an unpleasant odor. To overcome this problem, two approaches have been used. Enamels are pigmented with zinc oxide which reacts with the sulfur compounds to form white zinc oxide (these are known as the sulfur-absorbing enamels), or the enamels are pigmented with aluminium powder or white pigment to obscure any tin sulfide which might form. These are known as sulfur-resisting enamels.

Special Enamels

Enamels having additives such as waxes to assist the release of the product from the can, or enamels pigmented with aluminum powder or other materials are also used. The latter were described above as sulfur-resisting enamels, but they are also used in premium quality packs where sulfur staining is not a problem, simply to improve the appearance of the inside can surface.

Table 9.3: Main Types of Internal Can Enamel (Lacquer)

General Type of Resin and Components Blended to Produce it	*Flexibility*	*Sulfide Stain Resistance*	*Typical Uses*	*Comments*
Oleo-resinous (drying oil and or synthetic resins	Good	Poor	Acid fruits	Good general purpose natural range at relatively low cost
Sulfur-resistant oleo-resinous (added zinc oxide)	Good	Good	Vegetables, soups, especially can or as topcoat over epoxy-phenolic	Not for use with acid products
Phenolic (phenol or substituted phenol with formaldehyde)	Moderate	Very good	Meat, fish, soups, vegetables	Good at relatively low cost but film thickness restricted by flexibility
Epoxy-phenolic (epoxy resins with phenolic resins)	Good	Poor	Meat, fish, soups, vegetables, beer and beverages (top coat)	Wide range of properties may be obtained by modifications
Epoxy-phenolic with zinc oxide (zinc oxide added)	Good	Good	Vegetables, soups (especially can ends	Not for use with acid products; possible color change with green vegetables
Aluminized epoxy-phenolic (metallic aluminium powder added)	Good	Very good	Meat products	Clean but rather dull appearance
Vinyl solution (vinyl chloride-vinyl acetate copolymers)	Excellent	Not applicable	Spray on can bodies, roller coating on ends, as topcoat for beer and beverages	Free from flavor taints; sensitive to soldering heat and not usually suitable for direct application to tinplate

General Type of Resin and Components Blended to Produce it	*Flexibility*	*Sulfide Stain Resistance*	*Typical Uses*	*Comments*
Vinyl organosol or plastisol (high MW vinyl resins suspended in a solvent)	Good	Not applicable	Beer and beverage topcoat on ends, bottle closures, drawn cans	As for vinyl solutions but giving a thicker, tougher layer
Acrylic (acrylic resin, usually pigmented white)	Very good in some ranges	Very good when pigmented	Vegetables, soups, prepared foods containing sulfide stainers	Attractive clean appearance of opened cans
Polybutadiene (hydrocarbon resins)	Moderate–poor	Very good if zinc	Beer and beverage first coat, vegetables and soups if with ZnO	Costs and hence popularity depends on country

Two basic methods are used for the application of protective coatings to metal containers: roller coating and spraying, the former being the most widespread. Dipping was used in the early days but is now obsolete. Roller coating is used if physical contact is possible; thus, it finds use in the coating of material in sheet and coil form, and the external coating of cylindrical can bodies. Spraying techniques are used if physical contact is impossible or difficult; thus, it finds use mainly to coat the inside surface of can bodies, including two-piece D and I and sometimes DRD cans.

Because the coating is generally applied wet (*i.e.*, the resin is suspended in a carrier such as an organic or aqueous solvent for ease of application), it must be dried after application by solvent removal, oxidation and/or heat polymerization. This process (known as baking or curing) is usually carried out in a forced convection oven using hot air at up to 210°C for times of up to 15 minutes. However, recent developments include resin formulations which need lower temperatures and shorter curing times, often through the use of ultra-violet radiation to accelerate polymerization. Such UV-cured resins are virtually solvent-free; they contain photosensitive molecules which absorb the UV radiation and release free radicals which polymerize or cross-link the liquid resin to form a solid coating. Because the substrate is not heated, it can be handled immediately for further operations, and substantial savings (up to 50 per cent) in energy and space are possible.

Another recent development is that of power coating, where the resin is applied "dry" in the form of a fine powder, usually under the direction of an electrostatic field. It is used mainly where heavy coatings are required, such as in the protection of welded side seams where the bare metal that exists in the weld area (most of the tin is removed during the welding process) must be covered. Curing is usually by infrared radiation or high frequency induction heating, since hot air could disturb the uncured coating. Powdered coatings contain little or no volatile effluents and require low energy consumption for application and cure.

Electrophoretic deposition, a process originally developed for protecting automobile bodies, has been developed for coating two-piece cans. A resin film is deposited electrically from an aqueous suspension, giving a far more even distribution than that obtained by spraying. As well, its throwing power enables it to coat regions inaccessible to spray. The process overcomes the current wasteful spray techniques with almost 99 per cent coating utilization achieved. During the electrocoating operation the aluminum surface is passivated which helps adhesion and improves long term quality of the package.

Although the primary purpose in decorating the external surface of a metal container is to improve its appearance and assist its marketability, it also significantly improves the container's external corrosion resistance. Decoration of the external surface is similar in many respects to the process used to protect the internal surface, the constituents generally being dispersed in volatile solvents, applied on roller coating machines (apart from the printed image) and baked in tunnel ovens. When the container fabrication process involves severe deformation of the metal, the system used is likely to consist of a size (mainly epoxy or vinyl types), a base coat, a printed image containing several colors and a final varnish film to protect inks and base coatings from damage during fabrication and later use, and to and gloss to the print image. However, if high quality inks are used, a varnish coating may not be required.

Offset lithography has been used for over a century for decorating sheet metal. Since metal is non-absorbent, the coatings and prints cure by internal chemical reactions involving oxidation, polymerization or both. By adjusting the inks' tack properties, it is possible to print one wet ink on to the previous wet ink ("wet on wet") without the second blanket picking off the first layer. Methods used to "set" the ink in between single color presses in tandem include high-temperature air blasts, flame treatment and, more recently, the application of UV radiation to the print. The latter approach has necessitated a new approach to ink formulation. The print process is usually carried out with sheet stock prior to slitting into can body blanks or scroll shearing into end stock. With the advent of two-piece cans, less elaborate designs are used on beer and beverage cans because of the necessity to use presses that can print completely fabricated cans.

In recent years, the can manufacturing industry has been continuously changing. Besides the introduction of new materials and manufacturing processes, there has been a revolution in graphic and structural design. The traditional can, a common shape with few changes over time, has been limited to the application of easy-open closures and diverse shapes and sizes. Due to competition with other packaging materials, today's trend is toward new designs far from the traditional image. Quality and long shelf life are not enough to attract the consumer's attention. Design of cans has become a fundamental aspect due to the increasing pressure of aesthetics and innovation. Great efforts have been made to maintain can functionality while taking into account the new trend toward personalized designs. In line with this, the newest technologies allow for the development of cans with expanded bodies (*e.g.*, square body with circular ends). Also, the image of the traditional can is being replaced by designs in which the ends have a more reduced diameter than the body. In addition to this new look, innovation has resulted in a significant reduction of material and a more stable can piling. New easy-open closure systems also contribute to the new image of tinplate cans.

Chapter 10

The Use of Glass in Food Packaging

10.1 History

Glass is by definition an amorphous solid material madc by fusing silica with a basic oxide. Glass is called amorphous because it is neither a solid nor a liquid but exists in a vitreous, or glassy, state. It is also defined as an inorganic product that has been cooled to a rigid state, without experiencing crystallization. It is one of the oldest materials and closely intertwined with cultural history and cultural skills. It is used in art and architecture and to preserve and package the food we eat. The story of glass does not even begin in Mesopotamia, where master craftsman discovered how to make glass in the third millennium BC. Glass occurs naturally. It forms when, at extremely high temperatures, quartz sand melts and then the molten mass cools down. "Solidified liquid", that is glass. Even in the Neolithic period (Early Stone Age), about 7000 BC, people were using glass as a tool. They recognized the outstanding properties of the naturally formed glassy rocks. For example, they used the sharp cutting edges of obsidian to make wedges and scrapers. Glass packaging first begun to be used in 1500 B.C in Egypt. The Egyptians produced the first hollow glass containers, which they used for ointments and oils. Glass, first seen used as a pot, was being mixed with melted limestone, soda, sand, silicate and shaped into glass packaging.

It was the Egyptians who would most encourage the use of glass as decorative and household material for the conservation and storage of certain products. Until the Middle Ages the use of glass was in the hands of a privileged few, who kept their composition and manufacturing secret. Glass became an object of luxury used both for decoration and for the conservation, transport and storage of food, medicines, oils, *etc.* In fact, the glassmaker's trade was the only one to which the nobility could surrender in France during this time.

As a glass container, thanks to its specific qualities, together with ceramics, it is the oldest and most widely used material for the preservation and storage of its products. Thus, since the beginning of civilization it has served as a container for wines, oils, perfumes and medicines, but it is from the seventeenth century when its use is widespread, due largely to the cork stopper, which gives it one of its main qualities, the tightness.

A century and a half later, in 1790, the French Government announced that it would offer a grand prize to those who discovered a practical method of preserving food for some time to serve as a diet for Napoleon's soldiers, at war throughout Europe for those dates The researcher Nicolás Appert came up with the solution: he observed that certain foods packed in glass jars, sealed and subsequently heated, retained their intact characteristics and were not altered, and could be preserved indefinitely. The food industry was born at that time.

Until the twentieth century, the manufacture of glass containers was done by hand. In the first years of 1900, after numerous investigations, the first machine for the automatic manufacturing and mass production of glass containers was created. A few years later, in 1925, an "Individual Sections" machine was started up, which had four sections, going later to five and then to six. Currently, there are machines with 20 sections that allow to manufacture 800,000 bottles and jars in one day.

10.2 Glass Packaging

Glass packaging is a type of packaging commonly used in many foods ranging from heat treated or pressure-packed solid and powdered products to liquid products. Glass, which is usually transparent, is an amorphous (non-crytsalline) solid material which has a wide area of use for technological and decorative purposes. Unlike metals, glass does not have free and mobile electrons that absorb light energy. This is why it is transparent and translucent. However, colored glasses that selectively pass the light can also be made. The glass used in packaging industry is soda-lime glass. It has been stated that the first use of glass was around 3000 B.C., but that glass workshops were first established in the 16th century and that their development continued until the 19th century. It is stated that the production of glass packaging at workshops began in 1870. According to Aday (2014), the air blowing rods, which facilitate the production of glass packaging, were invented by the Romans. Glass packaging is generally in cylindrical shape. Among these cylindrical containers, those with narrow rims used in food packaging are called bottles whereas those with wide mouths are called jars. There are also glasses and demijohns, as well as vials and ampoules that are generally used in the field of pharmacy. In terms of use, the cylindrical shape is close to the spherical shape that provides maximum durability and is easily manufactured. The abundance in positive aspects of using glass as food packaging has made it widely preferred by consumers.

Glasses can also be manufactured in color. Colorants are used in little or suitable amounts within mixtures, such as chromium oxide for the production of green glasses; iron, sulfur and carbon for brown glasses; and cobalt oxide (zaffre) for the production of blue colored glasses. Decolorizers such as nickel and cobalt are added to the mix in order to mask the color emerging from trace amounts of impurities such as iron and sulfur found in the raw materials, when producing colorless (crystal) glass (Keles, 1998).

10.3 Glass Products

Glass containers are made either from the following types of glasses:

- Soda-lime glass: This is the most common commercial glass (90 per cent of total production), and also the least expensive. The chemical and physical properties of soda-lime glass are the basis for its widespread use.
- Lead glass: As the name already indicates, lead glass has a high percentage of lead oxide (at least 20 per cent of the batch). Lead glass has a relatively soft surface, making it especially suited for decorating using grinding, cutting, and engraving processes. This glass will not withstand high temperatures or sudden changes in temperature.
- Borosilicate glass: The third major group, borosilicate glass, is any silicate glass having at least 5 per cent of boric oxide in its composition. It demonstrates greater resistance to thermal changes and chemical corrosion. Thanks to its properties, borosilicate glass is suitable for use in industrial chemical process.

In general, the main parts of a glass container are mouth, neck, shoulder, body, heel, bottom and sting (concave surface inside the bottom). Depending on the mouth and the usefulness of the containers, two large families can be distinguished:

Bottles

Generally intended to contain liquid products, among which are:

- Waters
- Oils
- Snacks (with and without alcohol)
- Refreshing drinks
- Cavas
- Beers
- Dairy products
- Spirits
- Musts
- Cider
- Vinegars
- Wines

Jars

Designed for solid and semi-solid foods, for the following products:

- Olives and pickles
- Baby food
- Canned food (fish and vegetables)
- Dairy Products
- Spices
- Vegetables
- Jams and syrups
- Ready meals
- Desserts
- Sauces
- Juices

10.4 Composition of Glass

Glass is one of the most important packaging materials because of its high barrier and see-through properties. Architectural glass is made from three principal raw materials, easily found in nature: silica (SiO_2), lime ($CaCO_3$) and sodium carbonate (Na_2CO_3). The secondary materials are used to give special properties to glass or to facilitate the glass-making process. In the technical sense, glass is an inorganic substance formed from a mixture of sand (73 per cent), sodium oxide (13 per cent), and calcium oxide (12 per cent), with a proportion of broken glass or culler (15 to 30 per cent of total weight). Many glasses contain silica as their main component and glass former. Glass containers are produced by melting the mixture of silica (sand), sodium carbonate (melting agent), calcium carbonate and alumina (hardener) substances at high temperatures, casting them into the molds and cooling them.

The principal raw materials required for manufacture of glass are sand lime stone and soda ash. Screen printing, ceramic spray and plastic coating are used for decoration or printing on the glass.

Natural White Glass

- Silica (SiO_2) – 73 per cent
- Lime (CaO) – 11 per cent
- Soda (Na_2O) – 13-14 per cent
- Alumina (Al_2O_3) – 1.7 per cent
- Magnesia (MgO) – 0.3 per cent
- Potash (K_2O) – 0.3-0.4 per cent

- Iron oxide (Fe_2O_3) 0.05 per cent
- Sulphur Trioxide (SO_3) – 0.19 per cent

The content of soda decides the cost of the glass. To colour the glass appropriate materials are added for example:

- Green colour – Chromium and Iron,
- Amber – Iron, Sulphur and carbon
- Opal – Fluorine

Although glass is often regarded as a man-made material, it was formed naturally from common elements in the earth's crust long before the world was inhabited. The main components that are currently involved in the manufacturing process of glass containers come from nature. They are quality raw materials, which will give life to a container that adds all the remarkable characteristics of the ingredients that make it up. They exist in nature in large proportions and are easy to extract, ensuring minimal ecological impact. Natural materials such as obsidian (from magma or molten igneous rock) and tektites (from meteors) have compositions and properties similar to those of synthetic glass; pumice is a naturally occurring foam glass.

In general, the main raw materials used for the manufacture of glass containers are, in addition to the aforementioned glass, sand, limestone and soda hull, being able to be classified into the following groups:

- **Vitrifying agents:** These substances represent, as a whole, the main component and, basically, are responsible for the creation of the vitreous network.
- **Fluxes:** Components that favor the formation of glass, lowering its melting temperature and facilitating its elaboration.
- **Stabilizers:** These elements help reduce the tendency to devitrification.
- **Secondary Components:** In this section the tuners, dyes, bleaching, opacifying, *etc.* are framed.

For most glass, silica is combined with other raw materials in various proportions. Alkali fluxes (commonly sodium and potassium carbonates) lower the fusion temperature and viscosity of silica. Calcium and magnesium carbonates (limestone and dolomite) act as stabilizers, preventing the glass from dissolving in water. Other ingredients are added to give glass certain physical properties. For example, lead gives clarity and brilliance although at the expense of softness of the glass; alumina increases hardness and durability. The addition of about 6 per cent boron to form a borosilicate glass reduces the leaching of sodium (which is loosely combined with the silicon) from glass.

As a consequence of the sodium in glass being loosely combined in the silica matrix, the glass surface is subject to three forms of "corrosion": *etc.*hing, leaching and weathering. *etc.*hing is characterized by alkaline attack which slowly destroys the silica network, releasing other glass components. Leaching is characterized by acid attack in which hydrogen ions exchange for alkali or other positively charged, mobile ions. The remaining glass (principally silica) usually retains its normal integrity. Weathering, although not fully understood, is not a problem in commercial glass packaging applications since it may take centuries to become apparent. However, a mild form of weathering is commonly known as surface bloom and may occur under extended storage conditions.

Glass is neither a solid nor a liquid but exists in a vitreous or glassy state in which molecular units have disordered arrangement but sufficient cohesion to produce mechanical rigidity. Although glass has many of the properties of a solid, it is really a highly viscous liquid. During cooling, glass undergoes a reversible change in viscosity, the final viscosity being so high as to make the glass rigid for all practical purposes. Although glass at ambient temperatures has the characteristics of a solid, it is a supercooled liquid and will flow even at ambient temperatures over long periods of time, albeit extremely slowly. Evidence for this can be obtained by examining very old window panes which will be slightly thicker at the bottom than at the top.

Physically, glass has a random atomic structure in that the atoms are capable of arranging themselves in different orders. The basic structural unit is the silicon-oxygen tetrahedron in which a silicon atom is tetrahedrally coordinated to four surrounding oxygen atoms. However, although the silica atoms are always surrounded by four oxygen atoms, large groupings tend to be unordered. This amorphous structure, without slip planes formed by crystal boundaries that might allow deformation, is responsible for the stiffness and brittleness of glass.

10.5 History of Glass Manufacturing and its Process

Although the origin of the first synthetic glasses is lost in antiquity and legend, the first glass vessels were probably sculpted from solid blocks about 3000 B.C. In about 1000 B.C. the techniques of pouring molten glass or winding glass threads over a sand mold were developed, resulting in the formation of crude but useful glass objects. However, the real revolution in glassmaking came around 200 B.C. with the introduction of the blowing iron, a tube to which red-hot, highly malleable glass adheres. Blowing through one end of the iron causes the viscous liquid to balloon at the other end, leading to the production of hollow glass objects.

The basic ingredient of glass is silica derived from sand, flint or quartz. Silica can be melted at very high temperatures (1723°C) to form fused silica glass which has a very high melting point and is used for specialized applications including some laboratory glass.

By 200 A.D. articles of glass were in fairly common use in Roman households. During the following 1000 years, glassmaking techniques spread over Europe. However, glass remained expensive until improved techniques in the 18th and 19th centuries brought down the price of bottles and jars to a relatively affordable level.

Mechanization of glass container manufacture was introduced on a large scale in 1892 and several important developments occurred over the next few decades. These included the first fully automated machine for making bottles which was designed and built in 1903 by Michael J. Owens at the Toledo plant of Edward D. Libbey, and the development by the Hartford-Empire Company of there is (now generally taken to mean individual section, but actually named after its inventors Ingersall and Smith) blow-and-blow machine.

Further developments have occurred, resulting in the production of a wide range of glass containers for packaging. Generally, today's glass containers are lighter but stronger than their predecessors. For example, the weight of the pint (600 mL) milk bottle in England has been halved from over 567 g in the 1920s to 250 g today. Through developments such as this, the glass container continues to play a significant role in the packaging of food products.

The molten glass and conditioned in the feeders (glass heating/cooling channels) for the type of glass containers to be manufactured, and at a temperature close to 1000 degrees Celsius, is cut into drops (pieces of glass, of equal weight to the one of the containers) and distributed to the different molds that own the manufacturing machine, where by means of a series of pressure, pressing and vacuum processes, the new containers (bottles or jars) are shaped

Manufacturing Process

To better understand the work of the manufacturers of glass bottles and jars, it is important to know that the manufacturing process of these containers begins when the raw materials are mixed and automatically driven to the melting furnace, at temperatures close to 1,600 degrees Celsius, approximately. The ovens of these manufacturers of bottles and containers work continuously for 24 hours a day and 365 (366) days a year, during the process of their life, which is 8 or 9 years. In it, raw materials (including glass containers from recycling) enter at one end (enforne) and leave, once molten and conditioned, at the other end, through several arms (feeders) at the end of which and under them, are the machines, which produce the new containers.

Various steps involved in glass manufacturing process are discussed in detail as under:

Mixing and Melting

The largest constituent (68-73 per cent) of a typical soda-lime glass is silica; the second largest constituent (15-50 per cent) is cullet (*i.e.*, scrap or recycled glass), originating both as glass scrap from the factory and recycled glass from consumers. Although the use of cullet can cause problems with the production of some types of glass unless there is good separation of colored glass and removal of associated material such as labels, the use of cullet is economically desirable since less energy is required to melt cullet than new raw materials. Cullet also reduces the amount of dust and other particulate matter that often accompanies a batch made exclusively from new raw materials. Although the total primary energy use decreases as the per cent of cullet rises, the maximum energy saved is only about 13 per cent. Cullet is scrap or recycled glass

The raw materials are weighed, mixed and charged into a glass-melting furnace which is maintained at a temperature of approximately 1500°C. Here they are converted into molten glass that is chemically homogeneous and virtually free of gaseous inclusions (bubbles). The melting process consists of two phases: changing the solids into a liquid, and the fining or 'clearing up' of the liquid. During the refining process, gases (principally CO_2, SO_2 and water vapor) produced by the chemical reaction rise to the surface of the furnace and are removed. When the molten glass becomes free of gas (seed-free) it is then ready for forming into containers. It moves from the furnace into the working end of the furnace (mistakenly called the refiner) where thermal homogenization and cooling of the glass to the viscosity required for the particular operation begin. At this point the temperature of the melt has been lowered from 1250-1350°C to approximately 1100°C.

Forming Processes: The glass is carried from the working end of the furnace to the forming machine in a channel-like structure called a forehearth which is fired by a number of small burners, the aim being to ensure uniform temperature distribution throughout the depth of the glass. At the end of the forehearth is a gob-forming mechanism consisting of a rotating sleeve and vertical plunger. The glass exits in a continuous, viscous stream which is cut by rapidly moving, horizontal steel blades (shears) to form what is known as a 'gob', *i.e.*, a mass or lump of molten glass.

Precise control of temperature and shape during the formation of the gob is required for the high speed production of accurately formed containers, temperatures in the vicinity of 1100°C varying by no more than ±1°C being typical. The process of converting a cylindrically shaped gob of glass into a bottle or jar is called forming, and it is essentially a controlled cooling process. Various types of forming machines are used throughout the world, the predominant type being the IS (individual section) machine. As its name implies, it consists of up to 16 sections, each one an individually functioning, hollow glass machine. It performs two basic functions: it shapes the gob into a hollow container and simultaneously removes heat from the gob to prevent it from deforming significantly under its own weight. Two basic types of processes are used to make containers on the IS machine: the blow-and-blow (B and B) and the press-and-blow (P and B). A closure size of approximately 35 mm is the dividing line between narrow-neck B and B containers (*i.e.*, bottles) and wide mouth P and B containers (*i.e.*, jars).

Blow and Blow (B and B)

Bottles are normally produced by a two-step 'blow and blow' process whereby a gob of glass, accurately sheared in terms of weight and shape, is dropped into an externally air-cooled cast iron cavity known as the parison or blank mold. Some of the glass flows over a plunger in the base of the mold which is used to mold the finish of the container by means of ring molds. Compressed air is applied to force the glass down onto the plunger to form the neck ring. Sometimes vacuum is applied from the bottom as an alternative or additional procedure. When the finish molding is complete, the plunger is retracted and air blown in from the bottom, enlarging the size of the bubble until the glass is pressed out against the blank mold to form a hollow preform. This is then inverted and transferred to the blow mold where it elongates under its own weight. Air at about 200 kPa or vacuum is applied so that the glass is pressed against the metal surface of the blow mold which is air cooled to ensure rapid removal of heat. The mold is then opened and the fully blown parison (now at approximately 650°C) removed and held over a deadplate for a brief time to allow air to flow up through the deadplate and around the container to further cool it. It is then transported to the annealing lehr.

Press and Blow (P and B)

In the case of jars, a two-step 'press and blow' process is used. The body blank or parison is formed by pressing the gob of molten glass against the mold walls with a large plunger. When the cavity is filled, glass is then pushed down into the neck ring and the finish is formed. No baffle or counter-blow air is used in the formation of the parison cavity, the operation relying on the mechanical introduction of the plunger into the glass. The rest of the steps in the P and B process are identical to those in the B and B process.

Narrow Neck Press and Blow (NNPB)

In this more recent process for lightweight bottles, the gob is delivered into the blank mold and pressed by a metal plunger. The plunger and gob together have the same volume as the blank mold cavity. This enables the glass maker to decide exactly how the glass is distributed in the parison and, hence, to be able to more accurately control the uniformity of glass distribution in the finished container; weight savings of up to 30 per cent can be made. The second stage is similar to the B and B process. The parison is blown to a finished container having a more uniform wall thickness and, as a result, higher strength.

Annealing

The term annealing generally refers to the removal of stress, the annealing temperature or point being defined as the temperature at which stresses in the glass are relieved in a few minutes. The containers are transferred from the deadplate to a large oven known as a lehr which is equipped with a belt conveyer.

The function of the annealing lehr is to produce a stable product by removing any residual stresses resulting from non-uniform cooling rates during forming and handling. This is achieved by raising the temperature of the container to approximately 540°C (almost the softening point of the glass), holding it there for a few minutes and then cooling at a rate which is consistent with the removal of stress from a predetermined wall thickness. The critical area of temperature is between the upper annealing point (softening point) and the lower annealing point, after which they may be cooled at a rate which enables them to be handled as they emerge from the lehr.

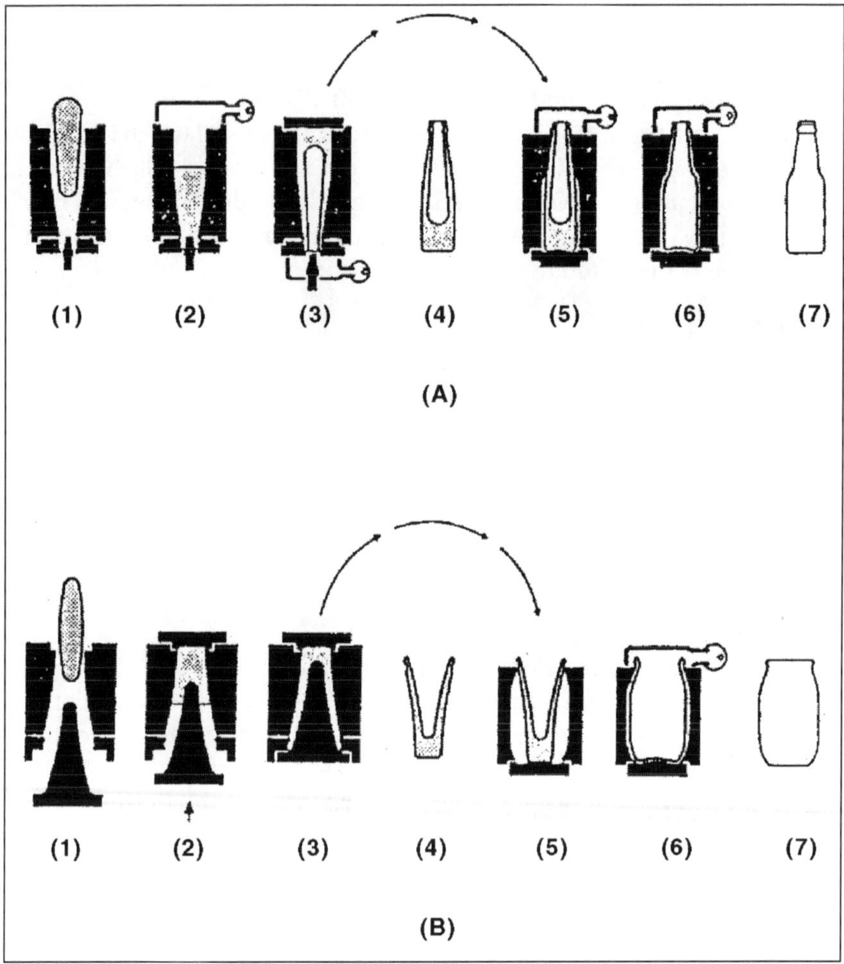

Figure 10.1: "Blow and Blow" and "Press and Blow" Processes for Glass Container Manufacture.

(A) (1) Gob dropped into blank mold; (2) Neck formed; (3) Blank blown; (4) Blank shape; (5) Blank transferred to blow mold (6) Final shape blown; (7) Finshed bottle; (B) (1) Gob dropped into blank mold; (2) Plunger presses blank shape; (3) Blank pressed; (4) Blank shape; (5) Blank transferred to blow mold (6) Final shape blown; (7) Finished jar.

During cooling the inside surface is hotter than the outside; this results in compression on the outer surface but tension at the inner surface. As mentioned earlier, glass fractures only in tension and usually at the surface. Sudden cooling introduces tensile stresses into the outer surfaces and compensating compressional stresses in the interior. Poorly annealed containers may be subject to breakage if the tension is high or the inner surface is bruised.

At the exit of the annealing arc, the packaging manufacturers treat them on their outer face to reduce the coefficient of friction between them, and allow easy sliding along the inspection lines and automatic packaging. At the end of the annealing arc, an aligner distributes the containers to the different inspection and transport lines.

Inspection

When glass container emerges from the annealing lehr, each one is inspected for faults. On these lines a series of electronic machines control 100 per cent of the containers and eliminate those that do not meet the expected quality requirements. Usually, they are optoelectronic machines that are responsible for carrying out dimensional and functional controls of the container:

- fragile or prevent its tightness.
- To eliminate those containers whose external profile does not correspond to the theoretical one.
- To eliminate containers whose minimum wall thickness does not meet the theoretical value, to ensure its resistance, against mechanical shocks.
- Reading systems of the mold numbers from an identification code to associate each defect with the corresponding mold.

All these machines used are linked to a computer, where all the information is stored and processed. Given the high number of containers produced annually by the manufacturers of containers and bottles, these machines carry out a control process, which would be impossible today to 100 per cent with people and with a much higher degree of effectiveness. The human control is carried out by highly specialized personnel, on the effectiveness at all times of the machines and this control and measure of their effectiveness, is what determines whether the containers produced are accepted or rejected.

Finally, and at the end of the inspection machines, there are the machines to pack the production (palletizers), which once palletized, is covered with a plastic (polyethylene) shrink sleeve and passed through a shrink tunnel or Retractable frame that contracts the cover on the containers, and with them protects them from a subsequent entry of dust or other element, during storage and transport.

Surface Treatments

The strength of a newly-made glass container can be rapidly reduced by moisture or abrasion, and some form of surface treatment to increase the strength is essential, since glass is non-lubricious. Two general types of surface treatment are applied to glass containers to modify mechanical properties.

i) **Hot End Treatment**: In this treatment (typically carried out while the glass container is 550°C) vapor containing tin or titanium (generally in the form of a tetrachloride) is brought into contact with the outside of the container, forming a thin film of metal oxide. This treatment prevents surface damage while the container is still hot, strengthens the surface and improves the adhesion of the subsequent cold end coating.

ii) **Cold End Treatment:** This treatment (typically carried out while the glass container is less than 100°C) is designed to protect the container surface and assist its flow through the filling line. Typically, it involves spraying an organic material in an aqueous base containing either waxes, stearates, silicones, oleic acid or polyethylene onto the outside of the container to increase its lubricity by providing a surface with a low coefficient of friction. It is important to check the compatibility of the cold end treatment with any adhesives used to attach labels. Sometimes only the cold end treatment is applied.

Shrink Sleeves

Although not strictly related to surface treatment of glass containers, shrink sleeves will be considered here since they can have an important influence on the formation of imperfections leading to container breakage due to surface contact. Most shrink sleeves are made of oriented plastic films that shrink around a glass container when heat is applied. Two types of protective labels are used on glass bottles in the form of a body sleeve: one constructed from thin, foamed PS (polystyrene), the other made from uniaxially oriented PVC (polyvinylchloride) or PS. The former offers some thermal insulation, while the latter (which can completely wrap the bottle from its neck to underneath the base if desired) contains the glass fragments and prevents shattered glass being scattered in all directions if the bottle is dropped.

10.6 Defects in Glass Containers

Some sixty defects can occur in finished glass containers, ranging from critical defects such as 'birds-wings' and 'spikes' (long, thin strands inside the container that would probably break off when the container was filled) to minor defects such as 'wavy appearance' (an irregular surface on the inside). Defects are classed as critical if they are hazardous to the user and make the container completely unusable; major if they reduce the usability of the container or its contents, and minor if they detract from its appearance or acceptability to the consumer. Accurate classification of defects in glass packaging and their commercial significance are areas of specific expertise.

Because of its amorphous structure, glass is brittle and usually breaks because of an applied tensile strength. It is now generally accepted that fracture of glass originates at small imperfections or flaws, the large majority of which are found at the surface. A bruise or contact with any hard body will produce on the surface of glass very small cracks or checks that may be invisible to the naked eye. However, because of their extreme narrowness, they cause a concentration of stress that may be many times greater than the nominal stress at the section containing them. Because of their ductility, metals yield at such points and equalize stress before failure occurs. Since glass cannot yield, the applied stress (when it is high enough) causes these flaws to propagate. Thus, it is the ultimate tensile strength of a glass surface which determines when a container will break.

In practice, a stress concentrator may be a small crack or check induced in the manufacturing process, or a scratch resulting from careless container handling. Therefore, the major step taken to make glass more break resistant involves the elimination of surface flaws (*e.g.*, microcracks) by careful handling during and after forming and annealing, since the condition of the surface has a great deal to do with its tensile properties.

The mechanical strength of a glass container is a measure of its ability to resist breaking when forces or impacts are applied. Glass deforms elastically until it breaks in direct proportion to the applied stress, the proportionality constant between the applied stress and the resulting strain being Young's modulus E. It is about 70 GPa for typical glass.

10.7 The Shape of the Glass Container

Usually, the shape of the container is determined by the nature of the product, each product group having a characteristics shape. Thus, liquid products generally have small diameter finishes for easier pouring; solid products require larger finishes for filling and removing the contents. As well as filling and emptying requirements, consideration must also be given to the nature and manner of labelling the container, and its compatibility with packaging and shipping systems.

The container finish (so-called because in the early days of glass manufacturing, it was the part of the container to be fabricated last), is the part of the container that holds the cap or closure, *i.e.*, the glass surrounding the opening in the container. It must be compatible with the cap or closure and can be broadly classified by size (*i.e.*, diameter), sealing method (*e.g.*, twist cap, cork, *etc.*), and special features (*e.g.*snap cap, pour-out, *etc.*).

The finish has several specific areas including the sealing surface which may be on the top or side of the finish, or a combination of the two; the glass lug which is one of several horizontal, tapering and protruding ridges of glass around the periphery of the finish on which the closure can be secured by twisting; the continuous thread which is a spiral projecting glass ridge on the finish intended to mesh with the thread of a screw-type closure; a transfer bead which is a continuous horizontal ridge near the bottom of the finish used in transferring the container from one part of the manufacturing operation to another; a vertical neck ring seam resulting from the joining of the two parts of the neck ring, and a neck ring parting line which is a horizontal mark on the glass surface at the bottom of the neck ring or finish ring resulting from the matching of the neck ring parts with the body mold parts. Not all glass containers have transfer beads or vertical neck ring seams.

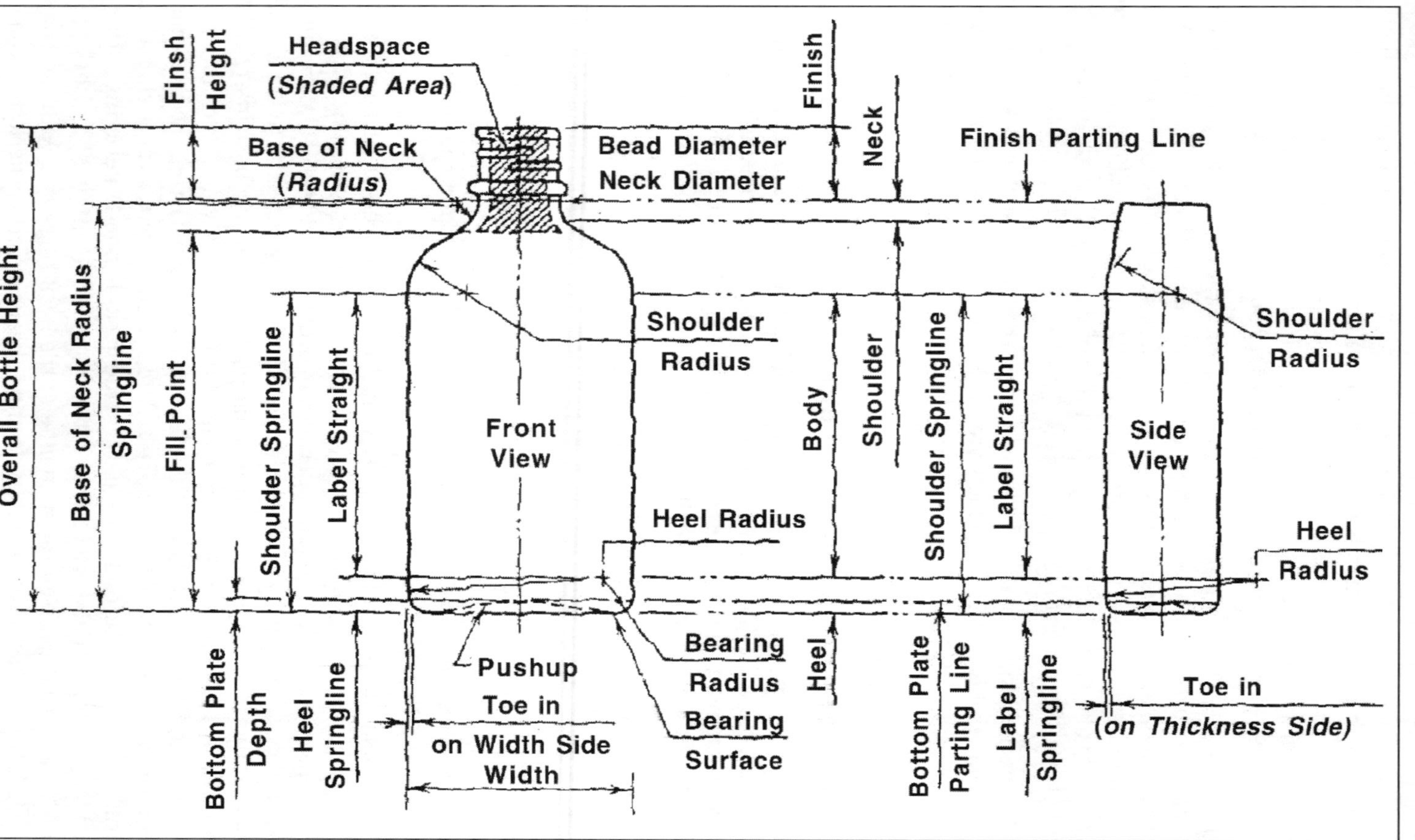

Figure 10.2: Glass Container Nomenclature

Although there are literally hundreds of different finishes used on glass containers, glass finishes are standardized and a specific set of dimensions, specifications and tolerances has been established for every finish designation by the Glass Packaging Institute (GPI) in the USA and equivalent bodies in other parts of the world.

Once a design has been accepted, the molds used in the manufacturing process must be made. They are usually constructed of cast iron and consist of three parts: a bottom plate, a body mold (divided vertically into two halves), and a neck or finish mold which is usually also split into two parts. Because of the high cost of mold manufacture, changes to container size and shape are usually only made if large quantities of the container are required. Generally, customers select their containers from the standard range provided by glass manufacturers unless they are extremely large users in which case the extra expense of customized designs is justified.

The GPI has established limits which are generally accepted as reasonable tolerances by most manufacturers. Allowance has been made for an increase in container size as a consequence of mold wear, as well as expected process capabilities of the manufacturer. Although closer tolerances can be met, this often incurs a higher cost since molds must be replaced more frequently.

10.8 Properties of Glass

Because of its amorphous structure, glass is brittle and usually breaks because of an applied tensile strength. It is now generally accepted that fracture of glass originates at small imperfections or flaws, the large majority of which are found at the surface. A bruise or contact with any hard body will produce on the surface of glass very small cracks or checks that may be invisible to the naked eye. However, because of their extreme narrowness, they cause a concentration of stress that may be many times greater than the nominal stress at the section containing them. Because of their ductility, metals yield at such points and equalize stress before failure occurs. Since glass cannot yield, the applied stress (when it is high enough) causes these flaws to propagate. Thus, it is the ultimate tensile strength of a glass surface which determines when a container will break.

In practice, a stress concentrator may be a small crack or check induced in the manufacturing process, or a scratch resulting from careless container handling. Therefore, the major step taken to make glass more break resistant involves the elimination of surface flaws (*e.g.*, microcracks) by careful handling during and after forming and annealing, since the condition of the surface has a great deal to do with its tensile properties.

The mechanical strength of a glass container is a measure of its ability to resist breaking when forces or impacts are applied. Glass deforms elastically until it breaks in direct proportion to the applied stress, the proportionality constant between the applied stress and the resulting strain being Young's modulus E. It is about 70 GPa for typical glass.

The principles of fracture analysis or diagnosis of the cause(s) of glass container breakage have been described by Moody. Four aspects are important:

Internal Pressure Resistance

This is important for bottles produced for carbonated beverages, and when the glass container is likely to be processed in boiling water or in pressurized hot water. Internal pressure produces bending stresses at various points on the outer surface of the container.

Vertical Load Strength

While glass can resist severe compression, the design of the shoulder is important in minimizing breakage during high speed filling and sealing operations.

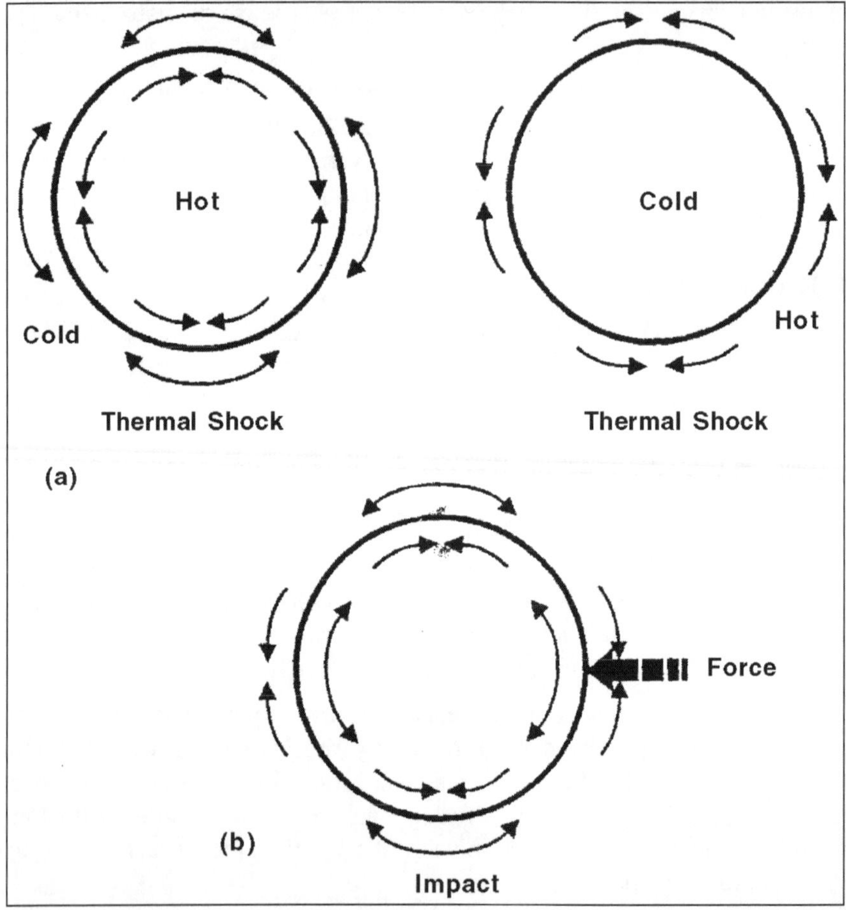

Figure 10.3: Cross Section of a Round Glass Container Illustrating Various Stresses on the Inside and Outside Surfaces.

Resistance to Impact

Two forms of impact are important: a moving container contacting a stationary object (as when a bottle is dropped), and a moving object contacting a stationary bottle (as in a filling line). In the latter situation, design features are incorporated into the sidewall to strengthen contact points. The development of surface treatments (including energy absorbing coatings) to lessen the fragility of glass when it contacts a stationary object has been very successful.

Resistance to Scratches and Abrasions

The overall strength of glass can be significantly impaired by surface damage such as scratches and abrasions. This is especially important in the case of reduced wall thickness bottles such as "one-trip" bottles. Surface treatments involving tin compounds (in conjunction with other treatments) provide scuff resistance, thereby overcoming susceptibility to early failure during bottle life.

Although the mechanical strength of a bottle or jar can increase with glass weight, this is at the expense of thermal strength which decreases with increasing glass weight. Considerable expertise is required by the glassmaker to determine the most appropriate design to satisfy the mechanical strength requirements and balance the thermal strength demands of the finished product.

The thermal strength of a bottle is a measure of the ability to withstand sudden temperature change. In the food industry, the behavior of glass with respect to temperature is of major significance, because relative to other forms of food packaging, glass has the least resistance to temperature changes. The resistance to thermal failure depends on the type of glass employed, the shape of the container, and the wall thickness.

Table 10.2: Coloring Agents Used in Glass

Effect	*Oxide*
Colorless, UV absorbing	CeO_2, TiO_2, $Fe2O_3$
Blue	Co_3O_4, $Cu_2O + CuO$
Purple	Mn_2O_3, NiO
Green	Cr_2O_3, $Fe_2O_3 + Cr_2O_3 + CuO$, V_2O_3
Brown	MnO, $MnO + Fe_2O_3$, $TiO_3 + Fe_2O_3$, $MnO + CeO_2$
Amber	Na_2S
Yellow	CdS, $CeO_2 + TiO_2$
Orange	CdS + Se
Red	CdS + Se, Au, Cu, UO_3, Sb_2S_3
Black	Co_3O_4 (+ Mn, Ni, Fe, Cu, Cr oxides)

When a glass container is suddenly cooled (*e.g.,* on removal from a hot oven), tensile stresses are set up on the outer surfaces, and compensating compressional stresses on the inner surface. Conversely, sudden heating leads to surface compression and internal tension. In both situations, the stresses are temporary and disappear when the equilibrium temperature has been reached. Because glass containers fracture only in tension, the temporary stresses from sudden cooling are much more damaging than those resulting from sudden heating, since the potentially damaged outside surface is in tension. It is found in practice that the amount of tension produced in one surface of a bottle by suddenly chilling it is about twice as great as the tension produced by suddenly heating the other surface, assuming the same temperature change in both cases.

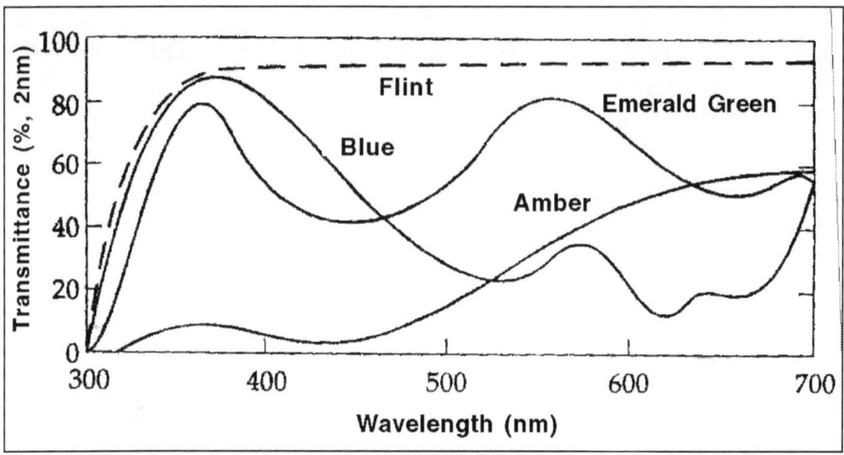

Figure 10.4: Typical Radiation Transmission of Common Container Glasses.

Thermal shock resistance cannot be calculated directly because the strength of glass containers is greater under momentary stress than under prolonged load. Therefore, empirical testing procedures are used. ASTM C 149 covers the determination of the relative resistance of commercial glass containers (bottles and jars) to thermal shock, and is intended to apply to all types of glass containers that are required to withstand sudden temperature changes (thermal shock) in service such as in washing, pasteurization or "hot pack" processes, or in being transferred from a warmer to a colder medium or vice versa. Resistance to breaking is determined by transferring glass containers which have been totally immersed in a hot water bath (typically at 63°C) for 5 minutes to a cold water bath (typically at 21°C) and observing the number of breakages.

Glass has no crystalline structure, so that when it is homogeneous and free from any stresses, it is optically isotropic. The optical properties of glass relate to the degree of penetration of light and the subsequent effect of that transmission, transmission being a function of wavelength. The spectral transmission of glass is determined by reflection at the glass surface and the optical absorption within the glass. In silicate glasses, transmission is limited by the absorption of silica at approximately 150 nm in the UV (ultraviolet) and at 6000 nm in the IR (infrared). Iron impurities further reduce transmission in the UV and near IR.

10.9 Advantages and Disadvantages of Glass Containers

Glass, as a packaging material has the following advantages:

- Part of abundant raw materials in nature.
- It is chemically inert against liquids and food products not posing compatibility problems.
- Hygienic that has easy cleaning and is sterilizable.
- It is odourless, does not transmit tastes or modify them.
- They have filling speeds comparable to those of cans.
- It guarantees the organoleptic and flavour properties of the food.
- It is transparent and display the contents.
- They are suitable for heat processing when hermetically sealed.
- Possibility of using anti-glass glass that prevents ultraviolet radiation from damaging the product.
- It is rigid and resistant to internal pressures, as well as high temperatures.
- Microwave compatible.
- They are reusable and recyclable.
- They are resealable.
- Impervious to moisture, gases, vapours and microorganisms, which guarantees the preservation and vitamins of the food even in prolonged storage.
- Mouldable, with versatility of shapes and colours.
- Preformed and customized containers.
- They are perceived by the customer to add value to the product.
- They can be printed on directly or by using paper labels.

Glass, which is perceived as healthy and thus preferred by people, also has its negative sides. The main disadvantages of glass as a packaging container are:

- Higher weight and hence higher transportation costs than other types of packaging containers.
- Lower resistance than other materials to fractures, scratches, and thermal shock.
- More variable dimensions than other containers.
- Additionally, energy resources such as fossil fuels and electricity are used in higher quantities during the production of glass. It causes 3 times more air pollution when compared to production of plastic

- Potentially serious hazards arise from glass splinters or fragments in foods.
- Especially the fact that it lets light pass-through results in the discoloration of the food.
- Permeability to UV light. This latter problem can be overcome by incorporating various oxides, sulphides, or selenides to colour glass and block out the incident UV radiation.
- Generates three times more air pollution than plastic packaging.
- It is costly in terms of shipping for long distances.

Chapter 11

Plastic in Food Packaging: A Modern Dilemma

History

With development of science and technology, alternative materials like plastics have emerged either to replace natural and traditional materials with a benefit of cost reduction purpose, to cater to certain applications which were not possible with the traditional materials, as well as for enhancing utility of traditional materials. In the new millennium, use of plastics both in terms of quantity and variety is perhaps the correct index of industrialisation of a country, the status of its economy and the standard of living of its citizens.

In today's world, life without plastics is incomprehensible. Every day, plastics contribute to our health and safety. The first man-made plastic was unveiled by Alexander Parkes at the 1862 Great International Exhibition in London. This material which the public dubbed Parkesine was an organic material derived from cellulose that once heated could be molded but that retained its shape when cooled. Parkes claimed that this new material could do anything rubber was capable of, but at a lower price. He had discovered something that could be transparent as well as carved into thousands of different shapes. But Parkesine soon lost its luster, when investors pulled the plug on the product due to the high cost of the raw materials needed in its production.

During the latter part of the 19_{th} century, a rush was on to find a replacement for ivory in billiards balls. John Wesley Hyatt, an American, finally came upon the solution in 1869 with celluloid. Hyatt, upon spilling a bottle of collodion in his workshop, discovered that the material congealed into a tough, flexible film. He then produced billiard balls using collodion as a substitute for ivory. But due to its highly brittle nature, the billiard balls would shatter once they hit each other. The solution to this challenge was the addition of camphor, a derivative of the laurel tree. This addition made celluloid the first thermoplastic: a substance molded under heat and pressure into a shape it retains even after the heat and pressure have been removed. Celluloid went on to be used in the first flexible photographic film for still and motion pictures.

Cellophane was discovered by Dr. Jacques Edwin Brandenberger, a Swiss textile engineer, who came upon the idea for a clear, protective, packaging layer in 1900. Brandenberger added viscose to cloth but the end result was a brittle material that was too stiff to be of any use. Yet Brandenberger saw another potential for the viscose material. He developed a new machine that could produce viscose sheets, which he marketed as cellophane. With a few more improvements, cellophane allowed for a clear layer of packaging for any product, the first fully flexible and water-proof wrap.

The 1920s witnessed a "plastics craze", as the use of cellophane spread throughout the world. DuPont, one of the industry leaders, became a hotbed for innovation concerning plastics. Wallace Hume Carothers, a young Harvard chemist, became the head of the DuPont lab. The company was responsible for the moisture-proofing of cellophane and was well on its way to developing nylon, which at the time they named Fiber 66. During the 1940s, the world saw the use of such materials as nylon, acrylic, neoprene, SBR, polyethylene, and many more polymers take the place of natural material supplies that were becoming exhausted.

Another important plastic innovation of the time was the development of polyvinyl chloride (PVC), or vinyl. Waldo Semon, a B.F. Goodrich organic chemist, was attempting to bind rubber to metal when he stumbled across PVC. In 1933, Ralph Wiley, a Dow Chemical lab worker, accidentally discovered yet another plastic: polyvinylidene chloride (better known as SaranTM). SaranTM was first used to protect military equipment, but it was later discovered that it was great for food packaging. SaranTM would cling to almost any material - bowls, dishes, pots and even itself; thus, it became the perfect tool for maintaining the freshness of food at home.

It was not until after the war, though, that the material became a tremendous hit with consumers and from that point on, its rise in popularity has been almost unprecedented. It became the first plastic in the United States to sell more than a billion pounds a year and it is currently the largest volume plastic in the world. Today, polyethylene is used to make such common items as soda bottles, milk jugs and grocery and dry-cleaning bags in addition to plastic food storage containers.

Time Line of Use of Plastic in Food Packaging

1862	Plastic made its first appearance at the Great International Exhibition in London, presented by Alexander Parkes. This material, called "Parkesine," was bio-based and had been derived from cellulose.
Early 1900s	Cellophane was created by Dr. Jacques Brandenberger, a Swiss textile engineer. The invention came about as Dr. Brandenberger was attempting to make clothing stain-resistant through the application of a flexible, clear film.
1930	Scotch Cellulose Tape, renamed Cellophane Tape at a later date, was invented by Richard Drew, an engineer working for 3M. This greatly benefited bakers and grocers, as they were able to keep their packages sealed in a more attractive and easy way.
1933	Saran™ was an accidental discovery by lab worker Ralph Wiley of Dow Chemical. This plastic material found its first use to be providing protection for military equipment, and was later used for the packaging of food.
1946	Earl Silas Tupper worked to create the well-known, highly-convenient plastic packaging we all know as Tupperware. Mr. Tupper chose to promote his polyethylene food containers through housewives who used Tupperware sales as a source of income. The invention of airtight plastic containers is considered among the most notable advances in the history of plastic packaging.
1946	"Stopette," a deodorant created by Dr. Jules Montenier, brought the spray bottle into the commercial world of plastics. Dr. Montenier created the spray bottle to dispense the deodorant when squeezed by consumers. Stopette was one of the sponsors of the popular program "What's My Line", and subsequently brought on a tremendous increase in the use of spray bottles.
1950	Canadians Larry Hansen and Harry Wasylyk are to thank for the modern garbage bag. Originally used commercially in places such as hospitals, this plastic product is found in abundance in virtually any public place you can imagine.
1954	Zipper-seal bags, created by Robert W. Vergobbi, were initially intended for use as pencil bags, but their versatility soon turned them toward food storage. Ziploc bags came about in 1968, and sandwich bags on a roll were later introduced.
1959	The first licensed character lunch box was created by Geuder, Frey, and Paeschke of Wisconsin. The tin box featured Mickey Mouse and a plastic handle. Plastic was used for the whole box in the 1960s.

1960	Sealed Air Corporation engineers Marc Chavannes and Alfred Fielding brought the world bubble wrap.
1986	Swanson TV Dinners came in microwaveable plastic trays, rather than the aluminum material used since the mid-1950s. These TV dinners were incredibly popular due to post-war fascination with time-saving devices, and the novelty of watching television.
1996	Bagged salad packaging reduced food waste and made buying fresh produce easier.
2000	Yogurt came in flexible plastic tubes for convenient snacking on the go.
2007	The two litre beverage bottle and the one gallon plastic milk jug reach a milestone in 'lightweighting'-both containers shed a third of their weight since they became widely used in the 1970s.
2008	Plastic bottles achieve a 27 per cent recycling rate, reclaiming 2.4 billion pounds of plastic. And polyethylene plastic bags and wraps achieve a 13 per cent recycling rate, reclaiming 832 million pounds of plastic. (The recycling rate for polyethylene plastic bags and wraps has doubled since 2005.
2010	Metallyte ™ films were introduced to help keep sharp cotents (coffee beans, grains, noodles, croutons) fresher by reducing packaging tears. The new films are lighter than foil-based designs.
2010	Heinz Dip and Squeeze was the first innovation for ketchup packaging in 42 years, allowing customers to choose whether to dispense or dip into the sauce.

Plastics Packaging in Modern Life

After the Second World War, the plastics industry underwent incredibly fast development. Plastics are one area that has seen major improvement in materials and their properties also. Polyethylene was one of the first plastics used widely for food packaging. There are several types of polyethylene in use today including low-density (LDPE), high-density (HDPE), linear low-density (LLDPE), and very low density (VLDPE). LDPE was the first to be developed by Imperial Chemical Industries in 1933. The company received a patent for production of the material that involved compressing ethylene gas and heating it to a high temperature. The first plastic sandwich bag on a roll was introduced in 1957. By 1966, over 25 per cent of all bread sold was in plastic bags made from LDPE. That package still is in wide use for most bread products sold. Some companies have gone back to the use of paper bags for bread to give it an artisanal feel; however, the paper bag does not keep the bread as well and the quality deteriorates much more quickly than when the bread is stored in a plastic bag. Although plastics have been more

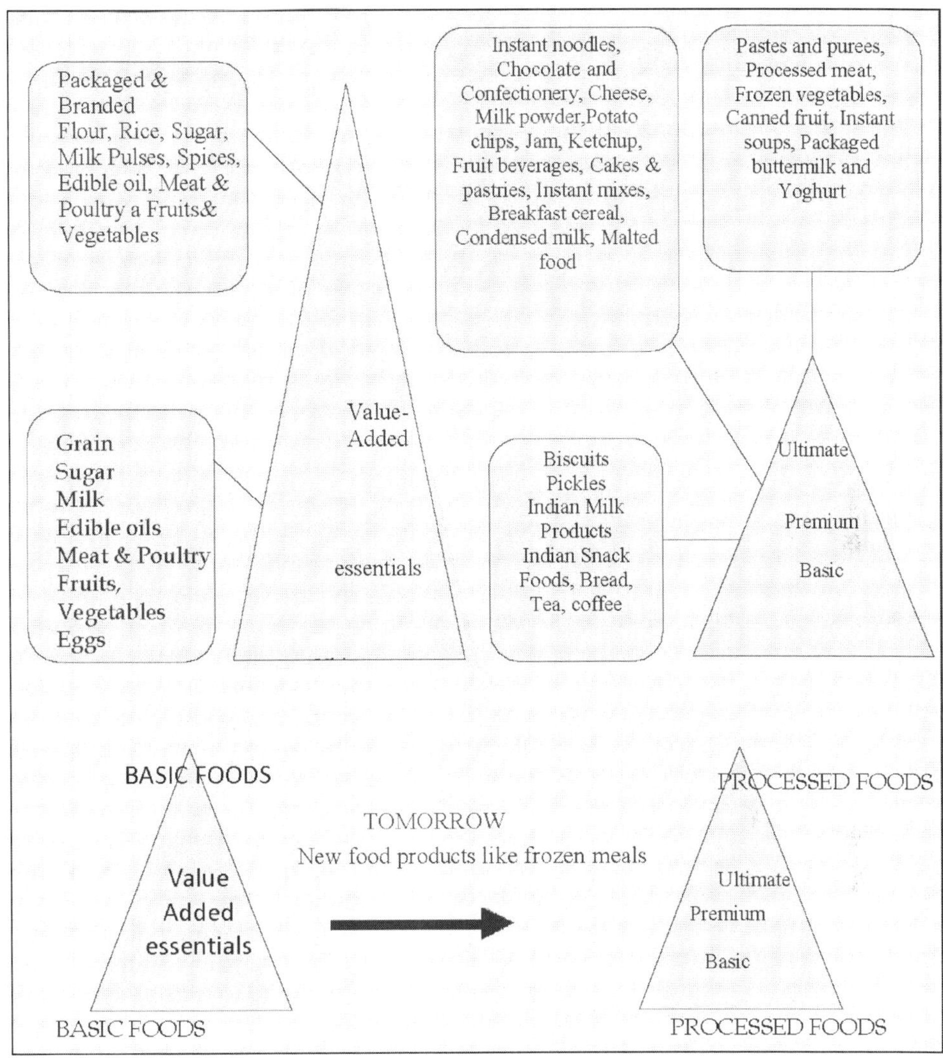

Figure 11.1: Today and Tomorrow.

widely used as food packaging materials in the past 50-60 years, new developments in plastics have helped to increase the usage. Professor Giulio Natta discovered isotactic polypropylene in 1954. The film is often oriented after the casting or forming process by first stretching the material in the machine direction and then stretching it in the crosswise direction to give oriented polypropylene (OPP). This stretching aligns the molecules, making a film with a better moisture vapor barrier, better clarity, and more stiffness. It is widely used as an overwrap for snack foods

The Packaging sector is the largest consumer of plastics now. Since the 1950s, plastics have grown into a major industry that affects all of our lives. In fact, since 1976, plastic has been the most used material in the world and was voted one of the top 100 news events of the century. There is probably no comparable sector of industry (apart perhaps from computing) which has grown so rapidly. None of the applications and innovations we take for granted would have been possible if it weren't for the early scientists who developed and refined the material. Those pioneers made it possible for us to enjoy the quality of life we do today.

Today plastics are the material of choice in packaging for the sectors such as FMCG, food and beverages, pharmaceuticals *etc.* Around half of all goods are now packaged in plastics, and yet thanks to constant innovation and achievement of resource efficiency, these plastics account for only 20 per cent by weight of all packaging materials. The 20th Century has correctly been described as the Century of Plastics. Plastics in packaging are the perfect tool to achieve sustainable development.

India is a growing market for plastics and consumes about 12.8 Million Metric Tonnes (MMT) of plastics annually against global consumption of 285 MMT per year. The plastics and polymer consumption is growing at an average rate of 10 per cent. The per capita consumption of polymers in India during 2014-15 was just 10.5 kg as compared to 109 kg in USA, 45 kg in China and 32 kg in Brazil. India is expected to be among the top ten packaging consumers in the world in coming years. In India, a large chunk of products that households buy for daily use are packaged in plastics. Plastics are used heavily for packaging due to innovative visual appeal for customer attraction and convenience. Additionally, they improve the hygiene quotient and shelf-life of the products especially in food and beverages segment. Overall, the Indian packaging industry is valued at over USD 32 Bn and offers employment to more than 10 lakh people across the country through ~10,000 firms. The industry is driven by key factors like rising population, increase in income levels and changing lifestyles. Growth prospects of end-user segments are leading to rise in the demand of the plastic packaging industry. Demand from rural sector for packaged products is being fuelled by the increasing media penetration through the means of internet and television. Further, India is emerging as the most favoured destination for organized retail destination in the world. And also the presence of E-commerce is expanding rapidly and is bringing around a revolution in the retail industry. Retailers are now leveraging digital retail channels thereby enabling wider reach out to customers with less amount of money spent on real estate. Therefore, organized retail and boom in e-commerce offers huge potential for future growth of retailing in India which in turn is pushing the growth of packaging sector.

Plastics Packaging Applications

Plastics, and plastics packaging, are now an essential part of our everyday life. The key to their success has been versatility. In packaging, plastics are used for many varied applications ranging from sterile storage of medical and pharmaceutical goods, to extending the shelf life of foodstuffs such as bread, meat and vegetables, and protecting sensitive technical products from damage. This means that plastics make a significant contribution to improving the quality of our life. At the same time they preserve valuable resources and help to save costs, as a result of their lower weight. Over time, plastics have become ever more sophisticated, lighter and more versatile due to innovative technologies and they have replaced traditional packaging such as glass and paper in many areas. Plastics act as an excellent barrier to oxygen, water and carbon dioxide. They are inert towards acids, alkalis and most solvents and therefore they ensure freshness and hygiene of the contents and high durability of the items packed. As per industry estimates, 35-40 per cent of the food products produced in India are wasted due to deficient infrastructure and lack of food processing capabilities. Plastics find applications in packaging which protects the longevity and quality of food and can reduce the post-harvest losses. Besides this, plastics can be reused and recycled. They have low energy requirements during production, hence considered to be energy efficient. They consume ~25 per cent less energy in production compared to other alternatives. They result in lower emission of CO_2 gas. Thus, when compared to glass or aluminium plastics results in lighter environmental footprint. As plastics possess versatile properties it can help us do more with less. One such property is light weight. As plastics are light in weight, they have a high product to package ratio which results in lighter weighed end product. For *e.g.*: only 1.5 pounds of flexible plastics can deliver ~60 pounds of beverage; compared to three pounds of aluminium or 50 pounds of glass. Thus, plastic packaging enables in shipping more products with less packaging material. And also brings down the fuel consumption and the overall transportation cost.

Various applications of plastic packaging include:

1. Plastic Bags and Plastic Sacks
2. Plastic Blow Moulded Containers
3. Plastic Food Packaging
4. Plastic Packaging for Soaps
5. Plastics for Detergents and Cosmetics
6. Plastics for Pharmaceuticals
7. Plastic Films
8. Plastics for Transit Packaging

Plastics packaging industry includes food packaging, medical packaging and flexible rigid packaging. Companies in the plastics packaging industry which use their expertise to develop user-friendly new products for the market will enjoy a strategic advantage over their rivals. If they also concentrate on market niches, their long-term commercial potential will be even better.

Packages can even be produced in shapes to relate to the product and assist marketing. A current pet food industry, pack, for example, is a plastic bag for cat food, shaped like a cat. Another is a chocolate pack shaped like a rabbit. Both have dramatic point-of-sale impact and have boosted sales of the product.

There is great scope also for technical innovation. Salad goods when unpackaged have only a very short life in the supermarket. But, when they are packed in a foil bag with a protective atmosphere, their shelf life can be increased by up to ten days. A film for meat packaging, for which a new barrier layer has been developed, now allows steaks to be kept in a special controlled atmosphere for over a month.

In addition to traditional packaging, the coffee industry is now demanding packaging without an aluminium layer. Packaging manufacturers have been asked to develop a film with the same barrier properties as films that contain aluminium, which preserve the aroma of coffee for over a year. Another example is beer in flexible plastic bottles sold in football stadiums. Cold-resistant films that remain flexible and impact-resistant in a deep freeze at a temperature of minus 30°C also provide excellent opportunities. Such solutions to problems provide the basis for market-success. In development of new products, those companies offering ever-lighter and easier to recycle packaging will have a competitive edge. The reason is the growing demand from consumers and legislators for less waste, greater material savings and more environmental protection. At the same time, less material helps to preserve valuable resources and reduce transport costs. Consequently, the industry is reducing packaging weight by using thinner and thinner films and thin-walled plastics containers.

Various Types of Packaging Plastics, their Applications and Benefits

Multiple types of plastics are being used as materials for packaging food, including polyolefin, polyester, polyvinyl chloride, polyvinylidene chloride, polystyrene, polyamide, and ethylene vinyl alcohol. Although more than 30 types of plastics have been used as packaging materials, polyolefins and polyesters are the most common.

Polyolefins

Polyolefin is a collective term for polyethylene and polypropylene, the 2 most widely used plastics in food packaging, and other less popular olefin polymers. Polyethylene and polypropylene both possess a successful combination of properties, including flexibility, strength, lightness, stability, moisture and chemical resistance, and easy processability, and are well suited for recycling and reuse. **Polyethylene:** The simplest and most inexpensive plastic made by addition polymerization of ethylene is polyethylene. There are 2 basic categories of polyethylene: high density and low density. High-density polyethylene is stiff, strong, less flexible, tough, resistant to chemicals and moisture, permeable to gas, easy to process, and easy to form. **It has higher softening temperature (121°C) and can therefore, be heat sterilised.** It is used to make bottles for milk, juice, and water; cereal box liners; margarine tubs; and grocery, trash, and retail bags.

Sacks made from 0.03 0.15mm high-density polyethylene have a high tear strength, penetration resistance and seal strength. They are waterproof and chemically resistant and are used instead of paper sacks. Low-density polyethylene is flexible, strong, tough, easy to seal, heat stable, inert, odour free and resistant to moisture **but has a relatively high gas permeability, sensitivity to oils and poor odour resistance. It is less expensive than most films and is therefore widely used.** Because low-density polyethylene is relatively transparent, it is predominately used in film applications and in applications where heat sealing is necessary. Bread and frozen food bags, flexible lids, and squeezable food bottles are examples of low-density polyethylene. Polyethylene bags are sometimes reused (both for grocery and nongrocery retail). Of the 2 categories of polyethylene, high-density polyethylene containers, especially milk bottles, are the most recycled among plastic packages.

Polypropylene

Harder, denser, and more transparent than polyethylene, polypropylene has good resistance to chemicals and is effective at barring water vapor. Its high melting point (160 °C) makes it suitable for applications where thermal resistance is required, such as hot-filled and microwavable packaging. Popular uses include yogurt containers and margarine tubs. When used in combination with an oxygen barrier such as ethylene vinyl alcohol or polyvinylidene chloride, polypropylene provides the strength and moisture barrier for ketchups and salad dressing bottles.

Polyesters

Polyethylene terephthalate (PET or PETE), polycarbonate, and polyethylene naphthalate (PEN) are polyesters, which are condensation polymers formed from ester monomers that result from the reaction between carboxylic acid and alcohol. The most commonly used polyester in food packaging is PETE.

i) **Polyethylene terephthalate:** It is formed when terephthalic acid reacts with ethylene glycol, PETE provides a good barrier to gases (oxygen and carbon dioxide) and moisture. It also has good resistance to heat, mineral oils, solvents, and acids, but not to bases. Consequently, PETE is becoming the packaging material of choice for many food products, particularly beverages and mineral waters. The use of PETE to make plastic bottles for carbonated drinks is increasing steadily. The main reasons for its popularity are its glass-like transparency, adequate gas barrier for retention of carbonation, light weight, and shatter resistance. The three major packaging applications of PETE are containers (bottles, jars, and tubs), semirigid sheets for thermoforming (trays and blisters), and thin-oriented films (bags and snack food wrappers). PETE exists both as an amorphous (transparent) and a semicrystalline (opaque and white) thermoplastic material. Amorphous PETE has better ductility but less stiffness and hardness than semicrystalline PETE, which has good strength, ductility, stiffness, and hardness. Recycled PETE from soda bottles is used as fibers, insulation, and other nonfood packaging applications.

ii) **Polycarbonate:** It is formed by polymerization of a sodium salt of bisphenol acid with carbonyl dichloride (phosgene). Clear, heat resistant, and durable, it is mainly used as a replacement for glass in items such as large returnable/refillable water bottles and sterilizable baby bottles. Care must be taken when cleaning polycarbonate because using harsh detergents such as sodium hypochlorite is not recommended because they catalyze the release of bisphenol A, a potential health hazard. An extensive literature analysis by vom Saal and Hughes (2005) suggests the need for a new risk assessment for the low-dose effects of this compound.

iii) **Polyethylene naphthalate (PEN)**: In the past 15-20 years, only one new plastic has been approved for food contact and that material is polyethylene naphthalene (PEN), which received FDA clearance in 2000. It is a condensation polymer of dimethyl naphthalene dicarboxylate and ethylene glycol. It is a relatively new member of the polyester family with excellent performance because of its high glass transition temperature. PEN's barrier properties for carbon dioxide, oxygen, and water vapor are superior to those of PETE, and PEN provides better performance at high temperatures, allowing hot refills, rewashing, and reuse. However, PEN costs 3 to 4 times more than PETE. Because PEN provides protection against transfer of flavors and odors, it is well suited for manufacturing bottles for beverages such as beer.

Although other new plastics have been developed, the process for clearance from FDA in the form of a letter of no objection is difficult to obtain as extensive safety and environmental evaluations must be performed. It should be noted that PEN has not seen any widespread use in food packaging, due mainly to the high cost of the material. There are a large number of new additives and processing aids that have been allowed.

iv) **Polyvinyl chloride (PVC):** An addition polymer of vinyl chloride, is heavy, stiff, ductile, and a medium strong, amorphous, transparent material. It has excellent resistance to chemicals (acids and bases), grease, and oil; good flow characteristics; and stable electrical properties. Although PVC is primarily used in medical and other nonfood applications, its food uses include bottles and packaging films. Because it is easily thermoformed, PVC sheets are widely used for blister packs such as those for meat products and unit dose pharmaceutical packaging. PVC can be transformed into materials with a wide range of flexibility with the addition of plasticizers such as phthalates, adipates, citrates, and phosphates. Phthalates are mainly used in nonfood packaging applications such as cosmetics, toys, and medical devices. Safety concerns have emerged over the use of phthalates in certain products, such as toys. Because of these safety concerns, phthalates are not used in food packaging materials in the United States (HHS 2005); instead, alternative nonphthalate plasticizers such as adipates are used. For example, di-(2-ethylhexyl) adipate (DEHA) is used in the manufacture of plastic cling wraps. These alternative plasticizers also have the potential to leach into food but at lower levels than phthalates. Low levels of DEHA have shown no toxicity in animals. Finally, PVC is difficult to recycle because it is used for such a variety of products, which makes it difficult to identify and separate. In addition, incineration of PVC presents environmental problems because of its chlorine content.

v) **Polyvinylidene chloride:** Polyvinylidene chloride (PVdC) is an addition polymer of vinylidene chloride. It is heat sealable and serves as an excellent barrier to water vapor, gases, and fatty and oily products. It is used in flexible packaging as a monolayer film, a coating, or part of a co-extruded product. Major applications include packaging of poultry, cured meats, cheese, snack foods, tea, coffee, and confectionary. It is also used in hot filling, retorting, low-temperature storage, and modified atmosphere packaging. PVdC contains twice the amount of chlorine as PVC and therefore also presents problems with incineration.

vi) **Polystyrene:** Polystyrene, an addition polymer of styrene, is clear, hard, and brittle with a relatively low melting point. It can be mono-extruded, co-extruded with other plastics, injection molded, or foamed to produce a range of products. Foaming produces an opaque, rigid, lightweight material with impact protection and thermal insulation properties. Typical applications include protective packaging such as egg cartons, containers, disposable plastic silverware, lids, cups, plates, bottles, and food trays. In expanded form, polystyrene is used for nonfood packaging and cushioning, and it can be recycled or incinerated.

vii) **Polyamide:** Commonly known as nylon (a brand name for a range of products produced by DuPont), polyamides were originally used in textiles. Formed by a condensation reaction between diamine and diacid, polyamides are polymers in which the repeating units are held together by amide links. Different types of polyamides are characterized by a number that relates to the number of carbons in the originating monomer. For example, nylon-6 has 6 carbons and is typically used in packaging. Nylon film is a transparent, non-heat stabilized plastic film. They absorb water. The higher the moisture content the more flexible they become. They offer excellent thermal and chemical resistance, with high tensile strength, and tear resistance. It has mechanical and thermal properties similar to PETE, so it has similar usefulness, such as boil-in bag packaging. Nylon also offers good chemical resistance, toughness, and low gas permeability.

viii) **Ethylene vinyl alcohol:** Ethylene vinyl alcohol (EVOH) is a copolymer of ethylene and vinyl alcohol. It is an excellent barrier to oil, fat, and oxygen. However, EVOH is moisture sensitive and is thus mostly used in multilayered co-extruded films in situation where it is not in direct contact with liquids.

Composite Films

There are many ways to create new applications for flexible and rigid packaging, and solve technical problems. Apart from new and modified polymers, one method is composite films, combining the benefits of individual layers to tailor-make the properties for the application. There are several processes:

Laminates

Plastic materials can be manufactured either as a single film or as a combination of more than one plastic. There are two ways of combining plastics: lamination and co-extrusion. Lamination involves bonding together two or more plastics or bonding plastic to another material such as paper or aluminum. Bonding is commonly achieved by use of water-, solvent-, or solids-based adhesives. After the adhesives are applied to one film, two films are passed between rollers to pressure bond them together. Lamination using laser rather than adhesives has also been used for thermoplastics. Lamination enables reverse printing, in which the printing is buried between layers and thus not subject to abrasion, and can add or enhance heat sealability.

Co-extrusion

One process that has improved overall properties of plastic films is coextrusion, developed in 1964 by Hercules. In this process a film with two or more layers of different types of plastic can be made in one step without the need to laminate the layers together with an adhesive, eliminates the use of solvents, and produces a film in one step instead in needing multiple steps. Because different types of resin or resin blends are being used, it requires careful control of the melt properties and viscosity to ensure the appropriate thickness of each distinct material. It is possible to make structures with much thinner layers than can be made when laminating. This process is more rapid (requires one step in comparison to multiple steps with lamination) but requires materials that have thermal characteristics that allow co-extrusion. Multiple-layer films offer better protection for products as some films are better moisture barriers and others offer better barriers to gases. One example is polyester film, which provides a better gas barrier, whereas polypropylene and ethylene vinyl alcohol (EvOH) films are better moisture barriers. These three can be combined readily in one structure to give protection from both moisture and oxygen permeation.

Because co-extrusion and lamination combine multiple materials, recycling is complicated. However, combining materials results in the additive advantage of properties from each individual material and often reduces the total amount of packaging material required. Therefore, co-extrusion and lamination can be sources of packaging reduction.

Extrusion Coating: Molten polymer is cast-coated onto the surface of another film with a higher melting point.

Water- or Solvent-based Coating: The coating is applied to a film and the solvent or water is removed using heat.

Vacuum Coating: The latest technology is to apply a very thin layer of an inorganic material such as aluminium to a film in a vacuum chamber.

Nano-composite Technology: Nano-composite technology is also being applied to the production of films with barrier properties. By incorporation of mineral particles in nano size (less than the wavelength of light), it is possible to create a form of "labyrinth" within the structure of the film, which physically retards the passage of molecules of (for example) gas. Already, nylon films have been commercialised with improved gas barrier properties by nano-scale additives, reducing the volume of material in the package and also simplifying recycling.

Barrier properties in other package forms are achieved by co-injection and co-extrusion of blow mouldings. To improve the gas barrier of PET bottles for the vast beer packaging market, coatings of PVDC have been used, but the latest developments use plasma technology to deposit a barrier inside the bottle, integrated with the bottle production line.

Protecting the environment is also becoming increasingly established for printing the films in addition to solvent-based printing inks, water-based ones are now becoming more common.

Use of Various Types of Packaging Plastics

In plastics, there are a variety of materials, which are used for packaging applications. The important plastics, which find large packaging applications are HDPE, PP, PVC, Polyester, Polystyrene, *etc.* These are used for building the body structure of the packages, while other polymers are often used, in less thickness, as coating to improve the functional properties of basic packaging. Examples of such speciality polymer coatings are Polyvinylidene Chloride (PVDC), Ionomer, Ethylene Vinyl Alcohol (EVOH) and others. These speciality coatings provide specific properties such as high barrier to oxygen, moisture aroma/flavour or it may impart sealability to a package even when its surface is contaminated with product. Sometimes, a speciality polymer coating may also provide a good hot tack resistance over a wider temperature range. In other situations, when rigors of handling and distribution demand higher mechanical strength, plastics can be used to enhance such properties. Thus, the advantage of plastics is in the range of mechanical barrier and other properties it offers and also the feasibility of tailoring a structure required to meet precisely the product protection, preservation and distribution needs.

PET (Polyethylene Terphthalate)

It is used in beverage containers, food containers, boil-in food pouches, processed meat packages *etc.* PET is more impermeable than other low-cost plastics and so is a popular material for making bottles for Coke and other "fizzy drinks," since carbonation tends to attack other plastics, and for acidic drinks such as fruit or vegetable juices. PET is also strong and abrasion resistant, and is used for making mechanical parts, food trays, and other items that have to endure abuse. PET films, trade-named "mylar," are used to make recording tape.

HDPE (High Density Polyethylene)

HDPE (High density polyethylene) is used in milk bottles, cereal box liners, detergent bottles, oil bottles, margarine tubs, toys, plastic bags *etc.*

PVC (Polyvinyl Chloride)

PVC (Polyvinyl chloride) is used in food wrap, vegetable oil bottles and blister packaging.

LDPE (Low Density Polyethylene)

LDPE (Low density polyethylene) used in shrink-wrap, plastic bags, garment bags, dry cleaning bags, and squeezable food bottles.

PP (Polypropylene)

PP (Polypropylene) used in margarine and yogurt containers, caps for containers, wrapping to replace cellophane, medicine bottles *etc.*

PS (Polystyrene)

PS (Polystyrene) used in egg cartons, fast food trays, disposable plastic silverware, cups, compact disc jackets.

To assist recycling of disposable items, the Plastic Bottle Institute of the Society of the Plastics Industry has devised the now-familiar scheme to mark plastic bottles by plastic type. A recyclable plastic container using this scheme is marked with a triangle with three "chasing arrows" inside of it, which enclose a number giving the plastic type: PETE, HDPE, PVC, LDPE, PP, PS, and OTHER.

The Table 11.1 depicts the applications of different plastics used across the industry segments and the attainable benefits of the same;

Use of Polyolefins in Food Packaging

The safe and reliable delivery of food has emerged as one of the major industrial concerns of our time that affects both the developing and western world. Surprisingly, some 30 per cent of food losses in developing countries are the direct result of improper food packaging and failed logistics.

Compared to western methods of packaging where plastic materials are widely used and where wastes represent less than 3 per cent from farm to supermarket, these statistics are highly indicative of the strong relationship between the type of packaging material used and its ability to minimise food wastes. The global challenge of food delivery has thus forced recognition of the importance of materials used for food packaging. Indeed, packaging materials must live up to tough requirements, safely preserving and protecting food contents without compromising the health of consumers. The materials must equally be viable commercial solutions, providing a unit cost reduction and delivering shelf appeal that is attractive to the end user.

Table 11.1: Applications and Benefits of different Plastic Products

Plastic	*Applications*	*Benefits*
PET	Food jars for jelly, jam and pickles	Excellent resistance to most solvents
	Plastic bottles for soft drinks, water, juice	High impact capability and shatter resistance
	Ovenable film and microwavable food trays	Clear and optically smooth surfaces
HDPE	Cereal box liners	Relatively stiff material with useful temperature capabilities
	Reusable shipping containers	Higher tensile strength
	Bottles for non-food items, such as shampoo, liquid laundry detergent, household cleaners, motor oil *etc.*	
PVC	Rigid packaging applications include blister packs and clamshells	High impact strength
	Packaging, film and sheet, and loose-leaf binders.	Brilliant clarity
	Flexible packaging uses include bags for bedding and medical	Excellent processing performance
LDPE	Container lids	Excellent resistance to acids, bases and vegetable oils
	Shrink wrap and stretch film.	Toughness, flexibility and relative transparency
	Squeezable bottles (*e.g.*, honey and mustard).	
PP	Medicine bottles	Low moisture vapor transmission
	Bottle caps and closures	Inertness toward acids, alkalis and most solvents
	Bottles for catsup and syrup.	
PS	Protective foam packaging for furniture, electronics and other delicate items.	Low thermal conductivity and excellent insulation properties
	Packing compact disc cases and aspirin bottles	Excellent moisture barrier for short shelf life products

Source: Industry reports, Analysis by Tata Strategic.

Although the packaging industry can choose from an array of materials for food packaging, polyolefins (polyethylene and polypropylene) present a flexible and robust solution for food packaging, offering a balance between performance, processability and cost to make them the industry's material of choice. Polyolefins is the first choice of material for advanced rigid packaging.

One of the most visible parts of the plastics invasion was Earl Tupper's "tupperware," a complete line of sealable polyethylene food containers that Tupper cleverly promoted through a network of housewives who sold Tupperware as a means of bringing in some money. The Tupperware line of products was well thought out and highly effective, greatly reducing spoilage of foods in storage. Thin-film "plastic wrap" that could be purchased in rolls also helped keep food fresh.

Polyethylene and polypropylene often offer the lowest unit cost of any packaging system when combined with an optimised pack design. Their low densities and high-performance levels, made possible with state-of-the-art high-performance grades, offer extreme light weighting.

Whether considering a full life cycle analysis or the packaging conversion process alone, polyethylene and polypropylene are materials that favourably address environmental concerns such as energy consumption, climate protection, and water conservation, compared to competing and traditional packaging materials.

Polyethylene and polypropylene intrinsically offer low densities which mean low packaging unit weight. This results in an important reduction of material used and minimises waste at end of life. The materials also significantly lower fuel consumption for transport and contributes to lowering our CO_2 emissions. In addition, advanced polyolefins are recyclable or can be turned back into energy to substitute fossil fuels in clean energy recovery processes.

Plastic Film

Plastic is any material made of polymeric organic compounds and additives that can be shaped by flow. This is the most common material used to make high quality and highly significant films.

In 1868 John Hyatt mixed celluloid with camphor and alcohol to find a substitute of billiard balls. Then came celluloid collars, cuffs, *etc.* After the turn of the century, cellophane was invented. The major growth for cellophane and all plastic films started during the late 1930s and after World War II when self-service shopping came into vogue. Many packages required transparency because people wanted to see what they were buying. Since then, many plastic films were added to the list of packaging films.

Plastic films are high performance materials, which play an essential part in modern life. They are mostly used in packaging applications along with some applications in agricultural, medical and engineering fields. Plastic films are perfectly stable, easy to work with and can be lighter than tissue.

Like plastic bottles and containers, film can be made with different resins, each of which has a unique combination of properties that makes it ideal for certain applications. Plastic film also can be clear or colored, printed or plain, single- or multi-layered and combined with other materials such as aluminum and paper. Thus, the only thing that all plastic film really has in common is that it is flexible, as used in grocery bags, as opposed to rigid, as used in soft drink bottles and butter tubs.

There are many material types used in plastic films ranging from single layer polymers to multilayer polymers with tie layers and copolymers. A number of plastics are used in applications such as nylon, polypropylene, cellophane, *etc.* Plastic films do not have voids in the surface that allow an ink or coating to penetrate into it. Certain plastics can be attacked or swelled by specific UV raw materials. These can be matched up with the type of substrate and allow the ink to penetrate the film.

Polypropylene Films

These are heat sealable films for flexible packing and have good clarity, high gloss and high tensile strength. PP film has excellent moisture barrier characteristics. The film can contain colorants, stabilizers, or other additives, and can be coated for the improvement of performance properties. The resin also has a high melting point, which makes it desirable in packages that require sterilization at high temperatures. PP film commonly is used to package such things as cigarettes, candy, snack foods, bakery products, cheese and sanitary goods. It also can be found in shrink wrap, tape, tobacco wrap, diaper coverstock and the sterile wrap used in hospitals and other medical care facilities. Because PP has only average gas barrier properties, it often is used in combination with such things as a PVDC coating or acrylic, which provide additional barrier properties.

Polyvinyl Chloride (PVC) Films

PVC film is gas permeable and necessary for packaging such things as red meat, which require a small amount of oxygen inside the package in order to remain fresh and maintain its bright red colour. PVC film also is a good barrier to oil and grease, is puncture resistant and has good cling and excellent clarity.

LDPE/LLDPE Films

Low density polyethylene (LDPE) film acts as a gas barrier, which is necessary for packaging such things as chicken, which would quickly spoil if exposed to oxygen.

These two polyethylene resins often are talked about as if they are one because they have similar properties; both have good clarity, are good moisture barriers and fair gas barriers, can be heat sealed and are strong and highly flexible. They both are also used in similar applications, including but not limited to stretch wrap; shrink wrap; bags for produce, bakery goods, candy and ice; bags-in-a-box; boil-in-bags; carton liners; bubble packaging; envelope film; industrial liners; overwrap; shipping sacks; textile bags; mattress bags; grocery sacks; garment bags; trash and can liners; and agricultural and construction film. They do, however, have some differences that make them preferable for different applications. For example, LDPE is often selected for its high clarity, ease of processing and high gloss. LLDPE, on the other hand, is selected for its tensile and impact strength (*i.e.*, its toughness) as well as its heat sealability. LDPE tends to be used more often in things like food and non-food packaging and shrink wrap; LLDPE is used more often in trash bags and stretch wrap.

Polyester Films

Polyester film features superior performance in printing, metalizing and cold seal applications. The film shows excellent scuff resistance, machinability and temperature characteristics. The adhesive can be clear or pigmented for ease of identification.

HDPE Films

Because it is part of the polyethylene family, HDPE film is found in many of the same applications as LDPE and LLDPE. For example, HDPE film is used in bakery bags, carton and box liners, cereal and cake mix bags, shipping sacks, industrial liners, retail bags, grocery sacks, T-shirt bags, trash bags and liners, agricultural film, construction film and envelope material (such as Tyvek), as well as many other products and packages. In recent years, it has made inroads into the film market mostly because of its down-gauging properties, which allow manufacturers to use less material (*i.e.*, source reduction) to make a package that can deliver an equal amount of product. HDPE also tends to be stiffer than other polyethylene films, which is an important characteristic for packages that need to maintain their shape. In addition, HDPE is strong and puncture resistant, has good moisture barrier properties and is resistant to grease and oils.

Nylon Films

Nylon, like PP, has a relatively high melting point, which makes it ideal for use in conventional and microwave cooking applications. It also has good oxygen barrier properties, which make it ideal for use with other materials (like paper) that do not have good gas resistance. Nylon film is a transparent, non-heat stabilized plastic film. They absorb water. The higher the moisture content the more flexible they become. They offer excellent thermal and chemical resistance, with high tensile strength, and tear resistance. Nylon is said to be making inroads into cheese packaging because it does not allow oxygen into the package (which would spoil the cheese) but does allow carbon dioxide out (which, if trapped inside the package, would cause it to balloon). Nylon also is easy to process, which may give it an advantage over other specialty film resins.

Cellophane Films

Cellophane film is made by the viscose process, a physical process for making regenerated rayon by treating cellulose with caustic soda, and with carbon disulfide to form cellulose xanthate, which can then be spun into fibers and reconverted to cellulose by an acid treatment.

EVA

EVA is used in some film applications. It has excellent adhesion, is inert and has good flex crack resistance and good heat-sealing properties. Typically, this resin is not used alone but in combination with other film resins (like PE). For example, many bag-in-box applications (like boxed wine, some juices and soft drink syrups) are made of an LDPE/EVA copolymer. EVA also is good for packaging such things as meat, poultry and ice.

Applications of Plastic Films

Plastics are considered to be one of the most valuable and versatile family of materials ever developed. New uses are developed for plastic films almost every day. Plastic film offers a number of significant benefits:

1. Plastic food wrappings prevent spoilage.
2. Plastic wrappings reduce food waste.
3. Plastic vapour and air barriers prevent moisture damage which in turn helps in supporting energy-conservation efforts in home construction.
4. Plastic stretch wrap and shrink wrap reduce packaging weight and bulk. This helps in reducing transportation and storage costs.
5. Agricultural plastic film helps in reducing weeds, keeping seedlings moist, wrap silage and increase crop yields. From the time these films are produced, to their use and reuse, and to their final recycling, or disposal, plastic films provide environmental benefits.
6. Plastic film can be coloured or clear, printed or plain, or laminated to aluminum, paper and other materials.

There are three types of packaging applications in which film is used: food, non-food and other. It is important to keep in mind that within each of these categories, plastic film can vary by resin and color; it also may be made of one layer of plastic or as many as ten layers depending on the complexity of the package. In addition, other materials such as aluminum or paper may be used in combination with plastic film in order to impart special properties. Therefore, even the following categories of film are themselves made up of many diverse types of film.

Food Packaging

Plastic films for packaging are available in the form of packaging pouch, packaging bags, packaging rolls, sheets, foils *etc.* Food packaging film is used in such things as in-store bags for produce (such as apples and potatoes); all nonfrozen baked goods (such as rolls and breads); bakery bread and bun bags; tray covers for institutional deliveries of bakery products; bags-in-a-box (film used to contain fluid in a supportive box, such as boxed wine); boil-in-bags (film used to contain food prepared by keeping it in the package and placing it in boiling water); candy and confection bags and wrappers; carton liners (for such products as cake mixes); and meat, poultry and seafood wraps (such as hot dog and bacon film); milk bags; and grocery bags. The only data available on the use of film in food packaging applications are restricted to the polyethylene family, despite the fact that other resins are used in food packaging as well. It is worth pointing out that LDPE is the polyethylene resin used most often in food packaging; it accounts for 65.5 per cent of the total, with LLDPE making up 25.6 per cent and HDPE making up 8.9 per cent

Non-food Packaging

Non-food packaging film refers to such things as industrial liners (film used to line supported structures such as gaylord boxes, frozen pork box liners and liners for shipments of nuts and bolts), shipping sacks (film used to protect and/or contain contents such as bark and mulch bags), bubble packing, mailing envelopes, dry cleaning bags, multiwall sack liners, diaper overwrap, stretch wraps and rack and counter bags. Again, data on film used in non-food packaging applications are confined to the polyethylene family. LDPE is the polyethylene resin used most frequently in non-food packaging applications. It composes 54.9 per cent of the polyethylene used in non-food packaging, whereas LLDPE composes 35.5 per cent and HDPE composes 9.6 per cent.

Other Packaging

The other types of packaging in which film is found are stretch and shrink wrap. Stretch wrap is a strong, highly flexible film that can be stretched to take the shape of a product or products. It is used in a variety of applications ranging from overwrapping fresh meats to securing shipping cartons to pallets. Stretch wrap usually is made of co-extruded LLDPE and LDPE, although it can be made from individual plastic resins, such as LLDPE, LDPE and PVC.

Shrink wrap, on the other hand, is a plastic film that is applied loosely around products, sealed by heating the seams and shrunk through a heating process to take the shape of the products. In shipping, it can be used to bind multiple packages of less than pallet size together (such as five 20-ounce cans of beans or three juice boxes) or used over an entire pallet of packages. In these applications, shrink wrap typically is made of LDPE, although it can be made from other resins as well, such as LLDPE and PP. In addition to shipping applications, shrink wrap also can be used for bundling purposes, such as bundling magazines and papers, and it can be used to protect and display such products as albums and compact disks. There are several ways to tell stretch wrap and shrink wrap apart: stretch wrap usually feels somewhat tacky to the touch and is very flexible, whereas shrink wrap may be more brittle (or crinkly) to the touch and does not stretch when pulled. In addition, stretch wrap usually is wrapped around products, whereas shrink wrap will enclose the product (that is, have a top and bottom cover), which makes it an attractive choice for shipping products in extreme weather conditions or for products that need extra protection.

Examples of Innovations for Advanced Packaging

Source Reduction/Light Weighting in Thin Wall Consumer Packaging for Yellow-fats, Dairy and Similar Markets

It is a new high stiffness packaging with high impact polypropylene block copolymer. The exceptional flow and strength allow processors to use injection moulding to achieve packaging with walls as thin as 0.3 mm without compromising on impact resistance.

Aesthetics/Appearance Improvements in Ice Cream Packaging

It is a transparent polypropylene for proven consumer appeal with impact properties designed for deep freeze storage and transportation at -20°C.

Processing Efficiency with Cycle Time Reduction in Pail Production

The high flow polypropylene block copolymer with high impact allows fast injection cycle time in multiple cavities and in more demanding shapes *e.g.*, rectangular paint pails. This allows unit cost reductions in both machine time and energy utilisation.

Convenience *e.g.*, Easy to Open in Sports-lock Bottle Closures

The design freedom of a random copolymer polypropylene allowing convenience features in Sports-lock' type bottle closures but with the outstanding taste and odour properties previously only associated with the mould design limited 'Organoleptic HDPE' grades.

Hot Fill Capability for Juice, Dairy, Sports-drink and Other Liquid Packaging

The market leading injection stretch blow moulding polypropylene (ISBM PP) grade that offers outstanding transparency, high productivity and light weighting capability, while also allowing improved hot fill capability versus competing HDPE and PET bottle solutions.

Plastic Packaging: A 'Service' Industry

If a packaging manufacturer offers complete solutions, the customer can transfer entire responsibility for the production process to the supplier. This outsourcing opportunity is increasingly used by customers.

It also includes innovative advice on packaging problems and reliable, fast deliveries, up to 'Just-In-Time' supply. The packaging industry is looking to cooperate more closely with suppliers of raw materials and machinery.

The packaging industry, raw material suppliers and machinery manufacturers are already cooperating globally in Europe, USA and in Japan. Fast methods of communication by e-mail and computerised design ensure that this cooperation is even more effective.

For example, the food industry needs re-sealable peelable packaging allowing products such as vegetables and meat to be removed, resealed and removed again later.

Laminates are another example of inter-sector cooperation, and here the packaging industry is working with manufacturers of adhesives and machinery to develop eco-friendly solvent-free laminates with a shorter curing time.

In other sectors, the potential of PET for high-quality clear rigid containers is still only at its beginnings: wide-mouth, hot-fill, high barrier solutions are moving from laboratories to production lines.

Advantages of Plastic Usage in Food Packaging

When the packaging questions are tough, plastics are often the answer. Sometimes they are the only answer, performing tasks no other materials can perform and providing consumers with products and services no other materials can provide.

Plastic packaging continues to have the wrap on consumer preference. Freshness, storage stability and ease of preparation are among the consumer goals driving the popularity of plastic food packaging. Offering safety, quality, convenience and savings, plastic packaging meets the needs of consumers. Different plastics offer different qualities, giving manufacturers and consumers the freedom to choose the type of plastic that best suits the application. Plastics can be rigid when protection is needed, or flexible for convenience's sake. They can be clear or opaque. And they can be molded into a wide variety of shapes and sizes. New ideas include plastic containers for cereal, coffee, spices and baby food, as well as squeeze bottles that allow portion control of juice concentrates and keep contents fresh in the refrigerator for up to five weeks. Freezer-to-oven-to-table plastic food packaging is now available for both microwave and conventional oven use. And plastic container design itself is participating in the cooking, with innovations such as tapered popcorn boxes that keep the kernels in the hot oil and microwavable cake mixes with reusable trays.

In other types of packaging as well, consumers and the hospitals, schools and other institutions that serve them increasingly are turning to plastics. Safe, sanitary, easy to use and economical, plastic packaging is the shape of the future. Various advantages of plastic packaging are:

Safety by Design

In the home, break-resistant, shatterproof and no-spill plastic bottles cut down on injuries and clean-ups anywhere the floor is hard and hands may be slippery.

Food-service outlets and their customers rely on plastic packaging to protect food products against contamination and retain desired temperatures longer. And single-serve plastic packaging for condiments not only preserves freshness and flavor, it also ensures the consumer a sanitary portion while cutting down on food waste.

Plastic packaging molds itself to modern lifestyles. Today's working parents and busy homes rely on its convenience and the services it provides. Microwave ovens have become a near necessity in homes, and plastic trays are the package of choice for consumers. Microwave cooking enables active people to eat well without spending their limited leisure time on food preparation. The elderly, too, benefit from the ease of microwavable food packaging.

Plastic packaging also preserves flavor and saves time in conventional cooking and storage. Squeeze bottles for condiments, boil-in-bag dishes, resealable bags for everything from shredded cheese to cereal, freezer bags that protect food against ice crystals, precooked foods that are microwavable in the package all contribute to quality meals in the home.

Extremely lightweight and molded to promote easy handling, plastic containers allow consumers to enjoy the savings of beverages, detergents and other products in the "large economy size." And plastic packaging, which can be transparent without being fragile, enables consumers to see what they're getting and to serve themselves.

In addition to saving space in today's smaller living quarters, plastic packaging can be as decorative as it is serviceable. Further, it won't leave rust rings on counters and fixtures. For a host of personal and home products, plastic packaging works well and looks good, too.

Differentiation through Design

Plastic packaging is unique in that it allows for a variety of design capabilities. Design innovation resulting in both aesthetic and functional value helps to differentiate end-market products on the shelf. Due to the large number of end-market products on the shelves today, design elements have become an increasingly important distinguishing characteristic in the eyes of the consumer.

A major trend in packaging design is the use of shrink-sleeve labeling. Shrink-sleeves are most often produced with PVC, but can also be produced from PETG (Polyethylene Terephthalate Glycol Co-monomer) and OPS (Oriented Polystyrene), which have different shrink characteristics and may be more appropriate for certain applications. Using this type of labeling provides 360-degree design capabilities that allow for maximum utilization of the container surface for optimal graphic visibility. In markets where product differentiation is limited, shrink-sleeves can offer a highend, attention-grabbing design that distinguishes the product from others on the shelf.

Packaging that enhances functionality of a product can also have an impressionable effect on the consumer, and plastic holds an advantage over some substitute materials, with its ability to be formed into any number of shapes that compliment or improve product usage. The production of dual-chamber containers, where two products mix as they are dispensed, is one example of a product design gaining popularity. In addition, packaging attributes that address the needs of on-the-go lifestyles such as single-use pouches, handles for easy carrying, and re-closable openings will continue to grow.

Use for Shipping, Storing, Savings

The use of plastics for shipping and storage will continue to grow. Strong, durable and tear resistant, plastic packaging saves energy, space and money. Plastic containers, which generally require less energy to manufacture than other packaging, also require less fuel to transport than heavier materials. Additional savings come from reductions in shipping damage and elimination of the need for additional packing materials, such as partitions between individual products. Strong enough for stacking and moldable into space-saving shapes, plastic containers can maximize warehousing room and lower storage costs.

Meeting unique packaging needs from anti-static protective packaging for electronic components to shelf-stable containers for products that once required costly cold storage are a specialty of plastics. Because they can be molded to fit contours, plastics provide the ultimate protection in packaging office machines, entertainment units, food products and other delicate products. Tough enough to withstand the stresses of transportation yet capable of screening out even the smallest particle of dust, plastic packaging delivers. These factors all add up to savings for producers and merchants and can result in lower prices for consumers.

Convenience

Consumer lifestyle changes have recently resulted in an increased emphasis on convenient, on-the-go packaging options, especially in the food and beverage industries. Characteristics which cater to these preferences include light weight, portability, and durability. Now that plastic has achieved a number of protective capabilities on par with glass and metal substitutes, it has begun to differentiate itself as the best option for new product innovation due to its wide range of applications.

Flexible plastic packages, such as stand-up pouches, have recently been gaining market share at the expense of some rigid packaging alternatives. Similar to films, less resin is used in manufacturing, which lowers production costs. In addition, flexible packaging requires significantly lower transportation costs and less customer shelf space than rigid packaging. Pouches also offer lightweight portability, effective single-serving options, and extensive marketing capabilities. Pouch packaging is not a new technology; in fact, it has been used since the 1960's by the U.S. Army. However, improved barrier properties, combined with current consumer life-styles and a growing demand for ready-to-eat (RTE) products have resulted in a surge of demand. Retort pouches are popular for these kinds of applications. In the retort process, the pouch is filled, sealed, and then retorted through a thermal process that cooks or sterilizes the product. Plastic pouches now have shelf-life comparable to that of metal or glass containers, and also have the advantage of being suitable for microwave preparation, which caters to a convenience driven consumer base. Companies who can effectively capitalize on the advantages of pouch packaging in the near future are expected to capture market share in this high growth space.

Barrier Protection

Recent trends in the plastic packaging industry have led flexible packaging producers towards the manufacture of thinner gauge films due to the material reduction advantages. Product pricing is an important competitive factor because most of these products are high volume, low margin items, which has led to commoditization and a lack of differentiation in most film, sheet, and bag. With resin costs accounting for approximately two-thirds of the cost of goods sold, keeping raw material costs to a minimum has become a priority.

Coinciding with the push towards thinner gauge plastics is increasing customer demand for improved performance and barrier protection. In addition, products used to package foods, beverages, and pharmaceuticals need to comply with legislative requirements of state and federal health and safety authorities. The food and beverage industry is focused on manufacturing thinner gauge film with high barrier protection, and the pharmaceutical industry is determined to realize continued improvement of child-resistant and tamper evident closures. Producing packaging that provides for better performance and protection is a major objective driving innovation in the space.

Advantages of Plastic

- They are light in weight, can be easily moulded and have excellent finishing.
- They possess very good strength and toughness. They possess good shock absorption capacity. · Advantages of plastic are corrosion resistant and chemically inert.
- They have low thermal expansion of co-efficient and possess good thermal and electrical insulating property.
- Advantages of plastic is very good water resistant and possess good adhesiveness.
- Plastic is strong, good and cheap to produce.
- Plastic is a recycling process and it does not decompose.
- Plastic bottles can be reused and restored over again and again.
- Plastic is one of the unbreakable.
- Plastic is an odourless.

- ☆ Plastic is used for building, construction, electronics, packaging (gladwrap) and transportation industries.
- ☆ Used to make - Water bottles, pens, plastic bags, cups *etc.*
- ☆ Advantages of plastic are Very cheap to make.
- ☆ Durability
- ☆ Strength
- ☆ Chemical resistance.
- ☆ Advantages of plastic are used to produce another product.
- ☆ Used to reduce soil and wind erosion.

Disadvantages of Plastics

- ☆ Plastic is a non-renewable resource.
- ☆ Most plastics do not degrade. Instead, they slowly fragment into smaller particles, referred to as microplastics, and, probably, nanoplastics. These particles, whether in the form of larger or smaller plastics, have profound detrimental consequences for ecosystems, biota, and the environment, but also for the economy and human health.
- ☆ Causes cancer
- ☆ Embrittlement at low temperature.
- ☆ Deformation under load.
- ☆ Low heat resistant and poor ductility.
- ☆ Combustibility.
- ☆ Produces toxic fumes when it is burnt
- ☆ It is a recycle process, but it is very costly.

Flexible films can be produced with very little resin, which reduces production costs when compared to other packaging materials. Co-extrusion technologies have enabled manufacturers to combine various resins in order to achieve a highly effective barrier, while at the same time producing thinner gauge films that cost less to the manufacturer. However, combining two or more materials in the package has created recycling problems for many co-extruded films.

Two specific packaging technologies that have been recently gaining momentum are modified atmosphere packaging (MAP) and nanotechnology. The MAP technique involves the practice of altering the composition of the internal atmosphere of a package in order to improve a product's shelf life. This trend is most common in foods, but is also used in pharmaceutical packaging. Nano-composites are plastics that have fillers dispersed throughout the resin which reduces permeability in the resulting film. Nanotechnology offers enhanced barrier protection, increased shelf life, and light-weight design. These developments have significant potential in improving the barrier performance of plastic packaging films and creating new markets for plastic packaging.

Globalization of Plastics Packaging Industry-Economic Factor

Plastics packaging came into widespread use with the introduction of polyethylene in the fifties. Before this, in addition to classical packaging materials such as paper, glass and wood, we used films of converted natural materials such as cellulose acetate and cellophane transparent cellulose film. The development of polystyrene, polypropylene, PVC, polyesters and polyethylene copolymers saw the start of the rapid increase in the use of plastics.

Plastics packaging is everywhere today. In spite of the size and economic importance of the industry, it is a fact that manufacturers of plastics packaging (predominantly medium-sized and small companies) are now open to a double dependency: on the one side the raw material suppliers dictate the prices of plastics, and on the other side there is massive downward pressure on prices by customers - particularly in the food industry.

In addition, there is increasing competition, especially from Eastern Europe, where manufacturers have capacities for high quality extrusion and printing at a lower cost. One can also expect increasing competition from the Far East, although manufacturers there are yet not as advanced as in Europe, particularly in the fields of barrier materials and printing technology. An additional pressure is that the packaging market is characterised by growing overcapacity, which is placing even more pressure on prices. Nevertheless, demand for packaging will continue to increase.

It is anticipated that growth rates over the last few years averaging 4-5 per cent per annum will remain at these levels, or increase. There also appears to be a slight trend in Europe towards flexible packaging systems.

As far as the individual plastic materials are concerned, very high growth rates are expected for PET, in the field of rigid packaging. The main application for PET is bottles for carbonated drinks and mineral water. In Europe and the USA, PET bottles are now also being tested for packaging beer.

Another application for PET which will gain in significance is food (such as preserves) that is hot-filled. Growth is also expected in the use of injection moulded polypropylene for large tubs and buckets that will gradually replace metal containers. PVC, which is under pressure from other materials in the food sectors, holds a strong position in 'bubble' packaging of pharmaceutical tablets and in 'display' packaging of products such as tools, ironmongery and hardware.

In terms of product areas, food packaging which is the largest single product area in the whole packaging industry will be the major growth market for plastics packaging. The growth of the market is assisted by demographic developments in the modern world, such as the steady increase in single and two-person households and the growing number of elderly people, which is fundamentally influencing consumer purchasing habits.

The market demand today is for practical, time-saving ready or deep-frozen meals in small microwaveable easy-to-open packaging or packaging that can be resealed. The trend towards these convenience products is enhanced by the fact that more and more people can now afford them. The flexibility of plastics, in protective properties and processability, has given them an excellent position for fulfilling the specific packaging requirements of this market.

Another new and growing special market is the packaging of pharmaceutical and medical products. Again, there is a strong and growing demand for these products in industrialised countries, where considerable consumer expenditure is available for healthcare. These products, such as medicines, prostheses and hygienic products place particularly high demands on packaging, for sterility, protection, appearance and security, which can well be met by plastics.

The increasing level of the globalisation of business processes has led to structural changes for raw material suppliers and customers of the packaging, which have far-reaching effects on plastics packaging companies. As for all large companies, it is essential for large purchasers of packaging today to have a global presence. They expect that packaging manufacturers will be capable of providing local deliveries to them, wherever they may be located. Those who manufacture products in China assume that their packaging supplier will also build a factory there.

Increasingly raw materials suppliers, such as BASF and Shell, are merging in full or part to remain competitive in the global market. The trend is moving towards raw materials supplies being restricted to pure plastics without any additives, in order to streamline production. Plastics processors will need to purchase additives such as stabilisers, pigments and lubricants separately and then mix them themselves to obtain the correct blend.

This means that the raw material suppliers will shift responsibility for raw materials to the plastics processing industry. If this trend becomes established it will be essential that the plastics processors acquire the necessary expertise and invest in new specialised machinery to meet this challenge. It is also unavoidable that small and medium-sized companies will have to join forces to form larger groups and operate on a global level.

This means that, on the one hand, they will form a counterbalance to the raw material suppliers, while on the other they will only be able to offer their customers a comprehensive range of packaging through consolidated purchasing with other plastics processors.

At the same time, wide-ranging joint ventures between companies will allow small and medium-sized companies to gain better financial control. On a political level, too, co-operation is urgently required. There is a whole series of very professional national and sector associations, such as the Plastic Packaging Industry Association in Germany, the British Plastics Federation and La Federation de la Plasturgie in France, to name but a few.

It is important that the industry speaks with one voice to ensure that it does not splinter. Already there is fruitful cooperation at European level with the European Plastics Converters Federation. This activity needs to be intensified and is particularly important in the fields of environmental protection and recycling. The European Commission together with the Packaging and Packaging Waste Directive of December 1994 imposed far-reaching regulations, including those for the reduction of packaging material and a high level of recycling. In Germany the packaging industry has suffered a loss of competitiveness on a European level as a result of the expensive waste separation system of the Dual System. Regulations need to be standardised in all countries to prevent this imbalance.

Plastics in the Environment

There is a growing awareness of the health and environmental consequences of food packaging, especially plastic packaging made from petroleum products. There are multiple costs to our increasing use of packaging for food and a conversation about this use is probably in order.

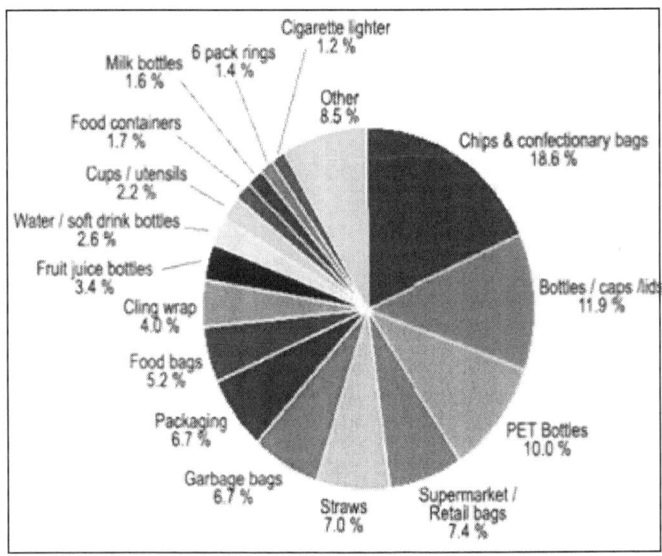

Figure 11.2: Source of Plastic Waste.

Causes of Plastic Pollution

- Trash dumps and landfills are unfortunate major problems, as they allow pollutants to enter the ground and affect wildlife and groundwater for years to come.
- As plastic is less expensive, it is one of the most widely available and overused item in the world today. When disposed, it does not decompose easily and pollutes the land or air nearby when burned in the open air.

- ☆ Burning plastic is incredibly toxic, and can lead to harmful atmospheric conditions and deadly illness.
- ☆ Therefore, if it is in a land fill, it will never stop releasing toxins in that area.

Plastic Pollution Facts

- ☆ Daily, 8 million pieces of plastic reach the oceans.
- ☆ Yearly, this translates into between 4.8 and 12.7 million tonnes.
- ☆ It is the equivalent of a garbage truck full of plastic dumped into the ocean every minute
- ☆ Of the total amount of plastics sent to landfills, 79 per cent is transported to the oceans, less than 10 per cent is recycled and 12 per cent is incinerated.
- ☆ 25 trillion macro- and 51 trillion microplastics litter the oceans.
- ☆ Of these, 269,000 tonnes float on the surface.
- ☆ This equates to 1345 blue whales and 500 times the number of stars in the Milky Way.
- ☆ Plastic has been found throughout the Globe, including in remote and isolated locations.
- ☆ Plastic in expected to increase 10 fold in the next 5 years

There are three possible environmental problems to be considered. Firstly, plastics are mostly made from oil, natural gas or coal, and these are all limited natural resources that must be conserved. Secondly, the manufacture of plastics produces a lot of harmful pollutants which manufacturing companies need to deal with properly. When plastics are used, recycled, or disposed of, or left in the environment as litter, they break down and release harmful chemicals. These pollutants include heavy metals such as cadmium and lead, and chemicals such as benzene, dioxins, and other pollutants, which all release harmful toxins into our air, water, and bodies. Thirdly, old and unwanted plastics are not always easy to dispose of. Buried in landfill sites they will take a very long time to rot. Right now, most plastic is being wasted sent to landfills or, more likely, incinerators. Burning plastic in incinerators releases toxic heavy metals and chemicals. Incinerators produce a variety of toxic discharges to the air, water, and ground that are significant sources of powerful pollutants, including dioxin and other chlorinated organic compounds that are well known for their toxic effects on human health and the environment. Many of these toxins enter the food supply and become more concentrated as they move up through the food chain. In addition to air and water emissions, incinerators create toxic ash—or slag—which contains heavy metals, dioxins, and other pollutants. This toxic ash must be land filled, and the pollutants present in the ash can then leach into groundwater.

In fact, garbage incinerators and medical waste incinerators are two of the largest sources of dioxin identified by the U.S. Environmental Protection Agency. Dioxin is the common name for a class of 75 chemicals. It is a toxic waste product formed when waste containing chlorine is burned or when products containing chlorine are manufactured. Dioxins are among the most potent synthetic chemicals ever tested, causing cancer and harming our immune and reproductive systems even at very low concentrations.

In landfills, leachate is produced when water picks up toxins as it seeps through the trash. This trash includes plastics of all types, even older plastics that have been proven to be toxic but are still in our landfills. Although landfills attempt to collect this toxic leachate, it also leaks into ground and surface water, releasing pollutants into the environment and causing health risks for humans and wildlife.

There were 359 million tonnes of plastics produced worldwide in 2018. Only a very small percentage of recycled bottles are used to make new bottles. Coca-Cola sources only 20 per cent of its packaging from recycled material, while Nestlé Waters North America uses just over 16 per cent recycled content. Reduction of the use of packaged and throw-away items seems the right course. Reitman-White and Doppelt (2006) are correct when they say that retailers like PCC and The Food Co-op can't change the market by themselves. Consumers also need to take responsibility for their own personal choices, by choosing more sustainable products, and by simply choosing to use fewer packaged products.

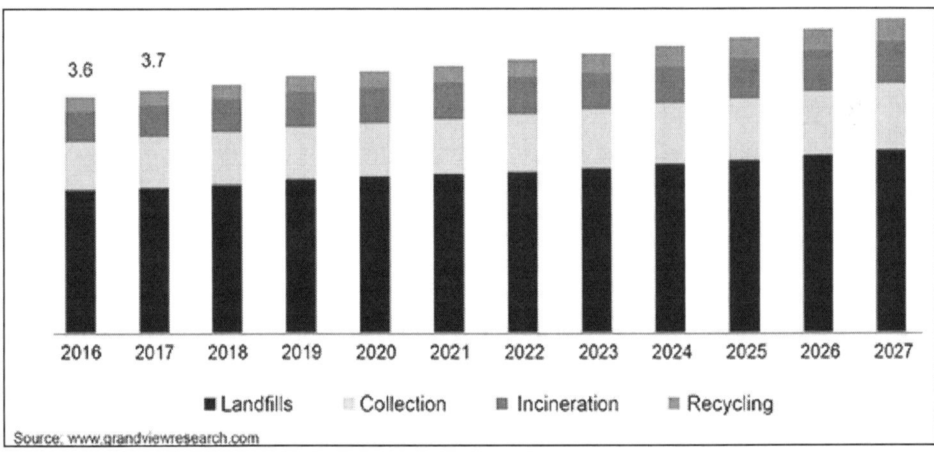

Figure 11.3: US Plastic Waste Management Market Size, by Service, 2016-2027 (USD Billion).

A Throw-away Society

There's no dispute that the Western World has a throw-away problem. 3.13 million tons of PET plastic were discarded in the U.S. in 2018. There were 1.9 million grocery bags and other plastics bags collected in the 2018 International

Coastal Cleanup. According to the EPA, 8.7 per cent of plastic material generated in the U.S. Municipal Solid Waste (MSW) stream was recycled in 2018. Another 15.8 per cent was combusted for energy, while 75.6 per cent was sent to landfills. The plastic waste that is thrown into seas every year can kill as many as 1 million sea creatures. Plastics cause serious environment pollution such as soil pollution, water pollution, and air pollution.

All this packaging contributes not only to the landfill, but also consumes a huge amount of resources including energy to produce. Despite a growing amount of postconsumer content, most paper products are made from trees cut from our dwindling forests. Plastics take up 6 to 8 per cent of the total oil and gas we use, a portion that will look more and more significant as we exhaust the world's oil and gas supplies. In addition to environmental concerns, food packaging raises health concerns. Many studies warn that potentially carcinogenic and hormone-disrupting chemicals in plastic food packaging can leach into food. Leaching increases when plastic comes in contact with oily or fatty foods, especially when heated, and from old or scratched plastic. Types of plastics shown to leach toxic chemicals are plycarbonate, PVC and styrene. A growing number of environmentally concerned consumers are creating a hot market for new, biobased products. Could these bio-plastics replace oil-based plastics?

PVC: The Toxic Plastic

Polyvinyl chloride, also known as vinyl or PVC, poses risks to both the environment and human health. PVC is also the least recyclable plastic.

1. Vinyl chloride workers have elevated risk of liver cancer.
2. Vinyl chloride manufacturing creates air and water pollution near the factories, often located in low income neighborhoods.
3. PVC needs additives and stabilizers to make it unsealable. For example, lead is often added for strength, while plasticizers are added for flexibility. These toxic additives contribute to further pollution and human exposure.
4. Dioxin in air emission from PVC manufacturing and disposal or from incineration of PVC products settles on grasslands and accumulates in meat and dairy products and ultimately in human tissue. Dioxin is a known carcinogen. Low level exposures are associated with decreased birth weight, learning and behavioural problems in children, suppressed immune function and disruption of hormones in the body.

Plastic as Green Packaging

Environmental concerns influence the packaging industry, as indicated by increasing pressure on regulatory agencies from lobbyist groups. Standards for recycling are becoming more stringent, and pressures to reduce the amount of material in packaging are mounting.

This pressure is an additional factor that is causing a shift away from rigid packaging in favour of flexible options. Flexible packaging uses less material for the same application, and frequently the resins used in flexible packaging are more easily recycled. A plastic pouch will typically use 50 per cent less material than a rigid plastic container or bottle for the same application, and up to 75 per cent less material (by volume) than a glass container or bottle.

Bioplastics are also beginning to make headway in the plastics packaging space as biopolymers have recently begun to gain significant attention industry-wide after years of undelivered promises surrounding biodegradable and eco-friendly materials. One of the most influential biopolymers currently marketed is NatureWorks' corn-based polylactic acid (PLA.). Major corporations and venture capitalists are banking on it. NatureWorks LLC, a subsidiary of Cargill Dow, opened a new factory in Nebraska in 2001 and is generating more than 300-million pounds of corn-based plastics per year using 40,000 bushels of Cargill corn each day in the process. This corn-derived polymer is used for soft-drink cups, salad and fruit containers, as fill in pillows and comforters, and even in donut boxes and gift card wraps. But is it really an environmentally friendly alternative?

PLA offers packaging manufacturers a cost-competitive, renewable option to replace traditional petroleum-based plastic materials. PLA uses 68 per cent fewer fossil fuel resources than traditional plastics in its manufacturing and is the world's first greenhouse-gas-neutral polymer. In addition, current research shows NatureWorks PLA can exist in the present North American infrastructure with the existing commercial systems for recycling PET and HDPE.

Spartech Corp. has recently announced the production of Rejeven8, made from 95 per cent NatureWorks PLA and is said to better match properties of traditional PET. Some manufacturers are already beginning to use PLA in thermoformed and shrink wrap packaging applications. By doing so, retailers are able to differentiate themselves on the shelves, by catering to an increasingly eco-friendly consumer base.

Packaging and Plastics: What's a Consumer to Do?

One of the most common and persistent concerns raised by PCC members is the need to reduce packaging, especially plastic packaging. Sustainable solutions seem elusive, though new possibilities are emerging. How can conscientious consumers support a market for more sustainable products? How do we distinguish between real, long-term solutions and short-term fixes that appear attractive but fail in the long run to protect our environment and health? One such conundrum involves the new biodegradable, plant-based plastics. Will these be the wave of the future? Informed consumers increasingly are concerned about the environmental and health risks of packaging particularly the packaging of our food.

Compostability and Recycling Issues

Bioplastics are currently used in disposable items like packaging, containers, straws, bags and bottles, and in non-disposable carpet, plastic piping, phone casings, 3-D printing, car insulation and medical implants. The global bioplastic market is projected to grow from $17 billion this year to almost $44 billion in 2022. There are two main types of bioplastics.

PLA (polyactic acid) is typically made from the sugars in corn starch, cassava or sugarcane. It is biodegradable, carbon-neutral and edible. To transform corn into plastic, corn kernels are immersed in sulfur dioxide and hot water, where its components break down into starch, protein, and fiber. The kernels are then ground and the corn oil is separated from the starch. The starch is comprised of long chains of carbon molecules, similar to the carbon chains in plastic from fossil fuels. Some citric acids are mixed in to form a long-chain polymer (a large molecule consisting of repeating smaller units) that is the building block for plastic. PLA can look and behave like polyethylene (used in plastic films, packing and bottles), polystyrene (Styrofoam and plastic cutlery) or polypropylene (packaging, auto parts, textiles). Minnesota-based Nature Works is one of the largest companies producing PLA under the brand name Ingeo. Advocates highlight that corn-based plastics are 100 per cent biodegradable as long as they're in a controlled environment with at least 140-degree temperatures and 90 per cent humidity for 40 to 60 days. Given these conditions, they degrade into carbon, hydrogen and oxygen.

Unfortunately, most consumers don't have access to the kind of industrial composting facilities needed to create these conditions. One composting specialist says few people know that corn-based plastic requires very specific conditions to break down; it will not degrade in backyard compost, nor will it degrade more quickly than other plastics in a landfill. Another common misperception is that corn-based plastic can go into residential recycling bins with other plastics. This is not true. There currently is not a large enough market to recycle this type of plastic.

Some professional recyclers express concern that PLA can contaminate the #1 plastic recycling stream because it's indistinguishable from other clear plastics and may be mixed in mistakenly. Others think recycling is unlikely to be affected adversely until PLA makes up a far greater percentage of the plastic stream.

PHA (polyhydroxyalkanoate) is made by microorganisms, sometimes genetically engineered, that produce plastic from organic materials. The microbes are deprived of nutrients like nitrogen, oxygen and phosphorus, but given high levels of carbon. They produce PHA as carbon reserves, which they store in granules until they have more of the other nutrients they need to grow and reproduce. Companies can then harvest the microbe-made PHA, which has a chemical structure similar to that of traditional plastics. Because it is biodegradable and will not harm living tissue, PHA is often used for medical applications such as sutures, slings, bone plates and skin substitutes; it is also used for single-use food packaging.

While bioplastics are generally considered to be more eco-friendly than traditional plastics, a 2010 study from the University of Pittsburgh found that wasn't necessarily true when the materials' life cycles were taken into consideration.

The study compared seven traditional plastics, four bioplastics and one made from both fossil fuel and renewable sources. The researchers determined that bioplastics production resulted in greater amounts of pollutants, due to the fertilizers and pesticides used in growing the crops and the chemical processing needed to turn organic material into plastic. The bioplastics also contributed more to ozone depletion than the traditional plastics, and required extensive land use. B-PET, the hybrid plastic, was found to have the highest potential for toxic effects on ecosystems and the most carcinogens, and scored the worst in the life cycle analysis because it combined the negative impacts of both agriculture and chemical processing.

When bioplastics are not discarded properly, they can contaminate batches of recycled plastic and harm recycling infrastructure. If bioplastic contaminates recycled PET (polyethylene terephthalate, the most common plastic, used for water and soda bottles), for example, the entire lot could be rejected and end up in a landfill. So separate recycling streams are necessary to be able to properly discard bioplastics.

Bioplastics are also relatively expensive; PLA can be 20 to 50 per cent more costly than comparable materials because of the complex process used to convert corn or sugarcane into the building blocks for PLA. However, prices are coming down as researchers and companies develop more efficient and eco-friendly strategies for producing bioplastics.

Alternatives to Curb the Wide Use of Plastic

1. Use paper or cloth bags instead of plastic bags.
2. Reuse plastic bags for many times to reduce consumption, and hence curbing the production of them.
3. Use glass containers such as Pyrex.
4. Use of plastic bags can be reduced through codification by imposing heavy taxes or fees on the production of them.
5. Compliance with proper criteria and standards on the production of plastic bags makes the bags thicker, more durable and biodegradable.
6. Reduce the use of plastic-wrapped products.
7. Mothers should be cautious not to use plastic bottles for feeding their babies.
8. Do not buy plastic toys which children can chew, and replace them with toys made of natural materials.
9. Do not buy or use plastic cups, plates or spoons, especially the disposable single-use ones.
10. If you select between different plastic products, choose the product that can be reused or recycled.
11. Do not take more plastic bags than you need when shopping.

12. In case you buy a product in a plastic can or bottle, reuse it for other purposes instead of disposal.
13. Avoid using plastic cups made of rigid polystyrene in drinking tea, coffee and other hot drinks. Use paper or glass cups instead.

GMO Issues

Even if consumers can get their corn-based plastic to a composting facility, many eco and health-conscious consumers still are critical of the raw material used to create the product. Corn-based plastics are made from non-organic corn, which already is at least 46 per cent of the U.S. corn planted. GE corn is so widespread that many companies, such as Patagonia and PCC, have chosen not to use corn-based plastics until a non-GE source of raw material can be guaranteed. Others have opted for the Source-Offset program offered by NatureWorks, whereby, for every pound of NatureWorks PLA delivered, a corresponding amount of non-GE field corn is purchased. Theoretically, NatureWorks PLA could be made from other high-sugar crops such as sugar beets (also genetically engineered). But the only economically viable raw material today is corn.

GE-corn Issues

GE corn is the most petroleum intensive crop except for cotton. Petroleum is used to power machines in cultivation and harvest, to make synthetic fertilizers and pesticides, to transport corn for manufacture into bioplastic, and to shred and heat the bio-plastic for composting. GE corn threatens organic farmers and a sustainable environment. Pollen drift and genetic contamination reduce biodiversity. Insects develop resistance to the engineered Bt toxin over time, threatening to render natural Bt sprays useless for organic farmers, a tool allowed as a last resort. GE corn kills the larvae of beneficial pollinators, including the Monarch butterfly, and other beneficial insects. It fails, for example, to kill the African Cotton Worm, but kills the beneficial Lacewing that eats Cotton Worms.

Why not GE Corn Plastic?

A society wanting sustainable solutions needs to avoid quick fixes that aren't sustainable and ultimately make things worse. From a systems perspective, GE corn plastic is a classic case of a "fix that fails" and "shifting the burden" dynamics.

GE corn plastic builds a market and provides justification to expand GE acreage. Expansion is certain to have negative economic, social and environmental consequences for PCC and many others. Crop diversity slowly but surely will be reduced, putting the entire food system at risk. Large industrial agricultural corporations will exert even greater control over production and distribution, which increasingly will threaten small farmers and organic food production. This will lead to increased prices because there'll be fewer suppliers controlling the market. In short, use of the product now may help in the short run, but harm PCC's business in the future.

While GE corn plastic might seem to make sense in the short term, its many negative consequences could make it harder or even impossible ever to return to sustainable solutions. Bottom line: it's no solution to replace one problem with another. Better to avoid GE bioplastics now and encourage better choices that won't harm people, the economy and the environment for a long time into the future. With this type of issue, "slower is faster and faster is slower."

Resource Conservation

Conserving resources means using less raw materials and energy throughout a product's life cycle from development through disposal or recycling. Plastics are derived from natural resources typically oil and natural gas. And yet, in part because of plastics' unique characteristics *i.e.,* lightweight, durability, formability *etc.* they can conserve more resources during a product's life when compared to some other materials.

After their intended use, plastic containers often can be used again for the same or a different purpose. Plastic grocery sacks can tote wet swimsuits home from the beach or garbage to the bin. In the hands of ingenious consumers, plastic milk jugs become planters and plastic soda bottles are converted to bird feeders.

EPA used data from the American Chemistry Council and the National Association for PET Container Resources to measure the recycling of plastic. While overall the amount of recycled plastics is relatively small *i.e.* three million tons for a 8.7 per cent recycling rate in 2018; the recycling of some specific types of plastic containers is more significant. The recycling rate of PET bottles and jars was 29.1 per cent in 2018, and the rate for HDPE natural bottles was 29.3 per cent in 2018. These figures have grown dramatically as new technologies, markets and collection systems are developed. Recycled plastic soft-drink bottles are being used to create new products, such as new bottles, fiberfill for winter clothing, carpeting and building materials. Recycled plastic milk jugs and soft-drink-bottle base cups are being used to create drainage pipes, buckets and plastic "lumber" for boat docks. And mixed plastics are being recycled into landscaping "timber" and outdoor benches.

Plastic packaging, which constitutes less than 4 per cent of all municipal solid waste by weight, also can be disposed of safely in landfills. And when incinerated, plastics - with their high energy content help the waste mix burn more efficiently, enhancing waste-to-energy conversion and leaving less ash for disposal.

Plastic Recycling and Beyond

Plastic waste management has assumed great significance in present day context. The plastics industry supports recycling that is sustainable, economical and environmentally responsible. Recycling of plastics is considered the next viable and technically feasible option to tackle plastic waste management. Various schemes are being implemented to mitigate the impacts of plastic waste worldwide. It makes rising sense environmentally as well as economically and current trends demonstrate a considerable increase in the rate of recovery and recycling of plastic wastes. Recycling requires participation from the public and, therefore, needs citizens to perform separation of waste materials at the source. Separating the plastic waste from other waste will prevent plastics to be land filled and will allow it to be recycled with other plastics of the same kind. Converting waste plastics to fuel is beneficial as it not only allows for waste plastic to be disposed of but also presents the opportunity of developing an alternative to fossil fuel. Plastic waste is also integrated with bitumen for laying roads. Non-recyclable plastic waste finds its application in the co-processing of plastic waste in cement kilns. This refers to the use of wastes in industrial pro

Plastic Bottles

When it comes to resource conservation, recycling is one area where consumers can make a big difference. Lightweight, shatterproof plastic beverage bottles are great to take to work, on errands or just about anywhere. Annual consumption of plastic bottles is set to top half a trillion by 2021, far outstripping recycling efforts and jeopardising oceans, coastlines and other environments. A million plastic bottles are bought around the world every minute and the number will jump another 20 per cent by 2021, creating an environmental crisis some campaigners predict will be as serious as climate change.

Plastic bottles are among the most readily recycled plastics. Most plastic bottles used for soft drinks and water are made from polyethylene terephthalate (Pet), which is highly recyclable. According to the EPA, 29.1 per cent of recyclable PET plastic bottles and 19.3 per cent of recyclable HDPE plastic bottles were actually recycled in 2018; there is a tremendous opportunity to recycle even more.

Before one pitches that empty water bottle into the trash, consider this: Every ton of plastic bottles recycled saves about 3.8 barrels of oil. Recycled plastic bottles are used to make hundreds of everyday products, ranging from fleece jackets and carpeting to detergent bottles and lumber for outdoor decking. Over 80 per cent of U.S. households have access to a plastics recycling program, be it curb side collection or community drop-off centers, yet, in most areas, the demand for recycled plastics exceeds the available supply. In recent years, the number of U.S. plastics recycling businesses has nearly tripled. Today, more than 1,600 businesses are involved in recycling post-consumer plastics.

Plastic Bags

From carrying groceries to taking out the trash, plastic bags are used and reused for dozens of tasks every day. In addition to being convenient and recyclable, plastic bags have many positive attributes from an environmental perspective:

Energy Efficient: Plastic grocery bags require 20-40 per cent less energy to manufacture than paper ones.

Lightweight and Compact: Lightweight plastic bags help save space and fuel in transport. For every seven trucks needed to deliver paper bags, only one truck

Table 11.2: Synthesis of Current Knowledge, Uncertainty and Recommended Actions Relevant to Environmental and Human Health Concerns Arising from Current Production, Use and Disposal of Plastics

	Established Knowledge	*Concerns and Uncertainty*	*Recommendations for Industry, Research and Policy*
Production and use	☆ Plastics are inexpensive, lightweight, versatile, water resistant and durable ☆ Annual growth in plastic production is approximately 9 per cent (currently >260 Mt yr^{-1}) ☆ Around 8 per cent of world oil production is used to make plastics ☆ Plastics bring extensive societal, human health and environmental benefits ☆ >33 per cent of production is used for disposable items of packaging	☆ Is our usage of hydrocarbons for plastics sustainable? ☆ To what extent could biopolymers replace oil-based plastics? ☆ Is there sufficient arable land for production of biomass (crops) required for biopolymers? ☆ To what extent does use of plastic powders as cleaning abrasives, and scrubbers results in direct release of particles to environment?	☆ Increase/incentivize material reduction and reuse ☆ Construct life cycle analysis of production, disposal/recycling of major polymers (including biopolymers, degradable and biodegradable polymers) and plastic products ☆ Develop alternative monomers, polymers and additives using green chemistry approaches ☆ Revise international standards for and introduce accurate/informative labelling of recyclable, 'degradable', 'biodegradable' and compostable polymers
Disposal: Waste management	☆ Plastics are a substantial part of domestic and industrial wastes in landfill ☆ Recycling of some polymers (*e.g.* PET) has increased considerably in recent years, but substantial quantities of plastic waste not compatible with recycling	☆ Are current disposal strategies sustainable—lack of space in landfill? ☆ To what extent do chemicals leach from plastic in landfill?	☆ Icrease/incentivize product design towards use of recycled feedstock and increased end-of-life recyclability ☆ Improve methods to collect and separate plastic waste for recycling

	Established Knowledge	*Concerns and Uncertainty*	*Recommendations for Industry, Research and Policy*
	☆ Biodegradable polymers typically require industrial composting and will not readily degrade in landfill ☆ Biodegradable plastics can compromise recycling	☆ Little is known about the degradability or environmental fate of additives used in biodegradable polymers	☆ Investment in/incentivize recycling operations standardize labelling so consumers can identify products with high end-of-life recyclability (traffic light system) ☆ Research and monitoring of leachates from landfills
Disposal: Littering and dumping	☆ Plastic debris is common in marine habitats worldwide, including poles and deep sea ☆ The abundance of plastic debris is increasing/stabilizing (not declining) ☆ Plastic debris is fragmenting, with pieces <20 μm on shorelines and in water column	☆ To what extent will breakdown of plastic debris increase the abundance of small fragments in the environment? ☆ Rates of accumulation of debris on land, in freshwaters and in the deep sea are not certain ☆ Do biodegradable or compostable plastics degrade in natural habitats?	☆ Education/incentives to promote the value of end-of-life plastics as a feedstock for recycling ☆ Education and associated enforcement on the wasteful and adverse ecological effects of plastic spillage, dumping and littering ☆ Develop standard protocols and monitoring to evaluate trends in the abundance of plastic debris across in natural habitats ☆ Cleaning programmes in natural, urban and industrial locations ☆ Research on breakdown of degradable and biodegradables

	Established Knowledge	*Concerns and Uncertainty*	*Recommendations for Industry, Research and Policy*
Issues relating to wildlife	☆ >260 species are known to ingest or become entangled in plastic debris ☆ Ingestion is widespread in some populations (>95 per cent of individuals) and can compromise feeding ☆ Entanglement in plastic debris can lead to severe injury and death	☆ Does ingestion of, or entanglement in, plastic debris have effects at the population level or can such effects combine with other stressors to do so? ☆ To what extent do plastics transport/release chemicals to wildlife? ☆ What are the consequences of the accumulation of small plastic particles (*e.g.* abrasives from cleaning applications) in the environment?	☆ Research to establish the distribution, abundance and environmental consequences of micro- and nano-plastic fragments research to establish potential for plastics to transport chemicals to food chain ☆ Research to establish population-level consequences of ingestion and entanglement ☆ Education, monitoring and cleaning
Issues relating to human health	☆ Some plastics contain potentially harmful monomers and additive chemicals, including flame retardants and plasticizers ☆ Adverse effects of additives evident in laboratory animals ☆ Measurable levels of chemicals used as additives/monomers are present in the human population ☆ Canadian government declared BPA a toxic substance. ☆ USA National Toxicology Program expressed concern for adverse health effects	☆ What are the effects of low-dose chronic exposure to chemicals or mixtures of chemicals used as plastic monomers or additives? ☆ Dose–response curves may not be monotonic and so should not be extrapolated in risk assessment	☆ Conduct cumulative risk assessment/management of plastic additives and monomers ☆ Biomonitoring of body burdens of additives/monomers ☆ Effects on susceptible subpopulations (babies, children) and on those with high-exposure risks ☆ Evaluate effects of exposure to mixtures of additives/monomers design/validate appropriate ☆ Species/protocols to assess chronic low dose exposures to additives/monomers by humans

Source: Thompson *et al.*, 2009. Plastics, the environment and human health: current consensus and future trends. *Philos Trans R Soc Lond B Biol Sci.* 2009 Jul 27; 364(1526): 2153–2166.

is needed for the same number of plastic bags, helping to save energy and reduce emissions.

Reusable: More than 80 per cent of today's consumers reuse plastic bags as liners for household wastebaskets, shoe totes and laundry or garment bags.

Recyclable: Recycled plastic bags can be made into a wide range of second-generation products, including new bags, exterior house siding, and plastic lumber for decking, park benches and picnic tables.

High Fuel Value: Plastics can help trash burn more efficiently in energy-recovery facilities, creating energy that can be used to make electricity in some communities.

Plastics Disposal

Despite the continuing growth of recycling, improvements in source reduction and increases in energy recovery, some waste will always require disposal. One should learn about plastics in landfills and their important role in protecting our environment and our groundwater from potential hazards posed by landfills.

Plastics and the Future

The speed of technological change is increasing exponentially such that life in 2030 will be unrecognizable compared with life today; plastics will play a significant role in this change. Plastic materials have the potential to bring scientific and medical advances, to alleviate suffering and help reduce mankind's environmental footprint on the planet.

In conclusion, plastics offer considerable benefits for the future, but it is evident that our current approaches to production, use and disposal are not sustainable and present concerns for wildlife and human health. We have considerable knowledge about many of the environmental hazards, and information on human health effects is growing, but many concerns and uncertainties remain. There are solutions, but these can only be achieved by combined actions (Table 11.2). There is a role for individuals, via appropriate use and disposal, particularly recycling; for industry by adopting green chemistry, material reduction and by designing products for reuse and/or end-of-life recyclability and for governments and policymakers by setting standards and targets, by defining appropriate product labelling to inform and incentivize change and by funding relevant academic research and technological developments. These measures must be considered within a framework of lifecycle analysis and this should incorporate all of the key stages in plastic production, including synthesis of the chemicals that are used in production, together with usage and disposal.

Chapter 12

Thermoplastic Polymers

All the plastic materials which can be softened and melted by heating, but they set again when cool are called thermoplastics. It is a plastic that can be melted repetitively and make another shape. They are solid materials at room temperature but viscous liquids when heated to temperatures of only a few hundred degrees. This characteristic allows them to be easily and economically shaped into products. They can be subjected to heating and cooling cycles repeatedly without significant degradation. Polyethylene, polypropylene, polyvinyl chloride, polystyrene, polybenzimidazole, acrylic, nylon and teflon are examples of thermoplastics.

12.1 Properties of Thermoplastic Polymers

Optical Properties

There are a number of optical properties of importance with thermoplastic polymers including clarity, haze, color, transmittance, reflectance, gloss and refractive index.

The clarity of a film indicates the degree of distortion of an object when viewed through the film, "see-through" clarity referring to the ability of the film to resolve fine details of fairly distant images viewed through the film. It should not be confused with haze as hazy materials often have quite good see-through clarity. In order to achieve a product with a high clarity, it is important that the refractive index is constant throughout the sample in the line of direction between the object in view and the eye. The presence of interfaces between regions of different refractive index will cause scattering of the light rays.

Amorphous polymers free from fillers or other impurities are transparent unless chemical groups which absorb visible light radiation are present. Generally, crystalline polymers are translucent, but they will be transparent when the crystal structures such as spherulites are smaller than the wavelength of light. However, if the structures are greater in diameter than the wavelength of light, the light waves will be scattered if the crystal structures have a different refractive index to that of the amorphous regions. The clarity of crystalline plastics can be improved by quenching or by random copolymerisation.

Since refractive index is dependent on density, it follows that where the crystalline and amorphous densities of polymers differ, there will be a difference in refractive index. Thus, thick polyethylene objects will be opaque since they cannot be quenched rapidly, and the spherulites formed have a significantly higher density (about 1010 kg m^{-3}) than the amorphous region (840–850). In the case of polypropylene, the difference is less marked (crystal density 940; amorphous density 850) and moldings are more translucent. Generally, the refractive index value for plastics is around 1.5.

Haze can be taken as a measure of the "milkiness" or "cloudiness" of an otherwise transparent polymer, and is often the result of surface imperfections, particularly in the case of thin films. The appearance of haze with consequent loss of contrast is caused by light being scattered by the surface imperfections, or by in homogeneities in the film due to voids, large crystallites, incompletely dissolved additives or cross-linked material. Internal haze does not arise with amorphous polymers; with crystalline polymers it increases with the degree of crystallinity and size of spherulites or other forms of crystal aggregates, as well as with the ratio of the density between the crystalline phase and the amorphous-phase.

Methods for measuring haze are specified in ASTM D 1003 and in BS 2782. Haze is defined in these standards as the percentage of transmitted light which, in passing through a specimen, deviates by more than 2.5° on average from an incident parallel beam by forward scattering from both surfaces and from within the specimen.

When light falls on a material, some is transmitted, some is reflected and some absorbed. The transmittance is the ratio of the light passing through to the light incident on the material, and the reflectance the ratio of the reflected light to the incident light.

The gloss (strictly speaking specular gloss) of a film has been defined as the degree to which a surface simulates a perfect mirror in its capacity to reflect incident light. Thus, it is a function of the reflectance and the surface finish of a material. Where transmittance and reflectance do not add up to unity, then some of the light waves are absorbed; if this absorption does not occur uniformly over the visible spectrum, the material appears colored. Gloss is determined by standard tests (ASTM D 523, D 1834 and D 2457, and BS 2782) using a device (known as a glossmeter) which measures the percentage of the light, incident at an angle (usually 45°) to the surface of the film, that is reflected at the same angle. The fraction of the original light that is reflected is the gloss of the sample.

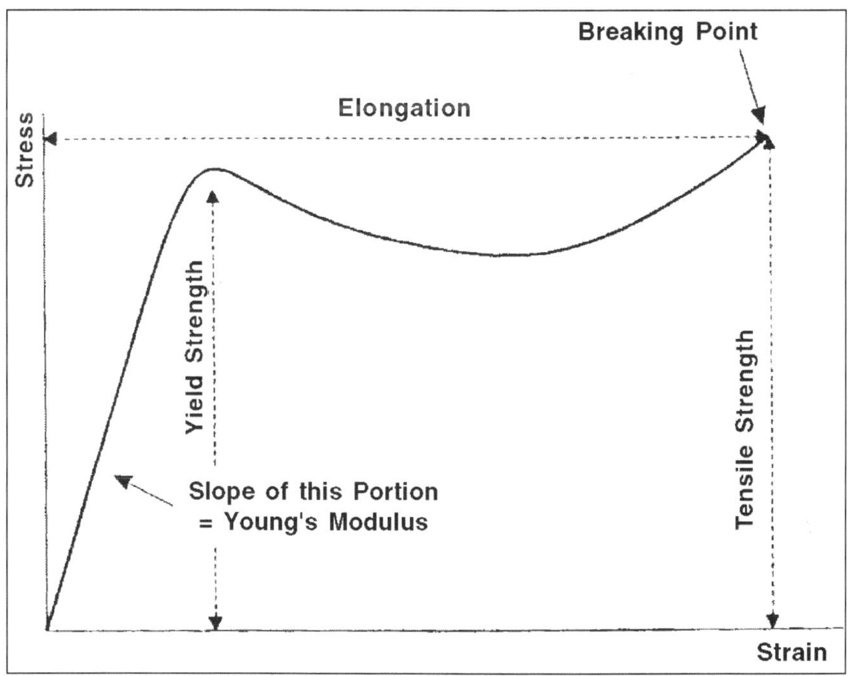

Figure 12.1: Typical Stress/Strain Curve for Thermoplastic Polymers Impact Strength.

Toughness of a plastic is measured by its **resistance to impacts**. It is the ability of a material to resist both fracture and deformation. To be tough, a material must exhibit both fairly good strength and ductility to resist cracking and deformation under impact loading.

The impact properties of polymeric materials are directly related to the overall toughness of the material, toughness being the ability of the polymer to absorb applied energy. The area under the stress-strain curve is directly proportional to the toughness of the material. The impact strength of a film is a measure of its ability to withstand shock loading.

The pendulum impact tester, also known as the Izod Test, can be used to measure impact strength. An impacting head on the end of a pendulum is swung through an arc into and through the sample. This test measures the difference between the potential energy of the pendulum at the maximum height of its free swing, and the potential energy of the pendulum after rupture of the sample. This difference in energy is defined as impact strength and is useful in predicting the resistance of a material to breakage from dropping or other quick blows.

Another method (ASTM D 2176 Method B) utilizes the M.I. T. Fold Endurance tester. It has an upper spring-loaded clamp which moves only up and down, an oscillating folding head which supports the smooth, folding surfaces, a driven device to provide rotary oscillating motion to the folding head, and a counter to register the number of folds. The sample is clamped into the folding head and the upper spring-loaded clamp depressed by a weight equivalent to the desired tension on the sample. The sample is then clamped into the upper clamp and the tension applied. The motor is started and the sample folded until it is severed at the crease.

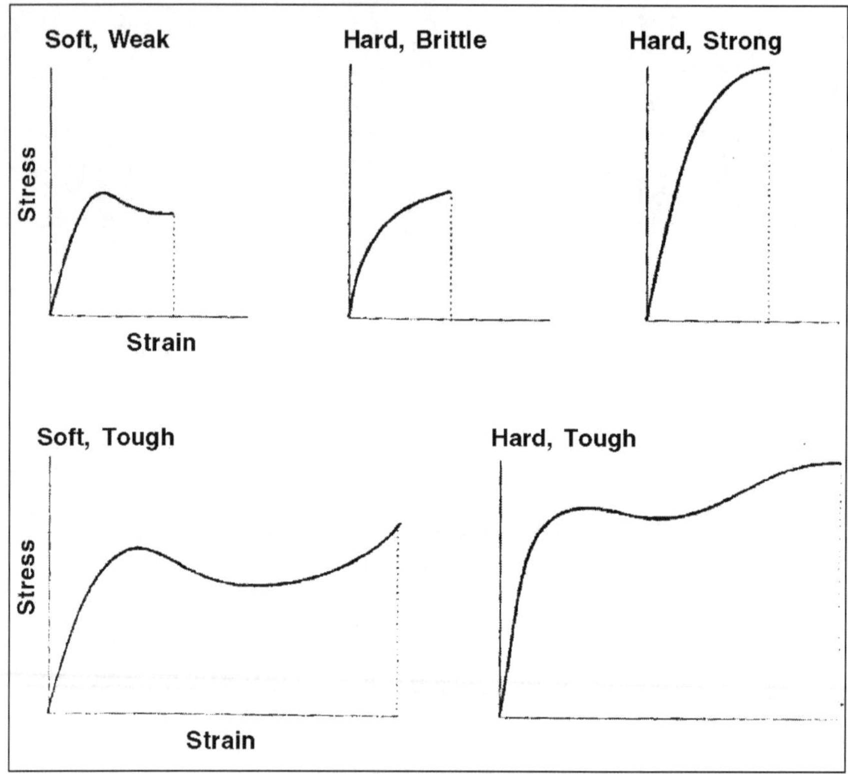

Figure 12.2: Typical Stress/Strain Curve Obtained with Plastic Polymers.

The above two tests may not always be applicable to polymeric films; the number of folds required to break the test specimen depends on its thickness. Thus, nylon film 2.54 μm thick may withstand over 200,000 double folds in the M.I.T. tester before failing, but 25.4 μm thick nylon may fail after fewer than 100,000 double folds.

12.2 Tensile Properties

The four properties tensile and yield strength, elongation and Young's modulus are considered under a single heading because the same equipment is used for measuring each of them. The basic principles of tensile testing (tensile meaning capable of being drawn out or stretched) can be found in any introductory physics textbook; a discussion relating to plastics is given by Shah. The tensile test is the simplest mechanical test to visualize; in it an increasing tensile stress is applied to a material and the resulting changes in length are monitored.

The *stress*, σ, is defined as the force per unit cross-sectional area of material. Thus, if the force is F Newtons and the cross-sectional area is A square meters, then the stress is given by:

$$\sigma = F/A \text{ N m}^{-2}$$

The *strain*, e, is defined as the fractional change in length of the material, *i.e.*, $\Delta l/l_0$, where l_0 is the initial length and Δl the change in length. It is expressed as a dimensionless ratio.

Tensile strength (or more accurately, ultimate tensile strength) is the maximum tensile strength which a material can sustain and is taken to be the maximum load exerted on the test specimen during the test, divided by the original cross-section of the specimen.

Yield strength is the tensile stress at which the first sign of a non-elastic deformation occurs and is the load at this point (known as the yield point) divided by the original cross-section of the specimen.

Elongation is usually measured at the point where the film breaks and is expressed as the percentage of change of the original length of the material between the grips of the testing machine. Its importance is as a measure of the film's ability to stretch, a large value for elongation indicating that the material will absorb a large amount of energy before breaking.

Young's modulus, or the modulus of elasticity, is the ratio of stress to strain over the range for which this ratio is constant, *i.e.*, up to the yield point. It is a measure of the force that is required to deform the film by a given amount and so it is also a measure of the intrinsic stiffness of the film.

Yield strength is usually more important than ultimate tensile strength, especially during the passage of the film through packaging or printing equipment where a sudden "snatch" could cause a non-reversible distortion and thus, for example, out-of-register printing.

Both yield point and elongation are important properties during the unwinding of plastic films. There is the danger of uneven stretching of the film if the elongation is high unless special handling techniques are used. Too low an elongation should also be avoided as any sudden unbalance in the unwind operation could lead to breaking of the film. Because a certain amount of tension is necessary during unwind, the possibility exists that films with a low yield strength could be stressed beyond their yield point.

One problem which arises when considering the strength of thermoplastic polymers is the time and temperature dependence of the changes which occur under mechanical stress, particularly in the case of thermoplastics. This time-dependence is well illustrated by the behavior of thermoplastics under tensile stress. This is measured by gripping the test material between fixed and moving clamps which are capable of separating at a range of speeds, the stress in the sample being measured continuously and recorded on a chart. At very slow speeds the molecules can readily disentangle, and the measured tensile strength depends largely on the magnitude of the weakest intermolecular forces. At faster speeds, on the other hand, there is little time for either disentanglement or slipping and the breaking point will not happen until the largest intermolecular forces have been overcome.

Methods for testing tensile properties of thermoplastic films are described in ASTM D 882 and BS 2782-301. It is usual when carrying out a tensile strength determination to plot the stress against the strain, *i.e.*, the load against the elongation. Such a stress/strain curve can give a great deal of information. For instance, brittle plastics will break at a much earlier stage of the curve, sometimes before the yield point has been reached.

A great deal of information about the material can be obtained from the shape of its stress/strain curve. In addition to the numerical values for tensile strength, Young's modulus, elongation, *etc.* it is possible to obtain some idea of the toughness of the material by measuring the area under the curve. This area is a measure of the energy needed to break the test specimen and hence is directly related to toughness.

12.3 The Transfer of Substances in Polymeric Materials

The solution and transport behaviour of low molecular weight substances in polymeric materials has become increasingly important in recent years, with the widespread use of polymer films and rigid plastics for food packaging.

Unfortunately, there are many examples of foods packaged with an apparent lack of proper consideration of the effects of the end use environment on properties, or of limitations imposed on performance due to unfavorable solution or transport characteristics. The plasticization of polymers by sorption of ambient vapors or liquids resulting in the decrease in mechanical properties and the loss of beverage components (*e.g.*, CO_2, flavour, *etc.*) are just two of many examples which could be cited. An objective of research in this field is to establish mechanisms and expressions relating solubility and transport with the molecular properties and characteristics of the components.

The protection of foods from gas and vapor exchange with the environment depends on the integrity of packages (including their seals and closures), and on the permeability of the packaging materials themselves. There are two processes by which gases and vapours may pass through polymeric materials:

1. A pore effect, in which the gases and vapours flow through microscopic pores, pinholes, and cracks in the materials; and
2. A solubility-diffusion effect, in which the gases and vapours dissolve in the polymer at one surface, diffuse through the polymer by virtue of a concentration gradient, and evaporate at the other surface of the polymer.

This "solution-diffusion" process (also known as "activated diffusion") is described as true permeability.

Most polymers when sufficiently thin exhibit both forms of permeability. Porosity falls very sharply as the thickness of a polymer is increased, reaching virtually zero with many of the thicker types of commercially available materials. True permeability, however, varies inversely as the thickness of the material and hence, cannot be effectively eliminated merely by increasing the material thickness.

This chapter is concerned with aspects of the solution, diffusion and permeation of gases and vapours ("permeants") in effectively non-porous polymeric materials. Mention will not be made of systems involving liquid solvents or higher concentrations of sorbed solvent vapours. Such systems often show behaviour which seemingly deviates strongly from that observed with gases or low concentrations of sorbed vapours. Such deviations are classified as anomalous diffusion, Case II, or Super Case II behavior and the interested reader is referred to a more detailed review.

The first recorded observation of the permeation of gas through a membrane appears to be that of Thomas Graham who in 1829 observed the permeation of CO_2 into a wet pig's bladder. In 1831 J.K. Mitchell, an American physician and the inventor of the toy rubber balloon, discovered that his balloons collapsed at different rates when they were filled with different gases.

The next major step came in 1855 when Adolf Fick proposed his law of mass diffusion by analogy with Fourier's law for heat conduction and Ohm's law for electrical conduction. In 1866 Graham postulated that the permeation process entailed dissolution of the penetrant, followed by transmission of the dissolved species through the membrane as though through a liquid, a process Graham called "colloidal diffusion". His concept is similar to that which is still used today, known as the solution-diffusion model. The subject was placed on a quantitative basis in 1879 by Von Wroblewski who showed that the solubility of gases in rubber obeyed Henry's law, and combined this with Fick's law to obtain the now familiar expression relating permeation rate and the area and thickness of the membrane.

Under steady state conditions, a gas or vapour will diffuse through a polymer at a constant rate if a constant pressure difference is maintained across the polymer. The diffusive flux, of a permeant in a polymer can be defined as the amount passing through a plane (surface) of unit area normal to the direction of flow during unit time, *i.e.*:

$$J = Q/A.\,t \qquad (12.1)$$

where Q is the total amount of permeant which has passed through area A during time t.

The relationship between the rate of permeation and the concentration gradient is one of direct proportionality and is embodied in Fick's first law:

$$J = -D \frac{\delta C}{\delta x} \tag{12.2}$$

where J is the flux (or rate of transport) per unit area of permeant through the polymer, c is the concentration of the permeant, D is defined as the diffusion coefficient and $\delta c/\delta x$ is the concentration gradient of the permeant across a thickness δx.

Consider a polymeric material X mm thick, of area A, exposed to a permeant at pressure p_1 on one side and at a lower pressure p_2 on the other. The concentration of permeant in the first layer of the polymer is c_1 and in the last layer c_2.

If x and $(x + \delta x)$ represent two planes through the *polymer* at distances x and $(x + \delta x)$ from the high-pressure surface, and if the rate of permeation at x is J mL per sec, and at $(x + \delta x)$ is $I + (\delta J/\delta x)\delta x$, then the amount retained per unit volume of the polymer is $(\delta J/\delta x)$. This is equal to the rate of change of concentration with time:

$$\frac{\delta}{\delta x}(J) = \frac{\delta c}{\delta t} \tag{12.3}$$

A negative sign is used because the concentration of permeant decreases across the material. Under steady state conditions, $\delta c/\delta t = O$ and J = constant. When the concentration gradient is zero (*i.e.*, $c_1 = c_2$) there will be no permeation.

If equation 12.2 is substituted into equation 12.3 then:

$$\frac{\delta}{\delta x}(J) = \frac{\delta}{\delta x} D \frac{\delta c}{\delta x} = \frac{\delta c}{\delta t} \tag{12.4}$$

and on rearrangement of the terms:

$$\frac{\delta c}{\delta t} = \frac{\delta}{\delta x} D \frac{\delta c}{\delta x} \tag{12.5}$$

and

$$\frac{\delta c}{\delta t} = D \frac{\delta_2 c}{\delta x_2} \tag{12.6}$$

Equation 12.6 is a simplified form of Fick's second law of diffusion and applies under circumstances where diffusion is limited to the x-direction and D is independent of concentration. There are extensive discussions of mathematical methods for solving the diffusion equation, together with solutions for many of the more common situations.

12.4 Permeant Relationships

For monocondesable gases, it has been predicted that the permeability ratio of a pair of gases will be relatively constant over a series of polymers.

It can also be seen that regardless of the film material, O_2 permeates about four times as fast as N_2, and CO_2 permeates about six times as fast as O_2 and 24 times as fast as N_2. There is no pattern for the ratios of water vapor to any of the three gases.

It may be thought to be strange that CO_2, the largest of the three gas molecules, has the highest permeability coefficient. In fact, it has the lowest diffusion coefficient of the three gases as would be expected from its relative size, but its permeability coefficient is the highest because its solubility coefficient S is much greater than that for the other gases.

Table 12.1: Diffusion, Solubility and Permeability Coefficients for Low Density Polyethylene Film at 25°C to CO_2, O_2 and N_2

	$D \times 10^6$ $cm^2\ s^{-1}$	$S \times 10^2$ $mL\ (STP)\ mL^{-1}\ atm^{-1}$	$P \times 10^{10}$ $[mL(STP)\ cm\ cm^{-2}\ s^{-1}\ (cm\ Hg^{-1})]$
CO_2	0.37	25.8	12.6
O_2	0.46	4.78	2.88
N_2	0.32	2.31	0.969

It has been argued that the permeability coefficient for a particular permeant/polymer system is a product of a factor F determined by the nature of the polymer, a factor G determined by the nature of the permeant, and an interaction function H. The latter is taken as unity where little or no interaction occurs, and becomes larger the greater the degree of interaction between the permeant and the polymer.

Thus, the permeability coefficient of a polymer x to a permeant k can be expressed as:

$$P_{xk} = F_x G_k H_{xk} \tag{12.7}$$

If H_{xk} is taken as unity, the ratio of the permeability coefficients of a polymer x to permeants k and j can be shown to be the same as the ratio between their respective G factors:

$$\frac{P_{xk}}{P_{xj}} = \frac{G_k}{G_i} \tag{12.8}$$

The above ratio would be almost independent of the nature of the permeant.

Similarly for the permeability coefficients of two polymers x and y and a permeant k:

$$\frac{P_{xk}}{P_{yk}} = \frac{F_x}{F_y} \qquad (12.9)$$

The above ratio would be independent of the nature of the polymer material.

From a knowledge of various values of P it is possible to calculate F values for specific polymers and G values for specific gases if the G value for one of the gases (usually nitrogen) is taken as unity. For non-condensable gases it has been predicted that the ratios of the permeability coefficients of a pair of gases will be relatively constant over a series of polymers.

Chapter 13

Important Plastics Processing Methods

Plastics are mainly organic polymers of high molecular mass, but they sometimes contain other substances as well. Plastics are usually synthetic, most commonly derived from petrochemicals, but many are partially natural. They are usually classified by their chemical structure of the polymer's backbone and side chains. Some important groups in these classifications are the acrylics, polyesters, silicones, polyurethanes, and halogenated plastics.

Plastic processing can be defined as the process of converting the plastics raw material into semi-finished products. There are a variety of methods used to process plastic. Each method has its advantages and disadvantages and are better suited for specific applications. Some of the important plastic processing methods are described as under:

Classification of Processing Methods

1. **Primary Processing Methods:** Injection, Extrusion, Blow, Compression and transfer moulding.
2. **Secondary Processing Methods:** Roto, Thermoforming, Coating, Casting, Fabrication and Calendaring *etc.*
3. **Tertiary Processing Methods:** Cutting, Drilling, Welding and Bending *etc.*

13.1 Extrusion

Extrusion is one of the most important plastics processing methods in use today. It is a high-volume manufacturing process in which raw plastic material is melted and formed into a continuous profile. Most plastic materials are processed in extruders and commonly pass through two or more extruders on their way from the chemical reactor to the finished product. All of the thermoplastics are formed into sheet or film by the process of screw extrusion.

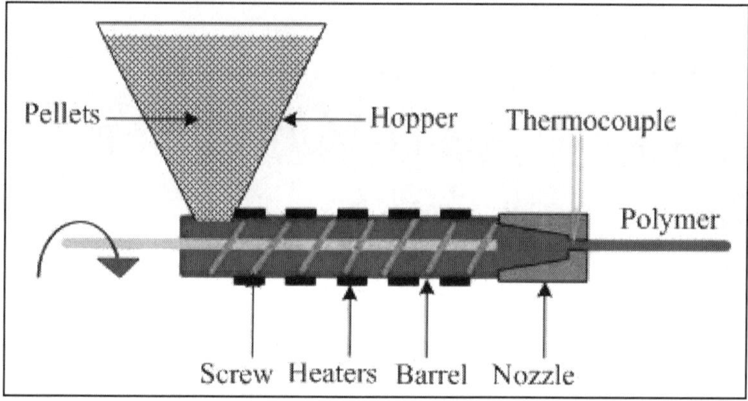

Figure 13.1: Extrusion of Polymers.

The heart of the extruder is the Archimedean screw which revolves within a close-fitting, heated barrel. It is capable of pumping a material under a set of operating conditions at a specific rate, depending on the resistance at the delivery end against which the extruder is required to pump. The extruder resembles a mincer into which granules are fed, heated and compressed until they fuse into a melt which is forced through a slot or circular die. Sufficient heat is produced by the energy of the screw to allow the process to continue with very little heat input.

The standard single screw extruder (Figure 13.2) receives cold powder or granules through a hopper to a throat at one end of the barrel and delivers it to the compression zone of the screw. In this section the diminishing depth of thread causes a volume compression and an increase in the shearing action of the material. It melts and is converted into a homogeneous mass by contact with the heated walls of the barrel and by the heat generated by friction. Generally external heating is only required at the start of the run, the frictional or exothermal heating being sufficient for steady-state operation. After the compression zone, the melt passes through the metering zone where the flow is stabilized, before being pumped through the die which determines its final form. The output from the die is known as the extrudate.

The screw is the most important component of the extruder and different designs are used for extruding different polymers. Extruder screws are characterized by their length to diameter ratios (commonly abbreviated to L/D ratios) and their compression ratios, this being the ratio of the volume of one flight of the screw at the infeed end to the volume of one flight at the die end. L/D ratios commonly used for single screw extruders are between about 15: 1 to 30: 1, while compression ratios can vary from 2: 1 to 4: 1.

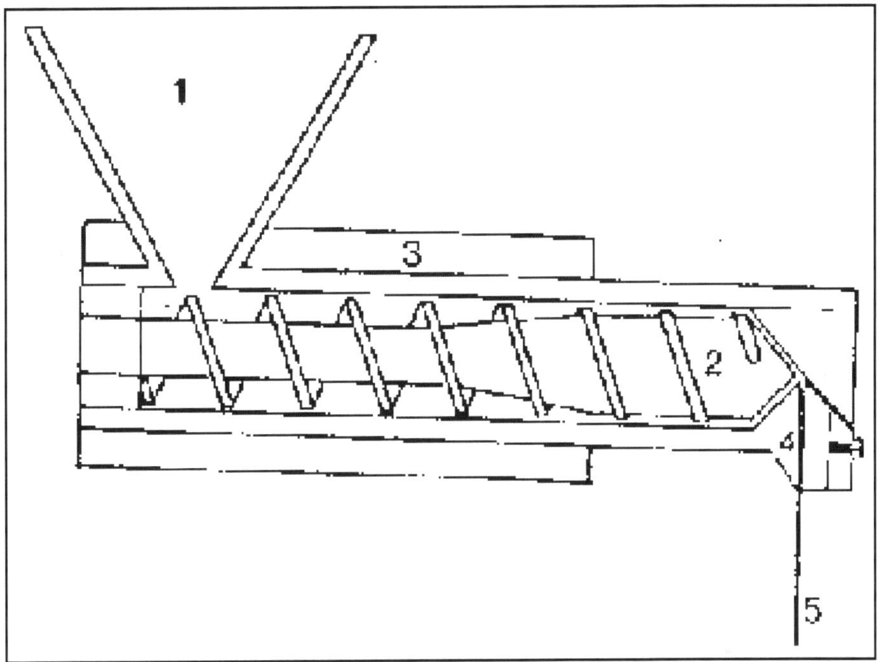

Figure 13.2: Single Screw Extruder
(1) Input hopper for resin; (2) Feeder screw; (3) Heating elements; (4) Slot extrusion die; (5) Extruded film.

There are basically two processes by which the extruded thermoplastic can be converted into film: the tubular process and the flat film process.

In the tubular (or blown film) process, a thin tube is extruded (usually in a vertically-upward direction), and by blowing air through the die head, the tube is inflated into a thin bubble. This is cooled, flattened out and wound up. The ratio of bubble diameter to die diameter is known as the blow-up ratio. Most low density polyethylene blown films used in packaging are made using blow-up ratios of between 2.0 and 2.5: 1. Blown film extrusion can produce defects such as variations in film thickness, surface defects, low tensile and impact strength, haze, blocking and wrinkling.

The properties of the film depend strongly on the polymer used and the processing conditions. The higher the density, the lower the flexibility and the greater the brittleness. The higher the molecular weight, the greater the tensile strength and resistance to film brittleness at low temperatures, but the lower the transparency.

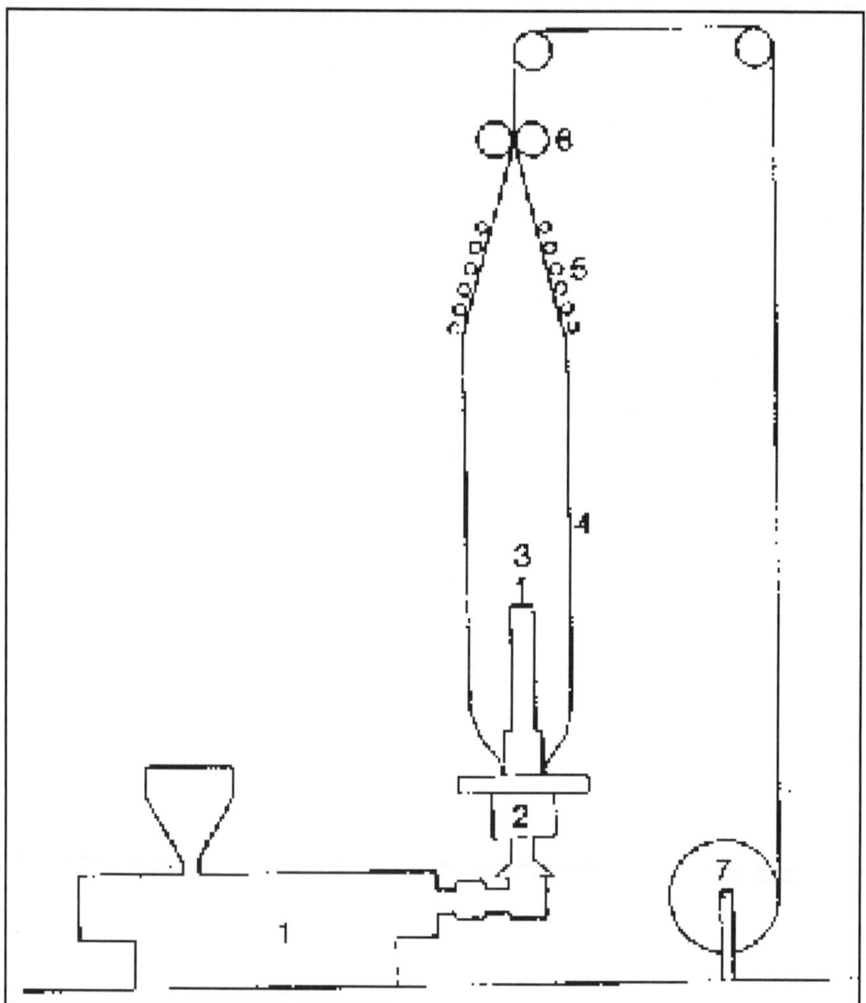

Figure 13.3: Blown Tubular Film Extrusion
(1) Screw extruder; (2) Circular die; (3) Air at constant flow; (4) Tubular film; (5) Guide frame; (6) Nip rollers; (7) Wind-up roll.

In flat film (also known as cast film or slit die) extrusion, the molten polymer is extruded through a slit-die into a quenching water bath or onto a chilled roller. In both cases, rapid cooling of the extruded film is most important. The ratio of the haul-off rate to the natural extrusion rate is referred to as the draw-down ratio. Draw-down ratios between 20: 1 and 40: 1 are typical.

A comparison of the tubular and flat film processes lists among the advantages of tubular film the fact that the mechanical properties are generally better, and that the process is easier and more flexible to operate. The cost for making wide tubular film is much lower than for wide cast film due to the cost of precision grinding long

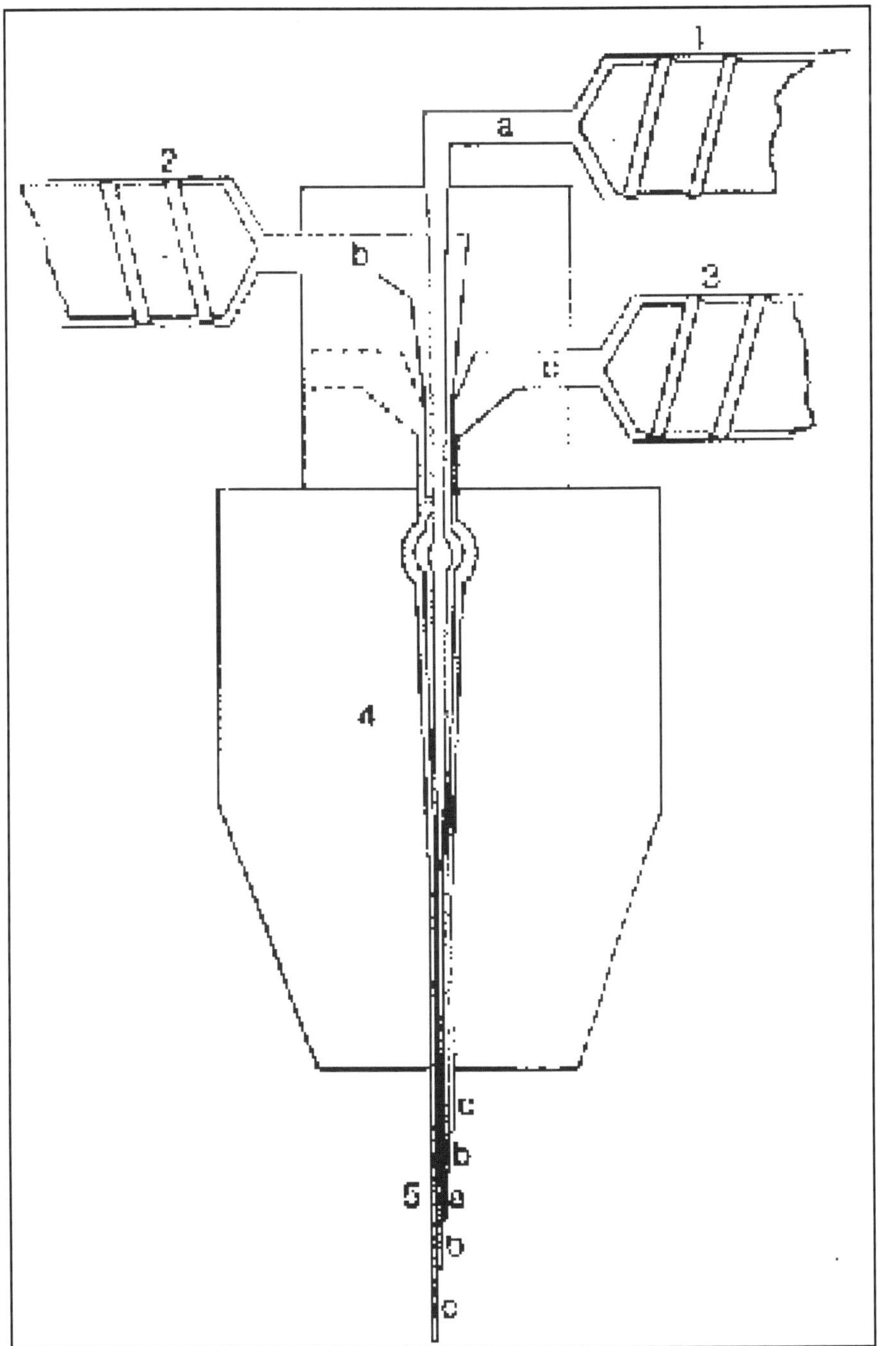

Figure 13.4: Three-component Slot Die
(1) Extruder screw (polymer a); (2) Extruder screw (polymer b);
(3) Extruder screw (polymer c); (4) Combining adaptor; (5) Five-layer film.

chill rolls. The advantages of the flat film process include less thickness variation, very high outputs and superior optical properties. This latter advantage is a consequence of the quicker cooling which can be achieved in the flat film process where cooling is by conduction, compared with the tubular film process, where cooling is by convection. Slower cooling permits the formation of more and larger crystals in the film, leading to haze which arises from light scatter between the crystal interfaces.

By using combining adaptors, it is possible to extrude simultaneously two or more different polymers which fuse at the point of film formation into a single web. Such a process is known as coextrusion and permits the production of a single web having, for example, barrier properties not possessed by anyone of the component polymers. A two-component slot die is capable of producing a two- or three-layer film from two materials, while a three-component die can produce a five-layer film from three materials.

Film thicker than 0.25 mm is normally defined as sheet, and is commonly thermoformed into objects that hold their shape such as trays or cups. The polymer is normally extruded horizontally into a nip formed by two hardened cooling rolls. These determine the final product thickness and surface finish. Sheets up to about 1.3 mm thick can be wound onto rolls, thicker sheet being cut to desired lengths.

13.2 Extrusion Blow Molding and Injection Blow Molding

Extrusion blow molding uses many arrangements for making and forming the parison, and the reader is referred to Irwin for a detailed discussion. In the simplest method, a mold is mounted under the die and the parison extruded between the open halves of the mold. When the parison reaches the proper length, the extruder is stopped and the mold closes around the parison.

A blow pin mounted inside the die head allows air to enter and blow the parison into the final container shape. The shape of the bottle or jar is defined, but the distribution of material (and thus wall thickness) is less well controlled. The cycle restarts after the part has cooled and the mold opened. To use the full capacity of the extruder, numerous systems are used.

One development is to use more than one mold, moving the filled mold away to cool while another is moved into position to receive the next section of extruded tube. The molds can be reciprocating ones, or can be mounted on a rotary table. In several food packaging applications such as fresh milk, the bottles are blow molded and filled on line in a continuous operation.

Extrusion blow molding is widely used with the following resins: high density polyethylene, polypropylene, poly(vinyl chloride) and acrylonitrile copolymers. The common grades of poly(ethylene terephthalate) cannot be extrusion blown.

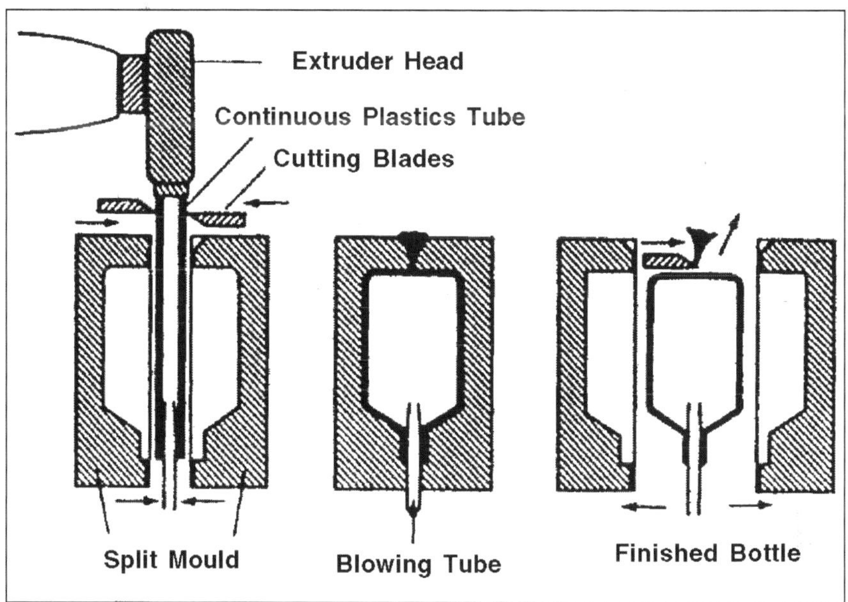

Figure 13.5: Extrusion Blow Molding of Plastic Bottles.

A related development has been the production of co-extruded bottles, where two or more extruders, each handling a different plastics material, produce a multi-layer parison having the desired properties. For example, a high-barrier, high cost material might be sandwiched between layers of a relatively low cost material to give a bottle with the desired barrier properties at an economical price.

Injection blow moulding is used for the production of hollow objects in large quantities. The main applications are bottles, jars and other containers. The Injection blow moulding process produces bottles of superior visual and dimensional quality compared to extrusion blow moulding. The process is ideal for both narrow and wide-mouthed containers and produces them fully finished with no flash.

Injection molding machine is similar to extruder. The main difference between the two machines is in screw operation. In extruder screw rotates continuously providing output of continuous long product (pipe, rod, sheet). Screw of injection molding machine is called reciprocating screw since it not only rotates but also moves forward and backward according to the steps of the molding cycle. It acts as a ram in the filling step when the molten polymer is injected into the mold and then it retracts backward in the molding step. Heating elements, placed over the barrel, soften and melt the polymer. The mold is equipped with a cooling system providing controlled cooling and solidification of the material.

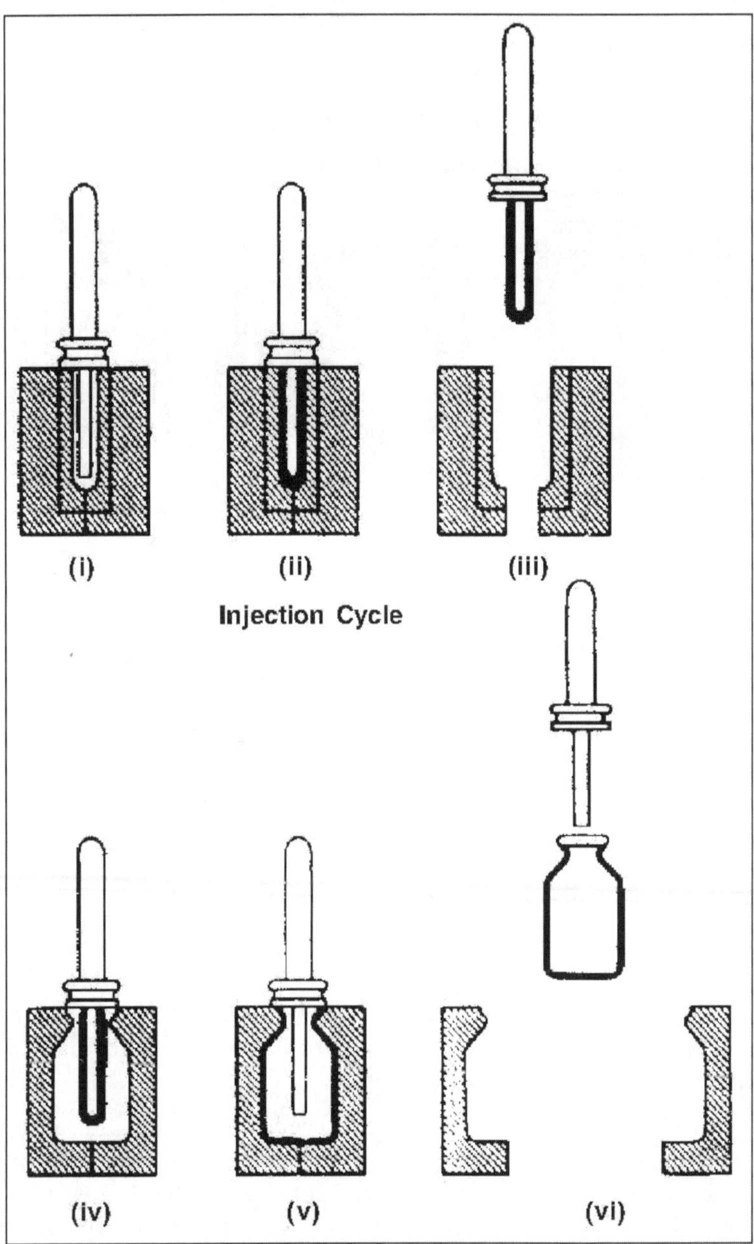

Figure 13.6: Injection Blow Molding of Plastic Bottles.

Injection blow molding is a non-continuous cyclic process most closely resembles the blowing of glass bottles. The parison is formed in one mold and then, while still molten, is transferred to a second mold where blowing with compressed air forms the final shape. After cooling, the mold is opened and the bottle ejected.

Several molds must be available if the injection molding machine is to operate near full capacity. The major advantage of injection molding over extrusion blow molding is that the process is virtually scrap-free, the finished parts usually requiring no further trimming, reaming or other finishing steps. In addition, the dimensions of the bottle (including the neck) show very little variation from bottle to bottle, and with some materials, improved strength and clarity are obtained due to the effect of a limited degree of biaxial orientation.

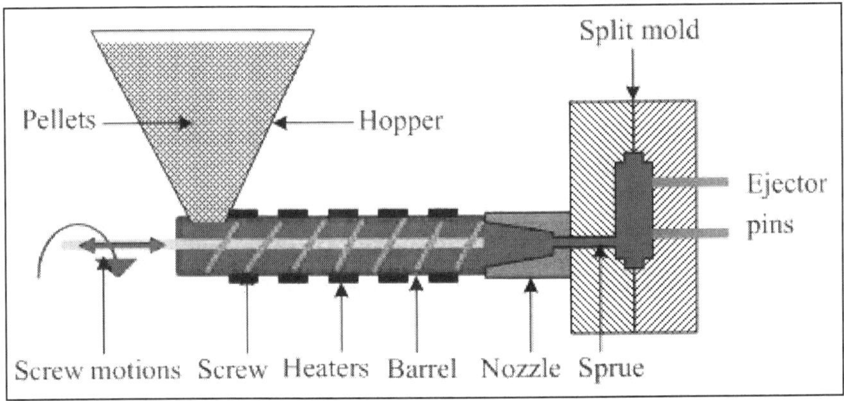

Figure 13.7: Injection Molding.

The resins most commonly used for injection blow molding are high density polyethylene, polypropylene, polystyrene and poly(vinyl chloride). Recently, poly(ethylene terephthalate) has been injection blow molded, and is likely to replace poly(vinyl chloride) in some applications.

In recent years, co-injection blow molding has been developed using two or three injection units working with one mold to produce either a part component or a perform which is later blow-formed using compressed air inside a mold to make a bottle or jar. The various component materials are metered into cavities in such an order that the barrier material flows through the main structural material to create a multi-layer structure. This process is used to produce five layer retortable containers from three materials, typically polypropylene as a structural layer and ethylene vinyl alcohol copolymer as a barrier layer with tie layers in between.

An alternative process for the production of retortable plastic containers has been developed by Hitek in Australia. The process is known as Cuspation-Dilation and was developed from the observation that a sheet of hot plastic could not be perforated by a metal needle; instead the sheet formed a cusp. Basically the process involves a set of metal fingers inside a female mold which stretch the molten plastic sheet into a series of cusps. The fingers then dilate to further stretch the material into the shape of the mold. Containers with very even cross section can be made using this process, without the thickness that would be required in an injection molded product.

13.3 The Orientation of Polymer Films

The process of orienting plastics involves modifying the structure of polymer films like polypropylene (PP), polyester (PET), and polyethylene (PE) by heating and mechanically stretching them with specialized orientation machines. The goal is to increase material yield by down-gauging the film's thickness while retaining its physical properties or to attain specific performance advantages in terms of mechanical, optical, barrier and porosity characteristics.

Orientation of polymer films is a means of improving their strength and durability in order to broaden their scope of application and make them serviceable in thinner gauges. Films may be oriented in either one direction (uniaxial orientation) or, more commonly, in two directions, usually at right angles to each other (biaxial orientation). Virtually all thermoplastics can be oriented to some extent, but amorphous films can be more readily oriented than crystalline films.

The first biaxial orientation process was developed in Germany about 1935, and during World War II oriented polystyrene film was made in Europe for capacitors and coaxial cable insulation. In 1948 Dow Chemical Company developed the Saran (vinylidene chloride/vinyl chloride copolymer) resins and Dewey and Almy (now the Cryovac Division of W.R. Grace) developed the film process and introduced Saran shrink film to the market. In 1952 E.I. du Pont de Nemours and Co., Inc., began selling oriented poly(ethylene terephthalate) film.

In essence orientation of thermoplastic film (whether uniaxial or biaxial) is a process of stretching the material in such a manner as to line up the molecular chains in a predetermined direction. Once lined up, the ordered arrangement is frozen in the strained condition. Biaxially oriented films possess superior tensile and impact strengths, improved flexibility, clarity, stiffness and toughness, and increased shrinkability. Gas and water vapor permeability may also be reduced, generally by 10 to 50 per cent depending on the type of polymer, and the degree and temperature of orientation.

Gas and water vapour permeabilities for amorphous polymers (*e.g.*, polystyrene and polyesters) appear to be nearly identical for both oriented and unoriented films. Crystalline polymers (*e.g.*, polypropylene and vinylidene chloride/vinyl chloride copolymer) show significant reduction in water vapor permeability when oriented. This difference is greatest at low degrees of crystallinity (10–15 per cent) and gradually becomes less as the degree of crystallinity increases, until at 40–50 per cent crystallinity, no differences are discernible. The gas permeability rates are largely dependent on the amporphous. The gas permeability rates are largely dependent on the amorphous content, which outweighs any effect introduced by orientation.

On the other hand, orientation generally has a detrimental effect on elongation, ease of tear propagation and the sealability of the film. The heat sealability range is narrowed and the film may vary in properties with age. Oriented film cannot be easily heat sealed because it shrinks and puckers at temperatures below the sealing temperature. Although various mechanical devices ('point sealing') have been designed to mitigate the shrinkage, a more suitable solution is to apply a surface coating of some thermoplastic having a lower melting point. For example, oriented polypropylene (OPP) may be coated with a dispersion of a copolymer of vinylidene chloride and vinyl chloride, or a copolymer of polypropylene with a small quantity of ethylene.

Among the more common commercially oriented films are poly (ethylene terephthalate), polyamide, vinylidene chloride/vinyl chloride copolymers, polypropylene and low-density polyethylene, the latter commonly being irradiated before blowing into film. Because radiation cross-links the molecules, the film can be stretched without becoming fluid at the melting point of a non-irradiated film, resulting in greatly improved tensile strength and shrink tension compared to non-irradiated low-density polyethylene. High density polyethylene is not oriented because its very rapid crystallization limits the extent to which it can be stretched. When the resin is blended 70: 30 with low density polyethylene, the rate of crystallization is slowed. The crystallinity imparts properties similar to those achieved through radiation cross-linking. The largest application of orientation techniques is in the manufacture of OPP which results in a considerable improvement in its barrier properties.

In the case of crystalline polymers, the action of orientation induces additional crystallization, with the crystalline structure aligned in the direction of stretching. The induced crystallinity is general and does not occur in spherulite form; therefore, oriented films usually have a high degree of clarity, because of the relative absence of spherulites which cause light scattering.

For many applications, shrinkage is not desirable and a greater degree of heat stability is required. Films can be annealed by application of heat to partially relax the forces while maintaining the film in a highly stretched condition. It is then cooled to room temperature and the restraint on the film released. Such a film is referred to as heat set and will not shrink if heated to below the annealing temperature. The procedure of annealing does result in some reduction in dimension in the stretched direction or directions.

The most common method used to orientate a thermoplastic film is to stretch it after it has been heated to a temperature at which it is soft. This temperature is below the flow temperature at which the molecules would glide readily past one another when the material is stressed, but above the glass transition temperature. As a result of this stretching, the direction of the molecules changes towards that in which the material is stressed, and the molecules are extended like springs. The temperature is then dropped below the softening point of the material while the molecules are held in this configuration, so that the molecules are frozen in the strained position.

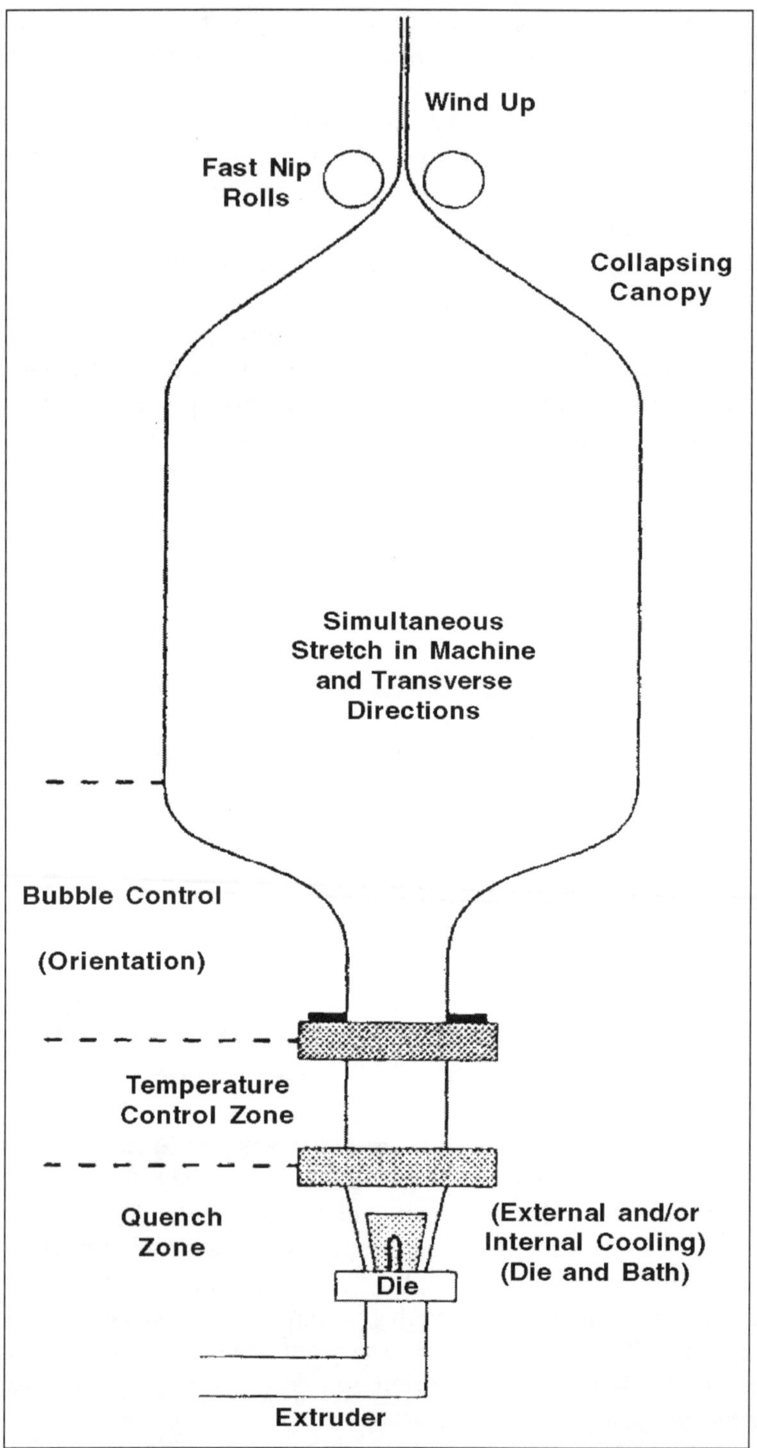

Figure 13.8: Orientation by the Bubble Process.

Films can be oriented using two processes–flat sheet and tubular. In the flat sheet or tentering process, thick (500–600 μm) cast film is fed to a system of differential draw rolls which are heated to bring the film to a suitable temperature below its melting point. The film is stretched in the machine direction and then fed to a tenter frame where a series of clips (mounted side by side on endless chains) grasp both edges of the film and draw it transversely as it travels forward at an increasing speed. Draw ratios in both directions normally vary between 4 : 1 and 10 : 1. After tentering, the film is passed over a cooling roller and reeled up. The two operations can also be carried out sequentially, *i.e.,* the film is stretched in the machine direction first, and then fed into a tenter frame where it is drawn transversely.

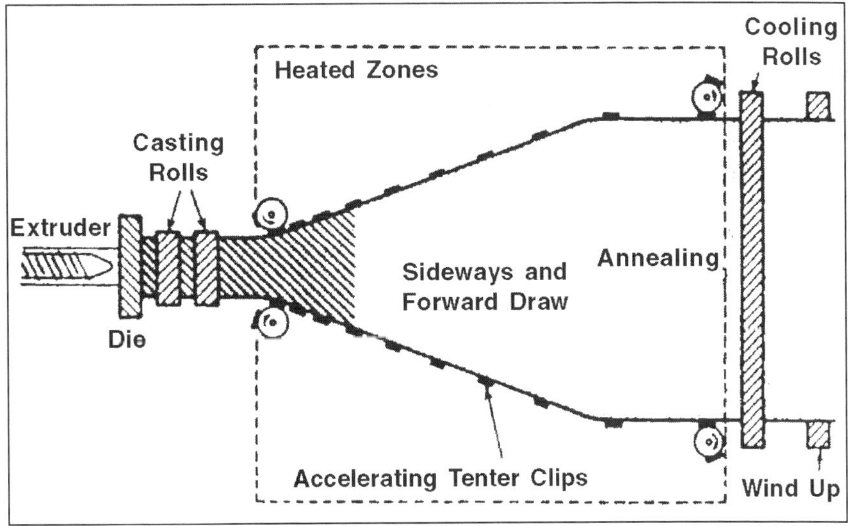

Figure 13.9: Simultaneous Orientation Process Using a Tenter.

In the tubular or bubble process, molten polymer is extruded from an annular die and then quenched to form a tube. The tube is flattened by passing through nip rolls and reheated to a uniform temperature. The air pressure in the tube is increased to expand the film transversely, the draw ratio being varied by adjusting the volume of entrapped air. Pinch or collapsing rolls at the end of the bubble are run at a faster speed than rolls at the beginning of the bubble, thus causing drawing of the film in the machine direction. The film is then wound-up.

The amount of orientation imparted to a film depends on the stretching temperature, the amount of stretching, the rate of stretching and the quench. Quenching is carried out either by extruding the web onto a chill roll or by passing it through a quench tank prior to orientation. Generally orientation is increased by decreasing the stretching temperature, increasing the amount of stretch, increasing the rate of stretch, and increasing the amount of quench. Films such as vinylidene chloride/vinyl chloride copolymer and polypropylene which have glass transition temperatures below room temperature show an appreciable crystallization rate even at room temperature, and therefore, have to be quenched and oriented immediately after extruding.

The potential energy stored in the extended molecules is the so-called "elastic memory" characteristic of oriented, non-heat set thermoplastics. When such a film is reheated to its orientation temperature, it shrinks as the molecules tend to return to their original size and spatial arrangement.

13.4 The Shrink and Stretch Films

The first shrink film was Saran and it was used initially for frozen poultry. The unfrozen birds were placed in a bag, a vacuum was applied to remove air and the bag was then sealed, typically with a metal clip. On immersion in hot water, the bag shrank tightly around the bird which was then frozen, commonly by passage through a glycol bath at –30°C. The close contour wrap ensured rapid heat transfer during freezing since there were no air pockets inside the package. This latter fact, coupled with the impermeability of the film, prevented freezer burn, *i.e.*, sublimation of water vapor from the surface of the frozen poultry.

Today a range of shrink films made from a variety of polymers is available. With the exception of Saran, most of these other shrink films require temperatures above 100°C to obtain a suitable degree of shrinkage, necessitating the development of hot air tunnels or heat guns.

Three properties of shrink films are important when selecting a film for a particular application. The first is the range of temperature over which a film will shrink. The lower the shrink temperature, the simpler and less expensive the shrink process. Films with a wide softening range are usually preferable since this makes temperature control of the heating equipment less critical.

The degree of shrinkage is also important, some applications requiring a higher degree of shrinkage than others. The amount of shrink can vary from 15 per cent to 80 per cent depending on the polymer composition and manufacturing techniques. Of particular interest is the relationship between degree of shrinkage and temperature, films with a steep shrink/temperature curve (*e.g.*, polypropylene, where a ±5°C variation in tunnel temperature could vary the degree of shrinkage by up to 20 per cent) being more difficult to handle because of the closer temperature control necessary.

Shrink tension, the stress exerted by the film when it is restrained from shrinking at elevated temperatures, is the third important factor. Generally, the lower the temperature at which orientation was carried out, the greater the shrink tension. Tension of 300 to 1000 kPa is desirable in order to provide a tight package after shrinking. With a shrink tension above 2000 kPa, care must be taken to prevent crushing and distorting of the package by limiting shrink temperature and time.

Balanced orientation is especially important for printed films, since uniform shrinkage is essential to avoid distortion of the print after shrinkage. Even a balanced biaxially oriented film may not shrink evenly in both directions if the product is of a very irregular shape. In these situations, it may be necessary to choose a print design which is not affected by such distortion.

The stretch films were first introduced in the early 1970s. In stretch wrapping the film is stretched around the article and the loose end 'wiped" against the underlying film, the film-to-film adhesion or cling being sufficient to hold it in place. Other less common ways of attaching the end are heat sealing, adhesives, mechanical fasteners and tying. Orientation of the polymer chains occurs on stretching to give a stiffer film, improving the tightness of the wrap and the stability of the load. The residual tension in the film gives a tight contour wrap. A simple analogy is that of an elastic or rubber band which can be stretched over an object and then assume a tight position.

The main films used in stretch wrapping are low density polyethylene, linear low-density polyethylene, poly (vinyl chloride), ethylene-vinyl acetate copolymer and polypropylene, the choice depending on such factors as appearance (*i.e.*, requirements for clarity, sparkle, *etc.*) and the protection required (gas and moisture barrier, and/or physical protection in preventing pallet loads from disintegrating).

There are four important mechanical properties of stretch films. The stretchability or elongation is a measure of how much a given film can be stretched without distorting the load, the greater the stretchability, the lower the material costs since less film is used for a given application. The stress is a measure of the static force exerted by the film on the load, and is highest at the corners of a straight, square load. This force can crush or distort the load if not controlled during the wrapping process. The restretch force or elasticity is a measure of the force available to restrict further movement of the load during transit as a result of vibration, sway and high impact forces. The breaking strength is a measure of the ultimate force that the film can withstand without failure.

Low density polyethylene is classed as a low stretch film with a stretchability of about 30 per cent. Under ideal conditions linear low-density polyethylene can stretch up to 400 per cent, although the practical limit is nearer 200 per cent. About 60 per cent of the initial stretch is retained by this film, the figure for low density polyethylene being about 70 per cent. Both are quite adequate for the stretch wrapping of large pallet loads. The trend is towards lighter gauge films, 20 μm being common, with some as thin as 13 μm.

An important property of stretch films is their cling with blown linear low-density polyethylene having much less cling than poly (vinyl chloride) or ethylene vinyl acetate copolymer, although cast linear low-density polyethylene approaches the others in this property. Additives such as glyceryl mono-oleate and polyisobutylene are sometimes added by film manufacturers to improve cling. Elevated humidity can sometimes enhance film cling because some cling additives function by attracting moisture from the atmosphere. As well, antioxidants, antistatic agents and antiblock agents (the latter to prevent the layers of film on a roll from becoming permanently bonded together) are also frequently added.

The advantages and disadvantages of stretch versus shrink film can be considered under various headings. From the energy point of view, stretch wrapping requires no shrink tunnel or hot air guns. As a consequence, stretch wrap systems can be used in cold stores. There are also savings in yield, since a thinner gauge film can be used for stretch wrapping. With a shrink film, typically 1 kg of film per apllet is used; this was reduced to 0.5 kg with the early stretch films. The new linear low density polyethylene films can be used at around 0.3 kg per pallet. Furthermore, stretch film will not stick to film-wrapped packs as will shrink film if used to overwrap pallet loads. Installation and commissioning costs are usually less for stretch film.

One early disadvantage of stretch wrapping was the lack of control of film tension, resulting in many instances in damage to the goods being wrapped due to excessive film tension. Now-a-days suitable equipment incorporating adjustable torque units is available to overcome this problem. One associated problem is that different films retain different stress, since all films start to relax immediately after they have been stretched. Most of the relaxation (about 99 per cent) occurs within 24 hours. The opposite of relaxation is stress retention which is defined as the capacity of a film to maintain the applied tension during stretch wrapping.

The percentage of *stress retention* should not be confused with the percentage of *stress recovery,* the latter being analogous to the behaviour of a rubber band. Stress recovery is important whenever the unit load decreases in size. Then stress relaxation is not due to film elongation with time, but to the reduction in the size of the load. In these situations, stretch wrapping has major advantages compared with most strapping methods since the latter provide very little stress recovery.

Blown low density polyethylene has about 85 per cent stress recovery while poly(vinyl chloride) has about 100 per cent, after 16 hours at 20 per cent initial elongation and 25°C. Thus, the former film would be more suitable for containing heavier loads with the latter being preferred when significant reduction in unit load size occurs. However, such a relationship may not hold at higher ambient temperatures since stress retention of PVC decreases much faster than LDPE as the temperature increases. Stress recovery must be taken into account when choosing a stretch film and setting the tension on the torque unit.

Recent developments include the introduction of pre-stretching, where stretching occurs between two rollers immediately prior to the film being applied to the load. This permits films to be stretched to a higher ratio and the weight of film per pallet is thus reduced. A further advantage with this system is that less 'necking' or narrowing of the film occurs as it is stretched, because the force is applied more evenly over a shorter distance.

An alternative approach to high stretch is the use of a pre-heat system where the film is heated before stretching, reducing the force required to stretch the film and thus, the risk of mechanical damage. A further development has been the introduction of power stretch systems where the driving force for stretch is not the rotating load but the driven rollers over which the film is drawn. The stretching force is thus isolated from the load which reduces the risk of film damage at high stretch.

13.5 The Types of Sealing

Heat Sealing

The heat sealability of a packaging film is one of the most important properties when considering its use, and the integrity of the resultant seal is of paramount importance to the ultimate package integrity.

A number of factors are involved in determining the quality of a heat seal. They can be conveniently summarized under three headings:

1. *Machine factors*: dwell or clamp time; temperature and pressure.
2. *Resin factors*: density; molecular weight and additives in the resin.
3. *Film factors*: gauge; style or form (*e.g.*, whether gussetted or not) and treatment for printing.

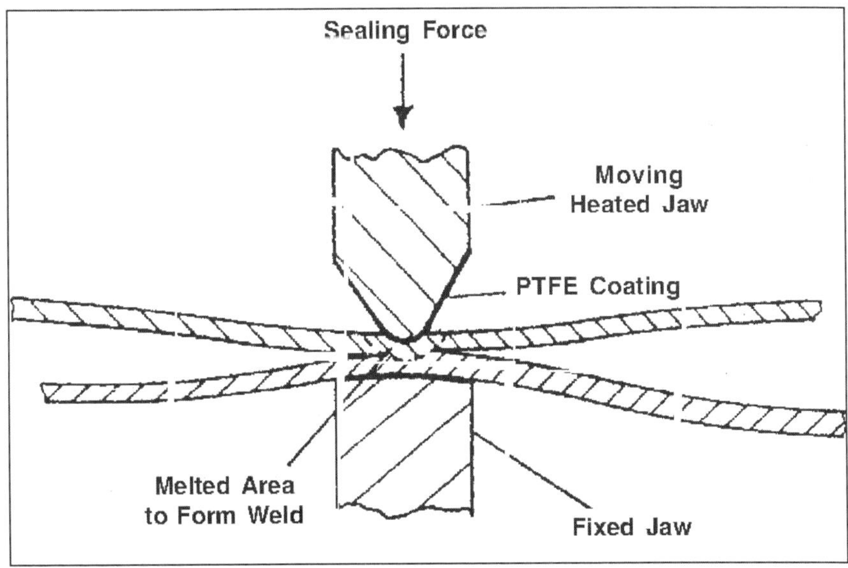

Figure 13.10: A Simple Heat Sealer

All of these factors tend to interact in a complex way. For example, the amount of heat available may be limited by the capacity of the heating elements, by the rate of heat transfer of the sealing bar and its coating, or the type of product being packaged. Increasing the dwell time (*i.e.*, the time during which heat is applied) will increase the heat available, but this may prove to be uneconomic since fewer units will be able to be handled per minute.

Heat sealable films are considered to be those films which can be bonded together by the normal application of heat, such as by conductance from a heavy heat-resistant metal bar containing a heating element. Nonheat-sealable films cannot be sealed this way, but they can often be made heat sealable by coating them with heat-sealable coatings. In this way, the two facing coated surfaces become bonded to each other by application of heat and pressure for the required dwell time.

Conductance Sealing

Conductance (also known as resistance) sealers are the most common type of heat sealers in commercial use and typically consist of two metal jaws (often patterned or embossed to give the seals extra strength), one of which is electrically heated, the temperature being controlled thermostatically. The second or backing jaw is often covered with a resilient material such as rubber to distribute pressure evenly and aid in smoothing out the film in the sealing area. Frequently the unheated jaw is water cooled, although in some situations it may be heated to the same temperature as the first jaw to enable sealing through sheets of film in exactly the same way each time.

Conductance sealers are used for coated regenerated cellulose films or any other materials with a heat seal coating such as foil or paper. However, they are not suitable for unsupported materials such as polyethylene film which would simply melt and stick to the jaws. Serrated jaws can be used to ensure that the two webs are stretched into intimate contact with high local pressure; they also improve appearance. For all sealing jaws a non-stick coating is desirable; Teflon-[poly(tetrafluoroethylene)] is commonly used, either as a cloth-reinforced film or, in the case of serrated jaws, as a powder bound in a heat-resistant vehicle.

Dwell time should be able to be controlled to fractions of a second and be easily adjustable. Likewise, the pressure between the jaws should also be easily adjustable. Both these factors will need to be changed when different materials are heat sealed.

A variation on this type of sealer is the band sealer, where the films travel between two endless bands of metal which are pressed together by heated bars. The heat passes through the bands and seals the films; the bands are then pressed together by chilled bars to withdraw heat from the seal. Band sealers are widely used for sealing pouches and have the advantage of being continuous.

Impulse Sealing

In this method the films (usually unsupported materials) are clamped between two cold metal bars and then fused by the effect of a short heat impulse; cooling occurs under pressure. Much lighter jaws than for resistance sealers are used. A high current is sent for a short period through a nichrome resistance wire or ribbon covered with Teflon tape. The current heats the wire to the desired temperature, the temperature being controlled with a transformer.

Dwell time of the heating impulse must also be controlled, as must the length of the cooling period. Thus, two timers are often found on impulse sealers, the second timer controlling the cooling of the resistance wire to allow the film to harden under pressure and prevent deformation of the film in the sealed area. Sometimes one of the jaws is water cooled to prevent excess heating and promote rapid cooling. If very heavy films are to be sealed, both jaws may contain resistance wires. To prevent the film from sticking to the heated jaws, it is usual to cover them with a slip sheet such as glass cloth impregnated with Teflon. Generally, the seals produced by impulse sealing are of excellent quality.

Ultrasonic Sealing

In this method of sealing, a generator feeds a 20 kHz signal into a transducer which transforms electrical energy into mechanical vibrations of the same frequency. The mechanical energy is converted into heat at the interface and produces an almost instantaneous weld. Little overall heating is produced, thus enabling oriented films to be sealed without any change in dimensions. This method is also useful for sealing multiwall paper bags which have an inner polyethylene liner, since the heat is generated only at the welding interface and there is no heat damage to the paper.

A conical welding head acts as a focussing tool for the vibrations which pass through the film to an anvil. The pressure, along with the dwell time (expressed in terms of the rate at which the film is fed between the welding head and anvil) and temperature, is adjusted to get optimum seal quality.

Dielectric Sealing

In this method, dielectric heating is used in which a high-frequency current (typically 50–80 MHz) is passed through two or more layers of film. The electrodes (usually made of brass) are the top and bottom jaws, the latter being the ground of the circuit. When the layers of film are in place between the closed jaws, a high-frequency current is passed between the jaws, heating and liquefying the films. The pressure which the jaws exert on the films helps to bring about thorough fusion and bonding.

This method of sealing is used principally with poly(vinyl chloride) and nylon 6/6 films as they are difficult to heat seal by direct means because they tend to degrade at temperatures close to their softening point. It is only applicable to materials which are polar and capable of forming a dipole moment.

Hot-wire Sealing

This method of sealing (which uses a thin wire or strip of metal with a, radiussed edge, heated with a low voltage current) has found application for the manufacture of polyethylene bags and pouches in tubular form, the cutting and sealing operations being carried out in the one step. It is only applicable to thermoplastic films which can tolerate high temperatures for a short time and which also have a low viscosity in the fused stage. When unsupported films are trim sealed by this method, they tend to form a strong bead in their seal areas due to surface tension and orientation. This method is also used to a limited degree with laminated constructions.

There is a tendency with this method of sealing to get what is known as "angel hair"–fine strands of polymer protruding from the seal area. This can be controlled with the use of proper temperatures and times. Generally, films thicker than 0.05 mm are difficult to seal through, especially if they have gussetted structures.

Chapter 14

Use of Nanotechnology in Foods and their Packaging

Introduction

The concept of nanotechnology was first introduced by Noble Laureate Physicist Richard Feynman in 1959. The word "nano" comes from the Greek for "dwarf". Eric Drexler popularized the term 'nano-tech' in 1980's during his works on these incredible devices.

A nanometer is a thousandth of a thousandth of a thousandth of a meter (10^{-9} m). Food nanotechnology has its history from Pasteurization process introduced by Pasteur to kill the spoilage bacteria (1000 nanometers), made the first step of revolution in food processing and improvement in quality of foods. Later, Watson and Crick's model of DNA structure which is about 2.5nm opened the gateway of applications in biotechnology, biomedical, agricultural and production processes. Further, the invention of carbon nanotubes "buckyball fullerene" which is 1nm in size served as the cutting-edge discovery to the world of innovation and led to the era of nanoscience.

Nanotechnology is an interdisciplinary study which allows us to develop new materials with new, interesting and useful properties. Now a days, nanotechnology has become one of the most promising technologies to revolutionize conventional food science and the food industry. Nanotechnology-assisted processing and packaging has proved its competence in food systems. According to a definition in a recent report (Nanotechnology in Agriculture and Food) food is "nanofood" when nanoparticles, nanotechnology techniques or tools are used during cultivation, production, processing, or packaging of the food. It does not mean atomically modified food or food produced by nanomachines.

Nanotechnology plays an active role in trying to solve problems such as food packaging and preservation. Research is in progress into their potential in food nanotechnology sector including food packaging, foods and supplements due to their unique functions and applications of nanomaterials. Nanotechnology is rapidly developing technology impacts every aspect of the food system from cultivation to food production to processing, packaging, transportation, shelf life and bioavailability of nutrients.

Scientists and industry stakeholders have already identified potential uses of nanotechnology in virtually every segment of the food industry, from agriculture (*e.g.*, pesticide, fertilizer or vaccine delivery; animal and plant pathogen detection; and targeted genetic engineering) to food processing (*e.g.*, encapsulation of flavour or odour enhancers; food textural or quality improvement; new gelation or viscosifying agents) to food packaging (*e.g.*, pathogen, gas or abuse sensors; anticounterfeiting devices, UV-protection, and stronger, more impermeable polymer films) to nutrient supplements (*e.g.*, nutraceuticals with higher stability and bioavailability). Applications of nanotechnology in food packaging sector are related to improve the properties of packaging and to mechanical, thermal and barrier, biodegradable, improve the durability of packed products. Tens of millions of dollars are being spent in a global race to apply nanotechnologies in food production, processing and packaging.

Nanotechnology enables the designers to alter the structure of the packaging materials on the molecular scale, to give the materials desired properties. Simple traditional "packing" is to be replaced with multi-functional intelligent packaging methods to improve the food quality, thanks to the application of nanotechnology in this field. This will make products cheaper, production more efficient and more sustainable through using less water and chemicals. Producing less waste and using less energy is a central concern of food manufacturers, and the drive towards production efficiency is likely to continue to boost nanotechnology funding (Asadi and Mousavi, 2007).

Nanoscale biotech and nano-bio-info will have big impacts on the food and food-processing industries. The future belongs to new products, new processes with the goal to customize and personalize the products. More than 180 applications are in different developing stages and a few of them are on the market already. In 2017, nanotechnology had a market value amounting to nearly 49 billion U.S. dollars worldwide, and was projected to increase to nearly 76 billion U.S. dollars by 2020. In terms of numbers, the USA leads, followed by Japan, China, and the EU. The U.S. Accounts for Over 29.5 per cent of Global Market Size in 2020, While China is Forecast to Grow at a 12.4 per cent CAGR for the Period of 2020-2027.The Nanotechnology market in the U.S. is estimated at US$16 Billion in the year 2020. The country currently accounts for a 29.53 per cent share in the global market.

China, the world second largest economy, is forecast to reach an estimated market size of US$22.1 Billion in the year 2027 trailing a CAGR of 12.4 per cent through 2027. Among the other noteworthy geographic markets are Japan and Canada, each forecast to grow at 11.4 per cent and 11 per cent respectively over the 2020-2027 period. Within Europe, Germany is forecast to grow at approximately 9.4 per cent CAGR while Rest of European market (as defined in the study) will reach US$22.1 Billion by the year 2027. More than 400 companies around the world today are active in nanotechnology research and development (R and D) and this number is expected to increase to more than 1000 within the next 10 years.

Application of Nanotechnology in Food Science and Processing

The potential benefits of Nanofoods–foods produced using nanotechnology are astonishing. Nanotechnology is having an impact on several aspects of food science, from how food is grown to how it is packaged. Advocates of the technology promise improved food processing, packaging and safety; enhanced flavour and nutrition; 'functional foods' where everyday foods carry medicines and supplements, and increased production and cost-effectiveness. Companies are developing nanomaterials that will make a difference not only in the taste of food, but also in food safety, and the health benefits that food delivers. Estimates of commercially available nanofoods vary widely between 150-600 nanofoods and 400-500 nanofood packaging applications.

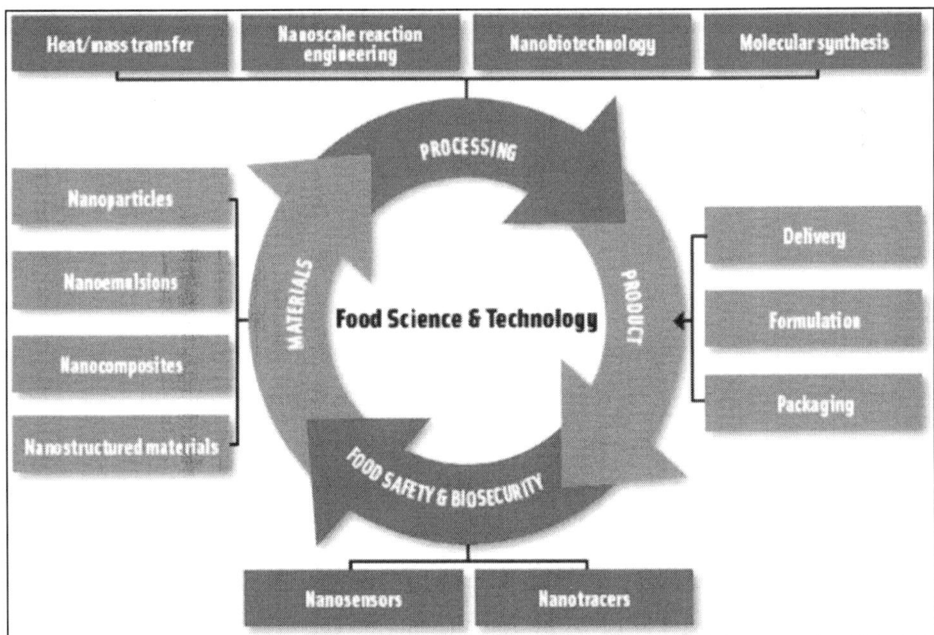

Figure 14.1: Nanotechnology Applications in Food Science and Technology.

In a world where thousands of people starve each day, increased production alone is enough to warrant worldwide support. For the past few years, the food industry has been investing millions of dollars in nanotechnology research and development. Some of the world's largest food manufacturers, including Nestle, Altria, H.J. Heinz and Unilever, are blazing the trail, while hundreds of smaller companies follow their lead. Yet, despite the potential benefits, compared with other nanotechnology arenas, nanofoods don't get a lot of publicity. The ongoing debate over nanofood safety and regulations has slowed the introduction of nanofood products, but research and development continue to thrive - though, interestingly, most of the larger companies are keeping their activities quiet. Although the risks associated with nanotechnology in other areas, such as cosmetics and medicine, are equally blurry, it seems the difference is that the public is far less apt to jump on the nanotechnology bandwagon when it comes to their food supply.

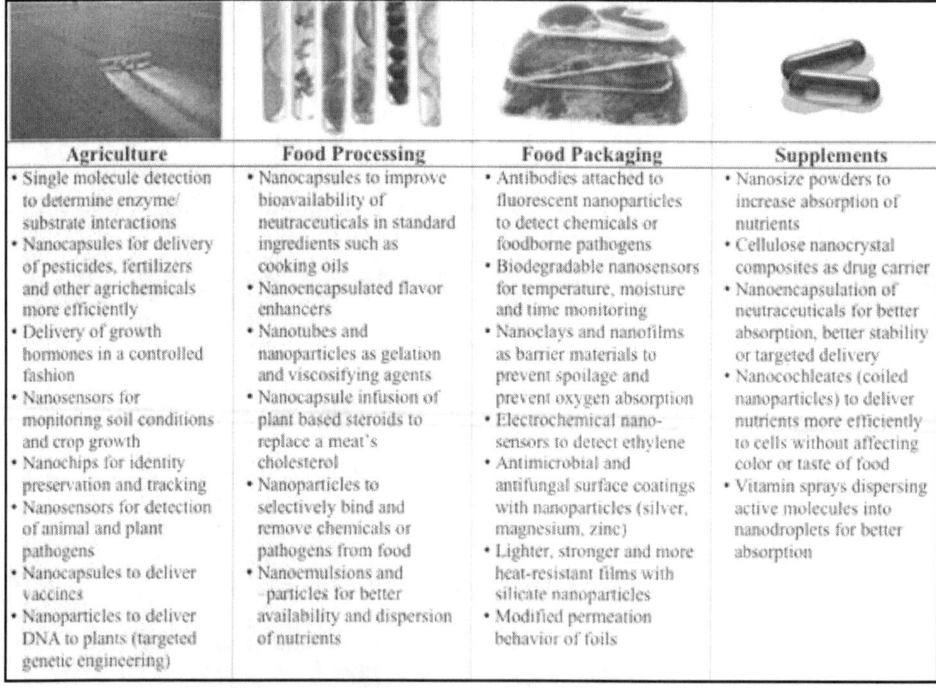

Agriculture	Food Processing	Food Packaging	Supplements
• Single molecule detection to determine enzyme/substrate interactions • Nanocapsules for delivery of pesticides, fertilizers and other agrichemicals more efficiently • Delivery of growth hormones in a controlled fashion • Nanosensors for monitoring soil conditions and crop growth • Nanochips for identity preservation and tracking • Nanosensors for detection of animal and plant pathogens • Nanocapsules to deliver vaccines • Nanoparticles to deliver DNA to plants (targeted genetic engineering)	• Nanocapsules to improve bioavailability of neutraceuticals in standard ingredients such as cooking oils • Nanoencapsulated flavor enhancers • Nanotubes and nanoparticles as gelation and viscosifying agents • Nanocapsule infusion of plant based steroids to replace a meat's cholesterol • Nanoparticles to selectively bind and remove chemicals or pathogens from food • Nanoemulsions and particles for better availability and dispersion of nutrients	• Antibodies attached to fluorescent nanoparticles to detect chemicals or foodborne pathogens • Biodegradable nanosensors for temperature, moisture and time monitoring • Nanoclays and nanofilms as barrier materials to prevent spoilage and prevent oxygen absorption • Electrochemical nano-sensors to detect ethylene • Antimicrobial and antifungal surface coatings with nanoparticles (silver, magnesium, zinc) • Lighter, stronger and more heat-resistant films with silicate nanoparticles • Modified permeation behavior of foils	• Nanosize powders to increase absorption of nutrients • Cellulose nanocrystal composites as drug carrier • Nanoencapsulation of neutraceuticals for better absorption, better stability or targeted delivery • Nanocochleates (coiled nanoparticles) to deliver nutrients more efficiently to cells without affecting color or taste of food • Vitamin sprays dispersing active molecules into nanodroplets for better absorption

Figure 14.2: Examples for Nanofood Applications (*Source*: Nanowerk).

Nanotechnology has potential applications in all aspects food processing, food packaging and food monitoring. Production of functional foods such as soft drinks, ice cream, chocolate and chips are marketed as healthy foods by reducing fat, carbohydrate or calorie content or by increasing protein, fibre or vitamin contents. Development of foods capable of modifying their colour, flavour or nutritional properties according to a person's dietary needs, allergies, or taste preferences, production of stronger flavourings, colourings and nutritional additives and lowering costs of ingredients. Increasing the shelf life of the products by using packaging materials, which can release antimicrobials or control air and moisture exchange with the environment can be done also by nanotechnology techniques. At the present time, numerous nanofood products are in development. Nanofoods which contain nanoscale ingredients and additives are already available on the supermarket shelves. Nestle and Unilever are reported to be developing nanosize emulsion-based ice cream with a lower fat content but that retains a fatty texture and flavour.

Many of the largest companies such as Heinz, Kraft foods and Nestle are investing in nanotechnology and they are on their way to commercializing the products. In the forefront of nanofood development is Kraft Foods, which took the industry's lead when it established the Nanotek Consortium, a collaboration of 15 universities and national research labs, in 2000. Kraft's focus is on "interactive" foods and beverages. These products are customized to fit the tastes and needs of consumers at an individual level. Possible products include drinks that change colours and flavours to foods that can recognize and adjust to a consumer's allergies or nutritional needs. Other large companies, such as Nestlé and Unilever, are exploring improved emulsifiers that will make food texture more uniform. These huge Western companies are responsible for the bulk of the food industry's research and development, however the nanofood industry is truly a global phenomenon.

Nanoparticles are added to many foods to improve flow properties, colour and stability during processing, or to increase shelf life. For example, aluminosillcate materials are commonly used as anticaking agents in granular or powdered processed foods, while anatase titanium dioxide is a common food whitener and brightener additive, used in confectionery, some cheeses and sauces.

Processing techniques, which produce nanoparticles, particles up to a few hundred nanometres in size, and nanoscale emulsions are used in the manufacture of salad dressings, chocolate syrups, sweeteners, flavoured oils and many other processed foods. The formation of nanoparticles and nanoscale emulsions can result from food processing techniques such as high-pressure valve homogenization, dry ball milling, dry jet milling and ultrasound emulsification. Although many food manufacturers may remain completely unaware that their foods contain nanoparticles, it is likely that, these processing techniques are used precisely because the textural changes and flow properties they produce are attractive to the manufacturers.

Nano-encapsulated active ingredients including vitamins and fatty acids are now sold commercially for use in processing and preservations of beverages, meat, cheese and other foods. For example, industrial sausage and cured meat production requires the addition of many additives to speed up the production process, stabilize colour and improve taste. Aquanova, a German Company, has developed a nanotechnology based on carrier system using 30 nm micelles to encapsulate active ingredients such as vitamins C and E and fatty acids, which can be used as preservatives. Aquanova markets these micelles as "NovaSol". The German industry magazine "Fleischwirtschaft" claims that NovaSol offers considerable advantages of faster meat processing, cheaper ingredients, higher colour stability and ready to use liquid form. These nanoadditives have been available to German manufacturers since 2006.

Table 14.1: List of Nanomaterial-based Biosensors with their Application in Food Science and Food Nanotechnology

Nanomaterials	*Analyte*	*Samples*
Silicon dioxide	Act as food colorant, hygroscopic, anticaking, and drying agent	Food preservation and packaging
Titanium dioxide	Used as whitener in dairy products (*e.g.*, milk and cheese)	Food preservation and packaging
Zinc oxide	Reduces the oxygen flow inside the packed containers	Food preservation and packaging
Silver nanoparticles	Acts as antibacterial agent, absorbs, and decomposes ethylene in fruit and vegetables	Food preservation and packaging
Inorganic nanoceramic	Used in cooking (frying)	Food preservation and packaging
Polymeric nano-particles	Used as bactericidal and efficient delivery mechanism	Food preservation and packaging
Chitosan	Used as coating agent for mandarin, strawberries, and fresh fruits	Anti-fungicide
Gold nanoparticles AuNPs	Integration of DNA or enzymes or antibodies with Au NP Pathogens Glucose	Food storage applications Meat and dairy industries Fruit juice
SWCNT (single-wall carbon nanotubes)	Integration with biomolecules Fructose Methyl parathion and chlorpyrifos	Wine Honey Phosphate-buffered solution
MWCNT (multi-walled carbon nanotubes)	Integration biomolecules Paraoxon Fructose	Food industry Phosphate-buffered solution
CdTe QDs (cadmium telluride quantum dot nanoparticles)	Integration biomolecules	Food industry
Cu and Au NPs	Pathogens	Surface water

Nanomaterials	*Analyte*	*Samples*
ZrO_2 NPs	Parathion	Phosphate-buffered solution
Exfoliated graphite nanoplatelet xGnPs	Glucose	Phosphate-buffered solution
Glyco-NPs	*E. coli*	Phosphate-buffered solution
Quantum dots, QDs	*Salmonella typhi*	Chiken carcass wash water
Gold nanorods	*Pseudomonas*	Sodium chloride
Silica particles coated with silver shells	*E. coli*	Water
Au NPs	*Mycobacteriumavium* subsp. *paratuberculosis*	Milk
CdTe QD	2,4 D (herbicide)	Phosphate-buffered solution
(CdSe)ZnS core shell QDs	Paraoxon (insecticide)	CH_3OH/H_2O (v/v) solvent
Au NPs	Paraoxon (insecticide)	Glycine buffer
Fe_3O_4 MNPs	Glucose	Acetate buffer solution
CdTe QDs	Glucose	Phosphate-buffered solution
CdSe@ZnS NPs	Maltose	Buffer solution
Silver zeolite	Antimicrobial agent	Preservations, disinfectors, and decontaminants
Al_2O_3, La, Nano	Water purification and soil cleaning	Oxidation of contaminants
Colloidal metals	Food supplements	Enhanced uptake
Graphene	Nanoplate-based nano-composites	Detects contaminants in food
Cellulose nanocrystals	Biocompatible high water uptake	Food packaging
Magnetic nano-particles	Large specific surface area	Pathogen monitoring
Carbon nanotubes	Optical, electrical, mechanical, and thermal conductivity	Food inspection and vacuum proof food packaging
Allyl isothiocyanate and carbon nanotubes	Antimicrobial packaging	Enabled effective storage of shredded cooked chicken meat
Nanolaminates	Food-grade film	Improve the texture properties of foods and serve as carriers

Nutritional additives are another growing source for the use of nanoparticles in foods. The Institute of Medicine, U.S National Academy of Sciences defines Functional Foods as foods that provide a health benefit beyond the traditional nutrients which they contain. Nano-encapsulation involves an active ingredient in a nano-size capsule. Dairy products, cereals, breads and beverages are now fortified with vitamins, minerals, probiotics, bioactive peptides, antioxidants and plant sterols. Some of these active ingredients are now being added to foods as nanoparticles or at particles of a few hundred nm in size.

Nanoparticles are being developed that will deliver vitamins or other nutrients in food and beverages without affecting the taste or appearance. These nanoparticles actually encapsulate the nutrients and carry them through the stomach into the bloodstream. Active ingredients including vitamins, preservatives and enzymes have recently been added to foods in microscale capsules. For instance, many of the commonly used Omega-3 food additives are micrometres in size, such as the 140-180 micron micro-encapsulated tuna fish oils, which are used by Nu- Mega Driphorm to fortify Australian bread. According to the manufacturer, nanocapsules of tuna fish oil added to Tip Top Bread provide valuable nutrients, while the encapsulation prevents the bread from tasting fishy. The same technology is also employed in yoghurt and baby foods.

Aquanova, Novasol range of nano-encapsulated bioactive ingredients in addition includes vitamins, co-enzymes, isoflavones, flavonoids, carotenoids, preserving agents, food colouring substances and other bioactive substances. These products are found in a wide range of food additives and in beverages additives, such as Solu tm E200 marketed by BASF, which is a vitamin E nano-solution, especially formulated for clear beverages like sports beverages, flavoured and enhanced waters.

The effectiveness of these ingredients depends on preserving and enhancing their bioavailability. Nano-sizing or nano-encapsulating active ingredients delivers greater bioavailability, improved solubility and increased potency compared to these substances in larger or micro-encapsulated form. The greater potency of nanoparticle additives may well reduce the quantities of the required additives, and also their greater chemical reactivity.

NutraLease Ltd Company, a start-up company of the Hebrew University of Jerusalem has developed novel carriers for nutraceuticals in food systems. Some of the nutraceuticals incorporated in the carriers include lycopene, beta-carotenes and phytosterols. These products are used in healthy foods, especially to prevent the accumulation of cholesterol. The nano-sized self-assembled structured liquids (NSSL) technology allows for encapsulation of nutraceuticals, cosmeceuticals, and essential oils and drugs in food, pharmaceuticals, and cosmetics. Another advantage to the NSSL technology is that it allows the addition of insoluble compounds into food and cosmetics. One of the first products developed with this technology, a healthier version of canola oil is already available to consumers in Israel.

Nanotechnology also promises to provide a means of altering and manipulating food products to more effectively deliver nutrients like protein and antioxidants for precisely targeted nutritional and health benefits. Shefer developed the encapsulated system, which resulted in nanospheres or microspheres. The major potential product applications for the nano-sphere/microsphere system are baked goods, refrigerated/frozen batters, tortillas and flat breads, processed meat products, seasonal confectionery, specialty products, chewing gums, dessert mixes, and nutritional foods.

Oilfresh Company introduced anti-oxidation device for the restaurant industry. This company has announced the introduction of Oilfresh1000 in the U.S. market, a catalytic antioxidant device for use in restaurant deep-frying machines, powered by nanotechnology. This product not only keeps frying oil fresh significantly longer, but also allows restaurants the flexibility of switching to more healthful vegetable oils. This results in better taste, crisper deep-fried foods, better consistency of product, lower costs and greater profits, and substantial benefits to health and environment.

In other parts of the world, nanotechnology efforts are focused on the agricultural side of food production. A joint effort among universities in India and Mexico is directed at developing non-toxic nanoscale herbicides. Researchers at Tamil Nadu Agricultural University in India and Monterrey Tech in Mexico are looking for ways to attack a weed's seed coating and prevent it from germinating.

The range of current nanofood research and development is as impressive as the industry's projected growth. Last August, UK-based Cientifica estimated that nanotechnologies in the food industry were currently valued at $410 million and would grow to $5.8 billion by 2012.

It has been suggested that the number of companies applying nanotechnologies to food could be as high as 400 (Cientifica 2006). A number of major food and beverage companies are reported to have (or have had) an interest in nanotechnology. It is also widely anticipated that the number of companies applying nanotechnologies to food will increase dramatically in the near future. Considering such rapid developments in this field, and the global setup of international food companies, it is not unreasonable to anticipate that more nanofood products will appear on the EU markets within the next few years.

Application of Nanotechnology in Food Packaging

Today, food packaging and monitoring are a major focus of food industry-related nanotech R and D. Packaging is enclosing the food material in a container to ensure the delivery of product in best condition to the consumer for final use. Therefore, good packaging plays a vital role in developing the image and brand of company product. Packaging protects the food and allows it to reach the consumer in hygienic and safe condition. The use of protective coatings and suitable packaging by food industry can certainly increase the shelf life of food products. The role of nanotechnology in smart/intelligent food packaging is increasing now a days.

Novel food packaging technology is by far the most promising benefit of nanotechnology in the food industry in the near future. Nanotechnology involves the study and use of materials at an extremely small scale at sizes of millionths of a millimetre and exploit the fact that some materials have different properties at this ultra small scale from those at a larger scale. One nanometer is the same as one millionth of a millimetre. The new nanocomposites developed by Ding and his team are assemblies of functionalised nanoparticles, hundreds of micrometres in size, capable of disintegrating in liquid into nanoparticles that attach to and kill microorganisms. Many scientists believe that such discoveries are pointing the way to what packaging will be like in the future. The Foundation for Scientific and Industrial Research at the Norwegian Institute of Technology (SINTEF) for example is already using nanotechnology to create small particles in the film and improve the transportation of some gases through the plastic film to pump out dirty air such as carbon dioxide. It is hoped that the concept could be used to block out harmful gases that shorten the shelf life of food. SINTEF scientists are looking at whether the film could also provide barrier protection and prevent gases such as oxygen and ethylene from deteriorating food.

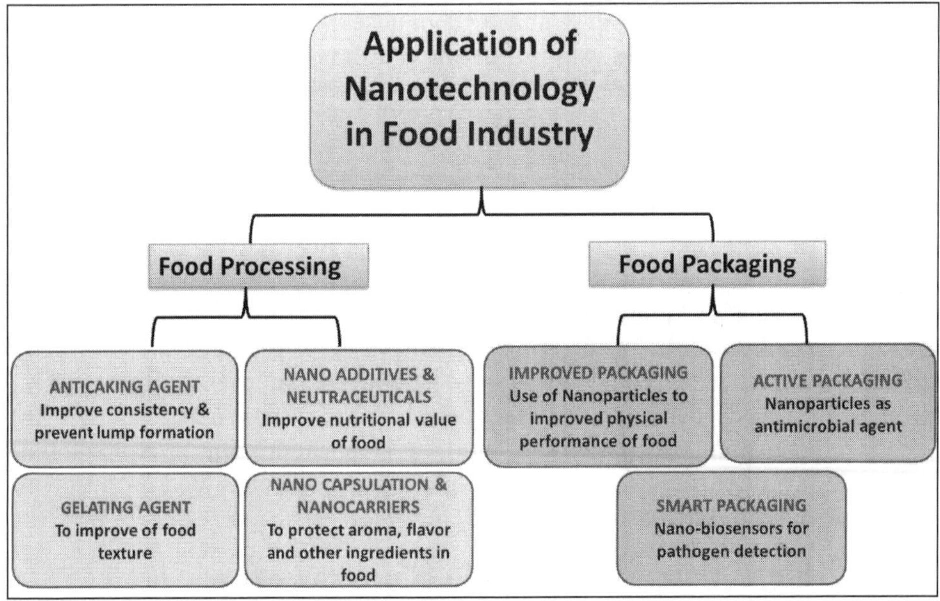

Figure 14.3: Schematic Diagram Showing Role of Nanotechnology in different Aspects of Food Sectors.

The crucial role of nanotechnology in food packaging process is considered as the largest commercial application in food sector. Nanotechnology is a technology that has the potential to revolutionize the food industry. Food packaging is considered to be one of the earliest commercial applications of nanotechnology in the food sector. El Amin had stated that rising consumer demand for improved quality, extended shelf life and environmentally-friendly products, along with enhanced regulations, EU is pushing for more innovation in the food packaging sector. He added that the food packaging manufacturers are responding to the consumer and regulatory trends by not only focusing on developing economic and effective packages for protecting the food products, but also on the aesthetic value of the packages. Customers today demand a lot more from packaging in terms of protecting the quality, freshness and safety of foods. He believes that the nanotechnology, which uses microscopic particles, is effective and affordable and will bring out suitable food packaging in the near future. He added that in the food packaging sector, nanomaterials are being developed with enhanced mechanical and thermal properties to ensure better protection of foods from external mechanical, thermal, chemical or microbiological effects. This would endow packaged foods with an additional level of safety and functionality. This technology would potentially increase the shelf life of foods.

A scientific and industrial research at the Norwegian Institute of Technology is using nanotechnology to create tiny particles in the film, to improve the transportation of some gases through the plastic film to pump out dirty air such as carbon dioxide. It is hoped that the concept could be used to block out harmful gases that shorten the shelf life of the foods. The scientists are looking at whether the film could also provide barrier protection and prevent gases such as oxygen and ethylene from deteriorating foods.

Waxy coating is used widely for some foods such as apples and cheeses. Recently, nanotechnology has enabled the development of nanoscale edible coatings as thin as 5 nm wide, which are invisible to the human eye. These edible nano-coatings could be used on meats, cheese, fruits, vegetables, confectionery, bakery goods and fast foods. They could provide a barrier to moisture and gas exchange, act as a vehicle to deliver colours, flavours, antioxidants, enzymes and anti-browning agents and could also increase the shelf life of manufactured foods, even after the packaging is opened. The U.S. Company Sono-Tec Corporation announced in early 2007 that it has developed an edible antibacterial nano-coating, which can be applied directly to bakery goods. It is currently testing the process with its clients.

Another trend in this respect is the chemical release nano-packaging. This technique enables food packaging to interact with the food. The exchange can be processed in both directions. Packaging can release nanoscale antimicrobials, antioxidants, flavours, fragrances or nutraceuticals into the food or beverages to extend its shelf life or to improve its taste or smell.

In many instances chemical release packaging also incorporates surveillance elements, where the release of nano-chemicals will occur in response to a particular trigger event. Conversely, nano-packaging using carbon nanotubes are being developed with the ability to pump out carbon dioxide that would otherwise result in food or beverages deterioration. Nano-packaging that absorbs undesirable flavours is also in development stage.

Industrial nanotech (OTC: INTK), a company that specializes in nanotechnology innovation and product development, had announced the successful application of the company, s8217, sNansulate protective coating to provide thermal insulation and corrosion protection of dairy processing equipment. The company reported that it has received a repeat order for Nansulate PT from a dairy plant located in Crowell, Texas. The Nansulate PT was used to coat dairy processing tanks and pipes in order to protect them against corrosion and insulates against heat loss, increasing efficiency of the manufacturing process by reducing both energy and corrosion-related expenses. The is increasing annual production at a rate of approximately 2.8 per cent per year. Given these massive production volumes and annual increases, energy conservation and equipment protection are becoming critical issues for the US dairy industry.

Hence, in recent years, there is more concern about research and innovation in food packaging materials ranging from films, carbon nanotubes, to waxy nano-coatings for some foods. In the dominion of packaging, nanotechnology is been adopted much more rapidly. The use of nanoparticles might help in production of new food packaging materials with improved mechanical, barrier and antimicrobial properties to increase shelf life. Beside antimicrobial characteristics, nanoparticles can be used as vehicle to deliver antioxidants, enzymes, flavours, anti-browning agents and other materials to extend shelf life, even after.

The application of nanoparticles is not limited to antimicrobial food packaging but nanocomposite and nanolaminates have been actively used in food packaging to provide a barrier from extreme thermal and mechanical shock extending food shelf-life. In this way, the incorporation of nanoparticles into packaging materials offers quality food with longer shelf-life. The purpose of creating polymer composites is to have more mechanical and thermostable packing materials. Many inorganic or organic fillers are being used in order to achieve improved polymer composites. The incorporation of nanoparticles in polymers has allowed developing more resist packaging material with cost effectiveness. Use of inert nanoscale fillers such as clay and silicate nanoplatelets, silica (SiO_2) nanoparticles, chitin or chitosan into the polymer matrix renders it lighter, stronger, fire resistance, and better thermal properties. Antimicrobial nanocomposite films which are prepared by impregnating the fillers (having at least one dimension in the nanometric range or nanoparticles) into the polymers offer two-way benefit because of their structural integrity and barrier properties.

About 400-500 nano-packaging products are estimated to be in commercial use at the moment, while nanotechnology is predicted to be used in the manufacture of 25 per cent of all food packaging within the next decade. The significant purpose of nano-packaging is to set longer shelf life by improving the barrier function of food packaging to reduce gas and moisture exchange and UV light exposure. A variety of nanomaterials such as silver nanoparticle, titanium nitride nanoparticle, and nano-titanium dioxide, nano-zinc oxide, and nanoclay are introduced as functional additives to food packaging.

Nanotechnology enabled food packaging can be divided into three main categories.

1. Improved Packaging

Nanoparticles are mixed with polymer chain to improve the gas barrier properties, as well as, temperature, humidity resistance of packaging. The use of nanocomposite in contact with food is approved by United States Food and Drug Administration.

2. Active Packaging

Application of nanocomposites as an active material for packaging and material coating can also be used to improve food packaging. Many researchers were interested in studying the antimicrobial properties of organic compounds like essential oils, organic acids, and bacteriocins and their use in polymeric matrices as antimicrobial packaging. However, these compounds do not fit into the many food processing steps which require high temperatures and pressures as they are highly sensitive to these physical conditions. Using inorganic nanoparticles, a strong antibacterial activity can be achieved in low concentrations and more stability in extreme conditions. Therefore, in recent years, it has been a great interest of using these nanoparticles in antimicrobial food packaging. An antimicrobial packaging is actually a form of active packaging which contacts with the food product or the headspace inside to inhibit or retard the microbial growth that may be present on food surfaces. Many nanoparticles such as silver, copper, chitosan, and metal oxide nanoparticles like titanium oxide or zinc oxide have been reported to have antibacterial property.

The use of nanomaterials is helpful to interact directly with food or environment to allow better protection of the product. There are many nanomaterials which are used for different food applications on industrial scale. Several nanomaterials like nanocopper oxide, nanosilver, nanotitanium dioxide, nanomagnesium oxide and carbon nanotubes can provide antimicrobial properties. Titanium dioxide having a particle size of less than 100 nm is widely used as food additive and antimicrobial agent for food packaging and storage containers. Interestingly, TiO_2 is widely used as a disinfecting agent as it generates highly reactive oxygen species (ROS) that are toxic to pathogenic microorganisms. Silver nanoparticles are used as antimicrobial agents in food packaging, storage containers, chopping boards and refrigerators and also as health supplement. Presently, the use of silver nanoparticles as antibacterial agents in food packaging is increasing. Due to the property of low volatility and high temperature stability, silver at the nanoscale is identified to be a good anti-microbial and anti-fungal agent. It is claimed to be effective against more than 100 different bacteria. Silver has several advantages compared to other antimicrobial agents. Silver is toxic and has broad spectrum to various strains of fungi, bacteria, algae, and certain viruses. On the other hand, other molecular antimicrobials are normally specific to organisms (to varying degrees). Silver is shelf stable for long time. It is also reasonably effective at biofilm penetration, which is not achieved by many molecular antimicrobials. Nano silver particles can disrupt their barrier components, like lipopolysaccharides and proteins, and pierce into membranes (outer and inner) of the cells, Silver nanoparticles have the capability to hinder enzymes in respiratory chain and obstruct the penetration of phosphate and protons across the membrane, thus reducing the level of ATP. This attributes to the antimicrobial activity of silver nanoparticles. The high catalytic movement of nanoparticles can result in toxicity and oxidative stress within the microbial cells due to the formation of reactive oxygen species. Conventional viewpoint is that silver atoms cause cellular destruction in the same manner (observed for conventional silver antimicrobials) by detaching from the silver nano particles surface. Silver nano particles are efficient media to convey a huge quantity of silver ions to the cell interior in a short period of time. Antimicrobial activity of silver nanoparticles against *E. coli* and *Bacillus cereus* spores is greatly enhanced if combined with titanium dioxide and carbon nanotubes, respectively. Silver-doped TiO_2 nanoparticles also inactivated *B. cereus* spores on aluminum and polyester surfaces and destroyed airborne bacteria and molds when incorporated into air filters. Stabilization of silver nanoparticles with SDS or PVP increases antimicrobial activity against *E. coli* and *Staphylococcus aureus*. Therefore, surfaces of refrigerators and storage containers are coated with silver nanoparticles to prevent growth of microorganisms.

Carbon nanotubes can be exploited for their antibacterial properties apart from improving the polymer matrix properties. Straight contact with CNT masses was found to be fatal for *E. coli*, because long and thin CNTs puncture the microbial cells causing irreversible damages. The application of carbon nanotubes is stopped at present since quite a few studies reported that when in contact to skin, CNTs are cytotoxic to human cells.

Zinc and zinc oxide are used as nutritional additives and also as antimicrobial agents in food packaging. Compared to nanosilver, ZnO and MgO nanoparticles is likely to present more safe and reasonable packaging of food in the future. ZnO displays increasing antibacterial activity with decreasing particle size. This activity can be stimulated even with visible light and does not necessitate the presence of UV light. Silicon dioxide and carbon have particles of a few hundred nm in size and are used as food additives and for food packaging. These could provide a cheap, safe alternative to nano-sized silver, which has good antimicrobial properties, but is expensive and as a heavy metal, is not suitable for human contact.

3. Intelligent/Smart Packaging

Several types of nanosensors used in the food packaging industries include nanoparticles-based sensors, electronic noses, array biosensors, nanocantilevers, nanoparticle in solution, and nano-test strips. Packaging with nano-sensors is useful to trace the external or internal conditions of food products, pellets and containers throughout the food supply chain. Packaging that incorporates sensors and sometimes nanosensors can be “smart,” which means that it can respond to environmental conditions or repair itself or alert a consumer to contamination and/or the presence of pathogens. The trademark of intelligent packaging systems is to increase the communication feature of a package, for instance to vigorously replicate the original food quality in real time, contrasting to the fixed “Use By” and “Best Before” dating approach. Nano sensors have the ability to react with microbial

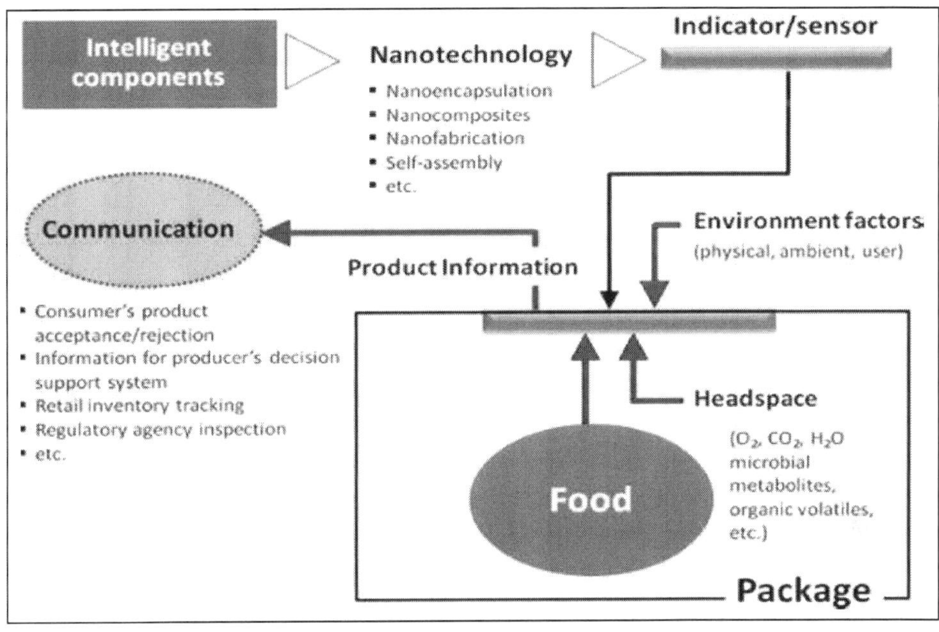

Figure 14.4: Intelligent Packaging.
(*Picture Courtesy* S.D.F. Mihindukulasuriya *et al.*, 2014, Trends in Food Science and technology).

contamination, changes in environment (such as humidity or temperature in storage rooms, oxygen exposure level) or degradation products. The sensor is made-up using appropriate nanotechnology. It can interrelate with external environmental factors and internal factors (food components). Consequently, response will be generated by the sensor (*e.g.*, electrical signal, visual cue) that correlates with the condition of the food product. Indicators of freshness for packaging of food are a vital prerequisite in the positive improvement of freshness indicators. It is the understanding about the quality-indicating metabolites. A freshness sensor should respond to the occurrence of these metabolites with the essential sensitivity. "Electronic Tongue" technology, an additional growth in the field, is made using sensor arrays to indicate the state of the food. This device includes an array of sensors tremendously sensitive to gases, which are released by microorganisms causing spoilage, producing colour changes that indicate whether the food is deteriorated. It is designed for sensing biochemical or microbial changes in the food. It can detect specific pathogen developing in the food or specific gases from food spoiling. Some smart packaging has been developed to use as tracing device for food safety. Packaging equipped with nano-sensors is designed to track either the internal or external conditions of food products, pellets and containers, throughout the supply chain. For example, such packaging can monitor temperature or humidity over time and then provide relevant information of these conditions, for example by changing colour. Some of these nano-sensors are under development as example in the Georgia Tech in the United State used modified carbon nanotube as biosensor to detect microorganisms, toxic substances and spoilage of foods or beverages. Another example, Opel, which makes Opal film incorporating 50nm carbon black nanoparticles was used as biosensor that can change colour in response to food spoilage. Titanium dioxide nanoparticles based, oxygen-sensing inks were used as Tamper proofing. Currently, Nestle, British Airways and MonoPrix Super market are using chemical sensors, which can easily detect colour change.

Nanosensors in plastic packaging can detect gases in food when it spoils and packaging itself changes the colour to alert the consumer. Further, film packed with silicate nanoparticles can reduce the flow of oxygen into the package and leaking of moisture out of package can keep the food fresh. It can prevent the growth of mould inside the refrigerator. Sensors have been developed to detect *Escherichia coli* contamination in packaged foods. Nanotechnology can reduce the packaging waste associated with processed foods and also support the preservation of fresh foods and thereby extending their self-life. The current technology can detect the microbial pathogens in food products in 2 to 7 days. In addition, it is used for detection of toxin, pesticide, and spoilage. Nanotechnology would be used to manufacture smart packaging for extending shelf life of the product and enable it to transport even further. Smart packaging containing nanosensors and antimicrobials is being developed to detect food spoilage and release Nano antimicrobials to extend shelf life, enabling supermarkets to keep food for even longer periods before its sale. It is recommended that nanopackaging should be designed in such a way to release antimicrobials, antioxidants, flavours, enzymes and nutraceuticals to extend the shelf life of the food products. There is need to undertake further toxicological and migration studies in order to ensure safe development of nanotechnologies in the food packaging industry. Attempts should be made to develop low-cost packaging by using nanotechnology.

Nanotechnology is also enabling sensor packaging to incorporate cheap Radio Frequency Identification (RFID) tags. The nano-enabled RFID tags are much smaller, flexible and can be printed on thin labels. This increases the tags versatility and thus enables much cheaper production. Other nano-based track and trace packaging technologies are also being improved. For instance, a United States company Oxonica Inc, has developed nano-barcodes to be used for individual items or pellets, which must be read with a modified microscope. These have been developed primarily for anti- counterfeiting purposes [44]. An ingestible nano-based track and trace technology is promised by pSiNutria, a spin out of nanotechnology company pSivida. Potential pSiNutria products include products to detect pathogens in food, for food tracing, and preservation and temperature measurements in food storage.

Another trend in the application of nano-packaging is the nano-biodegradable packaging. The use of nanomaterials to strengthen bioplastics (plant-based plastics) may enable bioplastics to be used instead of fossil-fuel based plastics for food packaging and carry bags.

Advantages of Nano-based 'Smart' and 'Active' Food Packaging over Conventional Packaging

A desirable packaging material must have gas and moisture permeability combined with strength and biodegradability. Nano-based "smart" and "active" food packagings confer several advantages over conventional packaging methods from providing better packaging material with improved mechanical strength, barrier properties, antimicrobial films to nanosensing for pathogen detection and alerting consumers to the safety status of food.

Using Clay Nanoparticles to Improve Plastic Packaging for Food Products

Use of nanomaterials in food packaging is already a reality. Packaging has been revolutionized due to the integration of nanocomposites, nanosensors, biodegradable nanocomposites for leakage proof, gases free, and pathogen less food packaging. They act as barrier for exchange of gases and maintain the quality of food using nanoclays, the formation of bacterial and fungal organisms or any kind of pathogens and toxins are terminated using antimicrobial packaging using silver, titanium oxide, zinc oxide and other bionanoparticles. Clay nanocomposites are being used to provide an impermeable barrier to gases such as oxygen or carbon dioxide in lightweight bottles, cartons and packaging films. Clay nanocomposites are being used in plastic bottles to extend the shelf life of beer and make plastic bottles nearly shatter proof.

Chemical giant Bayer produces a transparent plastic film (called Durethan) containing nanoparticles of clay. The nanoparticles are dispersed throughout the plastic and are able to block oxygen, carbon dioxide and moisture from reaching fresh meats or other foods. The nanoclay also makes the plastic lighter, stronger and more heat-resistant.

One example is bottles made with nanocomposites that minimize the leakage of carbon dioxide out of the bottle; this increases the shelf life of carbonated beverages without having to use heavier glass bottles or more expensive cans.

With different nanostructure, the plastics can obtain various gas/water vapor permeability to fit the requirements of preserving fruit, vegetable, beverage and other foods. By adding nanoparticles, people can also produce bottles and packages with more light resistance, stronger mechanical and thermal performance, and less gas absorption. These properties can significantly increase the shelf life, efficiently preserve flavour and colour, and facilitate transportation and usage. Further, nanostructured film can effectively prevent the food from the invasion of bacteria and microorganism and ensure the food safety.

Another example is food storage bins with silver nanoparticles embedded in the plastic. The silver nanoparticles kill bacteria from any food previously stored in the bins, minimizing harmful bacteria.

Nanotechnology in Plastics Packaging

The study published by Pira (2004) takes a detailed look at what these breakthrough technologies mean for plastics packaging. It looks at how the usability, durability and value-added of plastics packaging materials can be enhanced through nanotechnologies such as silicate nanoparticles, carbon nanofibres, carbon nanotubes, electrospun nanotubes, nanocapsules and self-assembled monolayers. It looks at all current and future nano-related research projects or products in these technologies or applications that could have an impact on the plastics packaging sector.

Nanoscale titanium dioxide particles could be the next breakthrough in plastic packaging, having the ability to prevent ultraviolet light from reducing shelf life.

Over time the sun's ultraviolet light can cause cracking, fading and other types of solar degradation to plastics. Ultraviolet light (UV) can also reduce the shelf life and effectiveness of products in transparent or semi-transparent packaging.

Finding better ways to block UV rays from passing through plastic packaging, while still allowing consumers to see the product inside, is a major goal within the industry (Elamin and Bird, 2007).

DuPont announced the release of its Light Stabilizer 210, a plastic additive designed using extremely small particles of titanium dioxide, which have been nanoengineered to absorb UV. He listed current potential applications for Light Stabilizer 210 as consumer packaged products; the company had also applied in the US for regulatory approval for use in the food sector. In testing, Light Stabilizer 210 blocked twice as much UV light as several classes of competitive products, the company claimed in a statement.

Light Stabilizer 210 works by absorbing UV rays and changing them into small amounts of heat which dissipate quickly without damaging the structure of plastic, DuPont stated. The key performance advantage of the new light stabiliser is that its extremely small particle size provides much more surface area for UV absorption, as a sizeable percentage of titanium dioxide particles in the product are nanoscale.

Creating a Molecular Barrier by Embedding Nanocrystals in Plastic for Improving Packaging

Until recently, industry's quest to package beer in plastic bottles (for cheaper transport) was unsuccessful because of spoilage and flavour problems. Today, Nanocor, a subsidiary of Amcol International Corp., is producing nanocomposites for use in plastic beer bottles that give the brew a six-month shelf-life. By embedding nanocrystals in plastic, researchers have created a molecular barrier that helps prevent the escape of oxygen. Nanocor and Southern Clay Products are now working on a plastic beer bottle that may increase shelf-life to 18 months. Zinc oxide nanoparticles can be incorporated into plastic packaging to block UV rays and provide anti-bacterial protection, while improving the strength and stability of the plastic film.

Using Nanotechnology Methods to Develop Antimicrobial Packaging and 'Active Packaging'

Kodak, best known for producing camera film, is using nanotech to develop antimicrobial packaging for food products that will be commercially available in coming time. Kodak is also developing other 'active packaging,' which absorbs oxygen, thereby keeping food fresh.

Using a Nanotech Bioswitch in 'Release on Command' Food Packaging

Researchers in the Netherlands are going on further to develop intelligent packaging that will release a preservative if the food within begins to spoil. This "release on command" preservative packaging is operated by means of a bioswitch developed through nanotechnology.

Embedded Sensors in Food Packaging and 'Electronic Tongue' Technology

Scientists at Kraft, as well as at Rutgers University and the University of Connecticut, are working on nano-particle films and other packaging with embedded sensors that will detect food pathogens, called "electronic tongue" technology, the sensors can detect substances in parts per trillion and would trigger a colour change in the packaging to alert the consumer if a food has become contaminated or if it has begun to spoil (Anonymous, 2004). With embedded nanosensors in the packaging consumers will be able to "read" the food inside. Sensors can alarm us before the food goes rotten or can inform the exact nutrition status contained in the contents. Nanotechnology is going to change the fabrication of the whole packaging industry. Self-assembly will reduce the fabrication costs and infrastructure. More flexible packaging methods will provide the consumers with fresher and customized products (Asadi and Mousavi, 2007).

Using Food Packaging Sensors in Defence and Security Applications

Developing small sensors to detect food-borne pathogens will not just extend the reach of industrial agriculture and large-scale food processing. In the view of the US military, it's a national security priority. With present technologies, testing for microbial food-contamination takes two to seven days and the sensors that have been developed to date are too big to be transported easily. Several groups of researchers in the US are developing biosensors that can detect pathogens quickly and easily, reasoning that "super sensors" would play a crucial role in the event of a terrorist attack on the food supply (Moraru *et al.,* 2003). With USDA and National Science Foundation funding, researchers at Purdue University are working to produce a hand-held sensor capable of detecting a specific bacteria instantaneously from any sample. They've created a start-up company called BioVitesse.

Regulating Nanotech in Food Industry

While there are lots of opportunities for using nanotechnology to improve food production, packaging, and quality, there is also some concern about how this will play out. The health implications of food processing techniques that produce nanoparticles and nanoscale emulsions also warrant the attention of food regulations. The potential for such foods to pose new health risks must be investigated in order to determine whether or not related new food safety standards are required. There is an urgent need for regulatory systems capable of managing any risks associated with nanofood and the use of nanotechnologies in food industry. Governments must also respond to nanotechnology's broader social, economic, civil liberties and ethical challenges. To ensure democratic control of these new technologies in the important area of food and agriculture, public involvement in nanotechnology decision making is also essential.

Many nanotechnology initiatives, commissions, or centers have been launched by governments, academia, private sectors in the United States, Europe, Japan, and some other countries around the globe to ensure rapid development and deployment of nanotechnology, promote economic growth, maintain global competitiveness, and improve the innovative capability. Some of them have also participated in proposing regulation to improve the protection of human health and the environment. These organizations or research centers mainly supported by the government sources play an essential role in performing or supporting nanotechnology researches, including the basic researches on nanotechnology, the applications of nanotechnology, safety assessment of nanomaterials, and the development of regulatory control. They also provide long-term coherence and platforms for interdisciplinary people or experts to promote nanotechnology to the public as well as to give an impetus to the nanotechnology industry. While the regulatory agencies may be making these efforts a little late, because some products are already available and development has been started on many more, one can hope that discussions would help consumers to benefit from improved and safe food products with a minimum of controversy (Earl Boysen).

In some countries (*i.e.*, the USA), some existing laws such as notably the Toxic Substances Control Act; the Occupational Safety and Health Act; the Food, Drug and Cosmetic Act, and the major environmental laws may at least provide some legal basis for regulating nanotechnology. Codex Alimentarius Commission, an intergovernmental agency established jointly by FAO and WHO, is promoting coordination of all food standards work undertaken by international governmental and non-governmental organizations. In order to prevent regulatory gap, Codex Alimentarius Commission, participating in the development of international food safety regulation, should take into account the use of nanoparticles and other nanoscale technologies in food and agriculture. Up to now, there is no international regulation of nanotechnology or nanoproducts. Only a few government agencies or organizations from different countries have established standard and regulation to define and regulate the use of nanotechnology.

In August 2006, the U.S. Food and Drug Administration (FDA) formed a Nanotechnology Task Force with goals that included:

1. Evaluate the effectiveness of the agency's regulatory approaches and authorities to meet any unique challenge that may be presented by the use of nanotechnology materials in FDA-regulated products.
2. Explore opportunities to foster innovation using nanotechnology materials to develop safe and effective drugs, biologics, and devices, and to develop safe foods, feeds and cosmetics.

EU legislation is in general considered more stringent than US legislation and more open to consumer opinions. For example, REACH requires that all chemicals, thus also NMs, must be registered, which includes the submission of safety data.

Table 14.2: A List of Recommendations on the Development of Regulations for Nanofoods

☆	Considering the criteria including particle size range, measurement methods, processing methods, physical and chemical properties, and safety concerns.
☆	Regulating nanofoods and food products containing nanoparticles, of which their larger forms may be already permitted materials, as a novel foods, ingredients, or additives.
☆	Several terms such as "micronized food", "ultrafine food", or "nanofood" can be considered. For the micronized food, a wider range of particle size (*i.e.*, up to a few micron) may be considered.
☆	Upper particle size limit for nanofoods and ultrafine foods may be loosened to a value above 100 nm depending on the product types.
☆	Not only focusing on sizes, but also concerning the changes in the bioactivity, physicochemical property, and functions of nanoscale materials in relation to their size reduction.
☆	Dividing nanofood products into different categories such as liquid, powder, aerosol, suspension, emulsion, and liposome for proper classification, management, and analysis.
☆	Requirement of food labeling to identify the presence of nanomaterials in product and provide possible particle size range and relevant safety information.
☆	Carrying out societal and ethical research for food nanotechnology.
☆	Nanotoxicity screening should include the physicochemical characterization of nanomaterials, in vitro assays (cellular and noncellular), and in vivo animal studies.
☆	Safety assessment or nanotoxicity study should be required for the nanoscale materials which are chemically modified.
☆	The nanoscale materials, which do not exist in nature, should be regarded as new substance and handled with more attention (*i.e.*, control measures, nanotoxicity study, personal protective equipment, and health monitoring).
☆	Potential toxicity of each individual nanoparticulate material needs to be considered on a case-by-case basis.

☆	Safety assessments and nanotoxicity study should be required for the nanoscale materials prepared by bottom-up synthetic approach, especially for those near or smaller than 100 nm even if their larger forms are proofed to be safe for consumption.
☆	Risk assessment should include the potential release of manufactured nanoparticles from the packaging materials into food materials.
☆	Addressing any regulatory gaps in the existing food and health related regulatory frameworks in order to control the use of ingredients in the form of manufactured, free nanoparticles in consumer products.
☆	Laboratory protocols for handling nanoscale materials and dusts are needed to protect nanotechnology workers and scientists from exposure to occupational health hazards. It is recommended to start with standard hygiene procedures, including gloving, protective clothing, and highly efficient respirators capable of removing nanoparticles, and to move ahead as new information becomes available.

Source: Chau *et al.,* 2007. Trends in Food Science and Technology 18: 269-280.

In Japan, the development of nanotechnology industry is highly supported by the Japanese government agencies including the Ministry of Education, Culture, Sports, Science, and Technology (MEXT), the Ministry of Economic, Trade, and Industries (METI), the Ministry of Health, Labor, and Welfare (MHLW), and the Ministry of the Environment (MOE). MEXT is responsible for research and development, and is building industry-academia-government cooperative research platforms to promote research and development in nanotechnology and materials science. METI works for the standardization of testing methods for safety evaluation of nanoparticles and MHLW develops the evaluation methods for health impacts of nanomaterials. Therefore, MEXT is responsible to perform researches on basic science for public acceptance of nanotechnology while METI, MHLW, and MOE are working on the regulatory science. Furthermore, the Nanotechnology Researchers Network Center (Nanonet), established by National Institute for Materials Science (NIMS), provides informational support for the development of nanotechnology.

In China food safety is regulated under the Food Safety Law, which does not include any NM specifications. The National Centre for Nanoscience and Technology (NCNST) and the Commission on Nanotechnology Standardisation are responsible for developing national standards in the nanotechnology area. One of these standards contains a definition for nano materials. Applications of nanominerals or NMs to be used as food ingredients have been rejected so far by the Chinese regulatory authorities (FAO/WHO, 2013).

In Taiwan a National Science and Technology Program for Nanoscience and Nanotechnology and "Nano Technology Industrialisation Promotion Plan" were initiated. In cooperation with the Standards Committee of TANIDA (Taiwan Nanotechnology Industry Development Association) a system for certifying nanoproducts, called nanoMark System was established with the aim to enhance the quality and image of nanotechnology products, protect consumer's rights, and

Table 14.3: Summary of Food Legislation in some Non-EU Countries

Country	*Responsible Organisation*	*Key Legislation*	*Online Resources*
USA	US Food and Drug Administration (FDA)	Federal Food, Drug, and Cosmetic Act (FFDCA)	http://www.fda.gov/regulatoryinformation/legislation/federalfooddrugandcosmeticactFDCAct/default.htm
	Environmental Protection Agency (EPA)	Federal Insecticide, Fungicide, and Rodenticide Act (FIFRA)	http://www.fda.gov/Food/default.htm http://www.epa.gov/oecaagct/lfra.html
Canada	Canadian Food Inspection Agency (CFIA) Public Health Agency of Canada (PHAC)	Food and Drugs Act	http://www.hc-sc.gc.ca/dhp-mps/nano-eng.php http://laws-lois.justice.gc.ca/eng/regulations/C.R.C. per cent 2C_c._870/
Australia and New Zealand	Food Standards Australia New Zealand (FSANZ)	Australia New Zealand Food Standards Code	http://www.foodstandards.gov.au/code/Pages/default.aspx http://www.nicnas.gov.au/communications/issues/nanomaterials-nanotechnology/nicnas-regulatory-activities-in-nanomaterials http://www.foodstandards.gov.au/consumer/foodtech/nanotech/pages/default.aspx
Switzerland	Swiss Federal Office of Public Health (FOPH) Federal Office for the Environment (FOEN)		http://www.bag.admin.ch/nanotechnologie/12171/12174/index.html?lang=en http://www.bag.admin.ch/nanotechnologie/12171/12176/index.html?lang=en http://www.bag.admin.ch/themen/lebensmittel/10380/index.html?lang=en
Russia	The Federal Service for the Protection of Consumer Rights and Human Well-Being of the Ministry of Health and Social Development (Rospotrebnadzor)	Sanitary Rules and Regulations ("SanPiN")	http://www.rospotrebnadzor.ru/en/deyatelnost/bilateral.php

Country	*Responsible Organisation*	*Key Legislation*	*Online Resources*
Japan	Ministry of Health, Labour and Welfare	Food Sanitation Law	http://www.mhlw.go.jp/english/policy/health-medical/food/index.html http://www.jetro.go.jp/en/reports/regulations/
India	Food Safety Standard Authority of India (FSSAI)	Food Safety and Standards Act, 2006	http://www.fssai.gov.in/AboutFssai/Introduction.aspx?RequestID=kHte14K1h8e3hHK4iHe_doAction=TrueWHO/FAO report, 2013
China	Ministry of Agriculture Ministry of Health National Institute of Metrology	Food Safety Law of China, 2009	http://en.nim.ac.cn/ http://en.nim.ac.cn/division/overview/924
Malaysia	Ministry of Science Technology and Innovation	Food Regulations 1985 The Food Act 1983	http://www.mosti.gov.my/index.php?option=com_content and view=frontpage and Itemid=27 and lang=en
Iran	Nanotechnology Committee of Food and Drug Organisation Iran Nanotechnology Initiative Council (INIC)		http://nanohealth.ir/pages/static_page.php?id=9 and site=1 and lang=2 http://irannano.org/nano/index.php?ctrl=section and actn=get_section and lang=2 and id=22
Thailand	Food and Drug Administration of the Ministry of Public Health	Food Act B.E.2522	http://eng.moph.go.th/index.php/safety
South Africa		Foodstuffs, Cosmetics and Disinfectants Amendment Act, 2007	http://www.sani.org.za/
Brazil	National Agency of Sanitary Surveillance (ANVISA) Ministry of Agriculture, Livestock, and Food Supply (MAPA) Ministry of Health (MS)		http://portal.anvisa.gov.br/wps/portal/anvisa-ingles http://www.agricultura.gov.br/

promote the development of a national nanotechnology industry. The Asia Nano Forum NEWSLETTER (Issue No. 24) reported that from 2004 to 2014, 39 companies and 1490 products passed the NanoMark certification.

In Australia, nanofood additives and ingredients are regulated by Food Standards Australia and New Zealand (FSANZ), under the Food Standards Code. Also Australian regulations appear to be struggling to stay abreast of the rapid expansion of nanotechnology into agriculture and food systems.

In India the key piece of regulation for food safety is the Food Safety and Standards Act (2006). The Government had launched in October 2001 a programme called the Nano Science and Technology Initiative (NSTI), followed by the programme "Nano Mission" in 2007. A series of research activities have been undertaken under this program and only recently some initiatives have started to address risk issues. Standardisation remains an area of concern, as India has only taken initial steps in addressing standardisation issues. As reported in some publications specifically addressing the topic of nanotechnology risk management in India, the nation does not have a legislation that takes in consideration nanoparticles as a hazard (Chugh, 2009), has a loose framework of legislation where nanotechnology risks can be addressed (Jayanthi *et al.,* 2012) and lacks resources and expertise to handle nanotechnology risks (Barpujari, 2011).

Overall, it is becoming increasingly important to have regulatory frameworks that properly address and specifically manage the potential risks of nanotechnology. Several countries over the world have been active in examining the appropriateness of their regulatory frameworks for dealing with nanotechnologies. Countries within the EU and outside EU follow different approaches in regulating nano-based products in agri/feed/food and nanotechnology in general. The EU, along with Switzerland, were identified to be the only world region where nano-specific provisions have been incorporated in legislation for agri/feed/food. In other regions nano-specific provisions are more implicit, building mainly upon guidance for industry.

Collaboration among countries around the world is required in order to exchange information and to ensure a high level of protection for humans and the environment, while not hampering the development of new beneficial products and their global marketing. From an international marketing perspective, a restrictive regulation may discourage the innovation and marketing of nanotechnology, leading to an impediment to the growth and development of nanofood industry. As an oversight of nanotechnology is to nurture beneficial technologies rather than stifle them, it is important to keep a balance between "too strict" and "too loose" in developing the standard, definition, and regulation for nanofoods.

Safety Issues

Over past years the popularity of the uses of structures on the nanometer scale in the food sector is increasing, therefore, interest and activities in this research area have greatly focused. Its applicability in the areas of food packaging and food safety are well known. Although the advances in nanotechnology are paving new paths day by day, there still persist many challenges and opportunities to improve the current technology and also issues about the consequences of nanotechnology that must need to be addressed in order to alleviate consumer concerns. Consumers are very doubtful in consenting new technologies in their foods and nanotechnology is not an exemption. The transparency of safety issues and environmental impact should be the priority while dealing with the development of nanotechnology in food systems and therefore, compulsory testing of nano foods is required before they are released to the market.

For example, the organizers of the Joint Symposium on Food Safety and Nutrition, organized by the Central Science Laboratory in the UK and the Joint Institute for Food Safety and Applied Nutrition at the University of Maryland, had chosen to focus their 2007 symposium on Nanotechnology in Foods and Cosmetics. They feel that nanotech materials both have "the potential for use in a vast variety of products and may pose new and unique safety issues".

According to the research, nanotech in food packaging will bring benefits in the form of:

1. Reduced usage of energy and raw materials (polymers)
2. Improvement of the mechanical properties of biodegradable polymers
3. Beneficial for reusable packaging due to improved mechanical strength and
4. Help to reduce the amount of packaging waste and save resources at the same time.

Besides a lot of advantages of nanotechnology to the food industry, safety issues associated with the nanomaterial cannot be neglected. The first type of the risk deals with inconclusive research on impacts and lack of effects data:

1. Knowledge about the effects of engineered nano-materials in the environment is still insufficient
2. Risk of environmental pollution by nanoparticles (persistence and unknown catalytic effects) whereas the rests of the risks are common risks also found in normal chemicals which relates to pollution and waste:
   (*a*) Release of nano particles into the environment during the production and disposal phases of a product,
   (*b*) Uncontrolled incineration of nanocomposites could cause emissions of nanoparticles into environment and

(*c*) Nanocomposite materials could disturb plastic recycling processes. Recyclers will have to deal with nanoparticular fillers, which eventually will be found in recycled materials.

In its February, 2007 meeting, the European Food Safety Authority Regulatory agency announced that it was forming a scientific panel to conduct a risk assessment of nanoparticles in food and food packaging. This panel should be able to draw input and expertise from across Europe. For example, Denmark's National Food Institute is working on a project to gather toxicology information on nanoparticles and the UK Food Safety Authority has put together a report that provides "an outline of potential areas for future regulation relating to the use of nanotechnology and nanomaterials in foods".

Many researchers discussed safety concerns associated with nanomaterial giving emphasis on the possibility of nanoparticles migrate from the packaging material into the food and their impact on consumer's health (Bradley *et al.*, 2011; Jain *et al.*, 2016). Although a material is being considered as GRAS (generally regarded as safe) substance, additional studies must be acquired to examine the risk of its nano counterparts because the physiochemical properties in nanostates are completely different from that are in macrostate. Moreover, the small size of these nanomaterials may increase the risk for bioaccumulation within body organs and tissues (Savolainen *et al.*, 2010). For example, silica nanoparticles which are used as anti-caking agents can be cytotoxic in human lung cells when subjected to exposure (Athinarayanan *et al.*, 2014). There are a lot of factors that affect dissolution including surface morphology of the particles, concentration, surface energy, aggregation, and adsorption. A model to study the migration of particles from food packaging has been developed by Cushen *et al.* (2014). They studied the migration of silver and copper from nanocomposites and observed that the percentage of nanofiller in the nanocomposites was one of the most crucial parameters driving migration, more so than particle size, temperature, or contact time. Since every nanomaterial has its individual property, therefore, toxicity will likely be established on a case-by-case basis (Mahler *et al.*, 2012). Further, regulatory authorities must develop some standards for commercial products to ensure product quality, health and safety, and environmental regulations.

Activist Concerns

Countering these developments are civil organisations such as the *etc.* group, which has expressed general concerns about the use of nanotechnology in food and agriculture and scrutinises socio-economic effects. In a report published last year the *etc.* group called for a moratorium, prohibiting the application of nanotechnology until the absence of health risks caused by nanoparticles will be scientifically evident.

This is a major warning sign that nanotechnology may soon hit a consumer roadblock once products come on the market. The report is part of a series of outlines being published this year as the European Commission develops a ten-year strategic plan for using the technology. More reports are expected to be published by the end of this year, the deadline.

The OCA (Organic Consumers Association), a grassroots non-profit public interest organization based in the U.S., is one of many vocal organizations calling for government regulation on nanofoods, at least until more safety testing is completed. These organizations argue that a lack of evidence of harm is not the same as reasonable certainty of safety, which is what food companies must demonstrate to the U.S. Food and Drug Administration (FDA) before introducing a new food additive.

"The OCA is focusing its efforts on educating the public about the potential risks of nanofoods and putting pressure on government agencies to increase oversight," says Minowa, adding that ever-tightening federal budgets, at least in the U.S., will make the latter a huge challenge. "There's a lack of consumer understanding, a lack of government oversight and a lack of labeling," says Minowa. "Combine these with a lack of testing and you have an equation for serious problems."

Although there is far less opposition to nanopackaging than there is to nanofoods, there are some who argue that the use of these devices will allow the food industry to further shirk their corporate responsibilities.

"While devices capable of detecting food-borne pathogens could be useful in monitoring the food supply, sensors and 'smart packaging' will not address the root problems inherent in industrial food production that result in contaminated foods: faster meat (dis)assembly lines, increased mechanization, a shrinking labour force of low-wage workers, fewer inspectors, the lack of corporate and government accountability and the great distances between food producers, processors and consumers," says the *etc.* Group, a conservation and sustainable advancement organization. "Just as it has become the consumer's responsibility to make sure meat has been cooked long enough to ensure that pathogens have been killed, consumers will soon be expected to act as their own meat inspectors so that industry can continue to trim safety overhead costs and increase profits."

Interestingly enough, the Environmental Protection Agency (EPA) in the U.S. declared on November 22, 2006 that it intends to regulate a large class of consumer items made with silver nanoparticles. The decision, which will affect not only washing machines but other consumer products such as odour-destroying shoe liners, food-storage containers, air fresheners, and a wide range of other products that contain nanosilver, marks a significant reversal in federal policy. Nanosilver containing consumer products that are applied to food packaging are not regulated by the EPA but by the FDA. The FDA is still considering whether it needs new rules for nanomaterials (Garber, 2007).

Chapter 15

Packaging of Cereals

15.1 Introduction

The term cereal is a derivative from Latin word 'cerealis' meaning 'grain' which is botanically, a type of fruit called a caryopsis, composed of the endosperm, germ, and bran. The cereals are fruits of cultivated grasses, common grass members of the grass family (a monocot family Poaceae, also known as Gramineae), which usually have long, thin stalks, such as wheat, rice, maize, sorghum, millet, barley and rye, whose starchy grains are used as food. The term cereal is not limited to these grains, but, also refers to foodstuff prepared from the starchy grains of cereal like flours, breads and pasta. Cereals have been important crops for thousands of years and the successful production, storage and use of cereals has contributed in no small measure to the development of modern civilization. Amazingly, the foods human beings eat most are grasses all around the world. Cereal grains are grown in greater quantities and provide more food energy worldwide than any other type of crops; they are therefore staple food crops.

Cereal grains were the first agricultural attempts by early man, and people still enjoy them today depending on where they live and what grows there well. Today cereals and cereal-based products are an important part of the diet in most countries, and each year new products based on cereals are developed and marketed to increasingly sophisticated consumers. In their natural form (as in whole grain), they are a rich source of vitamins, minerals, carbohydrates, fats, oils and protein. However, when refined by the removal of the bran and germ, the remaining endosperm is mostly carbohydrate and lacks the majority of the other nutrients. As human food, cereals are usually marketed in raw grain form or as ingredients of food products.

The cereals of commerce and industry are harvested, transported and stored in the form of grain. The anatomical structure of all cereal grains is basically similar, differing from one cereal to another in detail only.

The mature grain of the common cereals consists of carbohydrates, nitrogenous compounds (mainly proteins), lipids, mineral matter and water, together with small quantities of vitamins, enzymes and other substances, some of which are important nutrients in the human diet. Carbohydrates are quantitatively the most important constituents, forming 77–87 per cent of the total dry matter. The lipids in milled cereal products are liable to undergo two types of deterioration: hydrolysis from endogenous lipases, and oxidation from endogenous lipoxygenases or molecular oxygen. The products of lipid hydrolysis are glycerol and free fatty acids which give rise to unpleasant odours. The products of lipid oxidation cause the odour and flavour of rancidity. Damage to the grain and the fragmentation that occurs in milling promote deterioration by bringing the lipid and the enzyme together.

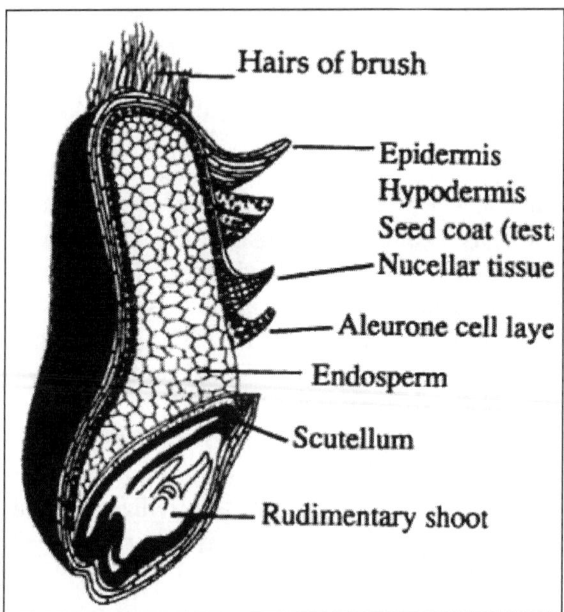

The processing of cereals for the manufacture of food may bring about alteration in the chemical composition in a number of ways:

1. Parts of the grain may be separated during processing and removed from the product, *i.e.*, the product may constitute only a fraction of the grain.
2. The various nutrients may be distributed non-uniformly throughout the various parts of the grain, so that certain nutrients are preferentially lost from or concentrated into the products when separation is made.
3. The processing treatments may bring about changes in the nutrients themselves. For example, chemical changes such as inactivation of

enzymes or hydrolysis of polysaccharides, and changes in distribution such as translocation of vitamins that occurs during the parboiling of rice.

The hazards to grain in storage are moisture, temperature, fungi, bacteria, insects and other pests. If the grain moisture content can be controlled, the hazards due to temperature-rise, fungi and insects can be largely avoided.

It has been known for many years that it is not the moisture content *per se* that is important for long term storage but rather the water activity *a*w, a "safe" moisture content for long term storage of grain and oil-seeds usually being accepted as one in equilibrium with 70 per cent relative humidity, *i.e.*, 0.70 *a*w. Above 0.75 *a*w molds will develop rapidly during storage and heating will occur, with subsequent deterioration in, and loss of, product. Moisture sorption isotherms for wheat, barley and maize have been published and all have a typical sigmoid shape. The moisture content corresponding to 0.70 *a*w varies quite considerably among the different grains, thus emphasizing the utility of expressing stability in terms of *a*w rather than moisture content.

Storage facilities for grains take many forms, ranging from piles of unprotected grain on the ground, underground pits or containers, and piles of bagged grain, to storage bins of many sizes, shapes and types of construction. It is essential that the grain has been dried to a moisture content corresponding to 0.70 *a*w or less prior to packaging and storage. Consumer packages for grain commonly consist of heat sealed pouches made from LDPE film; these provide a satisfactory moisture barrier and result in the required shelf lives for the grains.

15.2 The Storage of Wheat and Rice

The wheat grain is a living, respiring organism usually carrying endemic fungi. Respiration is slow at 14 per cent moisture content and 20°C, but rises as moisture content and temperature rise. The process of respiration generates heat (which is difficult to remove since wheat is a poor conductor) as well as CO_2 and water vapour, resulting in a loss in weight. Unless the grain is turned over to allow evaporation of the water, it will sweat and become caked in the bin.

Wheat at moisture contents between 16 per cent and 30 per cent can support fungal growth and there is thus, the risk of mycotoxin production. Above 30 per cent moisture content, wheat is susceptible to bacterial attack, leading to spoilage, excessive heat production and possibly charring. Insect life also becomes more active as the temperature rises and, because of their respiration, live insects in grain also raise the grain temperature. Deterioration during storage is aggravated by mechanical damage during harvesting since microorganisms attack damaged grains more readily than intact grains.

The shelf life for wheat as a function of moisture content and temperature is presented in Table 15.1. This table shows the importance of drying grain in that a drop of 3 per cent in moisture content increases the shelf life by a factor of four.

The major problems at these higher moisture contents include accelerated wheat enzyme activities and microbial spoilage.

Table 15.1: Safe Storage Life (Days) for Grains as a Function of Moisture Content and Temperature

Grain Temperature (°C)	*Grain Moisture, per cent*		
	14	*15.5*	*17*
10.0	256	128	64
15.5	128	64	32
21.1	64	32	16
26.7	32	16	8
32.2	16	8	4
37.8	8	4	2

Rice is the staple cereal in Japan and is also consumed in large quantities in many other countries. Brown (unmilled) rice is more nutritious than milled rice, but storage stability problems and a traditional consumer preference for whole (milled) rice have limited the quantities of brown rice packaged and sold for direct consumption. A major deterrent to greater user of brown rice is the accumulation of free fatty acids in rice stored under warm and humid conditions. Fatty acids can be released by lipase activity present in the rice aleurone (bran) layer of damaged grains and by high lipase-containing bacteria and fungi adhering to rice.

In a study to evaluate the effects of CO_2 gas flushing on the shelf life of brown rice, greater stability of the rice was obtained at refrigerated (4°C) storage temperatures when it was packaged in a laminate film (nylon-EVA Copolymer) bag rather than a regular (unspecified) plastic bag; gas flushing with CO_2 of the samples in the laminate bag improved stability even further. However, no differences between the three types of bags were found when stored at room (24°C) temperature. The gas flushed bags formed a hard, rigid pack similar to a vacuum packaged product and maintained appearance for at least 3 years. It was suggested that if bulk rice in warehouses could be stored as brown rice (minus hulls) instead of as rough rice (with hulls) in laminate, gas flushed bags, savings of at least 20 per cent by weight and 30–35 present by volume should result.

15.3 Breakfast Cereal Foods

Breakfast cereal foods can be classified according to:

1. The amount of domestic cooking required,
2. The form of the product, and
3. The cereal used as raw material.

All cereals contain a large proportion of starch which in its natural form is insoluble, tasteless and unsuited for human consumption. It must be cooked to make it digestible and acceptable. In the case of hot cereals, the cooking is carried out in the home, while ready-to-eat cereals are cooked during manufacture.

If the cereal is cooked with excess water and moderate heat as in boiling, the starch gelatinizes and becomes susceptible to starch-hydrolyzing enzymes in the human digestive system. If the cereal is cooked with a minimum of water (or without water) but at higher temperatures as in toasting, nonenzymic browning between protein and reducing sugars may occur and there may be some depolymerization of the starch.

Ready-to-eat cereals probably owe their origin to the Seventh Day Adventist Church whose members, preferring an entirely vegetable diet, experimented with the processing of cereals in the mid-nineteenth century. A granulated product called "Granula" and made by J.C. Jackson in 1863 may have been the first commercially available ready-to-eat breakfast cereal. A similar product called "Granola" was made by J.H. Kellogg by grinding biscuits made from wheatmeal, oatmeal and maize meal.

Ready-to-eat cereals comprise flaked, puffed, shredded and granulated products, generally made from wheat, maize or rice, although oats and barley are also used. The basic cereal may be enriched with sugar, honey or malt extract. All types are prepared by processes which tend to cause hydrolysis (dextrinization) rather than gelatinization of the starch.

Flaked products are made from wheat, corn, oats or rice. After cooking (often at elevated pressure) and the addition of flavourings such as malt, sugar and salt the cereal is dried to 15–20 per cent moisture content and conditioned for 1–3 days. It is then flaked, toasted, cooled and packaged.

Puffed products are prepared from conditioned whole grain wheat, rice, oats or pearl barley or a dough made from corn meal or oat flour with the addition of sugar, salt and sometimes oil. It is cooked for 20 min under pressure, dried to 14–16 per cent moisture content and pelleted by extrusion through a die. A batch of the conditioned grain or pelleted dough is fed into a heated pressure chamber which is injected with steam. The starch becomes gelatinized and expansion of water vapor on release of the pressure causes a several-fold increase in volume. The puffed product is dried to 3 per cent moisture content by toasting and then cooled and packaged.

Shredded products are made from whole wheat grains which are cooked to gelatinize the starch. After cooling and conditioning, the grain is fed through shredders. The shreds are baked for 20 min at 260°C, dried to 1 per cent moisture content, cooled and packaged.

Granulated products are made from a yeast dough consisting of wheat flour and salt. The dough is baked as large loaves which are then broken up, dried and ground to a standard degree of fineness.

Flaked or puffed cereals are sometimes coated with sugar or candy to provide a hard, transparent coating that does not become sticky even under humid conditions. The sugar content of cornflakes increases from 7 per cent to 43 per cent as a result of the coating process, and that of puffed wheat from 2 per cent to 51 per cent.

Deterioration

There are five modes of deterioration to be considered when selecting suitable packaging materials for breakfast cereals. They are:

1. Moisture gain resulting in loss of crispness.
2. Lipid oxidation resulting in rancidity and off-flavours.
3. Loss of vitamins.
4. Breakage, resulting in an aesthetically undesirable product.
5. Loss of aroma from flavoured product.

The shelf life of breakfast cereals depends to a large extent on the content and quality of the oil contained in them. Thus, products made from cereals having a low oil content such as wheat, barley, rice and maize grits (oil content: 1.5–2.0 per cent) have a longer shelf life than products made from oats (oil content 4–11 per cent, average 7 per cent). Although whole corn has a high oil content (4.4 per cent), most of the oil is contained in the germ which is removed in making grits.

Packaging

A detailed description of the materials used for the packaging of ready-to-eat breakfast cereals has been provided by Monahan, and a brief summary of the essential aspects is given below.

Loss of Crispness

Data on the permissible increase in moisture content or *a*w before loss of crispness occurs are scant and without this data, it is difficult to specify precisely the type of moisture vapor barrier required to achieve a given shelf life.

Packaging of breakfast cereals has traditionally been in fiberboard boxes with a super-calendered waxed glassine liner. More recently the glassine liner has been replaced by various plastic materials, in particular thin gauge HDPE which is usually folded rather than heat sealed. HDPE coextruded with a thin layer of EVA copolymer is a recent introduction, the EVA copolymer 'permitting a lower heat seal temperature and offering the consumer an appealing and peelable seal.

In a few cases where the cereal product is not hygroscopic and/or retains a satisfactory texture when in equilibrium with the ambient atmosphere, a liner may not be needed for moisture protection and may even serve to entrap rancid aromas. Where this is the case, either no liner or one which is vapor permeable may be used. Some shredded wheat products are in this category and are discussed below.

As well as providing a barrier to moisture vapor, the liner must also confine cereal aromas within and at the same time prevent foreign odours entering the packaged product. It should also be reclosable to protect the cereal remaining in the package.

Lipid Oxidation

The primary mode of chemical deterioration in dry cereals is lipid oxidation and two reasons have been advanced for this. First, the *a*w of dry cereals is at or below the monolayer which essentially stops all other types of deteriorative reactions. Second, unsaturated fats are required in lipid oxidation and the grains used in breakfast cereals have a high ratio of unsaturated to saturated fats.

To minimize oxidative rancidity it is important that the package excludes light. Excluding oxygen may be of limited assistance in extending shelf life although oxygen is almost never rate limiting. For this reason, most companies do not bother to use packaging which is a good oxygen barrier. In a study of the storage stability of a flaked oat cereal product packaged in materials of different oxygen barrier properties with and without the addition of an iron-based oxygen absorber, the absorber retarded or delayed lipid oxidation provided that it was used with a packaging material such as PVC/PVdC copolymer-coated PP-LDPE that was a good oxygen barrier.

The use of an anti-oxidant in the package liner has been shown to be successful in extending shelf life but is not generally permitted in most countries.

In shredded cereals, rancid odours tend to accumulate if shredded wheat is stored in an airtight container, due to the higher unsaturated fat content in this cereal. Therefore, the product is usually sold in breather boxes without inner or outer linings. When so packaged, the product is just as stable as any other prepared cereal, except that moisture absorption can take place much more readily. The net effect is a reduction in shelf life of these products.

Loss of Vitamins

The vitamin and mineral fortification of cereals is widely practiced in many countries and there are usually associated nutritional labelling requirements. The major factor influencing vitamin loss in packaged cereals is the temperature of storage. In a study on the effects of processing and storage on, micronutrients in breakfast cereals, it was concluded that micronutrient loss would not be a major factor in determining the shelf life of dry cereals. There was no substantial losses of added vitamins during normal shelf lives, with the possible exception of vitamin A and, to a slight extent, vitamin C. Vitamin A survived six months (the average distribution time) at room temperature with no measurable loss.

Mechanical Damage

The rigidity of the carton stock and the compression resistance of the finished carton together must provide the necessary resistance to product breakage throughout production line operations, warehouse storage and distribution from the manufacturer to the retailer and consumer. Rigidity also prevents the bulging of the carton. Protecting breakfast cereals from breakage does not appear to be a problem using currently available carton stock and carton designs.

Loss of Flavour

This can be a problem with certain cereal products to which fruit flavours have been added prior to packaging. In these situations, loss of flavour results in the product being considered to be at the end of its shelf life by the consumer. A study evaluating two typical cereal liner materials (HDPE and glassine) found that the permeability coefficients of *d*-limonene (a common flavour component in citrus products) in the HDPE liner were three to four orders of magnitude higher than those in glassine. It was also found that the solubility of *d*-limonene in the glassine liner was substantially lower than in the HDPE liner for the same vapor pressures. Hence equilibrium distribution of the limonene vapor between a product such as a fruit-flavoured cereal and the respective liners will result in a much lower limonene concentration within the glassine liner, and "scalping" of the cereal flavour can be assumed to be much more significant in the HDPE liner.

15.4 Pastas

The word "pasta" is traditionally associated with Italy and with wheat semolina, Italy cannot claim to have invented this popular food and semolina is not the original raw material. The Chinese invented pasta that was produced as noodles from rice and legume flours several thousand years B.C. Today pasta consumption is increasing in many countries to the extent that pasta can be classed as a truly international food.

Part of the appeal of pasta products (macaroni, spaghetti, vermicelli, noodles) is that they may be prepared from several raw materials according to countless formulations, and cooked and served in numerous ways to various tastes. Durum wheat semolina is considered to be the best raw material for pasta making because of the functional characteristics of its proteins and its high pigment content. The limited availability and high cost of durum wheat compared with other cereals has led to the use of flours and starches from rice, maize, barley, soft wheat, cassava and potato in pasta formulations in various parts of the world.

The original composition of pasta (water, durum wheat, semolina and egg) has altered considerably to include vitamin supplements, iron salts, powdered vegetables, tomato concentrate, milk protein, other cereal flours, and meat and cheese in filled pasta in order to satisfy the tastes and food habits of different populations. The introduction of modified atmosphere packaging has "enabled certain types of pasta (particularly the fresh and filled varieties) to progress from being a small-scale manufacturing operation to an established position in the food industry where, for example, fresh product is distributed across the USA.

Pasta can be subdivided into two categories. The first, macaroni, has come to represent a generic family of over 140 items in the United States and includes spaghetti, macaroni and vermicelli. This class of product is made from semolina, water, and in most cases, added vitamins and minerals, and contains about 1.5 per cent fat.

The second broad category of pasta is noodles which contain the same ingredients as pasta as well as egg solids at a minimum level of 5.5 per cent. The egg solids are usually in the form of yolks to enhance color and therefore, the lipid content of the noodles is about 4.6 per cent. Noodles may also contain additional protein from sources such as soy flour.

Dried pasta is produced by the reduction of dough moisture content from 30 per cent to about 11 per cent by means of a dehydration process, the length of which is determined by the temperature used. In recent years, the drying of pasta products at temperatures above 60°C has become widely accepted by European pasta manufacturers. Initial benefits realized from high temperature drying were control of bacteria in egg products and shorter drying cycles which permitted more compact drying lines for a given capacity. Currently two different approaches are used by manufacturers to realize the maximum benefits of high temperature drying. In one method high temperature is applied at the initial phase (predrier) of the drying process, with the remainder of the cycle performed at decreasing temperatures. The other method applies high temperatures during the final drying phase after predrying at conventional temperatures. A study which compared the quality characteristics of spaghetti dried by these two high temperature methods with conventional low temperature processes showed that the high temperature processes resulted in superior quality spaghetti.

If the drying stage of pasta, manufacture is not properly controlled; extensive growth of microorganisms such as *Salmonella* spp. and *Staphylococcus aureus* can occur, resulting in a potential hazard to public health. Extensive studies of the microbiology of pasta products have been reported for Canada and the United States; with a few exceptions, all samples met the overall criteria recommended by the International Commission for Microbiological Specifications for Foods.

Two modes have been identified for pasta product failure: moisture gain or loss and colour loss. The major mode for pasta failure is moisture gain, the optimum moisture content for pasta storage appearing to be 10–11 per cent. If the pasta moisture content increases to 13–16 per cent, mold growth (which makes the pasta unfit for consumption) and starch recrystallization or retrogradation (which makes the pasta unacceptably tough when cooked) occur. As with other dried products; the optimum moisture content for stability is derived from the maximum "safe" *a*w. Complete sorption and desorption isotherms have been obtained for macaroni at 15, 25 and 35°C, for egg noodles at 25 and 45°C and for vermicelli at 27°C.

The traditional packaging material for dried pasta was the paperboard carton, frequently with a plastic window so that the customer could view the contents.

Because of the problem of "dusting off" of pasta products and the static attraction of many plastic films for dust, the window was usually made from heavy gauge cellulose acetate as it has a very low static attraction for dust. It has been shown that the creases and end openings on a typical paperboard box increase the overall transmission rate by two times over that of paperboard alone, thereby reducing the shelf life to half under adverse (high temperature and humidity) conditions. In recent years pasta products have been packaged in plastic films such as OPP or coated LDPE.

The production of pasta involves kneading of the dough, followed by extrusion or lamination and then drawing. In the case of special pasta, this latter stage is accompanied by filling, using a cooked meat or vegetable-cheese mixture, thus resulting in a variety of potential microbiological flora. The Barilla style pasta contains salt which lowers its *a*w.

Fresh pasta products are not usually subjected to pasteurization but are refrigerated for retail distribution. The microbiological quality of fresh pasta will thus, depend on the quality of the raw materials, the cleanliness and hygiene of the processing environment and equipment, and the handling of the product during production and packing. In addition, the temperature of the product during storage, distribution and retailing is crucial with respect to microbiological quality. However, some companies are now pasteurizing the pasta after packing using hot air and/or microwaves.

These pasteurized fresh pasta products fall into the category of REPFEDs, *i.e.*, refrigerated, processed foods with extended durability.

Chapter 16

Packaging of Dairy Products

16.1 Packaging Materials

Basic role of dairy packaging, as well as for any other food product, is to provide a physical barrier to food in order to prevent the item from different damage (mechanical, physical, microbial contamination, *etc.*) and to maintain the best product quality. In the ideal case the packaging should constrain weight and nutrient losses as well as help in ex-tending shelf life of packed item. When choosing packaging material for dairy products, various important factors need to be considered such as toxicity, compatibility with the product, impact resistance, maintenance of sanitation, odour and light protection, chemically inactivity, shape and weight requirements, marketing appeal, printability and cost (Karaman *et al.,* 2015). The nature and the characteristics of the dairy product to be packaged define the selection of the appropriate packaging material and method. For example, if the product is susceptible to oxidation (such as butter) a selected material needs to have high barrier properties toward oxygen in order to enable the declared shelf life. Similarly, if the dairy product needs to be thermally treated after it has been packaged, the chosen material has to be heat tolerant. Some of the most frequently used materials for packaging of selected dairy products and their permeability properties are presented as under:

Metal Cans

The traditional method for packaging milk powders for consumers uses three piece tinplate cans where the atmospheric air is withdrawn from the powder and replaced with an inert gas such as N_2 prior to seaming the base onto the can. When correctly seamed, the can is essentially impermeable to oxygen, water vapor and light, and can be filled at high speeds. Its mechanical strength facilitates transport and handling, and the reuse possibilities of the empty can may contribute to its popularity in many parts of the world. It is usual for the top of the can to have a lid which can be levered off, and, in order to provide a gas-tight seal under the lid, an aluminum foil diaphragm is sealed to the rim of the can. This is punctured by the consumer immediately prior to use. The use of an easy-open lid incorporating a ring-pull made from aluminum and scored is now quite common; a plastic overclosure is supplied to provide a limited degree of protection once the metal end has been removed.

Laminates

In recent times aluminum foil/plastic laminates (sometimes containing a paper layer as well) have been introduced as a replacement for the tinplate can. The laminates can be formed, filled, gas flushed and sealed on a single machine from reel stock. Gas flushing is achieved by saturating the powder with inert gas but does not include the evacuation step used with cans. The main advantages associated with the laminated type of material are lower material cost and lighter material weight. The disadvantages are that such laminate packs do not have the mechanical strength and durability of rigid containers, and there can be difficulty in obtaining a satisfactory heat seal because of contamination of the heat seal area by powder during filling.

The physical strength of a laminate depends on the material of which it is composed. Aluminum foil 9 μm thick laminated to paper of weight 45 g m^{-2} and LDPE 25 μm thick will give a burst strength of 179 kPa. However, elimination of the paper and the introduction of a PET outer layer will change the physical properties considerably without greatly affecting the chemical properties. For example, 9 μm aluminum foil laminated with 12.5 μm PET and 64 μm LDPE increases the burst strength to 290 kPa.

Form-fill-seal (FFS) machines are usually designed to produce gusseted or ungusseted pouches. As there is less handling of the material with FFS machines during the production sequence, it is possible to omit the paper layer normally incorporated into premade bags. A typical material construction for milk powder on FFS machines is PET (17 μm)–LDPE (9 μm)–foil (9 μm)–LDPE (70 μm).

Fiber Cans

Fiber cans or composites manufactured by spiral winding of paper board strip are obtainable with a wide variety of liners. They can give a similar degree of protection to that obtained from aluminum foil/LDPE/paper bags and a strength similar to a metal can. They have the added advantages of being lighter than metal cans and not corroding, a problem which can occur with metal cans under high humidity conditions. Fiber cans are filled in the same way as metal cans but a disadvantage is that they cannot be cleaned with hot water. A typical specification for fiber cans for use with either whole or skim milk powder is 0.9 mm board and a foil coating of 5 μm with a nitrocellulose lacquer to protect the powder from the aluminum foil. An outer decorative label incorporating a fiber seal material gives increased protection against moisture penetration.

Occluded air trapped in vacuoles within powder particles is not removed during gas flushing of cans or laminate packs. Depending on the volume of occluded air in the powder, the equilibrium residual O_2 content of the pack could exceed the 0.02–0.03 mL g^{-1} normally tolerated. This problem can be alleviated by conditioning the powder under vacuum for 24–48 hours prior to gas flushing to remove occluded air.

Today many types of packaging techniques are employed, for the preservation of dairy products.

Vacuum Packaging

An alternative method of reducing the O_2 content of packaged whole milk powder is by compression of the powder. However, while such a procedure reduces the air content, it results in a significant increase in bulk density and (usually) a decrease in performance as measured by ease of solubility. Vacuum packaging is used to reduce the O_2 content, and it achieves this by evacuation of the interstitial air and some compression as a consequence the vacuumization. The main problem associated with vacuum packaging of milk powder is that of removing air from the package without removing powder fines which could damage the vacuum pump and contaminate the sealing area of the laminate bag. This problem can be avoided by applying the vacuum at a slow rate so that powder particles are not disturbed.

Gas Packing

The technique of gas packing is simple and, provided that the package is airtight, the O_2 content can easily be reduced from the 21 per cent present in air to 1 per cent or less immediately after the operation is finished. However, with spray dried powder this diminution in O_2 content is limited mainly to the atmosphere surrounding the hollow powder particles. Up to 28 days may need to elapse before equilibrium is established between the gases within and around the particles, although most of the change will usually be complete after 7–10 days. During this desorption period the O_2 content within the can may rise to as much as 5 per cent, considerably above the level of 1–1.5 per cent required to ensure optimal keeping quality. When O_2 levels of the order of 1 per cent are needed, there is no alternative in the usual gas packing technique but to store the powder after the initial gas packing long enough for desorption to take place and then to gas pack the cans a second time. This is costly and time consuming, requires considerable storage space, increases the handling which will remove the desorbed O_2 as it is liberated is likely to receive close scrutiny.

One such process was described in 1955, based on the removal of small amounts of O_2 by combination with H_2 in the presence of a palladium or platinum catalyst. The system involved flushing the package containing pellets and product with a combination of N_2 and H_2 gas. Initially the pellets were placed within gas permeable pouches and attached to the inside of the can lid. A comparison of conventional double gas packing with N_2 (with an interval of several days for desorption) with single gas packing with 90 per cent N_2, 10 per cent H_2 and the palladium catalyst for both whole and skim milk spray dried powders showed the new system was more effective in removing oxygen and maintaining an almost oxygen-free atmosphere within the cans. Assuming that the O_2 level within the can could rise to 10 per cent (an unlikely high value), the increase in moisture content caused by the reduction of this O_2 to water would be slightly less than 0.02 per cent. It was concluded that if the process was used on a commercial scale, both whole and skim milk powder would probably be usable after storage for 10 years at normal temperatures.

To avoid the possibility that someone might accidentally ingest the pouch, the American Can Company incorporated the palladium catalyst (in powder form on alumina) into the film construction. The multilayered film (known as Maraflex 7-F) consisted of PET/adhesive/aluminum foil/ionomer/catalyst/ionomer, an ionomer being chosen because the film between the catalyst and food had to be permeable to H_2 and O_2 as well as form a strong heat seal.

In trials using a gas flush of 8 per cent H_2 and 92 per cent N_2 with 100 g of whole milk powder in pouches, initial O_2 levels of 0.5–2.5 per cent were produced, these dropping to 0.1–0.2 per cent after one day. Two days after packaging O_2, readings were zero and an adequate amount of H_2 remained in excess of the quantity needed to scavenge the O_2. The moisture content increased from 2.42 per cent initially to a maximum of 3.34 per cent after one week and then declined to below the initial level after 2 weeks. Storage conditions were 21°C and 50 per cent RH. Similar results were obtained by other workers. Despite the apparent successes of the system, it has never been adopted commercially.

16.2 The Microorganisms in Raw Milk

In virtually all countries liquid milk for consumption must be pasteurized and cooled before it is packed. The primary purpose of the pasteurization is to destroy any pathogenic microorganisms present in the raw milk to make it safe for human consumption, while at the same time prolonging the shelf life of the milk by destroying other microorganisms and enzymes that might ruin the flavour.

In general, only 90–99 per cent of all microorganisms present in raw milk are destroyed by pasteurization, and pasteurized milk producers depend on refrigerated storage to achieve the required shelf life. Post-pasteurization contamination (PPC) with cold-tolerant Gram-negative rods is the limiting factor affecting the shelf life of commercial pasteurized milks at refrigerate storage temperatures. Inefficient sanitation of milk contact surfaces and contamination from the dairy plant atmosphere are the major causes of PPC, with the packaging machine fillers being considered to be the most important. Upgrading of pasteurized milk handling and packaging systems to ultra- clean standard appears to be an effective method for extending shelf life at refrigeration temperatures. In the absence of PPC, the shelf life of low acid pasteurized milk products depends on the activity of heat-resistant organisms which survive pasteurization, and their level of activity depends on the storage temperature. At 5°C a shelf life of 18–20 days would appear to b realistic for pasteurized milk products processed using an ultra-clean packaging system.

All fluid milk, except for a small quantity of "certified" raw milk, is pasteurized at either 63°C for 30 min (referred to as the Low Temperature Long Time (LTLT) method or sometimes as the holding method) or at 72°C for at least 15 s [referred to as the High Temperature Short Time (HTST) method] and then quickly cooled to 5°C or below. These heat treatments designed to destroy microorganisms that produce disease and to reduce the number of spoilage microorganisms present. They do not sterilize the product.

The disease-producing organisms of chief concern in milk are *Mycobacterium tuberculosis,* a non-spore-forming bacterium that causes tuberculosis and is frequently found in the milk of infected animals; Brucella species that cause brucellosis in animals and man; and *Coxiella burnetti* that causes a febrile disease in humans known as Q fever. The most resistant of these three organisms is *C. burnetti* which is characterized by a D65.6 of 30–36 s and a z of 4–5°C. *M. tuberculosis* is characterized by a D65.6 of 12–15 s and a z of 4–5°C, and *Brucella* species by a D65.6 of 6–12 s and a z of 4–5°C. It is left as an exercise for the interested reader to calculate the number of decimal reductions achieved for each of the above microorganisms by the LTLT and HTST pasteurization processes.

Now let us have look at the factors which affect the shelf life of pasteurized dairy products.

One of the most critical factors affecting the shelf life of pasteurized dairy products is the temperature of storage. Attention has been focussed on ways of predicting the effect of temperature on the growth of bacteria in foods. The "square root" equation has been found to provide a good description of the growth of a number of psychotrophic bacteria in dairy products.

The shelf life of pasteurized milk has been found to be determined mainly by the level of contamination with Gram-negative psychotrophic bacteria; the effect of temperature on growth of contaminants could be accurately determined by the "square root" plot but the conceptual minimum temperature for growth (To) varied. The variation was related to the quality of the pasteurized milk. The microflora of pasteurized milk varied significantly with storage temperature. Thus, while spoilage at refrigeration temperatures was mainly due to the growth of *Pseudomonas* spp., Enterobacteriaceae and Gram-positive bacteria assumed greater importance in the spoilage of milks stored at temperatures above 10°C. It has been suggested that 15°C may be a useful preincubation temperature for predicting the shelf life of pasteurized milk samples.

Effect of Light

During processing, distribution, storage and marketing, milk may be exposed to natural and artificial light. Flavour changes as well as loss of vitamins and other nutritional components are attributed to chemical reactions induced in the milk by light with wavelengths below 550 nm. This problem is quite widespread, a study in 1984 finding that 40 per cent of the milk samples in plastic purchased from stores showed light-induced off-flavours. Responses from taste panels have indicated that in the early stages of oxidation milk loses its naturally fresh flavour and becomes quite flat in taste without being objectionable as in the more advanced stages of oxidation.

Riboflavin (vitamin B_2) plays a central role as it is not only destroyed itself by light but, in addition, catalyzes the development of oxidized flavour and ascorbic acid oxidation by generating excited state (singlet) oxygen. Because of its absorption by riboflavin, light of wavelengths of 350–550 nm is the most damaging, the maximum damage occurring at about 450 nm. These wavelengths are contained in the emission spectra of white fluorescent tubes.

Riboflavin destruction by light has been shown to be greater in skim milk than whole milk, since light of 400–500 nm wavelengths can penetrate 40–50 per cent deeper into skim milk than whole milk. Destruction of nutrients by light could present legal problems in connection with nutrient standards and/or labels, since the milk after storage in supermarket cabinets may not meet regulatory requirements.

In a study on the relative destruction of vitamin A and riboflavin in skim milk when exposed to fluorescent light at an intensity of 2000 lumens m^{-2} for 24 hours, more than 75 per cent of the added vitamin A was destroyed in glass, clear polycarbonate and polyethylene containers. Paperboard containers provided the most protection, while gold-tinted polycarbonate which blocks light of 400–480 nm provided the second best protection.

Plastic sleeves made from poly (methyl methacrylate) with a yellow-green tint are available for installation over fluorescent tubes in display cases to shield products from light. Such tubes have 92 per cent transmission in the region from 800 to 440 nm, rapidly decreasing to zero per cent transmission at 385 nm and below. However, these shields would provide little protection from light in the damaging 400–550 nm wavelength region.

The extent of nutrient loss in a supermarket depends on the proximity of the containers to the light source, the number and wattage of light bulbs in the display case, the size of the exposed surface area, and the length of exposure. The detrimental effects of fluorescent light on milk could be alleviated by selection of packaging materials to minimize transmission of light; reduction of light intensities in display cases to 500 lumens m^{-2}; use of yellow or yellow-green lamps or filters in display cases; and responsible rotation at retail outlets to limit prolonged detrimental exposure to light.

It is clear that exposure of milk to light in the 400 to 550 nm wavelength region can result in the development of off-flavours and destruction of nutrients. It has been recommended that the packaging materials used for milk should not transmit more than 8 per cent of incident light at 500 nm wavelength and not more than 2 per cent at 400 nm. Exposure to direct sunlight should be avoided under all circumstances as this also tends to increase the temperature of the milk which accelerates microbial spoilage.

Effect of Oxygen

Oxygen plays an important role in the light-induced development of off-flavours in milk. Pasteurized milk at filling is generally saturated with O_2 (about 8 ppm), but if no additional O_2 can gain access, its content falls and the rate of the adverse reactions slows or stops. However, additional O_2 from a headspace or entering through a permeable container will maintain the O_2 content and keep the rate of oxidative reactions high. Oxygen has been shown to be limited in glass bottles because, unlike cartons and polyethylene bottles, they are impermeable and O_2 used up in oxidative reactions cannot easily be replenished.

Packaging Materials

Milk for retail sale is packaged in glass, coated paper and plastic containers of various compositions and constructions. Both single service and multiuse (refillable) containers are used, although the trend in recent years has been towards single serve packaging.

It has been demonstrated that the packaging material is a key to the protection of the flavour and nutritional qualities of fluid market milk. Numerous studies have shown that riboflavin is destroyed by light of the same wavelength as that which produces light-induced off-flavour.

Although the use of brown glass in milk bottles to reduce light damage was proposed more than seventy years ago it has only been adopted commercially in a few countries (*e.g.*, Sweden). The total amount of light passing through the container wall depends on the material from which the container is made and also on the colour either incorporated into the material (*e.g.*, the pigment in the glass) or used in printing it. The colour determines the wavelength of the light reaching the milk. Ruby glass is a most efficient barrier and amber is also very effective, but cost and marketing considerations militate against both.

Unpigmented polyethylene milk bottles in the 350 to 800 nm spectral region have been found to transmit 58 to 79 per cent of the incident light. Light transmission was reduced by pigmenting with titanium dioxide (1.6 per cent), the bottle being opaque below 390 nm. Use of a colorant with the titanium dioxide pigment further reduces transmission of light with wavelengths less than 600 nm. The unprinted area of a paperboard carton had less than 1.5 per cent transmission below 550 nm and was opaque to wavelengths below 430 nm.

In a study on light induced quality deterioration of milk, four common milk packaging materials were used: clear polyethylene pouch; coextruded laminate polyethylene pouch (outer white layer and an inner black pigmented layer);

standard paperboard carton, and returnable plastic jug. Off-flavour was detected in all containers except the laminate pouch.

A more recent study compared milk packaged in white polyethylene-coated cartons overprinted with blue, and polyethylene bottles. Cartoned milk was disliked by a flavour panel after about 17.5 h exposure to 4000 lux at 7°C and milk in the polyethylene bottles after 9 h exposure. The dissolved O_2 concentration dropped considerably in the bottled milk exposed to light, but only marginally in the cartons, light induced flavour development depending on O_2 availability. Because glass bottles are essentially impermeable, O_2 used up in oxidative reactions cannot easily be replenished and is likely to be a limiting factor.

From the quite extensive results now available in the literature, it would appear that the most efficient light barrier is the paperboard carton. This was not true of the early waxed cartons when wax penetrated the board, rendering it translucent. However, the modern polyethylene-coated board is very efficient and can be made even more so by overall printing, particularly in reds, yellows, blues, greens, browns and black in ascending order of efficiency.

Reuse of milk containers poses several problems to the consumer and the industry in that the industry has no way of controlling what a consumer may place in the container before it is returned for refilling, nor does the consumer have any way of knowing how a container may have been misused before being returned to the milk plant.

In a study where glass, polyethylene and polycarbonate multiuse milk containers were treated with 29 common household chemicals to simulate consumer abuse, glass was found to be the easiest to clean and was most resistant to retention of treatment contaminants. Neither the polyethylene nor the polycarbonate multiuse milk container appeared to comply with the provisions of the Grade "A" Pasteurized Milk Ordinance Recommendations of the US Public Health Service.

16.3 Flavoured Milk

Flavoured milk drinks are generally skimmed or semi-skimmed and contain less than 1 per cent fat. The package should be leak and tamper proof, should have sufficient wet strength and should not pass on any odour or taint to the product packed inside. The plastic based material used for sachets is octane LLDPE (O-LLDPE). OLLDPE when blended with 50 per cent LDPE provides excellent puncture resistance, excellent seal strength and hot tack. Co-extended multiple level low-density films with an outer opaque film and an inner black film for reducing the transmission of light are also used.

In India, flavoured milk drinks are available in sterilisable crown cork glass bottles, glass bottles with aluminium foil lid or snap-on plastic lid, plastic sachets and aseptic packs (Tetra bricks). Recently 200ml, translucent bottles of HDPE with an aluminium foil cap have also been introduced. Aseptic Packaging of Milk 178 Poly carbonate bottles with the leak proof screw cap are also used in place of glass bottles as these are unbreakable and are much lower in weight (1/6th the weight of glass bottle). PET containers are in use as they are light in weight, have good sales appeal and are strong.

16.4 The Packaging of Fresh Cream and Fermented Milks

Cream may be defined as that part of milk rich in fat which has been separated by skimming or mechanical separation. Cream is the concentrated form of milk fat. Cream contains a high per cent of milk fat and is very susceptible to spoilage. Cream may be free flowing and easily pourable, highly viscous and pourable only with difficulty, or spoonable but non-pourable. The method of production, the milk fat content and the system of treatment can have a marked effect on the physical characteristics of the cream. Moisture loss from cream can occur if not properly packed. It is prone to oxidative and lipolytic rancidity. It can absorb flavours

The fat content of cream may range from 10 per cent (half-cream) to 80 per cent plus (plastic cream); cream for butter manufacture would normally contain approximately 40 per cent fat. The United Nations FAO/WHO have suggested the following standards for Market Cream:

1. Pasteurized, sterilized and UHT treated cream > 18 per cent milk fat.
2. Half-cream, 10–18 per cent milk fat.
3. Whipping cream > 28 per cent milk fat.
4. Heavy whipping cream > 35 per cent milk fat.
5. Double cream > 45 per cent milk fat.

The packaging material for cream should;

1. Prevent light passage
2. Prevent water loss
3. Prevent oxygen transmission
4. Offer resistance for microbial contamination

For many years it was customary to package fresh cream in waxed paper cartons with press-in lids of the same material. However, on storage under refrigerated conditions the filled cartons showed a marked tendency to absorb moisture and become deformed. The waxed paperboard cartons were eventually superseded by plastic tubs (usually made from high-impact PS) of similar shape and sizes. These were initially covered with crimped-on skirted caps of aluminum foil, and later by aluminum foil heat sealed to the rim of the container. In addition, rectangular gable-topped waxed or polyethyle-coated paperboard packages are used, sometimes incorporating aluminum foil as a barrier in the laminate. The traditional glass bottle with a foil cap is still used in some parts of the world, but is increasingly being replaced by blow-molded polyethylene containers of similar shape, sealed with a close fitting plastic cap.

Whipped cream and synthetic formulations are sold in aerosol cans and polyethylene tubes. Imitation cream made from soybeans and vegetable oils is often marketed in wax coated paper board cartons. Ultra pasteurization has been applied to heavy and light creams. The product then goes for packing. A strong seal is necessary for product protection. PE extruded or wax coated paperboard tubs are used to pack single portion cream. Sterilized/UHT cream is packed in similar lines to that of UHT milk.

Fermented milks are products prepared from milks (whole, partially or fully skimmed milk, concentrated milk, or milk reconstituted from partially or fully skimmed dried milk) homogenized or not, pasteurized or sterilized and fermented by means of specific microorganisms.

Yoghurt, the most important of the fermented milk products, is a coagulated milk product obtained by lactic acid fermentation through the action of *Lactobacillus bulgaricus* and *Streptococcus thermophilus.* It may contain added fruit or fruit flavours as well as carbohydrate sweetening matter. Kefir has an alcohol content of 0.5–2 per cent and koumiss of 2–3 per cent, and both contain considerable quantities of CO_2 which is formed by the heterofermentative, aroma-forming lactic acid bacteria.

Table 16.1: Composition of Fermented Milks

Dry matter	14–18 per cent
Protein	4–61 per cent
Fat	0.1–10 per cent
Lactose	2–3 per cent
Lactic acid	0.6–1.3 per cent
Fruits and carbohydrate sweeteners	5–25 per cent
pH	3.8–4.6

Fermented milks can be classified into three consistency groups: set (firm), stirred (pasty) and liquid. The coagulum of firm products can be destroyed and whey expelled as a result of mechanical shear. Although this effect can be reduced by using packages which are conical in shape with the larger diameter forming the base, containers of this shape are unable to be stacked.

The protective effect of five different types of containers for packaging solid unflavoured yogurt stored at 8°C in the dark and under 2000 lux for 12 hours per day had been evaluated. The results are summarized in Table 16.2, with the rankings being based on the results of sensory analyses and the determinations of peroxide value, vitamins A and B2, and instrumental colour measurements. The transparent, brown-pigmented glass jar was found to offer superior protection against light and O_2. The authors recommended that at a storage temperature of 8°C, the shelf life of yoghurt package nonaseptically in the brown glass container was 16–18 days.

Table 16.2: Summary of Protective Effects of Different Packaging Materials for Yoghurt

Materials	*Protection Against*		*Rank in Order of Decreasing Total Protection*
	Light	*Oxygen*	
Transparent brown glass	Good	Perfect	1
Unpigmented glass	Moderate	Perfect	4
Transparent brown PS	Good	Moderate	3
Unpigmented PS	Bad	Moderate	5
Paperboard and PS	Excellent	Bad	2

The level of CO_2 in fermented milk products such as kefir contributes to the flavour and to the growth of the starter organisms. Thus, the influence of the packaging material on the CO_2 concentration in the product is clearly important. A study comparing regular LDPE-coated paperboard cartons with cartons made from a special board with a double thickness of LDPE on the inside and an aluminum foil lined board found that the CO_2 content dropped to 25 per cent or less of its original concentration in the first two types of containers after 13 days of storage, while it was maintained in the foil-lined cartons. It is also important to control the yeast content in kefir, as too high a content gives a pronounced yeasty flavour which is undesirable. After 13 days storage, the yeast content in the foil-lined carton had increased by only 60 per cent compared to a more than 6-fold increase in the LDPE-lined cartons.

16.4 The Packaging of Cheese

Cheese is the generic name for a group of fermented milk-based food products produced in at least 500 varieties throughout the world. Although some soft cheese varieties are consumed fresh, *i.e.,* without a ripening period, production of the vast majority of cheese varieties can be sub-divided into two well-defined phases; manufacture and ripening. Despite differences in detail in the manufacturing processes used for individual varieties, the conversion of milk into cheese generally comprises four stages.

Coagulation

Physicochemical changes in the casein micelles due to the action of proteolytic enzymes and/or lactic acid lead to the formation of a protein network called coagulum or gel.

Drainage

Separation of the whey, after mechanical rupture of the coagulum, by molding and in certain cases by pressure, to obtain a curd.

Salting

Incorporation of salt by deposition on the surface or within the body of the cheese, or by immersion in brine.

Ripening

Biochemical changes in the constituents of the curd brought about by the action of enzymes, mostly of microbial origin.

Cheese manufacture can be viewed as essentially a dehydration process in which the fat and casein in milk are concentrated between six- and twelvefold, depending on the variety. The degree of hydration is regulated by the extent and combination of the three steps listed above, in addition to the chemical composition of the milk. In turn, the level of moisture in the cheese, the salt content and the cheese microflora regulate the biochemical changes that occur during ripening and hence determine the flavour, aroma and texture of the finished cheese. Although the nature and quality of the finished cheese are determined in very large measure by the manufacturing steps, it is during the ripening phase that the characteristic flavour and texture of the individual cheese varieties develop.

Although some soft-cheeses are consumed fresh, most cheese varieties are not ready for consumption at the end of manufacture but undergo a period of ripening (also referred to as curing or maturation) which varies from about 4 weeks to more than 2 years. The duration of ripening is generally inversely related to the moisture content of the cheese, although many varieties may be consumed at any of several stages of maturity depending on the flavor preferences of consumers.

A great variety of microbial species are involved in the ripening of cheese, the total population generally exceeding 109 organisms per gram. The principal bacterial groups involved in ripening are the lactic acid Streptococci; Leuconostoc; Lactobacilli and Propionibacteria. Micrococci and Corynebacteria species may also be involved; they are aerobic and salt tolerant and thus, grow especially on the surface of cheeses.

Yeasts are widely distributed in nature and are found in raw milk and some cheeses, the basic flora in the majority of cheeses being species of the genus *Kluyveromyces.* Yeasts produce enzymes capable of degrading the constituents of the curd and so can contribute to modifying the texture of the cheese and to the development of flavor and aroma. They are capable of converting lactose into CO_2 and may also take part in lipid degradation. The role of yeasts in the ripening of certain cheeses may be complex and significant.

Among the fungal flora found in or on mold-ripened cheeses, species of the genus *Penicillium* are of particular importance. *Penicillium camemberti* is the original mold of Camembert and Brie and has a single habitat–the surface of a few cheeses and the environment of cheese factories. By contrast, *Penicillium roqueforti,* a microaerophilic mold, is widely distributed and is the mold of mold-ripened cheese; it grows very well at quite low O_2 levels (5 per cent). *Geotrichum candidum* is present on certain soft cheeses where it forms a characteristic grayish-white crust on the surface. The role played by molds in ripening is a major one for soft and mold-ripened cheeses.

The oxygen requirements of microorganisms are known to be variable. Some are strictly anaerobic and will grow only in the interior of the cheese, *e.g.,* Clostridia and Propionibacteria. Other microorganisms are microaerophilic and grow best in an environment with a low oxygen content, *e.g.*, lactic acid bacteria, especially Lactobacilli. Others are strictly aerobic and cannot grow actively except on the surface of the cheese, *e.g.*, molds, yeasts, micrococci and coryneform bacteria. Cheese ripened by the action of these microorganisms will necessarily be of small size. One exception is cheeses ripened by internal mold (*Penicillium roqueforti*), but in this case a certain aeration of the cheese is ensured by piercing, and the mold is able to grow in a relatively low O_2 atmosphere.

The composition of the atmosphere may also directly affect ripening. Thus ammonia, which is present in the ripening cellars of washed-rind soft cheeses, helps to neutralize the surface layer of the cheese. In this way it favours the growth and activity of the surface bacterial flora, *i.e.*, micrococci and coryneform bacteria. Very little is known about how the gaseous composition of the atmosphere in the factory, the packaging or the interior of the cheese influences the microbial, enzymic and chemical phenomena of ripening. Despite this, it is one of the important methods of achieving control over the process.

Water activity in cheese is determined by two basic parameters: water content and salt content. During ripening the *a*w is not stable but changes constantly, reducing until the *a*w of the surface is in equilibrium with the surrounding atmosphere. There is a gradient of *a*w from the moister central zone of the cheese to the drier zone near the surface. Overall, the *a*w of the whole cheese is the resultant of a set of complex physical phenomena (drainage; evaporation of water; diffusion of salt); chemical phenomena (interaction between the substrate and sodium chloride); and biochemical phenomena (proteolysis; lipolysis; glycolysis).

The salt concentration in the aqueous phase of most cheeses lies between 4 and 5 per cent, reaching 6–8 per cent in blue veined cheeses. These concentrations have a selective action at the microbial level and an inhibiting effect on enzyme activity. The selective action of salt on the growth of molds on the surface of soft cheese is well known: a reasonable degree of salting limits or even prevents the growth of *G. candidum,* without noticeably affecting that of *Penicillium.*

Fermentation of lactose to lactic acid during cheese manufacture reduces the pH of cheese curd to values that inhibit the growth of many pathogenic bacteria. The formation of lactic acid during manufacture and the metabolism of residual lactose during the initial stages of ripening reduce the pH of cheese to around 5 depending on variety. During ripening, the pH of the cheese rises due to the formation of alkaline N-containing compounds and or the catabolism of lactic acid.

The influence of pH on microbial growth and enzyme activity is particularly decisive. Only lactic acid bacteria, yeasts and molds can grow at pH values below 5. Enzymes are also very sensitive to variations in pH, most microbial proteases having greatest activity in the pH range 5.0–7.5, and lipases in the range 7.5–9.0. Curd at the end of drainage has a pH of less than 5.5 and is the site of active lactic acid fermentation. This acidic character of the curd is a necessity in that it slows down enzyme action and impedes the development of a harmful bacterial flora.

Soft cheese curd has a pH below 5, being around 4.5 for Camembert. Such a pH does not allow sufficient protease activity. Some neutralization of the curd is therefore necessary and this occurs either by the growth of molds and yeasts using lactic acid as a source of carbon in the case of semisoft surface molded cheeses of the Camembert type, or by the combined action of yeasts and the ammoniacal atmosphere of ripening rooms in the case of washed-rind soft cheeses. The pH of Camembert increases from 4.5 to over 6.0 in the interior and over 7.0 on the surface by the end of ripening (about 30 days).

Hard cheeses have pH values around 5.2–5.3 and neutralization of the cheese is undesirable. The pH tends to remain at about this level during ripening, partly because of the high degree of mineralization of the cheese which gives it a high buffering capacity.

The two key parameters contributing to the stability of cheeses are pH and *a*w. However, neither of these parameters are low enough to ensure complete stabilization of the product, with the result that cheeses as a class lie between perishable foods on the one side and intermediate moisture foods on the other. While the packaging will have no influence on the pH of the cheese, the *a*w of the surface (and ultimately the interior) of the cheese will be affected by the water vapor permeability of the packaging material.

Two other key factors which must be considered in the packaging of cheese (and indeed in the packaging of all dairy products) are the effect of light and oxygen. As discussed earlier in this chapter, light initiates the oxidation of fats, even at temperatures found in refrigerated display cabinets, giving rise in the case of unripened cheeses to off-flavours which have been described as "cardboardy" or "metallic". The oxidation reactions initiated by light may continue even if the cheese is subsequently protected from light, and any metallic ions contained in the packaging material will catalyze the reactions. In addition, the ingress of molecular oxygen through the packaging film is undesirable as it will contribute to the oxidation of fats and the growth of undesirable microorganisms.

Several attempts have been made to develop a classification system for cheese varieties and these have been reviewed by Fox. There is no one classification system applicable to all cheeses, but for the purposes of providing a framework to discuss packaging requirements, the following scheme is appropriate and will be used here: very hard and hard; semisoft and soft; fresh; and processed.

The Very Hard class of cheese (sometimes referred to as grating grade) is ripened by bacteria and is characterized by a moisture content (on a fat free basis) of <51 per cent, *e.g.*, Parmesan 42 per cent; Romano 31 per cent; Mozzarella 45 per cent. The Hard class of cheese is also ripened by bacteria and is characterized by a moisture content of 49–63 per cent, the range 49–56 per cent being categorized as hard, and the range 54–63 per cent being categorized as semi-hard. Cheeses in this class include Cheddar (still the cheese produced in the greatest quantities worldwide), Edam, Gouda, Cheshire, Gloucester, Derby and Leicester, as well as those with eyes such as Emmental and Gruyere. Also included in this class are Provolone, Mozzarella and Kasseri.

Table 16.3: Recommendations Regarding the Permeability of Packaging Materials[a]

Cheese Category	*Water Vapor Transmission Rate g m^{-2} day^{-1} at 25°C/75 per cent RH*	*O_2 Transmission Rate cm^3 m^{-2} day^{-1} atm^{-1} at 25°C/75 per cent RH*
Consumer-packaged	30	12,900
Ripening film for whole cheese	2.5	30
Gas or vacuum packaged	4.0	170

a: From British Standard BS 4855:1973.

In Cheddar and many of the other varieties in this class, lactose is metabolized to L-lactic acid which is then isomerized to D-lactate by non-starter lactic acid bacteria (NSLAB). In Swiss cheese, L-lactate and, to a lesser extent, D-lactate are converted to propionate, acetate and CO_2 when the growth of propionic acid bacteria is initiated on transfer of the cheese to the hot room.

Now let us study about the role of oxygen in the packaging of cheese.

The fate of oxygen in the cheese is still not completely understood. What is certain is that it is the gas present in the space between the packaging material and the cheese (either produced as a product of enzymic action in the cheese, left in the package after sealing, or diffusing through the packaging material) which determines whether or not microbial growth will occur on the surface of the cheese.

The secondary microflora of Cheddar cheese consists of lactic acid bacteria that gain access accidentally, usually as a result of contamination following pasteurization of the milk. These adventitious organisms may grow in the highly selective cheese environment such that certain non-starter lactic acid, bacteria (NSLAB) often attain high densities (> 107 colony-forming units per gram of cheese) which are maintained for months. Thus, adventitious bacteria are the dominant flora in cheese throughout much of the long ripening period.

In the presence of O_2 certain NSLAB convert lactate to acetate, CO_2 and water. The extent to which lactate oxidation occurs is dependent on the type of NSLAB present and their concentration in the cheese, which depends largely on the storage temperature and time. Also important are the surface area: volume ratio of the cheese and the O_2 permeability of the packaging material. The latter will vary with temperature and may vary with relative humidity, depending on the chemical nature of the material.

Packaging film for Cheddar cheese must be sufficiently impermeable to O_2 to prevent fat oxidation and mold growth. Another factor in mold growth may be the presence of O_2–consuming NSLAB in the cheese since these organisms will compete for O_2 permeating through the film, thereby reducing its concentration under the film. When there is a high density of NSLAB possessing the lactate oxidation system then the cheese will act as an O_2 sink and mold growth may be suppressed.

Since Cheddar cheese contains approximately 1.5 per cent w/w lactate, the, theoretical upper limit for the acetate concentration arising from lactate is about 1 per cent. In practice this level is never reached, mainly because of O_2 limitation. The major free fatty acid in Cheddar cheese is usually acetate and, it is considered by some to contribute to the flavor of the cheese, although concentrations greater than about 0.07 per cent w/w in young cheese give rise to off-flavors. The potential acetate production by NSLAB resulting from O_2 penetration into the cheese is another reason to use packaging materials with low O_2 permeabilities.

Data relating to the minimum partial pressure of O_2 necessary for the development of molds is scarce. In the absence of CO_2, molds will grow at O_2 partial pressures below 1 mm Hg. In the presence of CO_2 at partial pressures: of 65–110 mm Hg (8.5–15 per cent), microbial growth is inhibited up to an O_2 partial pressure of 15–20 mm Hg.

Table 16.4: Maximum O_2 Transmission Rate[a] of Packaging Material for Rindless Cheese[b]

	O_2 Transmission Rate	
	Non-shrink Film	*Shrink Film*
Mass 5 kg and heavier		
Cheddar type cheese	50	75
Gouda, Swiss and Edam type cheese	170	220
Mass up to 5 kg		
All types	120	300

a: $cm^3\ m^{-2}\ day^{-1}$ measured at 23°C and 93 per cent RH

b: From Australian Standard AS 2183-1978.

In a relatively early study to determine the maximum O_2 permeability requirements of plastic films that would ensure mold-free Provolone cheese (an Italian hard cheese chosen for the study because of its unusual sensitivity to mold growth), it was concluded that the O_2 permeability of the packaging material should not exceed 0.5 cm^3 100 m^{-3} day^{-1} atm^{-1} when stored under the customary conditions of 5°C and 76–85 per cent RH. Such a film was a laminate of LDPE (3.08×10^{-3} cm thick) and PET (1.27×10^{-3} thick).

Other published figures for the O_2 permeability requirements of packaging materials for hard cheese include those in Australian Standard AS 2183–1978. Without any indication as to how these values were arrived at, it is difficult to comment in detail. Also, without any mention of the degree of shrink expected or allowed for with the shrink films, it is not possible to assess whether or not the O_2 permeability requirements are identical for cheeses packaged in shrink and non-shrink film. The standard also contains a minimum value for the CO_2 transmission rate for Swiss-type cheeses of 680 cm^3 m^{-2} day^{-1} at 23°C and 93 per cent RH.

Table 16.5: Composition of the Films Proposed for the Packaging of Rindless Cheeses

Material	*Outside*		*Inside*
	First Layer	*Second Layer*	*Third Layer*
1	Copolymer of ethylene and vinyl acetate	Copolymer of vinylidene chloride and vinyl chloride	Irradiated copolymer of ethylene and vinyl acetate
	Thickness: 30 microns	Thickness: 8 microns	Thickness: 12 microns
2	Copolymer of ethylene and vinyl acetate	Copolymer of vinylidene chloride and vinyl chloride	Irradiated copolymer of ethylene and vinyl acetate
	Thickness: 15 microns	Thickness: 25 microns	Thickness: 12 microns
3	Nylon 6 (m.pt. 208°C)	Surlyn ionomer (sodium form)	
	Thickness: 30 microns	Thickness: 30 microns	
4	Mixture of 30% nylon 6 and 70% Surlyn ionomer (zinc form)	Copolymer of ethylene and 4.5 per cent vinyl acetate	
	Thickness: 49 microns	Thickness: 89 microns	
5	Surlyn ionomer (sodium form)	Copolymer of ethylene and 3.5 per cent vinyl acetate colored pale yellow by an inorganic pigment	
	Thickness: 29 microns	Thickness: 106 microns	

Material	Outside		Inside
	First Layer	*Second Layer*	*Third Layer*
6	Copolymer of ethylene and 3.5 per cent vinyl acetate colored pale yellow by an inorganic pigment	Surlyn ionomer (sodium form)	
	Thickness: 26.5 microns	Thickness: 108 microns	

Figures for the O_2 transmission rates of cheese bags used in 1986 to package 20 kg blocks of Cheddar cheese in New Zealand have been reported to be 27 cm^3 m^{-2} day^{-1} at 10°C and 100 per cent RH.

The utility of such published data is debatable, but it is likely that problems of mold growth on hard cheeses packaged in the films in common use today are a function of hygienic conditions in the packing room, the degree of vacuum inside the package and the integrity of the heat seal, rather than the O_2 transmission rate of the packaging material used.

Table 16.6: Permeability[a] of the Films Proposed for the Packaging of Rindless Cheeses

Material	*To Oxygen* $cm^3\ m^{-2}\ day^{-1}\ atm^{-1}$		*To Carbon Dioxide* $cm^3\ m^{-2}\ day^{-1}\ atm^{-1}$		*To Water Vapor* $g\ m^{-2}\ day^{-1}$ *at 38°C*
	0 per cent RH	*80 per cent RH*	*0 per cent RH*	*80 per cent RH*	*90 per cent RH*
1	155	155	765	765	17.0
2	250	250	1500	1500	29.5
3	25	125	85	455	6.2
4	600	595	2120	2325	4.2
5	955	865	3685	3295	2.3
6	440	640	1470	1565	4.4

a: According to DIN 53380 Standard.

The development of the technique of ripening under film (also known as rindless ripening) was motivated by various factors. One was the need to reduce blocks better suited to a high degree of mechanization. Another was to increase cost-effectiveness by replacing earlier techniques of coating, thus reducing handling during ripening and losses due to drying-out. The first techniques (which appeared around 1930 and were used for cheddar) involve coating the blocks with mineral oil. In 1950 J.B. Stine of Kraft Foods Company filed two patents concerning the use of a plastic packaging material for blocks of Swiss cheese; these patents have since passed into the public domain. A further factor was the development of prepacked, self-service, consumer portions of cheese for or by supermarkets, where the losses resulting from removal of the rind were considered unacceptable. This led to the idea of rindless cheese. There are three variables which may be used to adjust the process of ripening rindless cheese: the permeability of the packaging material; the ripening temperature, and the ripening time.

Processed cheese is usually packed in aluminium foil in cubes with different shapes. Tinplate cans are used for 200 gms and above quantity. PVDC coated plastic films are suitable for cheese packaging as they provide good oxygen and moisture barrier properties.

Cream cheese is packed in foil lined card board boxes of heat stable plastic packs. Saran is used as wrapping material for Neufchatel cheese. Air evacuation and gas flushing is used for cottage cheese, green cheese is packed by waxing and paraffining or alternatively vacuum packed in polyethylene or chyovac. Ripened cheese is packed in laminated cellophane film or pliofilm. Cheese consumer packs are generally consisting of Lacquered metal cans of laminated consisting of Nylon/ PVDC/Copolymer or polyster/PVDC/copolym or Nylon/Polyethylene. Processed cheese is packaged hot in metallic containers. Wax coated cellophane, A.P foil, polypropylene, PE, PVDE material is also used for packing processed cheese.

16.5 Butter milk

Butter milk is another most popular indigenous fermented milk product. It is packed in 200 ml polyethylene sachets. Recently some dairies are marketing butter milk in tetra packs.

16.6 Packaging of Ice cream

Ice-cream is a frozen dairy product prepared by mixing milk, cream, sugar, flavour and stabilisers, with the incorporation of air during the freezing process to provide a smooth texture. Since it is a frozen product, the packaging should be designed in such a way that the pack performs efficiently to contain the product at freezing temperatures. Super smooth ice creams are extremely delicate products and must be packaged with utmost care. Conventional form of packages includes paperboard cartons, paper cups and in some cases even metal containers. But, the above variety of packages are slowly but surely being replaced by other forms. Ice-cream needs to be protected against contamination, moisture loss and fluctuations in temperature. The package should be attractive, easy to open and dispose. The distribution and transportation of ice-cream is done in frozen condition and to maintain its frozen state, refrigerated trucks or dry ice is used. Dry ice pieces are wrapped in paper and placed around the ice-cream packs inside an insulated container, which is generally a corrugated fibre board box.

The various types of packages for ice-cream include:

- ☆ Paper board carton which is poly coated.

- ☆ Thermoformed/injection moulded plastic containers made from HIPS (high impact polystyrene), PP (Polypropylene) or HDPE (high density polyethylene). The materials used for the lids are LDPE (low density polyethylene) or PS (polystyrene). The lids are of snap on type.
- ☆ Laminates of BOPP (biaxially oriented polypropylene) or PET (polyethylene terephthalate) are used for candies.

The demand for thermoformed/injection moulded containers, tubs and cups are on the rise as this type of pack is hygienic and does not support the growth of microorganisms. This type of package is suitable to be used on automatic filling equipment, because of its comparative strength and rigidity as compared to paper based packs. Tight control of dimensions and weight in case of plastic containers (thermoformed) lead to a most cost-effective style of packaging, particularly when high speed filling is desired. Besides, they are easily available in the market today in a wide range of shapes, white, clear, translucent with reclosable lids or sealable closures. The reclosable tubs present the consumer with a great value pack, since it prevents crystallisation of the partly consumed contents, and, when empty, the package offers a wide range of reuse facilities. The scrap value of thermoformed packages is high and recyclability is preferred by collection and reuse of the polymer material for other applications.

16.7 Butter

Butter consists of fat, which is separated from the aqueous phase of milk. Butter should be kept under refrigeration. Butter usually contains 80-82 per cent fat, 15-16 per cent moisture and 0.5-2.5 per cent salt. Because of this high moisture content, butter, unlike solid fats is susceptible to mold growth. Flavour and odour are easily affected by absorption from other materials or through spoilage of butter due to rancidification, due to oxygen and light. The package should, therefore, be opaque and a high barrier against oxygen and foreign odours.

Butter is generally marketed in capacities ranging from 100 grams to 500 grams. The most commonly used butter wrap is the vegetable parchment paper of 45 gsm. Although parchment paper is greaseproof, it does not provide a sufficient barrier to oxygen. For superior product protection and for longer shelf-life, aluminium foil laminated to parchment or greaseproof paper is used for which specifications are given in IS: 7161- 1973 Specification for Vegetable Parchment or Greaseproof Paper/Al foil Laminate for Wrapping Butter. To avoid corrosion of aluminium foil (due to salt and lower pH of butter), it is coated with lacquer and a protective adhesive between the foil and paper is applied. The wrapped butter is placed in a plastic laminated paper-board carton to ease handling, distribution and storage at retail outlets under refrigeration.

Recently, embossed aluminium foil backed parchment paper has been introduced for UV light protection and sales appeal. A popular packaging style in some countries is to use plastic cups and plastic tubs with lids in different shapes and sizes. For such applications, PP (Polypropylene) and ABS (Acrylo-Butadiene-Styrene) are widely used. When butter is to be stored for a very long time and transported over long distances, hermetically sealed tinplate containers are used. Recent development is the use of smooth walled light-weight aluminium containers, which are convenient, colourful and hygienic. The containers are provided with printed or un-printed plastic or heat sealable foil lids. Butter packed in the metal containers has a shelf-life of one year and above.

16.8 Ghee

Ghee is a typically Indian product, known from almost vedic times. Ghee was almost exclusively a product made at home and consumed from home containers of glass, ceramics, earthenware or tinplate. Over a period of time, the manufacture of ghee has become a commercially well organised industry. Ghee is 100 per cent fat with an off- white colour, grainy texture and a characteristic flavour. The product has a very low moisture content (less than 0.5 per cent) and, therefore, unlike other dairy products, susceptibility to microbial deterioration is not very high. However, the product needs to be protected from chemical spoilage and rancidity caused by oxygen, light, heat, moisture and metal ions. A major portion of ghee was packed in lacquered or un-lacquered tinplate containers of capacities ranging from 250 grams to 15 kilograms. Since the product is very sensitive to oxygen, the tinplate containers are filled to the brim without an air gap. Ghee packed in tinplate containers is fairly stable and has a shelf-life of over one year.

Alternate packages, which are plastic based, are now gradually replacing tins. For shorter shelf-life, 200 grams, 500 grams and 1 kilogram capacity pouches made of polyethylene film are marketed, which are economical. Ghee is also marketed in lined cartons with flexible laminated plastics as inner liner materials and in tetrapaks. In both these packs long shelf-life is achieved. Laminated pouches of metallised polyester based films are also used. For packaging of ghee laminates of polyester, Nylon-6 and use of high barrier materials such as EVOH and EVAL can also be explored, as these materials could provide a fairly long shelf-life.

Chapter 17

Packaging of Fruits and Vegetables

17.1 Introduction

Growth, maturation and senescence are the three important phases through which fruits and vegetables pass, the first two terms often being referred to as fruit development. Ripening (a term reserved for fruit) generally occurs during the later stages of maturation and is considered the beginning of senescence, a term defined as the period when anabolic or synthetic biochemical processes give way to catabolic or degradative processes leading to aging and final death of the tissue.

Now let us study about respiration in fruits and vegetables.

Respiration involves the oxidation of energy-rich organic substrates normally present in cells such as starch, sugars and organic acids, to simpler molecules (CO_2 and H_2O) with the concurrent production of energy and other molecules which can be used by the cell for synthetic reactions. The greatest yield of energy is obtained when the process takes place in the presence of molecular oxygen. Respiration is then said to be aerobic. If hexose sugar is used as the substrate; the overall equation can be written as follows:

$$C_6H_{12}O_6 + 6O_2 \longrightarrow 6CO_2 + 6H_2O + \text{energy}$$

This transformation actually takes place in a large number of individual stages with the participation of many different enzyme systems. The water produced remains within the tissue but the CO_2 escapes and accounts for part of the weight loss of harvested fruits and vegetables, a range of 3–5 per cent of weight loss having been ascribed to respiration. When one mole of hexose sugar is oxidized, 36 moles of ATP (each possessing 32 kJ of useful energy) are formed. This represents about 40 per cent of the total free energy change, the remainder being dissipated as heat. Rapid removal of this heat is usually desirable and it is important that the packaging assists rather than impedes this process.

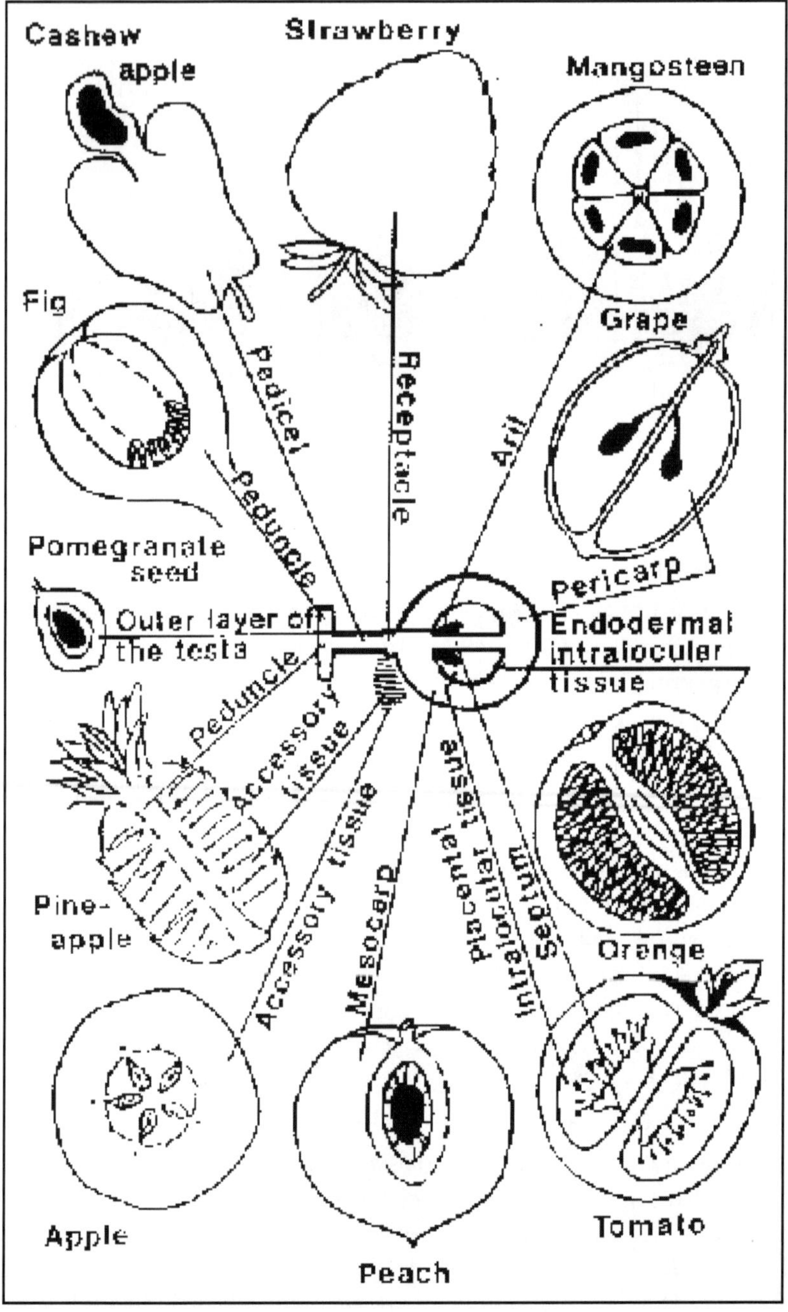

Figure 17.1: Derivation of some Fruits from Plant Tissue.

The rate of respiration is often a good index to the storage life of horticultural products: the higher the rate, the shorter the life, and the lower the rate the longer the life.

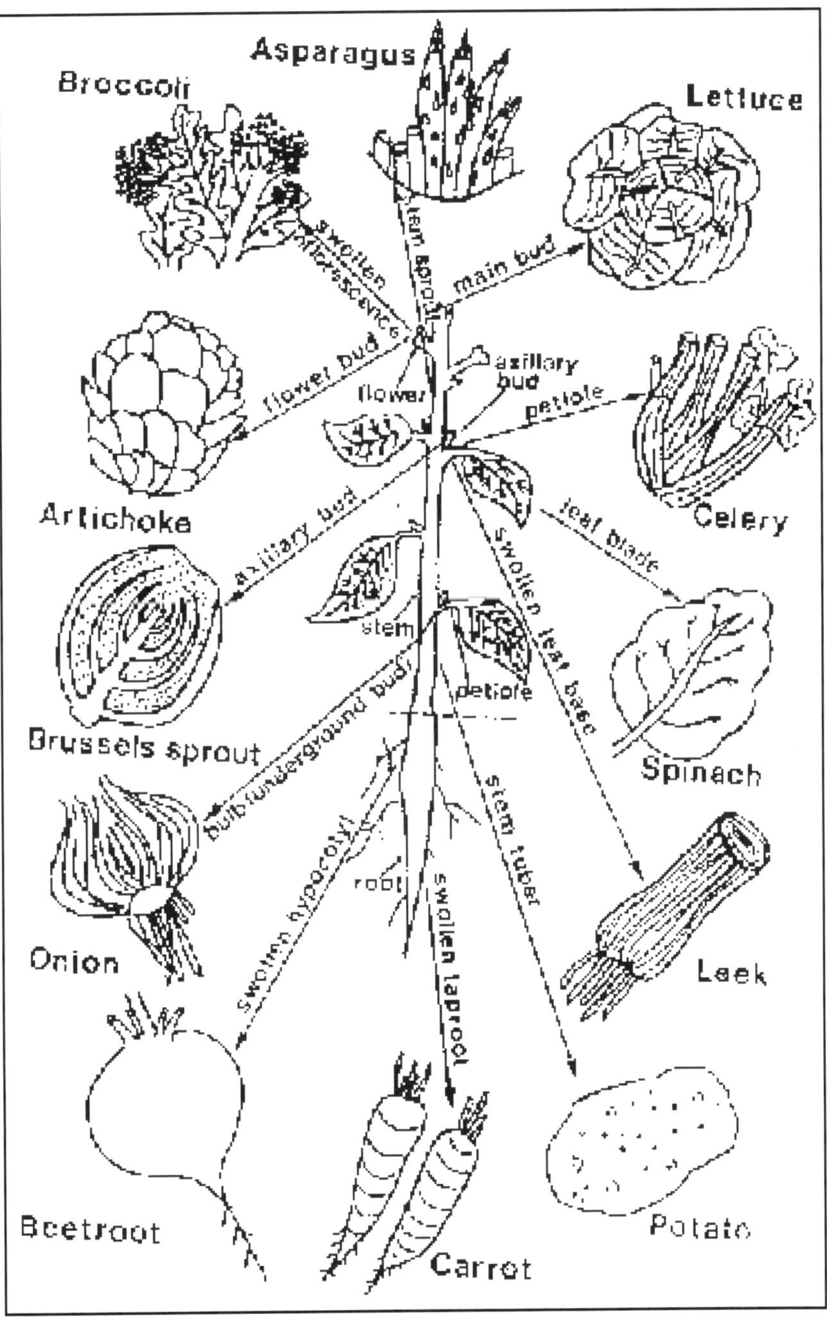

Figure 17.2: Derivation of some Vegetables from Plant Tissue.

From measurements of CO_2 and O_2 it is possible to evaluate the nature of the respiratory process. The ratio of the volume of CO_2 released to the volume of O_2 absorbed in respiration is termed the respiratory quotient (RQ). Values of RQ range from 0.7 to 1.3 for aerobic respiration, depending on the substrate being oxidized for carbohydrates, RQ = 1; for lipids RQ < 1; for organic acids RQ > 1.

Anaerobic respiration (sometimes called fermentation) involves the incomplete oxidation of compounds in the absence of O_2 and the accumulation of ethanol, acetaldehyde and CO_2. Much lower amounts of energy (2 moles of ATP) and CO_2 are produced from 1 mole of hexose sugar than that produced under aerobic conditions. As well, very little heat energy (approximately 5 per cent) is produced for a given amount of carbohydrate oxidation in anaerobic respiration compared to aerobic respiration. The oxygen concentration at which a shift from aerobic to anaerobic respiration occurs varies among tissues and is known as the extinction point. Very high RQ values (> 1.3) usually indicate anaerobic respiration.

Variations in the rate of respiration occur during organ development; as fruit increase in size, the total amount of CO_2 emitted increases although the respiration rate calculated on a per unit weight basis decreases continually. Three different routes are available for the exchange of gases between horticultural commodities and their surrounding atmospheres: lenticels and stomata, the cuticle, and the stem scars or floral (calyx) end. In leaves, the resistance of the epidermal layer to gas diffusion is mainly regulated through the stomatal aperture. A majority of studies on gas exchange in plant tissues indicate that the skin represents the main significant barrier to diffusion. Clearly the rate at which the three gases O_2, CO_2 and C_2H_4 diffuse through the tissue will have a significant effect on the rate of respiration.

Table 17.1: Classification of Vegetables According to Respiration Intensity

Class	*Respiration Intensity at 10°C mg CO_2 kg^{-1} h^{-1}*	*Commodities*
Very Low	Below 10	Onion
Low	10–20	Cabbage, cucumber, melon, tomato, turnip
Moderate	20–40	Carrot, celery, gherkin, leek, pepper, rhubarb
High	40–70	Asparagus (blanched), eggplant, fennel, lettuce, radish
Very high	70–100	Bean, Brussel sprouts, mushroom, savoy cabbage, spinach
Extremely high	Above 100	Broccoli, pea, sweet corn

17.2 Packaging of Fresh Horticultural Produce

As stated earlier, a simple consideration of equation would suggest that if the

CO_2 in the atmosphere were augmented (or the O_2 decreased), the respiration rate and storage life would be extended. The first scientific study on the effect of modified atmospheres on the shelf life of horticultural products was carried out by the French chemist Berard in the 1800s; he found that fruit in an anaerobic atmosphere failed to ripen. Research into the influence of O_2 and CO_2 levels on the chill storage life of apples, pears and berries was carried out in Cambridge, England in the 1920s by Kidd and West. Their results demonstrated that low O_2 and moderately high CO_2 levels around the fruit reduced the rate of deteriorative reactions.

Table 17.2: Classification of Horticultural Produce According to the Plant Organ Used

Class	*Commodities (Examples)*
Root vegetables	Carrot, celery, garlic, horseradish, onion, parsnip, radish, turnip
Tubers	Potato, yam, Jerusalem artichoke
Leaf and stem vegetables	Brussel sprouts, cabbage, celery, chicory, Chinese cabbage, cress, green onion, kale, lettuce, spinach
Flower vegetables	Artichoke, broccoli, cauliflower
Immature fruit vegetables	Bean, cucumber, gherkin, okra, pea, pepper, squash, sweet corn
Mature fruit vegetables	Melon, tomato
Reproductive organs	Most fruits

Historically, atmospheres surrounding fruits and vegetables have been altered in controlled atmosphere (CA) storage facilities where the levels of gases are continually monitored and adjusted to maintain the optimal concentration. Because CA storage is capital intensive and expensive to operate, it is more appropriate for those fruits and vegetables that are amenable to long-term storage such as apples, kiwifruit, pears and cabbages. Modified atmosphere (MA) storage typically involves some initial modification of atmospheric conditions which change further with time as a result of the commodity's respiration and the surrounding physical environment.

The recommended concentrations of O_2 and CO_2 for CA and MA storage of fruits and vegetables can be found in the published literature and will not be tabulated here. A very effective way of plotting this data has been presented with CO_2 concentration as the ordinate and O_2 concentration as the abcissa.

Since the 1960s attempts have been made to create and maintain modified atmospheres within plastic polymeric films. The availability of absorbers and adsorbers of O_2, CO_2, C_2H_4 and water provides additional foods for the packaging technologist to use to maintain a desired atmosphere within a package. Details of such systems will now be discussed.

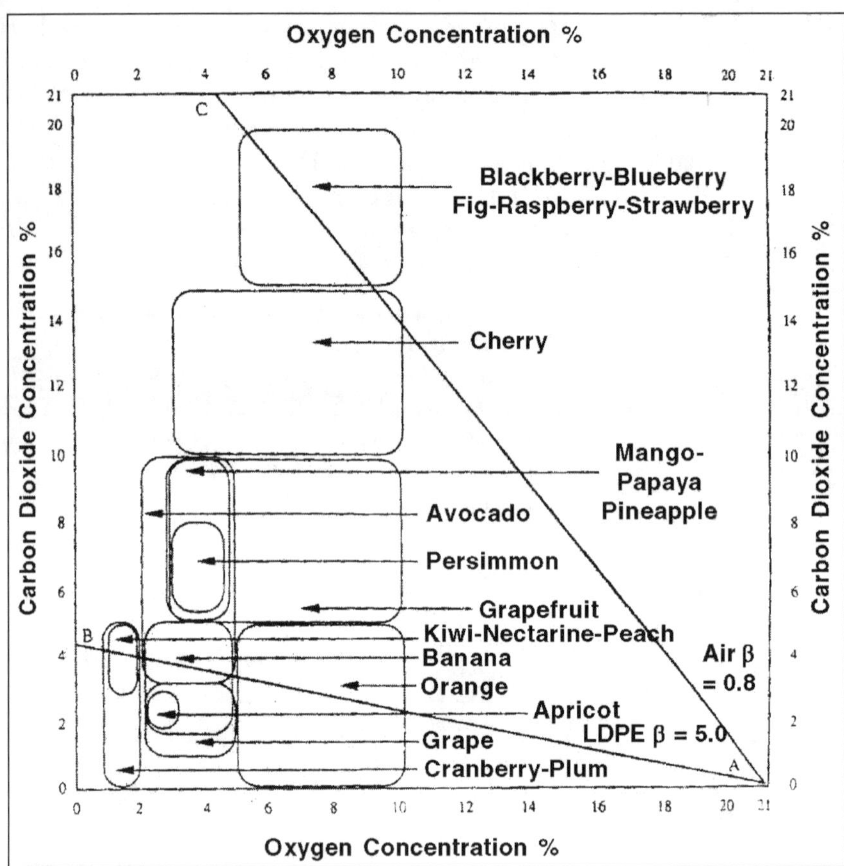

Figure 17.3: Recommended Modified Atmospheres for Storage of Fruits.

17.3 Package System for MAP

Polymeric films have been used for the packaging of horticultural products since the 1930s. Early work in the area stressed the primary role of packaging to reduce transpiration with many studies encouraging film perforation to avoid the development of injurious atmospheres in packages. Today interested centers on the development of non-perforated polymeric films which have the desired permeability characteristics so that modified atmospheres appropriate for the particular commodity and its storage temperature can be developed and maintained. In addition, the film must have the ability to cope with fluctuating storage conditions (*e.g.,* temperature, humidity and light) without injurious atmospheres developing. It is important to note that in the presence of light, green tissues undergo photosynthesis. In this situation, most of the CO_2 produced by respiration will be re-absorbed by photosynthesis, with the result that the O_2 concentration in the atmosphere around the produce will increase. This in turn will increase the rate of respiration.

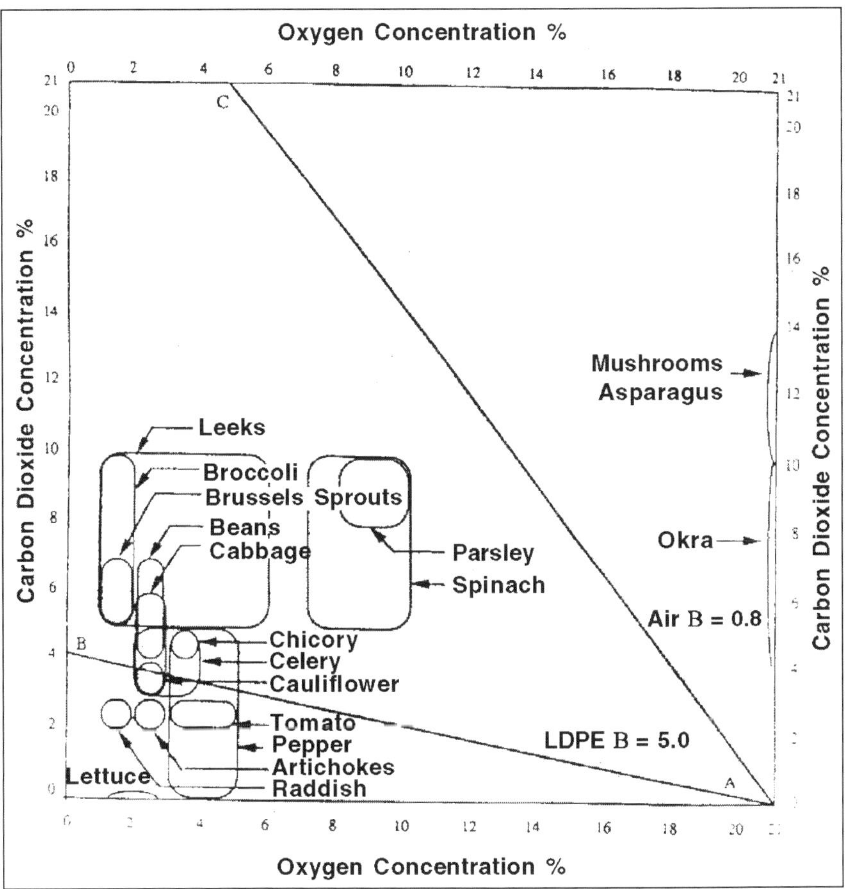

Figure 1.4: Recommended Modified Atmospheres for Storage of Vegetables.

Although many polymeric films are available for packaging purposes, relatively few have been used to package fresh produce. This is not surprising since few have gas permeabilities that make them suitable to use for MAP. Those most likely to be suitable include LDPE, plasticized PVC, PS and EVA copolymer. The O_2 and CO_2 permeabilities of these films is required at actual conditions of use, but regrettably such data is scanty, making the development of suitable packages very much a "pack and pray" exercise in many cases. It is to be hoped that detailed permeability data (including the activation energies for permeation, diffusion and solubility, as well as parameters such as degree of crystallinity) will become widely available to assist package developers.

There are some fruits and vegetables for which LDPE film will never be suitable if a MAP is required. What is needed is the presentation of data for other polymeric films in a similar format to that for LDPE to assist in the selection of appropriate films and avoid wasteful trials with films which will never give the MAs desired. There is also a need for modification of the permeability properties of the common polymeric films to make them more suitable for MAP; research in this field is currently being undertaken involving the incorporation of inert, inorganic material into the film during extrusion.

A variety of configurations is available, ranging from a plastic bag which is heat sealed (sometimes after gas flushing), to a rigid, largely impermeable tray (either thermoformed from typically PET, PVC or PS, or made from molded paper pulp with a thermoformed plastic liner) to which is sealed a polymeric film having the desired permeability characteristics so that either an optimum MA is created inside the package at equilibrium, or a gas mixture flushed through the package immediately prior to sealing is retained.

During 1960s individual seal packaging system was developed. It is known as UNIPACK. Individual seal packaging (ISP) consists of packaging individual units of fruit in a polymeric film (originally 10 micron HDPE but now coextruded polyolefins) which is then shrunk by blowing hot air over the package. The fruit may be cooled after passing through the hot air tunnel by rapid ventilation. Equipment for seal packaging has developed rapidly, from hand wrapping, through manual, semi- and fully-automatic sealers, to automatic form-fill-seal machinery capable of operating at 200 fruit per minute. It is also possible to code or brand each individual fruit. The use of stretch rather than shrink film has also been attempted but there appear to be greater problems with sealing and mechanization of the process.

ISP appears to function primarily by maintenance of a water saturated atmosphere that extends shelf life by reduction of shrinkage and weight loss. It also reduces disease severity by preventing cross contamination of produce in boxes and permitting easy removal of infected individual fruit. The modification of respiratory gases within ISP produce is generally not sufficient to cause significant atmosphere modification.

A marked reduction in shrinkage and weight loss without a deleterious effect on flavour has been shown to result from ISP. It doubles and at times triples the shelf life of the fruit as measured by appearance, firmness, shrinkage, weight loss and other quality attributes. ISP also delays various parameters of physiological deterioration better than cooling to optimal temperature.

ISP has been shown to reduce chilling injury in various citrus cultivars, and thus, when combined with cooling which reduces decay, fruit deterioration is reduced. ISP also contributes to the healing of wounds caused during harvesting and markedly reduces the incidence of blemishes on various fruit. Changes related to senescence such as deterioration of cell membrane integrity and softening are also delayed by ISP. Undesirable effects of ISP include germination of seeds inside grapefruit stored for a prolonged time, sprouting of carrots, and rooting of citron fruit.

The use of films for ISP which contain fungicides that are gradually released has been suggested as advantageous because most of the fungicides is in the film and is not measured when toxic residues in fruit are examined. New films which contain substances capable of absorbing ethylene would also be excellent for ISP since reduction of the ethylene concentration, around the fruit would extend shelf life by delaying the various reactions leading to sensecence. For the degreening of citrus fruit, the application prior to sealing of ethylene releasing agents ensures that ethylene accumulates to a level suitable for accelerating the degreening process. There are advantages over the conventional degreening techniques where ethylene gas aggravates the development of various blemishes resulting at times in the loss of more than half the fruit.

Despite all the advantages listed above, the commercial adoption of ISP has been slow, possibly due to the lack of availability of high speed, reasonably priced packaging and equipment. However, the technique appears to be used extensively in China and Japan where ISP is performed by hand.

We can notice number of differences between CA storage and MAP of horticultural produce.

Despite the gas composition around the produce being almost identical in both cases, the storage life of MAP produce is frequently less than that of produce under CA storage. The limiting factor is the appearance of molds and bacteria in MAP produce as a result of the high (near saturation) in-package relative humidity (IPRH), a consequence of the low water vapor transmission rate of the films typically used for MAP. Such a problem is not observed in CA storage where the flow system presumably results in lower relative humidity.

Although fungicides have been used in an attempt to control mold development in produce packages they have been only partially effective and, in some cases, have caused problems of their own. The incorporation of fungicides directly into films has not been fully effective either. However, even if fungicides were effective, they do not control bacterial development which is also associated with high humidity in MAP.

The optimum RH for storage of fresh produce has not been well established, and the generally recommended levels of 85–95 per cent, represent a compromise between preventing excessive weight loss while providing some control of microbial spoilage.

There are two possible approaches to controlling IPRH: perforation of the package which precludes the possibility of achieving MA conditions within the package, and the use of in-package water-adsorbing compounds which can maintain a target RH. The feasibility of the latter approach to produce predetermined humidity within fresh produce packages has recently been described using predried crystals of compounds possessing a type III sorption isotherm according to the BET classification system.

The use of xylitol, potassium chloride, sodium chloride and sorbitol produced and held the IPRH at 78–79, 84–85, 73–76 and 72–74 per cent respectively, of a model package containing tomatoes and an LDPE film for gas exchange. No shrivelling of the fruits was observed provided that their weight loss was less than 7 per cent. As well, because of the reduced RH and a_w inside the packages, microbial growth will not occur. Clearly, the use of potassium chloride would not be desirable since at that RH (0.84–0.85 a_w), most of the tomato fruit pathogens would be capable of growth.

In commercial packages, it is suggested that the salts be placed inside sachets of spunbonded olefin films which have small holes that readily permit the movement of water vapor but completely block the movement of liquid water. The critical feature of such a system is to have sufficient salt present such that when the desired quantity of water vapor has been absorbed, the salt remains saturated, for if the salt solution becomes unsaturated, the IPRH will rise. However, if the salt absorbs too much water vapor from inside the package, shrivelling of the produce will result. It should be possible to design sachets having the quantity of salt which will achieve the optimum results.

17.4 Minimally Processed Fruits and Vegetables

Minimally processed fruits and vegetables (MPFV) are products that have the attributes of convenience and fresh-like quality, their forms varying widely depending on the nature of the unprocessed commodity and how it is normally consumed. The primary spoilage mechanisms are the metabolism of the tissue and microbial growth; both will cause deterioration of the tissue and must be controlled to maintain tissue viability.

Mechanical damage to cells during processing is a major limitation to shelf life of minimally processed fruits and vegetables. It has been shown that keeping shredded lettuce cold, using sharp slicing knives, minimizing cellular damage, eliminating cellular fluids, removing liquid containing active enzymes from the surface by centrifugation, and reducing the initial microbial population all contribute to longer shelf life. However, other workers have reported that washing shredded lettuce in deionized water and then spin drying before packaging did not improve the appearance over lettuce which had not been so treated.

Of key importance with MPFV is the control of enzymes, either endogenous from the produce itself or exogenous from the invading microorganisms to maintain the firm, crisp texture and bright, light color. The enzymes can be controlled either by inactivation or chemical or physical means. Methods that are currently used have been discussed and include temperature, sulfite solutions, modified atmospheres, ascorbic acid compounds, divalent ions very high pressures, additives such as L-cysteine (which inactivates polyphenoloxidase), vanillin (which inhibits polygalacturonase), mannose (which reduces ethylene production, respiration and softening in pears), and gases such as SO_2, CO and ethylene oxide. However, it is unlikely that the widespread use of the latter three gases would ever gain regulatory approval given the current trend towards additive-free food.

The use of a gas-exchange process for extending the shelf life of raw foods, and in particular, minimally processed apples and potatoes, was heralded in 1980 but appears not to have progressed to the commercial stage. The process consisted of the replacement of intra-tissue gases in particulate raw foods with one or more gases, in appropriate sequence, that stabilized or at least significantly extended the like-fresh shelf life of the foods at ambient and/or chilled temperatures. Major savings in energy usage for the process of more than 75 per cent of that required for canning or freezing were also claimed.

Microbial deterioration of MPFV can be controlled by several methods including the use of chill temperatures, reduction of the total microbial population by the use of heat or irradiation, and the use of antagonistic organisms that control growth of undesirable microorganisms.

Very little has been published on the packaging of MPFV, but generally the same packaging processes as for fresh produce are used. However, allowance has to be made for differences in the respiration rate of produce which has been processed in some way, for example, the respiration rate of shredded lettuce is more than double that of whole lettuce quarters at 10°C, due to increased exposure of respiring tissue surface and physical damage causing a sound response. Thus, shredded lettuce will obviously need a considerably more permeable film than whole lettuce to result in aerobic equilibrium atmospheres.

The addition of an O_2 scavenger into an anaerobic package containing apricot and peach halves reduced textural loss from structural polymer breakdown, although a similar effect was not observed with pear halves. Scavenging the headspace O_2 from the package retarded textural breakdown to an even greater extent than the use of MA, although no details of the atmosphere were given. The optimum treatment for peeled and halved peaches and apricot halves consisted of a 2 per cent calcium chloride/1 per cent zinc chloride dip followed by anaerobic packaging and storage at 0–2°C; similar treatment did not retard degradation of pears.

The use of an ethylene absorbent (charcoal with palladium chloride) prevented the accumulation of ethylene and was effective in reducing the rate of softening in peeled and sliced kiwifruit and sectioned bananas, as well as chlorophyll loss in spinach leaves but not in broccoli.

Clearly, there is likely to be an increasing trend away from heavily processed towards minimally processed fruits and vegetables and considerable research will be required to develop the most effective packaging systems.

Packaging materials for frozen fruits and vegetables must protect the product from moisture loss, light and oxygen, of which the former is the most important. Freezer burn or sublimation of water vapor from the surface of frozen foods results in their becoming dehydrated with a concomitant loss in weight and their visual appearance deteriorates. All the common polymeric films have satisfactory water vapor transmission rates at freezer temperatures.

The earliest form of packaging material for frozen fruits and vegetables was waxed carton board, often with a moisture-proof regenerated cellulose film overwrap. These were then replaced with folding cartons having a hot melt coating of PVC/PVdC copolymer and the ability for the flaps to be heat sealed. Although still used to a small extent for low production volumes, the majority of frozen fruits and vegetables today are packaged in polymeric films based on blends of polyolefins, the major component of which is LDPE. Sufficient plasticizer is added to ensure that the films retain their flexibility, at low temperature. It is also common for the film to contain a white pigment to protect the contents from light which could oxidize the pigments. The film is usually supplied in roll form from which it is converted into a tube, then filled and sealed continuously in a form/fill/seal type of machine. Premade bags are used for low volume packaging operations.

The thermal processes used for canned fruits and vegetables differ markedly depending on the pH of the product: low acid products, *i.e.,* productions with a pH greater than 4.5 (which includes most vegetables) require a full 12D processes, typically 60–90 min at 121°C. In contrasts, those products with a pH less than 4.5 need only a mild heat treatment, typically 20 min in boiling water. Some products are acidified to lower the pH below 4.5 and thus, avoid the more severe heat treatment. For details about the processing procedures for individual fruits and vegetables, the reader is referred to a standard text in the area.

Canned fruits and vegetables are packaged either in tinplate cans or glass jars. The former must have the correct internal enamel applied to avoid corrosion of the tinplate. As with juices, it is important that all the air is removed from the product prior to packaging to minimize corrosion of the tinplate. For acid fruits containing anthocyanin red/blue pigments such as raspberries, the enamel coating must be particularly rigorous since the pigments act as depolarizers, accelerating the rate of corrosion. With some fruits only the ends of the can are enamelled, and for pineapple a plain can is used; as the tin dissolves from the tinplate and reacts with certain constituents of the pineapple, a yellow colour develops. White aluminum-pigmented epoxy resin enamels are used with fruits in some countries.

Many vegetables contain sulfur compounds which can break down during heat processing to release hydrogen sulfide. This can react with the tin and iron of the metal can to form black metallic sulfides which cause an unsightly 'staining of the can and also of the contents. While this process is encouraged and indeed is essential for the production of the desired flavour and colour in canned asparagus, it is avoided with other vegetables by the use of special enamels which contain zinc oxide. This reacts with the hydrogen sulfide to produce barely detectable white zinc sulfide on the inner surface of the can. White aluminum-pigmented enamels based on epoxy resins are also used for cans containing vegetables.

Glass containers are still used for packaging some commercially processed fruits and vegetables, generally for products at the premium end of the market. This is largely because the production rates for glass containers are much lower than those possible for metal cans. Cylindrical, widemouth glass jars are commonly used with either a twist-off or pry-off cap made from mild steel or aluminum coated with a suitable lining material. Considerably greater operator skills are required to retort glass jars compared to metal cans, since failure to control the overpressure correctly can result in either shattered containers or the loss of pry-off caps.

Retortable pouches made from laminates of plastic film generally with an aluminum central layer can also be used for the packaging of fruits and vegetables which are preserved by the use of heat.

The packaging of dehydrated fruits and vegetables requires the use of a package which will prevent or at the very least minimize the ingress of moisture and, in certain instances, oxygen. For example, products which contain carotenoid pigments (*e.g.*, carrots and apricots) can undergo oxidative deterioration, and dehydrated potatoes are liable to develop stale rancid flavors unless oxygen is excluded. Vacuum or inert gas packaging may be used if the product is particularly sensitive to oxidation.

Many vegetables, for example green beans, peas and cabbage, are treated with sulfur dioxide prior to drying to retard nonenzymic browning (the principal cause of deterioration in dehydrated vegetables) and increase the retention of ascorbic acid. Sulfur dioxide also has a useful antimicrobial effect during the initial stages of drying and, by varying the form in which it is introduced (sodium sulfite or metabisulfite), it can be used to control pH, which in turn influences the color and subsequent handling and drying characteristics of the product. Concentrations of sulfur dioxide in the dried product normally range from between 200 and 500 ppm for potatoes to between 2000 and 2500 ppm for cabbage.

For the packaging of dehydrated fruits and vegetables, the material normally used consists of one or more polymeric films having the desired barrier properties. This implies that the material must be a very good barrier to water vapor and, depending on the particular product, a good barrier to O_2 and maybe SO_2 and certain volatiles.

Chapter 18

Packaging of Meat and Meat Products

18.1 Introduction

There is a continuous search for improved methods of transporting food products from producers to consumers. With the increasing urbanization of society, the problems associated with the keeping quality of fresh flesh foods have become more accentuated. Large livestock slaughter and processing facilities have developed in areas where livestock production is highly concentrated, and such areas are, not surprisingly, *well* away from the centers of population density. As well, there is a major world trade in fresh and preserved flesh foods. In all these situations, packaging has a key role to play in protecting the product from extrinsic environmental influences and giving the food the shelf life required in the particular market.

In the packaging of red meat, there are two factors of major importance: colour and microbiology. Both of these will be considered in some detail before specific packaging materials and systems are discussed, since an understanding of red meat colour and microbiology is an essential prerequisite to the development of successful packaging for red meat.

Since the water activity of chilled meat is very high, unprotected meat will lose weight by evaporation and its appearance will deteriorate. Further weight loss will occur when meat is cut, since the exposed surfaces exude liquid which detracts from the appearance of packaged meat. This can be overcome by including an absorbent pad in the base of the package. Although efficient chilling can reduce the quantity of exudate, a certain amount will always be present when meat cuts are held for retailing. This unattractive bloody exudate found in vacuum packaged beef is referred to as "purge", "weep" and "drip"; 1–2 per cent purge is considered acceptable, while 4 per cent is considered excessive. Values of 2–4 per cent drip can have substantial economic implications if not controlled.

The importance of colour as a marketing attribute of red meat is well established, especially for self-service retailing. Consumers, used to seeing bright red meat prepared for sale, associate this colour with good eating quality, although there is little correlation between the two. This association of the colour of red meat (both in the chilled and frozen form) with freshness has been the dominant factor underlying retail meat marketing. The loss of this bright red colour, known as loss of "bloom" in the industry, is affected by many factors, although the consumer will usually relate the colour loss to bacterial growth. Because of the importance of meat colour, various methods of transportation, distribution and packaging of primals and subprimals have evolved which optimize the maintenance of a desirable meat colour.

The colour of meat as perceived by the consumer is primarily determined by such factors as the concentration and chemical form of the meat pigment myoglobin, the morphology of the muscle structure and the ability of the muscle to absorb or scatter incident light.

Myoglobin is the principal pigment of fresh meat and the form that it takes is of prime importance in determining the colour of the meat. The myoglobin molecule consists of a heme nucleus attached to a globulin type protein component. The heme group consists of a flat porphyrin ring with a central iron atom which has six bonding points or coordination links. Four of these are linked to nitrogen atoms; one is attached to the globin molecule; the remaining linkage is free to bind to other substances, usually water or oxygen. The colour of myoglobin depends on at least three factors:

1. The oxidation state of the iron atom: it may be reduced or oxidized;
2. The nature of the group at the sixth bonding point of iron;
3. The state of the globin: it may be native as in raw meat, or denatured as in cooked meat.

The colour of fresh meat depends chiefly on the relative amounts of the three pigment derivatives of myoglobin present at the surface: reduced myoglobin (Mb), oxymyoglobin (O_2Mb) and met myoglobin (MetMb). Reduced myoglobin (Mb) is purple in colour and predominates in the absence of oxygen. Oxymyoglobin (O_2Mb) is bright red in colour and results when Mb is oxygenated or exposed to oxygen; this is commonly known as "bloom." Metmyoglobin (MetMb) is brown in colour and exists when the oxygen concentration is between 0.5 per cent and 1 per cent, or when meat is exposed to air for long periods of time. The brown metmyoglobin cannot bind oxygen even though it is oxidized by the same oxygen that converts myoglobin to the bright red oxymyoglobin.

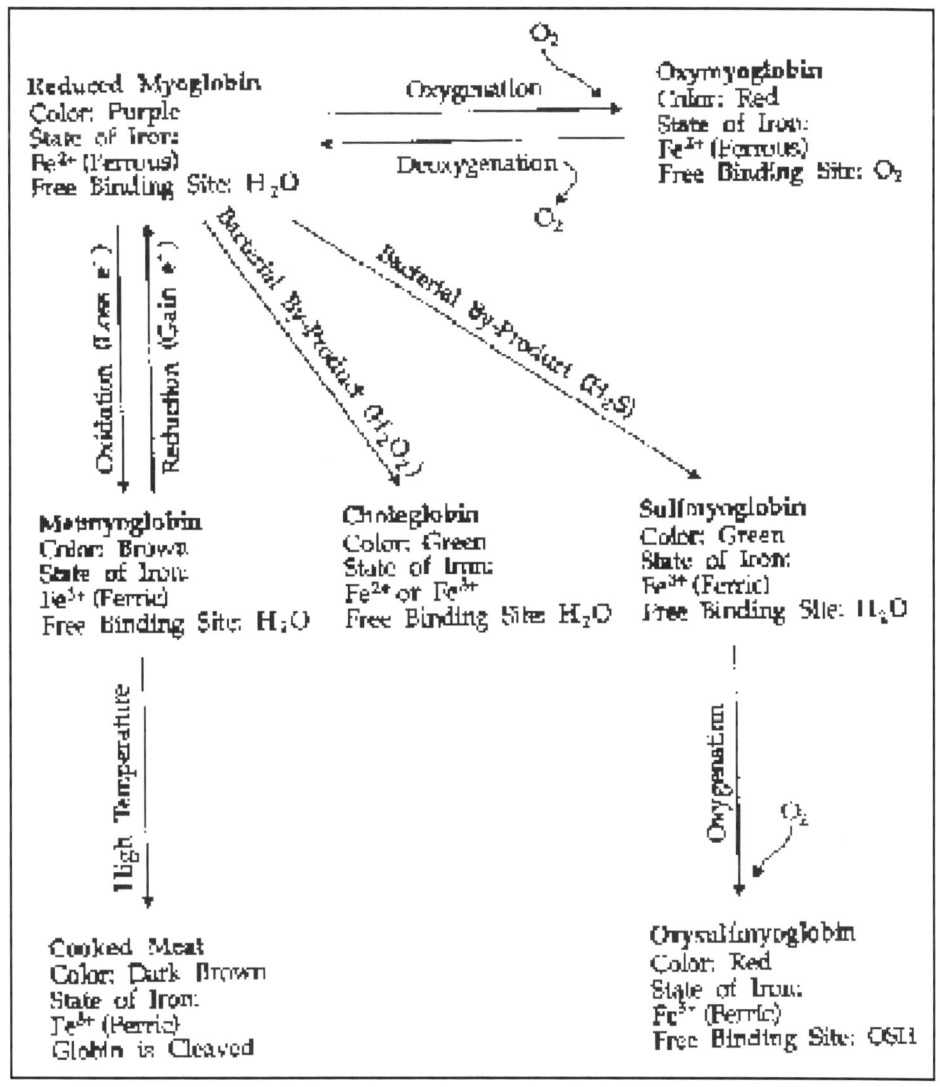

Figure 18.1: Primary Color States of Myoglobin.

The colour reactions of myoglobin are all reversible and dynamic with respect to the three primary forms of myoglobin: Mb, O_2Mb and $MetMb_+$. It is rare when all the myoglobin colour forms in meat are in the same form; generally two or more of the pigments will be present but the predominant pigment will be the most noticeable. $MetMb_+$ cannot take up oxygen but enzymes present in fresh meat are capable of reducing $MetMb_+$ to Mb which can then take up oxygen to form O_2Mb. As meat ages, the substrate for these enzymes is gradually used up and MetMb can no longer be reduced.

18.2 The Change of Colour of Meat

For meat which has been exposed to air for several hours the penetration; depth may be 6–7 mm. The bright red colour of O_2Mb will predominate and be apparent to the observer from the outside in to the point where the ratio of oxymyoglobin to myoglobin is about 1: 1, *i.e.*, about 84 per cent of the total depth of O_2 penetration. Since different muscles have different inherent surviving respiratory activities, X for different muscles will vary by a factor of 2–3 under a given set of conditions.

Due to a lack of oxygen, the colour of the interior of fresh red meat is purple because of the presence of reduced myoglobin. The outer most layer exposed to air is red due to the presence of oxymyoglobin. A thin brown layer of metmyoglobin can be observed between the red and purple regions due to a low concentration of oxygen. This layer thickens within for 1 or 2 days and becomes apparent, first by darkening the translucent surface tissue and later by breaking through to the surface. At lower oxygen partial pressures, the MetMb layer will be nearer the surface until at the critical partial pressure it is at the surface. The rate of these changes is strongly influenced by the temperature.

Oxygenation of purple reduced myoglobin to red oxymyoglobin is rapid, the surface of beef in air appearing red within half an hour at 5°C. In contrast, oxidation to metmyoglobin is slow, appearing first as a thin brown layer at the limit of oxygen penetration. The red oxymyoglobin is stable as long as the heme remains oxygenated, although the oxygen is continually associating and dissociating from the heme.

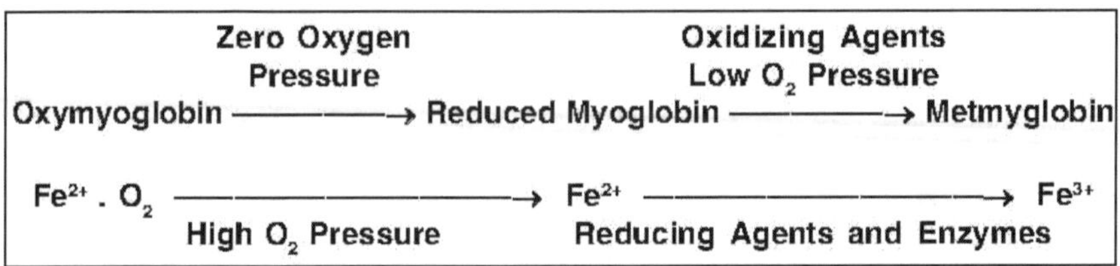

Figure 18.2: Schematic Diagram of Interrelationships between Three Major Meat Pigments.

The dissociation of oxygen from the heme (known as deoxygenation) is caused by conditions such as low pH (less than pH 5.4), high temperature, ultraviolet light, salts and especially low oxygen tensions which cause denaturation of the globin moiety of oxymyoglobin. The deoxygenation of red oxymyoglobin results in myoglobin (which is very unstable) which then becomes oxidized to brown MetMb. Surface dessication increases the salt concentration and promotes the formation of MetMb.

The conditions which cause deoxygenation of oxymyoglobin to myoglobin are also responsible for the conversion of unstable reduced myoglobin to brown metmyoglobin. Since the globin moiety is denatured, the oxidation (often called autoxidation) proceeds spontaneously and involves the loss of an electron from the iron of the heme. The actual oxidation is caused by the activity of oxygen-utilizing enzymes indigenous to muscle tissue. Through a process of respiration, the O_2 is converted to CO_2. When the O_2 has been substantially reduced, the MetMb pigments are reduced to the purple Mb.

Low storage temperatures suppress the activity of these enzymes, but with increasing storage temperatures, their activity increases thus, lowering the oxygen tension at the meat surface. Lamb muscles have the highest concentration of these enzymes compared to beef and pork and therefore, a very high oxygen demand; this explains the comparatively short retail shelf life of lamb. As well as encouraging enzyme activity, high storage temperatures will also decrease the depth of oxygen penetration into the meat tissues, moving the brown met myoglobin layer closer to the surface.

The maximum rate of metmyoglobin formation has been reported at oxygen partial pressures of 7.5±3 mm Hg at 7°C and 6±3 mm Hg at 0°C for beef. Earlier results which are still frequently quoted are less relevant since they were obtained using pure pigments and at temperatures well above those likely to be encountered in practice. For example, the rate of autoxidation for pure myoglobin solutions was reported as maximal at about 1 mm Hg of oxygen at 30°C and pH 5.7, while the maximal rate for ox blood hemoglobin at 25°C and pH 5.7 was reported to be 20 mm Hg of oxygen. Clearly, any value is very dependent on temperature, although independent of pH. If meat is held under high O_2 partial pressures (*i.e.,* above 30 mm Hg), autoxidation of Mb is minimized. These results have led to the development of systems allowing transport of meat at high O_2 partial pressures.

Metmyoglobin reduction to myoglobin in *post rigor* meat (*i.e.*, a change in the valence state of the iron molecule from ferric to ferrous) is primarily enzymic in nature, with an electron being donated by reducing enzyme systems. This process is often called "MRA" or "metmyoglobin reducing activity." Some muscles will remain bright red for longer periods of time than other muscles because any metmyoglobin formed is reduced back to reduced myoglobin and immediately oxygenated back to red oxymyoglobin; such muscles are said to have a higher MRA. Loss of reducing activity in meat *post mortem* is due to a combination of factors including fall in tissue pH, depletion of required substrates and co-factors, and ultimately complete loss of structural integrity and functional properties of the mitochondria.

It has been reported that bacteria cause discoloration of meat in their logarithmic growth phase, apparently due to the high oxygen demand of aerobic bacteria reducing the oxygen tension at the meat surface and causing the formation of metmyoglobin.

Some bacteria also produce by-products which oxidize the iron molecule. The most common are hydrogen sulfide (H_2S) and hydrogen peroxide (H_2O_2) which react with unstable myoglobin to produce sulfmyoglobin and choleglobin, respectively. H_2S production causes green discoloration on vacuum packaged meat, it generally being found only on meat having a high pH (pH > 6). On opening a vacuum packaged cut that has green discolorations on it, the green sulfmyoglobin becomes oxygenated to oxysulfmyoglobin which is red in appearance. However, the color change does not cause any diminution in the H_2S off-odour which resembles rotten eggs. H_2O_2 can either cause greening due to oxidation of heme pigments to form choleglobin or degradation beyond porphyrins to bile pigments.

The colour intensity of meat is determined by *ante mortem* factors such as species, stress, sex and age of the animal, the *post mortem* pH rate of decline and the ultimate pH of the meat. During *post mortem* glycolysis the pH of normal tissue falls from an *in vivo* value of 7.2 to an ultimate value of 5.4–5.6. As well as influencing the colour, the rate of fall of pH and the ultimate pH value also influences the water holding capacity and the texture of the meat. Differences in colour intensity between species are primarily caused by differing concentrations of myoglobin. Thus, beef which has the highest concentration is the darkest of the meat species, with lamb being intermediate in colour and myoglobin concentration. Pork has the lowest concentration of myoglobin and as such is the lightest in colour. Male animals usually produce darker meat than females due to a greater concentration of myoglobin in meat derived from male animals.

The colour intensity of meat is also influenced by pH and morphology of the muscle structure. The pH of a muscle is largely influenced by the conditions which exist immediately prior to or just after slaughter. Short term, violent excitement immediately before slaughter, or the slow cooling of carcasses after slaughter, can result in meat having a low ultimate pH. This is caused by a buildup of lactic acid which is formed as a metabolic end product in the anaerobic breakdown of glucose and glycogen. This condition is known as PSE since it produces pale, soft and exudative meat as a result of an abnormally rapid fall in pH immediately after slaughter when the combined effects of low pH and high temperature in the muscle lead to denaturation of sarcoplasmic and myofibrillar proteins. Because muscle glycolysis is relatively more rapid in hogs than bovines, PSE is rarely observed in beef animals. However, there are appreciable differences in beef with respect to both color and water-holding capacity duc to differences in the temperature/pH profiles postmortem. It would appear that PSE meat spoils in the same manner as nonnal meat.

PSE meat causes problems in packaging: the color is abnormally pale due to denatured protein, the physical structure increases light scattering making the meat more opaque and autoxidation increases causing colour fading. In addition, because of its low water-holding capacity resulting from protein denaturation, PSE meat produces excessive drip.

Biochemical conditions directly opposite to those producing PSE meat give rise to another type of abnormal meat from all meat animals but more especially beef cattle and hogs. This condition has been described as even more troublesome than PSE from a packaging point of view. The problem meat, known as dark-cutting or dark, firm dry (DFD), is translucent and sticky to touch, and unacceptable for retail packaging because of its dark purple colour. This meat has a high pH due to a low residual glycogen level remaining in the muscle at slaughter, following excessive stress (such as an extended transit haul) or exercise prior to slaughter. Such treatment causes depletion of the muscle glycogen before death. The low residual glycogen means that insufficient lactic acid is formed by glycolysis during the rigor process to lower the pH.

The muscle fibers of DRD meat are swollen and tightly packed together forming a barrier to the diffusion of oxygen and the absorption of light. As well, the high pH of the meat accelerates respiratory activity of the meat tissue, resulting in a very thin layer of red oxymyoglobin with the underlying purple reduced myoglobin more visually apparent. Moreover, the high pH encourages the growth of putrefactive microorganisms, thereby significantly reducing the keeping quality of this meat. For example, vacuum packaged beef of normal pH can be stored at chiller temperatures for periods in excess of 10 weeks, whereas DRD beef held under the same conditions will generally spoil within about 6 weeks. However, both normal and high pH cuts of beef packaged under CO_2 were unspoiled at 15 weeks.

The effect of carbon dioxide on meat colour is questionable. It has been reported that meat stored in high concentrations of CO_2 often develops a greyish tinge supposedly because of the lowering of the pH and subsequent precipitation of some sarcoplasmic proteins. It was recommended that CO_2 not be used in concentrations exceeding 20 per cent, and the claim was made that no MetMb formation occurred in atmospheres containing high concentrations of CO_2 providing that the O_2 partial pressure exceeded 5 per cent. A later report indicated that 50–80 per cent CO_2 is often found in residual air spaces in vacuum packages with no associated detrimental effect on meat colour.

More recently, the realization that discoloration in high CO_2 atmospheres must be due to oxygen leading to the use of 100 per cent CO_2 atmospheres on the grounds that meat colour would not be adversely affected if oxygen was rigorously excluded from the packages. In the case of raw red meats, the colour was actually improved by enzymic reduction of pigment that oxidized before the meat was packaged. However, the stability of raw meat colour on exposure, to air decreased as storage life increased, the colour stability at display reaching a minimum value after 12 weeks storage. After that time, acceptable meat colour at display was maintained for about 40 per cent of the time that fresh meat retained acceptable colour. This was claimed to be an inevitable consequence of prolonged chill temperature storage and not an effect of the 100 per cent CO_2 atmosphere.

Carbon monoxide (CO) has the potential to retard MetMb formation and fat oxidation. CO combines with myoglobin to form the bright red pigment carboxymyoglobin (COMb) which is spectrally very similar to oxymyoglobin (O_2Mb). Since COMb is much more stable toward oxidation than O_2Mb by virtue of the stronger association of CO to the iron-porphyrin site on the myoglobin molecule, the addition of CO at low levels can act to negate the detrimental color changes associated with the high levels of CO_2 required to maintain wholesomeness over prolonged transit times.

Significant retardation of rancidity in frozen pork sausage stored under CO-containing atmospheres has been reported, and it has been shown that the presence of CO increases the proportion of Mb present in a reduced state.

Despite these advantages, CO has not been approved by regulatory agencies for commercial use with meats. This is largely because of safety considerations, although it has been shown that less than 0.09 ppm of CO residue was left on cooked ground beef which had been exposed to CO storage.

Lighting is an important factor when presenting meat and meat products for retail sale. A high percentage of red in the light gives a particularly intense, red impression, and products that are already beginning to turn a greyish-red in daylight appear saturated with red when exposed to reddish lighting.

Incident light is a contributory factor in the discoloration of both fresh and frozen meat as, for example, in a retail supermarket display case. The extent of the effect depends on such factors as the wavelength and intensity of the light, temperature, oxygen, partial pressure, meat pH and storage time. Provided that UV wavelengths are avoided, the effect of light is small under usual refrigerated conditions.

The colour of frozen meat initially depends on the rate of freezing and the resultant size of ice crystals in the surface layer. Slow freezing produces large ice crystals with poor light scattering properties, giving the meat a dark, translucent appearance. Fast freezing, on the other hand, results in the formation of small ice crystals which scatter light and make the meat surface pale and opaque.

In frozen red meat, the principal deterioration during storage is photo-oxidation of the pigment. Whereas under direct illumination, chilled meat oxidation begins in the subsurface layer and progresses towards the surface, frozen meat oxidizes from the surface inwards. Loss of redness is detrimental to the marketing of frozen meat as well as chilled meat.

Considerable improvements in the colour shelf life (in the order of 5–10 times) of pre-packaged frozen beef have been reported when a light intensity of 100 lux was used compared to the normal retail cabinet intensity of 1000–2000 lux. Display lighting affected a greying and loss of saturation of the colour of frozen lamb chops.

18.3 Packaging of Fresh Meat

In the USA prior to 1967, whole beef carcass shipments were traditionally made to retail outlets where carcasses, were fabricated into retail items by butchers in individual stores. The "boxed beef" concept was introduced in 1967, dramatically changing beef processing, distribution and retail fabrication, without affecting retail presentation to the consumer. The basis for the boxed beef concept was vacuum packaging in low oxygen barrier materials which provided a good barrier to oxygen.

It is now a common practice to butcher beef carcasses at the slaughter house into 2–9 kg primal joints which are then vacuum packaged into plastic bags with low permeabilities to gases. In these circumstances, the atmosphere around the meat becomes depleted in O_2 (often < 1 per cent v/v) and enriched in CO_2 (> 20 per cent v/v), resulting in microbial changes quite different from those observed during aerobic storage. Vacuum packaged boxed beef is then distributed to retail outlets where these primals and subprimals are fabricated into consumer units, overwrapped in oxygen permeable film on PS foam or PVC trays and displayed for sale.

Boxed beef developed rapidly and changed beef distribution dramatically in the USA. In addition to beef, vacuum packaged boxed pork and lamb followed in the 1970's. In 1986 approximately 84 per cent of all beef processed in the USA was vacuum packaged.

Vacuum packaging achieves its preservative effect by maintaining the product in an oxygen-deficient environment. In anoxic conditions, potent spoilage bacteria are severely or totally inhibited on low-pH (< 5.8) meat. However, their growth on high-pH muscle tissue, or extensive fat cover of inevitably neutral pH, will spoil relatively rapidly in a vacuum pack. Vacuum packaging can therefore, extend the shelf life of primal cuts composed largely of low (normal) pH muscle tissue such as beef and venison by about fivefold over that achieved in air. For other meats and small cuts, only a twofold extension of shelf life can be safely anticipated.

The introduction of vacuum packaging for the distribution and storage of chilled beef has been hailed as the greatest innovation in meat handling during the last 25 years. Vacuum packaging is defined as "the packaging of a product in a high barrier package from which air is removed to prevent growth of aerobic spoilage organisms, shrinkage, oxidation, and color deterioration". The plastic film used for vacuum packaging must have a high resistance to gases and water vapour with perfect seals and good mechanical strength. The most commonly used film for fresh meat vacuum packages is PVDC. This technique is used for packaging of primal and sub-primal cuts of buffalo meat.

Of the 11 million tonnes of beef slaughtered in the USA in 1988, approximately 87 per cent moved into distribution in the form of boxed beef for industrial, restaurant and institutional use. Beef carcasses are broken down into primal and subprimal cuts, separated into boneless and bone-in cuts, and then vacuum packaged. Since only about two-thirds of a beef carcass is usable meat, there are clear advantages in the form of reduced refrigerated space for transportation and storage and less packaging material when boneless beef rather than bone-in or carcasses are stored and distributed. Another advantage is that the tenderness of the beef can be improved by aging without the evaporative weight loss incurred when carcasses are hung in the conventional manner.

Vacuum packaging involves enclosing large boneless joints (typically 3 to 15 kg in weight) in flexible plastic containers (usually bags) to prevent moisture; loss and exclude oxygen from the meat's surface. Packing under a vacuum reduces the volume of air sealed in with the meat. The plastic materials used for vacuum packaging must have low moisture and gas permeabilities and be strong enough to hold heavy beef joints.

In packaging fresh meat primals, many cuts contain bones which are often sharp and abrasive and readily puncture the flexible plastic materials used in vacuum packaging. To overcome this bone puncture a BONEGUARD material (approved by the FDA and USDA) consisting of a wax-impregnated and coated cotton scrim is employed.

Reported oxygen permeabilities of packaging films are usually measured at ambient temperatures and moderate humidities (typically 23°C and 50 per cent RH); but both temperature and humidity can affect the rates at which gases are transmitted through films. Data on the oxygen permeabilities of packaging films at chill temperatures are sparse, and those that do exist often do not include a complete specification of the film under test, or the test conditions. This is further complicated by the variety of test methods and units used. The oxygen permeabilities at sub-zero temperatures of two plastic films used for the vacuum packaging of meat have been reported. One film was a polyamide-polyethylene laminate, while the other was an EVA copolymer-PVC/PVdC copolymer laminate. Their oxygen permeabilities at –1°C were reported as 2.0 and 0.6 mL m^{-2} 24 h^{-1} atm^{-1} respectively, approximately one fiftieth of the values obtained at 23°C and 90 per cent RH.

As has been pointed out, once a piece of meat has been vacuum packaged in an oxygen barrier material having a gas permeability (at 23°C, and 75 per cent RH) below 50 mL m^{-2} day^{-1} atm^{-1} and adequately sealed to prevent air reentry, the shelf life of the meat is very much the same regardless of the packaging material. Therefore, the significant differences between the packaging materials and/or systems are not in the structure so much as in the physical properties, the production speeds of the system and the abuse resistance of the package itself.

When meat is vacuum packaged the contaminating flora are exposed to an atmosphere containing 20–25 per cent CO_2 and less than 10 per cent O_2. Both the high CO_2 and low O_2 tension depress the growth of Pseudomonads and facultative anaerobes predominate: lactic acid bacteria, *B. thermosphacta* and *Enterobacteriaceae.* The antimicrobial activities of the Lactobacilli, coupled with, the low storage temperature, combine in a synergistic manner to extend the shelf life of vauum packaged meat.

The oxygen in the small volume of residual air inside a vacuum package will be quickly consumed by meat respiration so that within 2 days the oxygen partial pressure at the surface of the meat will have dropped below 10 mm Hg. At these very low partial pressures the penetration limit of oxygen is very near the surface and the thin brown layer of MetMb which develops cannot conceal the underlying Mb; therefore the visible colour of vacuum packaged beef is purple.

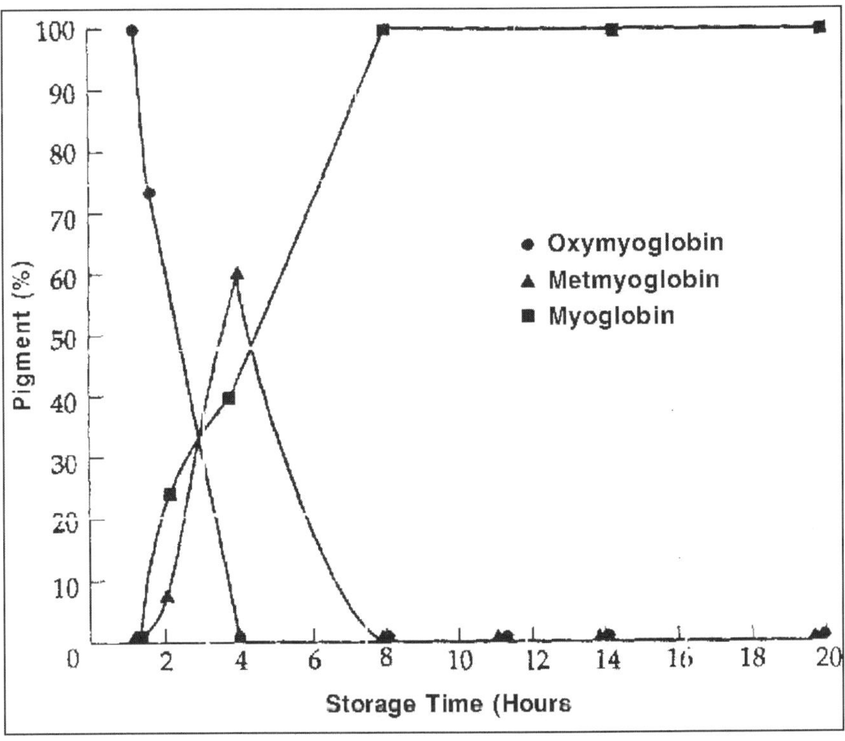

Figure 18.3: Initial Pigment Changes at 3.3°C of Anaerobically Packaged Beef.

The efficacy of vacuum packaging depends on there being close contact between a film of low gas permeability and all product surfaces. If there are vacuities within the pack, these will develop an oxygen-containing atmosphere as gases permeate into the package during storage. Bacterial growth will accelerate and product colour will deteriorate because of oxidation of Mb at meat surfaces exposed to such an atmosphere. Also, when the meat surface is large relative to the meat mass, an extensive film surface is presented for oxygen permeation and colour deterioration can also occur. Consequently, vacuum packaging is relatively ineffective for preserving products such as carcasses whose shapes preclude close application of the packaging film to all surfaces, and small cuts of any shape. Furthermore, because the anoxic conditions within the package result in the formation of purple Mb, vacuum packaging for retail display packages of red meat has not met with wide success because of consumers' negative perception of the colour purple.

18.4 The Shelf Life of Packaged Meats

In any discussion of the shelf life of packaged meats, it must be borne in mind that comparisons of published data are difficult and the making of generalizations foolhardy since there are a large number of variables which interact to determine the actual shelf life. The most important of these variables is temperature, and the statement that samples of vacuum packaged meat were held at a particular temperature is not especially helpful unless the range of temperatures encountered by the samples during storage is specified. Generally, it can be said that laboratory scale trials control temperatures over a much smaller range compared to commercial scale trials, and this is frequently the reason for the longer shelf life obtained in laboratory scale trials over that obtained in commercial production.

Other important variables include the microbiological status of the meat at the time of packing and the method used to determine the end of shelf life of the meat. These methods range from objective assessments of microbial counts to subjective assessments by taste panels consisting of trained or untrained members. Clearly, attempting to draw generally applicable conclusions from numerous published reports where the magnitude of these variables differs (or is not even specified) would only be misleading.

Vacuum packaged beef of normal pH can be stored at chiller temperatures for periods in excess of 10 weeks. However, high pH, dark, firm, dry (DRD) beef held under the same conditions will generally spoil within about 6 weeks. The degree of vacuum has been reported as having no effect on sensory panel ratings of vacuum packaged beef cuts. There are also advantages in aging beef in vacuum packages as opposed to carcass aging, including less loss due to water evaporation, less necessity for trimming of exposed surfaces and more efficient use of refrigerated space.

Although for many years vacuum packaging was only applied to chill beef, it has been applied in recent years to lamb and pork. Because of their relatively small size, pork and lamb carcasses are only partially boned before packaging, and the presence of bone can lead to puncturing of the package unless precautions are taken, for example, the use of BONEGUARD material (a woven cotton scrim impregnated with wax and having dead fold characteristics) or strong EVA copolymer film in the package structure.

In contrast to beef cuts, much of the surface of lamb cuts is adipose, rather than muscle tissue. Adipose tissue has pH values close to neutrality and has no significant respiratory activity. Packaged lamb can therefore, present a heterogeneous environment for microbial growth. This different microbial environment of high pH and, possibly, relatively high oxygen concentration, probably accounts for the shelf life of only 6 to 8 weeks reported for vacuum packaged lamb whereas 11 to 12 weeks is routinely attainable with beef. The longer shelf life of vacuum packaged beef is due to its flora being dominated by Lactobacilli which overgrow other spoilage organisms in the low temperature, low pH, low O_2 and high CO_2 environment of the package.

Laboratory and commercial trials in New Zealand seeking to extend the chill temperature shelf life of lamb cuts showed that after 12 weeks' storage at –0.5°C, lamb loins vacuum packaged in a PET-aluminum foil-LDPE laminate having an immeasurably low gas permeability developed floras of Lactobacilli that had not caused spoilage; meat colour was also much improved by the exclusion of oxygen. Commercial trials where the packaged cuts were held for 6 weeks at 0°C and then at -0.5°C were very acceptable at 9 weeks but most were spoiled by off-flavours at 12 weeks, the next sampling time. Since then a high CO_2 MAP has been developed in New Zealand which gives a chill temperature shelf life for lamb of at least 16 weeks; there is now less interest in extending the chill temperature shelf life of vacuum packaged lamb.

Various conflicting reports have appeared in the literature concerning the chill temperature shelf life of pork joints. For example, figures of little more than 2 weeks at 1°C and 3–4 weeks at 2°C were reported in an extensive review. However, it appears that limited use is being made of vacuum packaging for wholesale cuts of pork, where the attraction is likely to be convenience in butchery rather than savings in weight loss.

Recently a comparison of three packaging treatments [MAP (25 per cent CO_2; 75 per cent O_2); VSP and the mother bag system (100 per cent CO_2)] on the shelf life of fresh pork stored at 0°C was reported. All three treatments were equally efficient for the first 4 days of retail display, but the mother bag system gave the most promising shelf life results (21 days). The MAP shelf life was 14 days and the VSP 7 days.

The chill temperature shelf life of vacuum packaged lamb, mutton and pork has been extended using a process developed in Australia which involves application of foodgrade acetic acid at a concentration of 1.5 per cent and a temperature of 53–55°C (higher temperatures may cause discoloration). This procedure results in an immediate reduction of 90–99.5 per cent in the bacteria on the exposed surfaces and it has a residual action. A 3 per cent solution (higher concentrations may cause adverse color changes) at 23–25°C has a similar effect. An increase in storage life of about 50 per cent was achieved using this procedure prior to vacuum packaging. Because the 1.5 per cent acetic acid solution at 55°C may cause irritation to personnel and possibly corrosion to metal surfaces, the more expensive 3.0 per cent solution used at the lower temperature is preferred. Both processes have the approval of the FDA and the USDA, but not the European Community.

18.5 Recent Innovations in Meat Packaging

Active packaging, intelligent packaging, edible coatings/films and biodegradable packaging, and nanotechnology are the major recent innovations in the food packaging industry that have shown promising advanced properties in extending shelf life, improving food safety and quality, and protecting our natural environment.

Active packaging is intended to influence the packed food. The packaging absorbs food-related chemicals from the food or the environment within the packaging surrounding the food; or it releases substances into the food or the environment surrounding the food such as preservatives, antioxidants, and flavourings. The "releasing active materials and articles" are those designed to deliberately incorporate components that would release substances into or onto the packaged food or the environment surrounding the food; and "released active substances" are those intended to be released from releasing active materials and articles into or onto the packaged food or the environment surrounding the food and fulfilling a purpose in the food. Most important active packaging systems applied to meat and meat products are antioxidant and antimicrobial packaging, carbon dioxide emitters, and oxygen scavengers.

Antimicrobial active packaging is one of the most important concepts in active packaging because meat provides excellent nutrients for the growth of microorganism. Spoilage microorganisms including bacteria, yeast and molds, and pathogenic micrograms, specifically Salmonella spp., *S. aureus*, *L. monocytogenes*, *C. perfringens*, *C. botulinum*, and *E. coli* O157:H7 are the major concerns leading to quality deterioration and food safety issues in meat. The aims of using antimicrobial active packaging are to extend shelf-life and to ensure food safety of meat and meat products.

There are three basic categories of antimicrobial packaging:

1. Incorporation of an antimicrobial substance into a sachet/pad connected to the package from which the volatile bioactive substance is released during further storage;
2. Direct incorporation of the antimicrobial agent into the packaging film; and
3. Coating of packaging with a matrix that acts as a carrier for antimicrobial agents so that the agents can be released onto the surface of food through evaporation in the headspace (volatile substances) or migrate into the food (non-volatile additives) through diffusion.

For the effective application of antimicrobial active packaging in meat and meat products, the selection of an effective delivery method and antimicrobial substance, and minimal impact on the sensory properties of packaged product are the key points that should be considered.

High levels of oxygen present in meat packaging can facilitate microbial growth, lipid oxidation, development of off flavours and off odours, colour changes and nutritional losses. Lipid oxidation not only results in the development of off-flavours (rancidity), but also the potential formation of toxic aldehydes and the loss of nutritional quality because of polyunsaturated fatty acid (PUFA) degradation. Therefore, control of oxygen levels in meat packaging is important to limit the rate of such deteriorative and spoilage reactions. Antioxidant active packaging can be used as a means of improving product quality and extending shelf life of meat and meat product through controlling the level of oxygen.

18.6 The Poultry Meat and Eggs

Although most poultry meat is sold in the form of whole, oven-ready carcasses, there is now an increasing demand for cut-up portions and a variety of other further processed products, both raw and cooked.

Raw poultry meat is a perishable commodity of relatively high pH which readily supports the growth of microorganisms when stored under chill or ambient conditions. The shelf life of such meat depends on the combined effects of certain intrinsic and extrinsic factors, including the numbers and types of psychrotrophic spoilage organisms present initially, the storage temperature, muscle pH and type (red or white), as well as the kind of packaging material used and the gaseous environment of the product.

The main pathogenic organisms associated with poultry and poultry products are *Salmonella* spp.; *Staphylococcus aureus,* and *Clostridium perfringens.* Most studies on the extension of shelf life using CO_2 in modified atmospheres have concentrated on the suppression of spoilage organisms rather than the survival and growth of pathogens.

The vacuum packaging of poultry carcasses, cuts and other manufactured products can extend shelf life provided that the product is held under chill conditions. During storage at 1°C in either O_2 permeable film or vacuum packs, extensions in shelf life from 16 to 25 days in the case of breast fillets (pH 5.9–6.0) and from 14 to 20 days for drumsticks (pH 6.1–6.3) were observed for the vacuum packaged products. Somewhat surprisingly; deleterious flavour changes tended to precede the development of off-odours in vacuum packs of both types of muscle.

Although vacuum packaging may be used to encourage the development of an atmosphere around the product which delays microbial spoilage, other oysters in use involve an initial addition of about 20 per cent CO_2, to either individually packaged items or bulk packs of varying size in an O_2 impermeable plastic film. The MAP of poultry and related products has been reviewed.

The most comprehensive study of the use of CO_2-enriched atmospheres for extending the shelf life of poultry meat (chicken portions) was carried out in the USA in 1951. The maximum usable CO_2 concentration was 25 per cent since above this the meat became discoloured; even at 15 per cent a loss of bloom was sometimes noted.

Transferring CO_2-stored poultry meat to normal atmospheric conditions of cold storage gave an extended shelf life which was intermediate between the keeping time in air alone and that in the CO_2 atmosphere. Whether this effect is due to a lag phase in microbial growth induced by a change of gaseous, environment or merely to the slow diffusion of CO_2 from the tissues and hence a "residue" effect has not been established.

As was mentioned earlier in this chapter, CO_2 levels of 20 per cent or more extend; the shelf life by inhibiting the growth of many organisms at temperatures of –1°C to 7°C, the inhibitory effect being critically influenced by temperature. This was demonstrated graphically by a study into the effect of 80 per cent CO_2 (balance air) compared to air on the survival and growth of microorganisms most often associated with spoilage and foodborne disease in poultry carcasses at 2, 7 and 13°C. The CO_2 atmosphere substantially retarded the growth of the total bacterial load in uninoculated ground chicken meat and parts at all temperatures when compared to air; however, temperature had a larger overall effect than atmosphere. The study concluded that MAP of refrigerated poultry in elevated CO_2 atmospheres does not increase the microbial hazards when compared to air at the same temperature.

Another study found that 100 per cent CO_2 markedly reduced the growth rate of microorganisms whereas the growth rate in 20 per cent CO_2 was only slightly less than that in vacuum packs. No adverse effects of high CO_2 concentrations on meat quality were reported.

Although there is evidence that CO_2 packaging can be more effective in extending the shelf life than vacuum packaging, most studies in which an 'appreciable shelf-life extension has been demonstrated have involved CO_2 concentrations above the 25 per cent limit suggested by the early American work.

The optimum modified atmosphere to limit microbial growth and ensure organoleptic quality to 27 days of storage at 1.1°C of whole chickens in a polyamide/ionomer pouch with a minimum replacement of 95 per cent of the air required a usage level of 722 mL of CO_2 per kg of whole chickens. Although atmospheres enriched with CO_2 have been advocated for extending the shelf life of processed poultry products, the precise composition of the atmospheres necessary to achieve the desired result have yet to be clarified in the published literature.

While the number and types of microorganisms found on stored poultry is an important factor when determining the shelf life, the real determinant is the sensory quality of the raw and cooked product. Unfortunately, most published studies have not included sensory tests. One study which did evaluate the quality of raw and cooked poultry that had been stored under MA and refrigeration for up to 5 weeks. Their data indicated that MAP (80 per cent CO_2) poultry would be quite acceptable to consumers for up to 4–6 weeks depending on the temperature of storage. It was noted that commercial poultry processors may not get as long a shelf life because of difficulties in controlling the packaging process and temperature under production conditions.

In a study of broiler carcasses packaged under vacuum in film of low O_2 permeability, or under CO_2 in gas-impermeable packages, shelf life was a function of storage temperature, packaging and O_2 availability. Putrid spoilage in gas-impermeable packages after 7 weeks storage at 3°C or 14 weeks storage at –1.5°C was attributed to Enterobacteria. In vacuum packages with oxygen transmission rates of 30–40 mL m^{-2} day^{-1}, putrid odors were detected after 2 weeks storage at 3°C and 3-week storage at –1.5°C.

Concerning the safety of MAP chicken, the possible problem organisms would be *C. jejuni* which may be able to survive better in a MAP product, and *L. monocytogenes* and *A. hydrophila* which may, because of the extended storage lives of the MAP products, have additional time to grow to potentially high numbers. Although *C. perfringens* may be able to survive better in some MAs as compared to air, it would not be able to grow at the chill temperatures commonly used for MAP products. Thus, it is unlikely to be a health hazard in a MAP product unless the product is temperature abused, because high numbers of the organism must be ingested to cause illness.

Handling poultry products during a postcook packaging operation can contribute more to bacterial spoilage than the packaging operation itself. It has been reported that turkey product packaging had only a minor effect on numbers of bacteria, the significant increase in bacterial count being due to the handling operation, probably from equipment surfaces.

Infertile eggs from hens classified as *Callus domesticus* are used almost exclusively for human consumption, the average weight of shell eggs being about 58 g. Shell eggs consist of 8–11 per cent shell, 56–61 per cent albumen, and 27–32 per cent yolk. The pores in the shell permit gaseous diffusion as well as moisture loss and the entry route for microorganisms which might infect the egg. Egg products are commonly marketed in the following forms: (a) refrigerated, (b) frozen, and (c) spray dried.

The factors associated with the loss of shell egg quality are time, temperature and humidity. During the storage of shell eggs, the pH of albumen increases at a temperature-dependent rate from about 7.6 to a maximum value of about 9.7, the rise resulting from a loss of CO_2 through the pores in the shell. The rise in pH of albumen causes a breakdown in the gel structure of the thick white.

Several methods of altering the environmental conditions around eggs have been used to prolong their shelf life, the major one being refrigeration. Coating of the shell with mineral oil has also come into common usage to increase the shelf life by reducing the rate of CO_2 and moisture loss. Spray oiling has come into common usage for the shells of eggs, light mineral oils of food quality normally being used. It is essential that the treatment be applied to the eggshell within a few hours of production since the rate of CO_2 loss is very high during the first few hours, and weight loss during the first few days can be significant.

Clearly the rate of CO_2 loss at any particular temperature will be a function of the partial pressure of CO_2 in the external environment. Thus, it is not surprising that attempts to extend the shelf life of shell eggs have involved storing the eggs in an atmosphere containing CO_2 or alternatively packaging eggs in a CO_2-impermeable package. A comparative study evaluating the quality and shelf life of fresh shell eggs stored at room temperature with four different treatments (not packaged; packaged in air; packaged with 15 per cent CO_2 and oil coated) concluded that controlled atmosphere was the most efficient method of preserving egg quality at room temperature for a period of 7 weeks.

Spoilage of flesh foods such as fish and shellfish results from changes brought about by chemical reactions such as oxidation, reactions due to the fish's own enzymes, and the metabolic activities of microorganisms. The chemical composition and microbial flora of seafood vary considerably between species, different fishing grounds and seasons.

Both salt water and fresh water fish contain comparatively high levels of proteins (about 18 per cent) and other nitrogenous constituents. Non-fatty fish such as cod and haddock have a lipid content of less than 1 per cent in contrast to fatty fish such as herring and mackerel which can have lipid contents of up to 30 per cent. The spoilage of salt and fresh water fish appears to occur in essentially the same manner.

It is generally accepted that the internal flesh of healthy, live fish is sterile; microorganisms that exist on fresh fish are generally found in the gills, the outer slime, and the intestines. The *post mortem* changes leading to spoilage depend principally on the chemical composition of the fish, its microbial flora and subsequent handling, processing and storage.

Immediately *post mortem* a whole series of tissue enzyme reactions begin the process of autolysis (basically self digestion of the fish muscle) which leads eventually to spoilage. The autolytic enzyme reactions predominate for 4–6 days at 0°C after which the products of bacterial activity become increasingly evident with the appearance of undesirable odours and flavours. The rate of the autolytic changes are determined by many factors with the most important being temperature, pH, availability of O_2 and the physiological condition of the fish before death.

As spoilage proceeds there is a gradual invasion of the flesh by bacteria from the outer surfaces. Because bacteria can generally only use very basic nutrients as food, bacterial spoilage does not usually commence in whole fish until autolysis is well advanced. Breakdown of the muscle structure only occurs after spoilage has proceeded well beyond the point of rejection. The development of objectionable slimes, odours and flavours results mainly from bacterial activity.

The major spoilage organisms found on spoiled fish include *Pseudomonas, Moraxella, Acinetobacter, Flavobacterium,* and *Cytophaga* species. From a microbiological safety standpoint, the organisms of greatest concern when dealing with MAP fish products are the nonproteolytic *C. botulinum* type E.

Chapter 19

Sterilization of Packaging Material

19.1 Introduction

Aseptic processing is a thermal process in which the product and container are sterilized separately and brought together in a sterile environment. It involves pumping, deaeration, and sterilization of a food product, followed by holding it for a specified period of time (in a holding tube — required to have a 1/4" rise per foot length of tube), cooling it, and finally packaging it in a sterile container. The use of high temperature for a short period of time (in comparison with conventional canning) in aseptic processing yields a high-quality product. Care should be taken to ensure that all process calculations are performed after the deaeration stage and not based on the initial raw product.

Year	*Landmarks in History of Sterilization of Food Packaging*
1917	Dunkley of U.S. — sterilization of cans and lids by saturated steam; fill pre-sterilized product
1933	American Can Company developed a filling machine called the heat-cool-fill (HCF) system which used saturated steam to sterilize the cans and lids; sealing was done in a closed chamber which was kept pressurized with steam
1942	Avoset process – steam injection of the product coupled with retort or hot air sterilization of packages such as cans and bottles
1948	Dole-Martin aseptic process – product sterilization in a tubular heat exchanger, metal container sterilization using superheated steam at temperatures as high as 450/F since dry heat requires higher temperature than wet heat, followed by aseptic filling and sealing of cooled product in a superheated steam environment

Year	Landmarks in History of Sterilization of Food Packaging
Early 60s	Form-fill-seal package — tetrahedron packageTetra Brick aseptic processing machine
1961:	Alpura and Tetra Pak (Sweden) developed an aseptic carton system for milk
Late 60s	Tetra Brick aseptic processing machine
	Combibloc (blank carton) aseptic system; aseptic filling in drums and bag-in-box fillers
1981	Hydrogen peroxide was permitted for sterilization of food packaging surfaces
1997	Tetra Pak received a no-objection letter from FDA for aseptic processing of a low acid product containing large particulates

The required count reduction (number of D values) for the sterilization of the food contact packaging material surface is determined by the type of product, its desired shelf life and likely storage temperature. For non-sterile acidic products of pH <4.5, a minimum of four decimal reductions (4D's) in bacterial spores is required. For sterile, neutral low acid products of pH >4.5, a six decimal reduction is required. However, if there is the possibility that *Clostridium botulinum* is able to grow in the product, then a full 12D process should be given.

The non-sterility rate or error rate E_r is the number of non-sterile or faulty packages as a proportion of the total number of packages processed over a given period. It can be calculated using the following equation:

$$E_r = \frac{N}{R} A$$

where N is the number of microorganisms on, and A the area of, the food contact surface of the packaging material and R is the count or number of decimal reductions obtained in the sterilization process. Smaller containers with a smaller food contact surface area will have correspondingly less initial contamination and a less severe sterilization process will be needed to give a certain non-sterility rate E. Conversely, larger containers will require a more severe sterilization process. However, since container volume varies with the cube of the linear dimensions while surface area varies only with the square of the dimensions, the variation in the sterilization process with container size is less than might be expected.

It has been suggested that only 3 per cent of the total number of microorganisms on the package surface are spores. An upper value of 1000 microorgarlisms m^{-2} (30 spores m^{-2}) has been assumed for plastic films and paperboard laminates on reels, and 3000 microorganisms m^{-2} (90 spores m^{-2}) for prefabricated cups. Using these values, the non-sterility rate for a variety of different package sizes and types has been calculated and is reproduced in Table 19.1.

Table 19.1: Non-Sterility Rate Er for different Package Types, Package Sizes and Count Reductions D

	From the Reel				*Prefabricated Plastic Cups*
	Carton Package			*Plastic Cups*	
	200 mL	*500 mL*	*1 L*	*15 mL*	*150 mL*
Spore Count N	*30*	*30*	*30*	*30*	*90*
Surface Area A	*0.025*	*0.045*	*0.075*	*0.0037*	*0.018*
$R = 10^3$					1.6×10^{-3}
$R = 10^4$	0.75×10^{-4}	1.4×10^{-4}	2.3×10^{-4}	1×10^{-4}	1.6×10^{-4}
$R = 10^5$	0.75×10^{-5}	1.4×10^{-5}	2.3×10^{-5}	1×10^{-6}	1.6×10^{-5}
$R = 10^6$	0.75×10^{-6}	1.4×10^{-6}	2.3×10^{-6}	1×10^{-7}	1.6×10^{-6}
$R = 10^7$	0.75×10^{-7}	1.4×10^{-7}	2.3×10^{-7}	1×10^{-8}	1.6×10^{-7}

In order to attain a non-sterility rate of 10^{-6} in a 1 liter paperboard package of UHT milk, the sterilization process would have to result in 6.4D; for smaller packages, 6 decimal reductions would suffice. Table 19.1 indicates that the smaller the package and therefore the smaller the food contact surface, the lower the non-sterility rate E_r resulting from the packaging material sterilization process.

19.2 Sterilization of Food Contact Surface of Packaging Material

Because of the importance of the initial level of microbial contamination of the packaging material, steps should be taken to ensure that it is as low as possible. Thus, it should be produced, transported and stored under conditions which are as free from microorganisms as possible.

Three main sterilization processes for packaging material are in common use, either individually or in combination: irradiation, heat and chemical treatments. These will each be considered in turn.

Aseptic filling systems provided that the irradiated materials are smooth, UV-resistant and free from dust particles to avoid a "shadowing" effect on the surface. As well, it is important that the irradiation intensity is uniform and adequate for sterilization over the whole of a container which may be of complex shape. It is generally only used commercially in combination with hydrogen peroxide.

Infra-red Radiation

Infrared is part of the electromagnetic spectrum that is located between the visible region and microwaves, and its wavelength ranges from 0.5 to 100 μm. Infrared penetration causes vibrating movement of water molecules at a frequency of 60,000–150,000 MHz and thereby, causes heating. The dissipation of radiant energy in the form of heat produces a warming of the surface of the food and the penetration depth depends on the thickness of the product, the water activity, the product components and the wavelength of the infrared radiation. Infra-red (IR) rays are converted into sensible heat when they contact an absorbent surface, resulting in an increase in the temperature of the surface. Infrared technology is characterized by high energy efficiency, less water consumption, and environmental friendliness. In addition, it is also characterized by the homogeneity of heating with high heat transfer rate, low heating time, low energy consumption, and improving product quality and food safety. It also features low energy costs and the size of infrared equipment is small as well as controlling factors with high accuracy As with UV irradiation, IR irradiation is only applicable on smooth, even surfaces. IR rays have been used to treat the interior of aluminum lids which have been coated with a plastic lacquer material. Because of the possible softening of the plastic lacquer, the maximum temperature must not exceed 140°C.

Ionizing Radiation

In the field of packaging, ionizing radiation is used to sterilize packaging materials that cannot withstand the high temperatures used for thermal sterilization. The advantage of irradiation is not only to eliminate and/or reduce harmful human pathogens, which helps to ensure food safety, but also to improve the quality and extend the storage life of foods. Sterilization of the packaging materials by ionizing radiation, hence, has become a more common treatment especially for aseptic and pharmaceutical products. This is because of its potential to provide benefits over other traditional methods such as heat and chemical treatments which ordinarily cause nutritional loss in foods, damage heat-sensitive products and some plastics, or leave a chemical residue.

Particle irradiation techniques using gamma rays from cobalt 60 or cesium 139 have been used to sterilize the interior of sealed but empty containers, especially those made of materials which cannot withstand the temperatures needed for thermal sterilization, or which, because of their shape, could not be conveniently sterilized by other means. The bags made from plastic laminates (typically having a molded plastic filling connection) for use in aseptic bag-in-box systems are a good example of the latter category. They are given a radiation dose of 25 kGy (2.5 Mrad) or more which is sufficient to ensure sterility. They are sealed into microbial-proof containers prior to the irradiation treatment. A 20 kGy dose of high-energy electrons was found to sterilize 9 mm of a polyethylene strip infected with approximately 10_5 spores of *B. stearothennophilus.*

It is also possible to use low-energy (100 keV), large area electron beams for the surface sterilization of packaging materials and containers.

Saturated Steam

Saturated steam is a form of most heat. Sterilization of metal cans and lids by saturated steam under pressure was used as early as 1920 in the USA and is used today for sterilizing thermostable plastic cups. To attain adequate destruction in the short time available in high-speed packaging units, the surface temperature of the material must reach 135°C. Saturated steam is a preferred method for sterilizing metal food contact surfaces, downstream from the hold tube, including sterile hold tank, homogenizers, fillers and the aseptic packaging zone.

The most reliable sterilant is without doubt wet heat in the form of saturated steam. However, there are three major problems in the use of saturated steam. The first is that, in order to reach temperatures sufficiently high to achieve sterilization in a few seconds, the steam (and thus, the packaging material with which it comes into contact) must be under pressure, necessitating the use of a pressure chamber. Secondly, any air which enters the pressure chamber with the packaging material must be removed otherwise it would interfere with the transfer of heat from the steam to the package surface, Thirdly, condensation of steam during heating of the packaging material surface produces condensate which may remain in the container and dilute the product.

Despite these problems, saturated steam under pressure is being used to sterilize plastic containers. For example, immediately after deep-drawing, molded plastic cups of polystyrene and the lid foil are subjected to steam at 165°C and 600 kPa for 1.4 s (cups) and 1.8 s (lids). In order to limit the heating effect to the internal surface of the cups, the exterior of the cups is simultaneously cooled. This process has been shown to achieve a 5–6D reduction in *B. subtilis* spores.

Superheated Steam

Dry heat delivered by superheated steam or hot air is not as effective as moist heat at the same temperature for sterilization, therefore higher temperatures are required. Superheated steam was the method used in the 1950s for the sterilization of tinplate and aluminum cans and lids in the Martin-Dole aseptic canning process. The cans are passed continuously at normal pressure under saturated steam at 220–226°C for times of 36–45 s depending on the material of construction, aluminum cans having a shorter heating time because of their higher thermal conductivity.

When paper-based packaging materials are used, hot air is preferred as a sterilant over superheated steam. Combicans of cardboard laminates (*i.e.*, aluminium and paper) are sterilized in hot air at 145°C for 3 min. This method has only been found suitable for products that have a pH of less than <4.6.

When sterilizing by hot air or superheated steam, there is no effect on packaging material quality. Furthermore, the temperature within the entire aseptic filling chamber may be maintained at a level lethal to microorganisms, thus assuring asepticity. The high temperature of the headspace gas at the time of sealing also reduces the level of oxygen in the sealed package and can contribute towards prolonging product shelf life. Hot air sterilization units are commercially available for juice and beverages. Dry heat sterilization is slow and not suitable for heat-sensitive materials like many plastic items.

Hot Air

Dry heat in the form of hot air has the advantage that high temperatures can be reached at atmospheric pressure, thus simplifying the mechanical design problems for a container sterilization system. Hot air sterilization units are commercially available for juice and beverages. Dry heat sterilization is slow and not suitable for heat-sensitive materials like many plastic items.

Hot Air and Steam

A mixture of hot air and steam has been used to sterilize the inner surfaces of cups and lids made from polypropylene which is thermally stable up to 160°C. In this process, hot air is blown into the cups through a nozzle in such a way that the base and walls of the cup are uniformly heated.

Extrusion

Aseptic packaging system utilizing heat coextrusion process for sterilization is a form–fill–seal packaging system that relies on the temperatures reached by thermoplastic resin, during the co-extrusion process used to produce multi-layer packaging material, to produce a sterile product contact surface. During the extrusion of plastic granules prior to blow molding of plastic containers, temperatures of 180–230°C are reached for up to 3 min. However, because the temperature distribution inside the extruder is not uniform and the residence time of the plastic granules varies considerably, it is not possible to guarantee that all particles will achieve the minimum temperature and residence time necessary to result in sterility. The high temperature produced by extrusion produces sterile surfaces.

During production, the multi-layer package material is fed into the machine where it is delaminated under sterile conditions. This removes a layer of material and exposes the sterile product contact surface. The container material is then thermoformed into cups. The lid material that is also delaminated is then sealed onto the cup after filling. The sterility of the forming, filling and sealing areas is maintained by sterile air under positive pressures. Reports as to the efficacy of this heating method vary considerably. Therefore, aseptic filling into extruded containers appears only suitable for acidic products with pH less than 4.6. For products with pH greater than 4.6, extruded containers should be post-sterilized with hydrogen peroxide or with mixtures of peracetic acid. It would be prudent to validate any application of this method on a case-by-case basis.

Chemical Treatment

Hydrogen Peroxide

Hydrogen peroxide is the overwhelming choice for use as a chemical sterilant. Other chemicals which have been used as sterilants, primarily for use in systems for acid food, include various acids, ethanol, ethylene oxide and peracetic acid. Hydrogen peroxide is not an efficient sporicide when used at room temperature. However, the sporicidal activity increases substantially with increasing temperatures. Therefore, most aseptic packaging systems use hydrogen peroxide (at concentrations of 30 to 35 per cent) as a sterilant for packaging materials followed by hot air (60°C to 125°C) to dissipate residual hydrogen peroxide.

Overall, processing includes dipping, spraying, rinsing processes, or combined with UV-C or heat; At least 80° C and 30 per cent concentration is required; residual peroxide should be less than 100 ppb at time of filling and must decrease to 1 ppb within 24 hrs. Since it is hard to detect peroxide in foods, containers filled with water are run through the machine, initially.

Peracetic Acid

Produced by oxidation of acetic acid by hydrogen peroxide; effective even at 20/C (1 per cent solution will eliminate 7-8 logs of the most resistant spores in 5 min at 20°C; maximum usable temperature is 40°C).

Ethylene Oxide

It is a toxic gas and can penetrate porous materials; thus, used for presterilization of paperboard-based packaging materials (particularly preformed carton blanks which are to be assembled in an aseptic filler).

19.3 Verification of Sterilization

This is done by inoculation of the surface of the web, cup, or lid stock with the proper concentration of the test organism and allowing it to dry. The system is then run as in a commercial run and the finished containers are filled with an appropriate growth medium and observed for growth. Two of the most important factors affecting the success of the tests are the choice of the indicator organism and the physical state of the microorganisms used. The indicator organisms used are: B. stearothermophillus — strain 1518 (Superheated steam, peroxide + steam, extrusion), *B. polymyxa* — PSO (dry heat), *B. subtilis* strain A (peroxide + UV), *C. sporogenes* — PA 3679 (ethylene oxide), and *B. pumilus* (gamma radiation).

19.4 Types of Packaging Systems

1. The Cup Systems Used in Aseptic Filling

a) Performed Plastic Cups

This is one of the most common types of container for aseptic filling. Although the cups are usually made from high impact polystyrene (HIPS) or polypropylene lit, they can be made from coextruded, multilayered material if improved barrier properties are required. A typical example of the latter cup would be an outer layer of HIPS, then a laminating adhesive; a barrier layer of PVC/PVdC or EVOH copolymer; a laminating adhesive; and finally, LDPE. Such a construction with a PVC/PVdC copolymer barrier layer is claimed to give a non-refrigerated shelf life for pudding of more than twelvemonths.

The cups are fed onto a conveyor which is inside a sterile tunnel supplied with sterile air. The cups are sprayed inside with 35 per cent H_2O_2 solution (35 mg for a 250 mL cup size) and after about 3 s the solution is removed with compressed hot air at a maximum temperature of about 70°C, depending on the material from which the cups are made. The inside surface of the cups reaches a temperature of about 70°C which completes surface sterilization and reduces the peroxide residues to acceptable levels.

Cups can also be sterilized by carrying them through a 35 per cent peroxide bath at 85–90°C before heating and passing through a water bath. The cups then enter a sterile chamber where sterilization is completed by spraying with sterile water and drying with hot air.

The sealing material (usually aluminum foil with a thin coating of a thermoplastic to provide sealability) is typically sterilized with a 35 per cent peroxide solution which is then removed either by radiant heat, hot sterile air or by passing the material over a heated roller. In some systems UV radiation is used, either alone or in conjunction with peroxide.

b) Form-Fill-Seal Cups

The plastic material (commonly polystyrene because it is easily thermoformed) in the form of a web is fed from a roll into a thermoformer to give multiple containers still in web form. More complex coextruded multilayer materials incorporating a barrier layer of either PVC/PVdC or EVOH copolymer can also be treated in this way. Sterilization of the web is done prior to forming using a hydrogen peroxide bath. It then passes through a tunnel where it is heated to 130-150/C to prepare

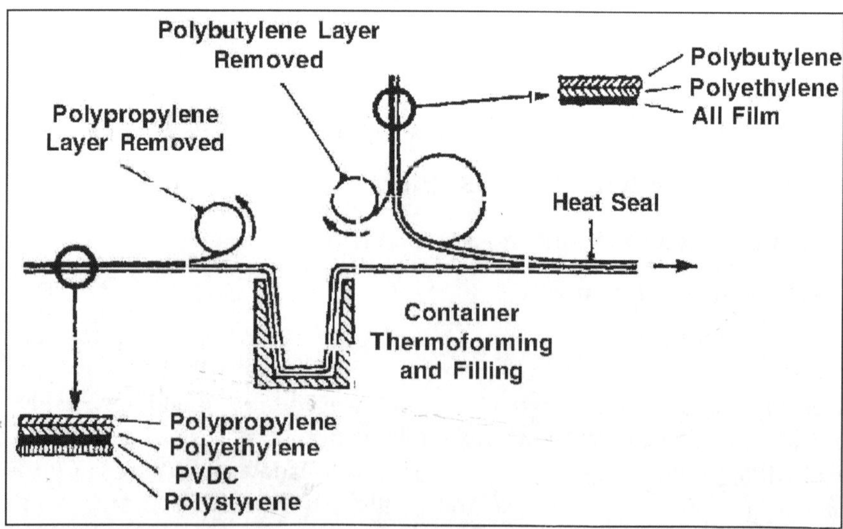

Figure 19.1: Schematic Representation of a Form-fill-seal System Using Peelable Coextruded Multilayer Films.

it for thermoforming. If an aluminum foil layer is incorporated into the laminate, mechanical forming rather than thermoforming is used. Mechanical force and compressed air are used to form the container in a water-cooled mold below the web.

The advantages in thermoforming cups from a reel compared to using premade cups include a favourable price ratio; simplified handling since the constant reloading of magazines is avoided; higher output by utilizing multiple tools; lower storage volume requirements for the packaging materials; and maximum sterility of the cups (both inside and outside surfaces) as well as the lidding material by running flat material through sterilizing baths.

2. The Carton Material

The carton material normally consists of layers of bleached and unbleached paperboard coated internally and externally with polyethylene, resulting in a carton which is impermeable to liquids and in which the internal and external surfaces may be heat sealed. There is also a thin layer of aluminum foil which acts as an oxygen barrier. The structure of a typical paperboard carton is shown in Figure 19.2. The functions of the various layers have been described as follows:

1. The outer polyethylene protects the ink layer and enables the package flaps to be sealed;
2. The bleached paperboard serves as a carrier of the decor and gives the package the required mechanical rigidity;
3. The laminated polyethylene binds the aluminum to the paperboard;
4. The aluminum foil acts as a gas barrier and provides protection of the product from light;
5. The two inner polyethylene layers provide a liquid barrier.

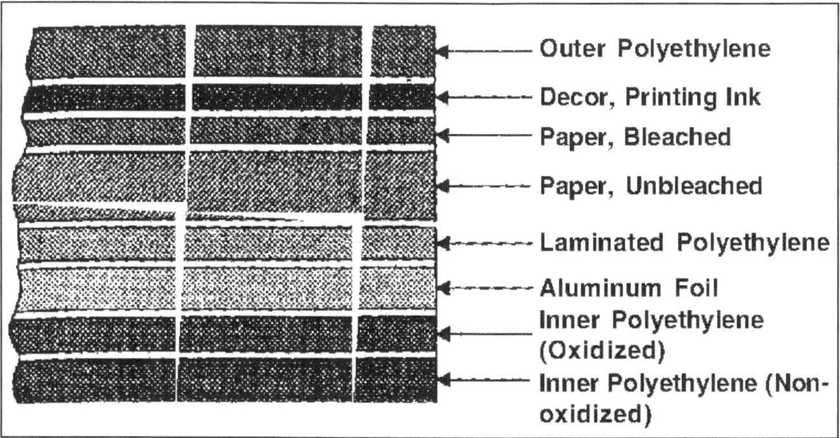

Figure 19.2: Typical Structure of a Paperboard Laminate Carton for Aseptic Filling.

a) Form-Fill-Seat Cartons

The packaging material is supplied in rolls which have been printed and creased, the latter being necessary to ease the forming process. A polyethylene strip is sealed to one edge (the reason for this is described later) and the packaging material sterilized using one of two procedures: a wetting system or a deep bath system.

In the wetting system, a thin H_2O_2 film (15–35 per cent concentration) containing a wetting agent to improve the formation of a liquid film is applied on the inner packaging material surface. The material then passes through a pair of rollers to remove excess liquid and is formed and sealed into a cylinder. It then passes under a tubular electric heater which heats the inside surface to about 120°C and evaporates the H_2O_2.

In the deep bath system, the packaging material is fed through a deep bath containing H_2O_2 (35 per cent concentration) at 78°C, the residence time being 6 s. After squeezer rollers have removed much of the peroxide, air at 125°C is directed through nozzles onto both sides of the material to heat it to increase the sterilizing effect and to evaporate the peroxide.

The sterilized packaging material is fed into a machine where it is formed into a tube and closed at the longitudinal seal by a heat-sealing element. In the process the strip of polyethylene which was added prior to sterilization is heat sealed across the inner surface of the longitudinal seal. It provides protection of the aluminum and paperboard layers from the product which could corrode or swell the layers if such a strip were absent. Product is then filled into the tube and a transverse seal made below the level of the product, thus ensuring that the package is completely filled. Alternatively, the packages may be produced with a headspace of up to 30 per cent of the total filling volume by the injection of either sterile air or other inert gases. The sterilization, filling and sealing processes are all performed inside a chamber maintained at an over-pressure of 0.5 atmos with sterile air.

The sealed packages are then pressed by molds into rectangular blocks, after which the top and bottom flaps or wings are folded down and heat sealed to the body of the package using electrically heated air.

b) Prefabricated Cartons

In systems of this type, prefabricated carton blanks are used, the cartons being die-cut, creased and the longitudinal seam completed at the factory of origin. The cartons are delivered to the processor in lay-flat form ready to be finally shaped in the filler and the top and bottom seams formed and bonded. Advantages cited for prefabrication include that the fact that the critical longitudinal seam can be flame welded (thus, ensuring good integrity of the seal), and that the same machine can be use used to produce different capacity cartons (provided that the same cross section is used) simply by adjusting the height.

The partly assembled lay-flat blanks are shipped in boxes to the processor, ready for loading into the filler. Stacks of blanks are loaded into a magazine from which they are individually, removed by suction pads opened up into a rectangle and placed on a mandrel. The polyethylene sealing surface at the bottom of the carton is softened by blasts of hot air. As the mandrel wheel turns on towards the bottom pressing station, the bottom is folded by means of rotary transverse and longitudinal folders. The bottom is then pressed and sealed against the end face of the mandrel, after which the carton is transferred to a pocket on the conveyor where the top is prefolded. Although all of the above operations take place under nonsterile conditions, steps are taken to avoid recontamination.

The carton then passes to the aseptic area which consists of several separate functional zones where operations are carried out in sequence. Sterility is maintained in each zone by a slight overpressure of sterile air. The inside surface of the carton is sterilized with a 35 per cent solution of H_2O_2, delivered either as a fine spray or as peroxide vapor in hot air so that the vapor condenses as liquid peroxide on the carton surface. The peroxide is removed by a jet of hot air at 170–200°C. Alternatively, the inside of the carton can be sprayed uniformly with a 1–2 per cent solution of H_2O_2 and then irradiated for about 10 s with high intensity UV radiation. The peroxide is then heated and removed by hot-air jets. Because the total quantity of peroxide used in this latter process is 20–30 times less than in the former, the problems of residual peroxide in the carton and peroxide contamination in the surrounding atmosphere are more easily dealt with.

The next stage of the process is filling. A certain amount of headspace is always advisable to ensure that the package can be opened and poured without spilling. When the contents require shaking (as is the case with flavored milk drinks and pulpy fruit juices), a headspace is essential. For products such as fruit juice it is advantageous to fill the area between the product and the top of the package with steam or an inert gas such as nitrogen. If steam is used, the headspace volume is reduced as a result of a condensation of the steam. A headspace is also crucial when it is not possible to seal the filled package through the product such as when it contains particulate solids. A headspace ensures that sealing can occur above the product line, thus preventing solid material from getting caught up in the top seam where its presence would lead to loss of sterility.

After filling, the carton top is folded and closed, the seal being made either by heating or ultrasonic welding. Production and date codes are added afterwards by ink jet printing or burning into the top seam. Although conventional cartons are of the "gable-top" type, cartons of the brick type can also be produced using basically similar equipment, with the additional step that the gables resulting from the top seam sealing are formed into a flat top. The protruding flaps or "ears" on each side are folded down and sealed to the package with hot air. The finished cartons are then discharged from the pocket chain onto a conveyor belt, ready for the final packaging process.

3. Sachet and Pouch Systems

a) Form-Fill-Seal Systems

A form-fill-sealing system is a self-contained packaging and sealing machine. It takes a V-fold or flat roll of material, forms it into a bag, fills it with the product and then seals the bag closed. Pillow pouches are usually used for packaging of milk; three-sided sealed pouch, however, is suitable also for aseptic packaging of particulates up to particle sizes of 12µ and bag sizes from 1-5 litres. A vertical form-fill-seal machine operates in a sterile chamber. The packaging material is passed through hydrogen peroxide and then drained and dried.

For standing pouches, a Japanese machine uses closed pouches from a reel with sterile interior surfaces, the exterior of which is sterilized in a hydrogen peroxide bath when the web with pouches enters the aseptic cabinet. The bags are then cut from the web, filled and sealed.

b) Layflat Tubing

This system uses a blown film polymer in the form of a layflat tubing so that only a transverse seal is required to form the bag. It is assumed that the inside of the tubing is sterile due to the temperature achieved during the extrusion process. The tubing is fed from the reel into a sterile chamber in which an over-pressure of air is maintained. The sachets are sealed at the bottom, cut, and moved into a filling station. After filling, they are sealed at the top and leave the chamber through a water seal.

4. Can Systems

Pioneered by Martin in the late 1940s; first system commissioned by Dole Corp. (CA) in 1950 for soups; uses superheated steam at 225°C for up to 40 s to sterilize can and ends; temperature should not exceed 232°C since tin flow underneath the enamel can occur, resulting in blister formation. During seaming of lid (the lining compound is still at ~ 220°C and is plastic; thus, seamed can should be transported in vertical position for at least 15 s after seaming to allow compound to settle and hermetically seal the can). For composite cans consisting of spirally wound body made from laminations of foil, plastics, and paper with metal ends, hot air at 143°C for 3 min. is used to sterilize packaging materials. Steam would cause swelling of the paper layers.

5. Bottle Systems

a) Glass and Jars

Glass bottles may be aseptically filled with food containing small particles, for instance for baby food. Jars may be filled with larger particles - 12mm cube size or larger - if one dimension is smaller. In a recent development, returnable bottles are filled aseptically, which up to now were applied only for UHT – treated milk. Saturated steam or dry heat (when dry heat is used, extended cooling by sterile air is required to minimize thermal shock when cool product is filled in it; no commercial unit, yet).

Basically, the same products can be filled into plastic bottles and jars as into glass containers. Closing is usually done by heat-sealing aluminium lids. For this reason, much attention has to be paid to avoid contamination of heat-sealing rims

b) Plastic

Plastic Cans: An aseptic machine for filling and closing of two-piece plastic cans, 'gourmet cans', was recently developed. Cans and lids with easy opening feature consist of PP/EVOH/PP. They are sterilized with hydrogen peroxide, UV radiation and heat-sealed inductively. The can is presently offered for liquids only for example coffee.

Non-Sterile Bottles: After blowing, the plastic bottles are conveyed into a sterile chamber which is kept at a slight over-pressure of sterile air. The bottles are inverted and sprayed inside with hydrogen peroxide and passed through a hot air tunnel to evaporate the residual peroxide. The bottles are rinsed with sterile water and then filled. A chemically sterilized, heat sealable closure, such as a plastic film or cap is then applied.

Sterile blown bottles: Bottles are extruded, blown with sterile air, and sealed under conditions that ensure internal sterility of the container. The sealed bottles are introduced into a sterile chamber (maintained at a slight positive pressure) where the outside surfaces are sterilized by hydrogen peroxide sprays. The closed top of the bottle is cut away, the neck trimmed, the bottle filled, and a foil cap or heat sealable sterile closure applied.

Single station blowing, filling, and sealing: This is a complex system. The separate operations of parison extrusion, blow molding, bottle filling, and sealing all take place in sequence in a single mold. Sterility of the inside surface of the container is ensured by the high temperature (164 to 234 °C) of the plastic material during extrusion of the parison, and the use of sterile air for blowing. After filling, the tube projecting from the bottle mold is vacuum-formed or sealed with jaws into a cap which closes the bottle. No special arrangements to ensure sterility are required since filling and sealing are carried out within the closed mold.

6. Bulk Packaging Systems

a) Metal Drum

Two major systems are in use and both use a 55 gallon metal drum constructed from steel with an electrolytically coated tin lining outside. The ends are double-seamed onto the body of the drum during manufacture and filling takes place through a threaded hole in which a cap is swaged after filling.

b) Bag-in-Box

In this system, the product is filled into a plastic bag which when full is put into an outer container such as a drum or a paperboard box. For large containers, filling occurs after the bag is placed in the box.

Chapter 20

Shelf Life of Packaged Food

20.1 Quality of Food

The demand for high quality food has constantly increased during recent decades, as has the interest in the food quality issue both in response to market pressure (such as requests from increasingly demanding and knowledgeable consumers) and also as a reaction to other factors, for example health and environmental concerns. The consumer of the 21st Century is a highly demanding one, exhibiting greater concern about quality, safety and health benefits with respect to products he/she buys.

Food quality can be defined as combinations of attributes or characteristics of a product that have significance in determining the degree of acceptability of that product to the consumer. In other words, it is a measure of the degree of excellence, a high standard or degree of acceptability by the consumer. It can also be defined as "summary of features and characteristics of a product or service that bear on its ability to satisfy stated or implied needs". In simple words, the product should have attributes to "satisfy the wants/needs of the consumer or conformance with the user's requirements". It also covers the safety and value for money.

The important components of food quality include food safety, sensory characteristics and nutritional value. Safety of food is a basic requirement of food quality. "Food safety" implies absence or acceptable and safe levels of contaminants, adulterants, naturally occurring toxins or any other substances that may make food injurious to health on an acute or chronic basis. Besides safety, quality attributes include nutritional value; organoleptic properties such as appearance, colour, texture, taste; and functional properties. Hence, food quality includes external factors as appearance, texture, and flavour; factors such as federal grade standards and internal. To have a quality food product, it is essential that besides ingredient quality, the manufacturer should ensure that the food processing environment is as clean as possible in order to produce the safest possible food for the consumer. Shelf life of the product should also be good.

20.2 Shelf Life of Food Products

Shelf life has a number of connotations and definitions. The National Food Processors Association in the USA recommended in 1978 as a working professional definition for internal industry use the following definition: "A product is within its shelf life when it is neither misbranded nor adulterated; when the product quality is generally accepted for its purported use by a consumer; and so long as the container retains its integrity with respect to leakage and protection of the contents."

The Institute of Food Technologists in the USA has defined shelf life as the period between the manufacture and the retail purchase of a food product, during which time the product is in a state of satisfactory quality in terms of nutritional value, taste, texture and appearance. An alternative definition is that shelf life is the duration of that period between the packing of a product and its use, for which the quality of the product remains acceptable to the product user.

To the general consumer, shelf life is the time limit, given typically on the best-before label, of how long the particular food can be kept before it should be thrown away or put another way, shelf life represents the length of time before the food is considered to be unsuitable for human consumption.

The shelf life of a food can be defined as the time, under defined storage conditions, during which food remains safe, retains desired sensory, chemical, physical and biological characteristics and acceptable quality to the consumers as well as complies with any label declaration.

The consumer can generally only judge the quality of the product at the time of consumption, not at the time of purchase. Thus, the shelf life refers to the time for which a food can remain on both the retailer's and consumer's shelf before it becomes unacceptable.

Generally, the quality of foods and beverages decreases with storage or holding time. Exceptions include distilled spirits (particularly whiskeys and brandies) which develop desirable flavor components during storage in wooden barrels, some wines which undergo increases in flavor complexity during storage in glass bottles, and many cheese varieties where enzymic degradation of proteins and carbohydrates, together with hydrolysis of fat and secondary chemical reactions lead to desirable flavours and textures in the aged cheeses.

However, for the majority of foods and beverages in which the quality decreases with time, it follows that there will be a finite length of time before the product becomes unacceptable. This time from production to unacceptability is referred to as shelf life.

Shelf life depends on the degradation mechanism of the specific food product. Most can be influenced by several factors: exposure to light, heat, and moisture; transmission of gases; mechanical stresses; and contamination by things such as microorganisms. Oxidative reactions are the most frequent event leading to the end of the shelf life of microbiologically stable foods, such as ambient stable and frozen foods. Since quality is a dynamic condition, continuously moving to lower levels, the shelf life of these products depends on the oxidation rate and the acceptability limit. The latter is the oxidation level above which the product becomes unsuitable for consumption.

Just like any other food, frozen foods deteriorate during storage by different modes or mechanisms.

Freezing foods prevents spoilage and preserves taste, but some frozen foods have a specific and defined shelf life. Foods that have exceeded their freezer shelf life are usually not dangerous, but they lack taste, nutrients, or both, or exhibit an unnatural texture when defrosted.

Shelf life after chill storage of perishable produce is often variable as it depends as much on the state of the product at harvest, the storage conditions utilized (*e.g.*, air or controlled atmosphere), packaging, length of storage, and the environmental conditions during the shelf-life period.

Storage studies are an essential part of food product development, with the manufacturer attempting to provide the longest shelf life practicable consistent with costs and the pattern of handling and use by distributors, retailers and consumers. Inadequate shelf life will often lead to consumer dissatisfaction and complaints. At best, such dissatisfaction will eventually affect the acceptance and sales of brand name products, while at worst it can lead to malnutrition or even illness. Therefore, food processors give considerable attention to determining the shelf lives of their products.

Although the shelf lives of foods vary, they are generally determined routinely for each particular product by the manufacturer or processor. Product shelf life must be set to ensure that this is the case over the entire product life, allowing sufficient product life for the distribution chain and retail turnover of product and some life for the consumer. Manufacturers, who are responsible for setting product shelf life must be able to justify the validity of the shelf life assigned. They are also under considerable pressure to produce products to meet the shelf life requirements of retailers and often this will dictate whether or not a product is stocked. Achieving the desired product shelf life is a powerful driver for product and packaging innovation to extend product life. Many new packaging materials such as modified atmosphere packaging (MAP) have been developed to complement developments in new preservation techniques. The role of packaging in the maintenance and extension of shelf life cannot be over emphasised.

It has been standard practice in almost all food manufacturing and processing establishments to put a "closed" code (so called because only those with knowledge of the coding system can interpret the code) onto the packaged product, sometimes on the primary package, but frequently on the secondary package. Generally, this code indicates the time of processing and packaging, *e.g.*, day and year, or shift, day and year, or hour, day and year. For many canned foods, it has long been mandatory to include such information on the end of the container in an embossed form, together with a code for the product itself to aid identification of unlabelled cans in the factory.

Since the advent of the consumer movement in the early 1970s, many different types of open dating systems have been proposed as part of the consumer's "right to know." An open date on a food product is a legible, easily read date which is displayed on the package with the purpose of informing the consumer about the shelf life of the product. Several types of dates can be used.

Pack Date: The date on which the product was packed into its primary package, *i.e.,* the immediate container in which it is sold. It does not provide any specific information as to the quality of the product when purchased or how long it will retain its quality after purchase.

Display Date: The date on which the product was placed on the shelf by the retailer.

Pull Date or Sell by Date: The last date on which the product should be sold in order to allow the consumer a reasonable length of time in which to use it.

Best Before or Best if Used by Date: The last date of maximum high quality.

Use by Date or Expiration Date: The date after which the food is no longer at an acceptable level of quality.

This form of dating is used infrequently because quality changes generally occur slowly and it is simply not possible to state that a food will be acceptable one day and unacceptable the next. An exception is with baked goods such as bread.

There is no uniform or universally accepted open dating system used for food throughout the world. In some countries, mandatory open dating of all perishable and sometimes semi perishable foods is required, while in other countries such requirements are voluntary. Arguments can be advanced both for and against the open dating of foods. However, there has been an obvious increase in the quantity of open dated food on sale throughout the world, and this trend is likely to continue.

The European Community (EC) has adopted a "minimum durability date". This is the length of time for which the product may be stored (the storage conditions may be specified, *e.g.,* temperature not to exceed 7°C) before its quality has deteriorated to such an extent that it would be regarded as unacceptable by the manufacturer or the consumer. This concept (basically equivalent to the "best before" date defined above) allows the processor to set the quality standard of the food, since the product would still be acceptable to many consumers after the "best before" date had passed.

20.3 Factors Affecting Product Quality and Shelf Life

For many foods, the product shelf life is limited by specific or key attributes that can be predicted at the time of product development. This is either on the basis of experience with similar products or observations of them, or from a consideration of:

- The make-up of the product (intrinsic factors)
- The environment that it will encounter during its life (extrinsic factors)
- And the shelf life limiting processes that this combination of intrinsic and extrinsic factors is likely to result in.

Intrinsic factors are the properties resulting from the make-up of the final product and include the following:

- Water activity (a_w) (available water)
- pH/total acidity; type of acid
- Natural microflora and surviving microbiological counts in final product
- Availability of oxygen
- Redox potential (Eh)
- Natural biochemistry/chemistry of the product
- Added preservatives (*e.g.*, salt, spices, antioxidant
- Product formulation
- Packaging interactions (*e.g.*, tin pickup, migration).

Selection of raw materials is important for controlling intrinsic factors, since subsequent processing can rarely compensate for poor quality raw materials.

Extrinsic factors are a result of the environment that the product encounters during life and include the following:

- Time–temperature profile during processing
- Temperature control during storage and distribution
- Relative humidity (RH) during storage and distribution
- Exposure to light (UV and IR) during storage and distribution
- Composition of gas atmosphere within packaging
- Consumer handling

20.4 The Alteration of Shelf Life

The alteration of product shelf life is controlled by following factors:

1. Product characteristics;

2. The environment to which the product is exposed during distribution; and
3. The properties of the package.

The shelf life of a product will be altered by changing its composition and form, the environment to which it is exposed, or the packaging system.

Based on the nature of the changes which can occur during storage, foods may be divided into three categories–perishable, semi-perishable and non-perishable or shelf stable, which translate into very short shelf life products, short to medium shelf life products, and medium to long shelf life products.

Perishable foods are those that must be held at chill or freezer temperatures (*i.e.*, 0 to 7°C or –12 to –18°C, respectively) if they are to be kept for more than short periods. Examples of such foods would include milk, fresh flesh foods such as meat, poultry and fish, and many fresh fruit and vegetables.

Semi-perishable foods are those which contain natural inhibitors (*e.g.*, some cheeses, root vegetables and eggs) or those that have received some type of mild preservation treatment (*e.g.*, pasteurization of milk, smoking of hams and pickling of vegetables) which produces greater tolerance to environmental conditions and abuse during distribution and handling.

Shelf stable foods are considered "non-perishable" at room temperatures. Many unprocessed foods fall into this category, and are unaffected by microorganisms because of their low moisture content (*e.g.*, cereal grains and nuts, and some confectionery products). Processed food products can be shelf stable if they are preserved by heat sterilization (*e.g.*, canned foods), contain preservatives (*e.g.*, soft drinks), are formulated as dry mixes (*e.g.*, cake mixes) or processed to reduce their water content (*e.g.*, raisins or crackers). However, shelf stable foods only retain this status if the integrity of the package which contains them remains intact. Even then, their shelf life is finite due to deteriorative chemical reactions which proceed at room temperature independently of the nature of the packaging, and the permeation through high barrier polymer packages of gases, odours and water vapor.

Bulk Density

The free space volume of a package (V) is directly related to the bulk density (p_b) and the true density (p_p) of the product as follows:

$$V = Vt - Vp = \frac{W}{p_b} - \frac{W}{p_p}$$

where, V_t = total volume of the package; V_p = volume of the product, W = weight of the product.

Thus, for packages of similar shape, equal weights of products of different bulk densities will have different free space volumes, and as a consequence, package areas and package behaviour will differ. This has important implications when changes are made in package size for the same product, or alterations are made to the process, resulting in changes to the product bulk density.

While the true density of a food depends largely on its composition and cannot be changed significantly, the bulk density of food powders can be affected by processing and packaging. Some food powders (*e.g.*, milk and coffee) are instantized by treating individual particles so that they form free-flowing agglomerates or aggregates in which there are relatively few points of contact; the surface of each particle is thus more easily wetted when the powder is rehydrated. Instantization results in a reduction of bulk density, *e.g.*, for skim milk powder from 0.64 to 0.55 g mL^{-1}. A wide range of bulk densities is encountered in food products, from around 0.056 g mL^{-1} for potato chips to 0.96 for granulated salt.

The free space volume has an important influence on the rate of oxidation of foods, for if a food is packaged in air, a large free space volume is undesirable since it constitutes a large oxygen reservoir. Conversely, if the product is packaged in an inert gas, a large free space volume acts as a large "sink" to minimize the effects of oxygen transferred through the film. It follows that a large package area and a low bulk density result in greater oxygen transmission.

In many foods such as those containing whole tissue components, or where the reacting species are partially bound as in membranes, structural proteins or carbohydrates, the concentration varies from one point to another, even at zero time. In other words, there are initial inhomogeneities and discontinuities for each compound. Furthermore, since most of these compounds will have little opportunity to move, the concentration differences will get greater as the reactions proceed out from isolated initial foci. This has been described as the "brush-fire" effect and is especially important in chain reactions such as oxidation.

As well, there may be several different stages of the deteriorative reaction proceeding at once, and the different stages may have different dependence on concentration and temperature, giving disguised kinetics. Such a situation is frequently the case for chain reactions and microbial growth which have both a lag and a log phase with very different rate constants.

The point to be taken from this is that for many foods it may be difficult to obtain kinetic data of use for predictive purposes. In such situations, use of sensory panels to determine the acceptability of the food is the recommended procedure.

20.5 The Transfer of Mass and Heat through the Package

The deterioration in product quality of packaged foods is often closely related to the transfer of mass and heat through the package. Packaged foods may lose or gain moisture; they will also reflect the temperature of their environment, since very few food packages are good insulators. Thus, the climatic conditions (*i.e.*, temperature and humidity) of the distribution environment have an important influence on the rate of deterioration of packaged foods.

In the case of mass transfer, it is the exchange of vapours and gases with the surrounding atmosphere which is of primary concern. Moisture vapor and oxygen are generally of most importance, although it may be necessary to consider the exchange of volatile aromas from the product or to the product from the surroundings. As well as oxygen, transmission of nitrogen and carbon dioxide may have to be taken into account in packages where the concentration of these gases inside the package has been modified from ambient to inhibit or slow down deteriorative reactions in the food.

Generally, the difference in partial pressure of the vapor or gas across the package barrier will control the rate and extent of permeation, although transfer can also occur due to the presence of pinholes in the material, channels in seals and closures, or cracks which result from flexing of the packaging material during filling and subsequent handling. Since the gaseous composition of the earth's atmosphere is essentially constant at sea level, the partial pressure difference of gases across the package barrier will depend on the internal atmosphere of the package at the time the package was sealed.

In contrast to the common gases, the partial pressure of water vapor in the atmosphere varies continuously, although the variation is generally much less in controlled climate stores. The magnitudes of these variations in external climate are detailed in, for example, British Standard BS 4672, and in various publications of the U.S. Army Natick Laboratories.

The general pattern is for temperature to rise and relative humidity to fall towards midday, and for relative humidity to rise and temperature to fall through the afternoon and evening. Obviously, this pattern would be modified by rainfall and vary with the seasons. The detailed climatic statistics of global maximum and mean temperatures which are available in many countries are of great assistance. Despite meteorological and secular trends, the daily and even the annual cycle of temperatures can be normalized to a standard cycle with a standard frequency distribution derived from the mean and range at many places. This is because of the sinusoidal trend of diurnal (earth's rotation) and seasonal (earth's revolution) solar radiation intensity. To summarize, mass transfer depends on the partial pressure difference across the package barrier, and on the nature of the barrier itself.

One of the major determinants of product shelf life is the temperature to which the product is exposed during its life time from production to consumption. Without exception, food products are exposed to fluctuating temperature environments during this time, and it is important (if an accurate estimate of shelf life is to be made) that the nature and extent of these temperature fluctuations is known. There is little point in carefully controlling the processing conditions inside the factory and then releasing the product into the distribution and retail system without some knowledge of the conditions which it will experience in that system. Such knowledge is essential in the case of products containing a "best before" or "use by" date.

The storage climates inside buildings such as warehouses and supermarkets are only broadly related to the external climate as reported by weather stations, and climatic variations in temperature and humidity can differ as much between different building constructions as between seasons on one site. Data for temperature in six major cities in the United States has also been presented.

If the major deteriorative reaction causing end of shelf life is known, simple, expressions can be derived to predict the extent of deterioration as a function of available time-temperature storage conditions. The basic types of deteriorative reactions which foods undergo were discussed later, together with the rates of these reactions and the factors controlling these rates. These reactions and their rates will now be analyzed in relation to product shelf life.

Fundamental to such an analysis is that the particular food under consideration follows the "laws" of "additivity" and "commutativity". Additivity implies that the total extent of the degradation reaction in the food produced by a succession of exposures at various temperatures is the simple sum of the separate amounts of degradation, regardless of the number or spacing of each time-temperature combination. Commutativity means that the total extent of the degradation reaction in the food is independent of the order of presentation of the various time-temperature experiences.

20.6 Shelf Life Testing of Food Products

The shelf life testing of food products can be studied under three categories:

1. Experiments designed to determine the shelf life of existing products,
2. Experiments designed to study the effect of specific factors and combinations of factors such as storage temperature, packaging materials or food additives on product shelf life; and
3. Tests designed to determine the shelf life of prototype or newly developed products.

There are several basic approaches to determine the shelf life of a food product:

Literature Study

The shelf life of an analogous product is obtained from the published literature or in-house company files.

Turnover Time

The average length of time which a product spends on the retail shelf is found by monitoring sales from retail outlets, and from this the required shelf life is estimated. This does not give the "true' shelf life of the product but rather the "required" shelf life, it being implicitly assumed that the product is still acceptable for some time after the average period on the retail shelf.

End Point Study

Random samples of the product are purchased from retail outlets and then tested in the laboratory to determine their quality; from this a reasonable estimate of shelf life can be obtained since the product has been exposed to actual environmental stresses encountered during warehousing and retailing.

Accelerated Shelf Life Testing

The basic assumption underlying accelerated shelf life testing (ASLT) is that the principles of Chemical kinetics can be applied to quantify the effects which extrinsic factors such as temperature, humidity, gas atmosphere and light have on the rate of deteriorative reactions. Laboratory studies are undertaken during which environmental conditions are accelerated by a known factor so that the product deteriorates at a faster than normal rate. This method requires that the effect of environmental conditions on product shelf life can be quantified.

Regardless of the method chosen or the reasons for its choice, sensory evaluation of the product is likely to be used either alone or in combination with instrumental or chemical analyses to determine the quality of the product. Not surprisingly, many food scientists and technologists in industry attempt to replace human judgment with instrumental or chemical analyses since the latter are not prone to fatigue or subject to the physiological and psychological fluctuations which characterize human performance. However, since it is human judgment which is the ultimate arbiter of food acceptability, it is essential that the results obtained from any instrumental or chemical analysis correlate closely with the sensory judgments for which they are to substitute.

That such is not always the case has been well documented. In chemical and physical tests, analytical parameters are isolated so that a single signal is monitored, whereas sensory responses are more complex due to the integration of multiple signals due to the interdependence of appearance, texture, aroma and flavour of a food. Hedonic (*i.e.*, like/dislike) responses, and to a lesser extent intensity (*i.e.*, stimulus strength such as firm/very firm) judgments are subject to many experimental influences such as past exposure to the product and those created by the actual test protocol (*e.g.*, number of samples and order of presentation). The potential fallacy of correlating hedonic responses with physical and chemical measurements has been discussed using as examples the colour, mouth feel and sweetness of specific foods.

Different methods are used to measure whether "fresh" samples are different from stored samples, or "control" samples from test samples. These methods require trained or experienced panelists, and three experimental designs are commonly used for the purpose of shelf life testing the paired comparison test; the duo-trio test, and the triangle test. Further details about these tests can be found in standard texts on sensory evaluation. Descriptive methods are used to measure quantitative and/or qualitative characteristics of products and require specially trained panelists. Effective methods are used to evaluate preference, acceptance and/or opinions of products and do not require trained panelists.

The selection of a particular sensory evaluation procedure for evaluating products undergoing shelf life testing is dependent on the purpose of the test. Acceptability assessments by untrained panelists are essential to an open-dating program, while discrimination testing with trained panelists might be used to determine the effect of a new packaging material on product stability. For shelf life determinations for new products and in situations where freshness standards cannot be maintained for the duration of the storage period, descriptive profile analysis is often employed.

In shelf life testing there can be one or more criteria which constitute sample failure. One criterion is an increase or decrease by a specified amount in the mean panel score. Another criterion is microbial deterioration of the sample to an extent which renders it unsuitable or unsafe for human consumption. Finally, any physical changes such as changes in odour, colour, texture, flavour, *etc.* that render the sample unacceptable to either the panel or the consumer are criteria for product failure. Thus, sample failure can be defined as the condition of the product which exhibits either physical, chemical, microbiological or sensory characteristics that are unacceptable to the consumer, and the time required for the product to exhibit such conditions is the shelf life of the product.

However, a fundamental requirement in the analysis of data is knowledge of the statistical distribution of the observations, so that the mean time to failure and its standard deviation can be accurately estimated, and the probability of future failures predicted. As has been pointed out, the length of shelf life for food products is usually obtained from simple averages of time to failure on the assumption that the failure distribution is symmetrical. If the distribution is skewed, estimates of the mean time to failure and its standard deviation will be biased. Furthermore, when the experiment is terminated before all the samples have failed, the mean time to failure based on simple averages will be biased because of the inclusion of unfailed data.

In order to improve the method of estimation of shelf life, knowledge of the statistical distribution of shelf life failures is required, together with an appropriate model for data analysis. Five statistical models normal, lognormal, exponential, Weibull and extreme-value distributions were fitted to failure data using the method of hazard plotting which provided information about the adequacy of fit of the observed data to the proposed model, the mean or median time to failure, and the probability of future failures. The Weibull distribution was suggested as the most appropriate shelf life model, and examples of its use to predict end of shelf life for food products have been presented.

The nature of product failure over time is commonly represented by a 'bathtub' curve which has many applications in the actuarial and engineering sciences.

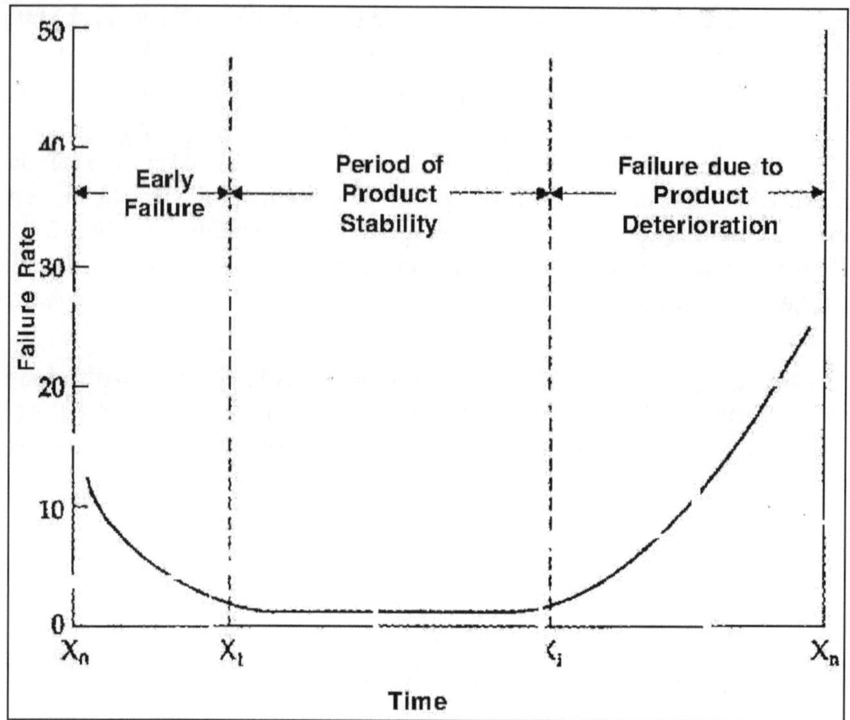

Figure 20.1: Bathtub Curve Showing Failure Rate as a Function of Time.

During the time between X_0 and X_1, early failures may occur due to faulty packaging (*e.g.*, pinholes or poor seals) and product abuse. However, the early failures should not be taken as true failures relative to the shelf life of the product. From X_1 to X_i, no product failures (barring random fluctuation) would be expected. From the time Xi to the termination time X_n, the hazard (failure) rate increases, this time representing the true failure due to deteriorative changes within the product. The length of shelf life is determined between the times X_0 and X_n, and the hazard function plays a central role in the analysis of failure data.

One problem with shelf life testing is to develop experimental designs which minimize the number of samples required, thus minimizing the cost of the testing while still providing reliable and statistically valid answers. Three experimental designs for use in shelf life testing of foods (partially staggered; staggered, and completely staggered) have been presented by Gacula and the interested reader should consult this paper for further details.

More recently, workers at the U.S. Army Natick R and D Center described a storage study of a military ration (called Meal, Ready-to-Eat) using sensory data, with the purpose of finding the effects of time and temperature on shelf life estimates, taking into account their variance. Four algorithms were used in the calculations: the contingency-table, logit, linear and nonlinear regressions. This paper should be consulted for further details.

20.7 ASLT and Dehydrated Products

The basic assumption underlying accelerated shelf life testing (ASLT) is that the principles of chemical kinetics can be applied to quantify the effects which extrinsic factors such as temperature, humidity, gas atmosphere and light have on the rate of deteriorative reactions. These basic principles and the way in which they can be applied to foods have been described later. By subjecting the food to controlled environments in which one or more of the extrinsic factors is maintained at a higher than normal level, the rates of deterioration will be speeded up or accelerated, resulting in a shorter than normal time for product failure. Because the effects of extrinsic factors on deterioration can be quantified, the magnitude of the acceleration can be calculated and the "true" shelf life of the product

The need for ASLT of food products is simple: since many foods have shelf lives of at least one year, evaluating the effect on shelf life of a change in the product (*e.g.*, a new antioxidant or thickener), the process (*e.g.*, a different time/temperature sterilization regime) or the packaging (*e.g.*, a new polymeric film) would require shelf life trials lasting at least as long as the required shelf life of the product. Companies cannot afford to wait for such long periods before receiving knowing whether or not the new product/process/packaging will give an adequate shelf life, because other decisions (*e.g.*, to construct a new factory; order new equipment; or arrange contracts for the supply of new packaging material) have led times of months and/or years. Some way of speeding up the time required to determine the shelf life of a product is necessary, and ASLT has been developed for that reason. Such procedures have long been used in the pharmaceutical industry where shelf life and efficacy of drugs are closely related. However, the use of ASLT in the food industry is not as widely used as it might be, due in part to the lack of basic data on the effect of extrinsic factors on the rates of deteriorative reactions, in part to ignorance of the methodology required, and in part to a healthy skepticism of the advantages to be gained from using ASLT procedures.

Dehydrated Products

In dehydrated vegetables, lipid and hydrolytic oxidation, together with nonenzymic browning and (in the case of green vegetables) chlorophyll degradation, are the major modes of deterioration. In dehydrated fruit, the major modes of deterioration is nonenzymic browning. Samaniego used temperatures of 30° and 40°C to accelerate deterioration in sliced green beans and onion flakes and found that, for example, at 40°C, the shelf life of onions was 11 times shorter, and at 30°C, 3.5 times shorter than at 20°C when the a_w was 0.56.

Canned Foods

It is generally assumed that if good manufacturing practices are followed, microbial deterioration of canned foods will not be a problem. Where there is thermophilic spoilage when canned foods are stored at elevated temperatures, this is more than likely due to inadequate cooling of the cans following thermal processing. Microbial spoilage at ambient temperatures is generally the result of 'leaker' spoilage, so called because the microorganisms are drawn into the can during cooling; chlorination of cooling water according to good manufacturing practice will alleviate this problem. Thus deteriorative reactions in canned foods will normally be limited to organoleptic changes such as loss of color, development of undesirable flavors, and nutrient degradation.

Labuza quotes a producer of canned meat products as stating that the major mode of deterioration is hydrogen gas production resulting from internal corrosion of the can. Samples were stored at 37.8°C to accelerate this deterioration; the shelf life at 37.8°C was considered to be 40 per cent of the shelf life at 4.4°C, corresponding to a Q_{10} of 1.3.

Starches

The use of accelerated tests for long shelf life products (*i.e.*, those which retain all their key characteristics after 18 months storage) which contain starches has been described. Such products include sterilized products packaged to exclude recontamination, and products formulated to inhibit spoilage. In these situations, temperature is used as the accelerating factor and a typical scheme. Each product should be tested in the actual package which is to be used to confirm that there are no problems with product/package compatibility.

A product which retains its desirable qualities after being subjected to the tests has a high probability of having a minimum shelf life of 18 months in normal distribution. However, caution is still necessary since interpretation of the results requires experience of similar product systems. This is particularly true of the 55°C tests where acid hydrolysis, color changes, *etc.* may be severe and yet are not reproduced at ambient temperature even after two years storage.

ASLT methodology for fatty foods has been reviewed with particular emphasis on the testing of antioxidant effectiveness. In all the classical ASLT methods, temperature is the dominant acceleration factor used, and its effect on the rate of lipid oxidation is best analyzed in terms of the overall activation energy E_A for lipid oxidation. An inherent assumption in these tests is that E_A is the same in both the presence and absence of antioxidants, although indications are that it is in fact considerably lower in the latter case.

Other acceleration parameters which are used for shelf life are the oxygen pressure, reactant contact and the addition of catalysts. The effect of these factors is generally much less important than that of temperature. An exception would be high-fat products packaged in metal containers where metal contamination of the product may be the most important factor in limiting shelf life.

20.8 Time-Temperature Records of Foods

Time-temperature recorders have been used for many years to provide a record of the temperature history to which a food has been subjected. Until recently they have tended to be relatively expensive, but with the commercialization of microchip technology, temperature recorders at quite modest cost are now available. They are in common use to record the temperature of frozen foods and fresh horticultural produce during transportation by sea. When the shipment arrives at a port, the temperature history of the product is inspected. If the product has been subjected to too high a temperature for too long a time, the load may be rejected without necessarily inspecting the product. This action is done on the basis that quality deterioration is cumulative and the product has lost enough of its shelf life to expect consumer dissatisfaction at a later stage. Alternatively, the decision may be made to dispose of the product quickly by discounting its price or some other marketing strategy.

The limitations of time-temperature recorders lie in the fact that they are comparatively expensive, can only monitor the specific area of the load where their sensors (typically thermistors or thermocouples) are located, and cannot follow the product through all phases of the distribution system unless a sensor is placed on each individual case or pallet. A further limitation concerns the interpretation of the data from a recorder; if the product has been subjected to a period at high temperature, how much of its shelf life has it lost? The answer to such a question is not simple, and often requires a value judgement by a qualified food technologist. In practice, such decisions need to be made by agents or brokers who are on the scene when the product arrives so that appropriate action can be taken immediately.

For the above reasons, time-temperature recorders have not, and will not, find widespread use in the food industry for monitoring the shelf life of foods. Instead, the use of time-temperature indicators is likely to become quite widespread over the next few years, due largely to an increasing product quality consciousness and awareness of product liability, especially in the USA.

A time-temperature indicator is a device which responds to the combined effects of time and temperature over a period. Many devices which can be attached to food packages to integrate the time and temperature to which the package is exposed have been developed, and the patent literature contains designs of more than 60 such devices. Initially the majority of these devices were developed specifically for frozen foods, but there is now widespread interest in shelf life devices for most categories of food, especially those foods where the rate of quality deterioration is highly temperature sensitive. Overviews of the major types of time-temperature indicators have been presented.

There are two classifications of time-temperature indicators devices which respond only after predetermined threshold temperature has been exceeded (these are said to be "partial history" indicators), and devices which respond to all temperature conditions (these are classified as "full history" indicators). Partial history indicators are intended to identify abusive temperature conditions and thus, there is no direct correlation between food quality change and the response of this class of indicator.

Chapter 21

Importance of Eco-friendly Packaging and its Sustainability

In modern societies, on the present level of economic development, to satisfy our needs, we have to rely on the services of the society. To be able to obtain consumer goods, we need among others the service of packaging. Service means that packaging consumer goods, helps their distribution, and gives an access to goods otherwise not accessible. It is most evident in food packaging. History of packaging began in the times when people used materials from nature for transporting food, through trade and industry development, when the need for better ways of protection and transport of goods emerged, until today when packaging is everywhere in modern society.

Conserving perishable food prevents early spoilage; invention of aseptic packaging prolongs shelf-life and makes possible distribution also to greater distances. This fact is more and more important in the present way of increasing urbanisation. Thirty-five years ago, only one third of the world's population was urban. Today, 60 per cent of the world's population lives in urban areas, a proportion that is expected to increase to 68 per cent by 2050. Projections show that urbanization, the gradual shift in residence of the human population from rural to urban areas, combined with the overall growth of the world's population could add another 2.5 billion people to urban areas by 2050, with close to 90 per cent of this increase taking place in Asia and Africa. This means that more people will live in cities than occupied the whole planet many years ago. What makes it worse, by 2021, the world has 37 'megacities'; in 2020 there were 1934 metropolises with more than 300,000 inhabitants representing approximately 60 per cent of the world's urban population. At least 2.59 billion people lived in metropolises in

2020 which was equivalent to one third of the global population. Greater Tokyo already has 37.39 million people, Sao Paulo, Brazil 22.04 million; and Bombay, India, 20.41million. (World resources, 2021). In such level of urbanisation, distribution of goods, especially food, is crucial, and the role of packaging is enormous and is becoming an important part of our lives today.

In the rapidly urbanising world, packaging is substantial to obtain basic consumer goods. Packaging protects the goods from mechanical, climate, chemical and micro-biological influences, but it also protects the surroundings from potentially harmful goods. Today, packaging is everywhere in modern society to such an extent that it represents a threat to the environment. No doubt that packaging should fulfil its function but should be used effectively with a minimum overall resource use. Packaging effectivity means that packaging fulfils its function with minimal use of resources and minimal overall wastage.

Traditional Food Packaging Technologies

Food Systems Population drift from rural to urban areas has caused drastic changes in the food supply network from farm to the consumer in many emerging nations. One traditional belief which can no longer be sustained is the old saying that, "There are always fish in the rivers and lagoons, and rice, taro, or sweet potato in the fields; therefore, let there be no concern for the next meal". It is more likely that the farmer has gone to the city or even overseas, and is earning a labourer's wage to keep his family in food during the off-season. So food is now brought to the market by many and various means, and redistributed to these new consumers. There are new vistas for traditional food markets where the technologies are tested beyond their limits (Hicks, 2002).

Given the circumstances in which many developing countries are today, the challenge for their traditional technologies is that often they do not contribute sufficiently to meeting socio-economic imperatives. This is true also of those food technologies where many of the processing methods have remained unchanged for centuries, and are becoming inadequate to cope with modern needs, because they are too labour-intensive and depend now too much on natural environmental conditions. It is now clear that there is a need to lessen the dependence on nature, reduce the drudgery, shorten the time of the work involved and upgrade the preparation, quality, packaging, presentation, and shelf-life of these traditional foods and their packaging.

The Need for Packaging

It is difficult to answer whether products do need packaging, or the consumer does? For whom is the package made? The product, or the consumer? The package is tailor-made for the product, but one shall not forget that all the products are made for consumers. From the consumer's point of view the package's function is to protect the product. In the case of bulk goods, or if the product cannot be used without the support of the package, the package must help the use of the product. Without the service of packaging, most of the goods, especially food, couldn't reach the consumer. Living in the countryside, one can buy goods, for example milk directly from the farmer, but a can is needed to carry it home. This can also be considered as a refillable package, with almost an unlimited number of uses. In developed countries, milk can only be bought from the store, packed in aseptic carton box. It is not possible to avoid it, unless one goes daily to the dairy, and drink from the tap.

From the packager's point of view, the most important function of the packaging is the protection of the usage and aesthetic values of the product from damages, and get the product sold to the consumer. For the producer, the package is also a value-creating media of the product. With the help of the package, the product can be sold to the consumer. For example, a barrel full of toothpaste has almost no purchase value, since who wants to buy 100 litres of toothpaste? Most probably nobody would buy even a handful of it in the grocery, but one certainly does buy 50 grams of toothpaste in a tube.

Till the middle of this century, groceries were a meeting point: people talked there. Most of the products were sold in bulk, one could have a look at its quality, discuss with the shopkeeper, asked information about them. The big supermarket stores are meant for shopping. One picks up the goods, pays and leaves. One does not see people talking, or ask somebody: Did you try this product? How is it like? The function of the shopkeeper is taken by the packages, and clearly in the competition of the goods wins that one, which can use the most efficiently those seconds one has for viewing them. The package has to arouse the attention, attract the consumer into arm's-length and influence him or her in buying decision, and in the other hand prepare the next purchase based on the imprinting in the mind.

Modern packagings are an expressive form of the consumer-lifestyle. Over the protective function, packagings are giving character, "personality" to the product. They are following the product to the consumer, giving practical as well as aesthetic value. The proportion of the usage and aesthetic functions is important. The quality of packaging reflects our universal culture. All those products that appear in the shops and offer themselves for purchase, apart of that - with very few exceptions - their usage function is primary, are creating the surrounding material environment.

Minimum Packaging Technology for Processed Foods

Technologies are called traditional if, unaffected by modernization, they have been commonly applied over a long period of time. In general, traditional technologies tend to be cheap, easy to produce, apply, maintain, and repair. They are generally labour-intensive, which can be economically beneficial, but as far as food packaging technologies are concerned, the final products are often hygienically sub-standard and they usually have a short shelf-life.

Many traditional foods have nonetheless remained unchanged in process or package for centuries, due to the fact that they developed in a particular location and are deep rooted in the natural, cultural, religious, and socio-economic environment. Some have disappeared without a trace as a result of modern influences, while others have expanded on a global scale, becoming household products, *e.g.*, soysauce, now a multimillion-dollar industry. The reasons for this phenomenon need to be examined (Hicks 1983).

The role of packaging in waste reduction is the most evident at food packaging. When food is processed and packaged, the food residues are often used as fuel, animal feed or some economically useful by-product. In absence of packaged processed food, the residues become garbage in the household. Another reason why food packaging reduces waste is that it reduces spoilage. In developing countries food wastage is between 20-50 per cent because of poor, or the lack of packaging. In Europe, where packaging is used in handling, transport, containment and storage, food wastage is approximately 2-3 per cent. As the use of packaging materials increases, the fraction of food waste decreases. Overall, for every 1 per cent increase of packaging, food waste decreases by about 1.6 per cent.

Satisfaction of human needs generally takes place with the use of natural resources. The social development has considerably accelerated in this century. The driving force of this is the demand for goods. Further, the population of the earth is increasing almost exponentially, which multiples the effect of humans on the nature. Neglecting the fact, whether it is possible, if the whole population had the consumption scale of the present maximum level (that of US for example), the effect of it would be catastrophic. It is crucial thus to satisfy our needs in the most optimum way.

Packaging as any other industrial activity, has positive and negative impacts on the environment. The negative impacts include the resource use, and the ecological effects of packaging related wastes and emissions. In a well functioning society, these negative effects shall not be neglected. The effectiveness and efficiency of packaging should be supervised and our consumption habits should be examined.

Food Preservation Principles and their Integration with Food Packaging

Food Unit Operations

The surplus foods grown in the village have a need to be more carefully harvested, protected from spoilage and damage, packaged, and transported by various means to these markets. Unless the goods are sold with minimum spoilage and at their peak flavour, appearance, and nutritional value and presented in an attractive way, they may not be eaten at all. This is a worse situation than if the crop had never been grown and can represent serious loss and waste to a community. Very little investment has been made so far in developing traditional technologies or in applying scientific knowledge in most of the developing countries; meanwhile the more expensive products of imported technologies have further slowed the development of indigenous technologies. It has been increasingly recognized that the time has come when these traditional technologies must be upgraded through scientific application of packaging principles and then integrated with other packaging functions such as marketing and advertisement. In addition, careful environmental considerations need to be given to minimum packaging forms to avoid pollution problems and ensure sustainability.

Before food can be packaged, there are many unit operations involved after harvesting the raw materials, including cleaning, grading, disposal of unwanted material, then stabilization of the enzymatic, biochemical and microbial spoilage. If a study of the preservation and packaging of foods is undertaken, a key question is, "What factors cause spoilage and deterioration in foods". The main factors are microorganisms (bacteria, yeast, and moulds), as well as enzymes, temperature, and biochemical changes in the foods. Food preservation techniques are designed to prevent these spoilage changes and impart a keeping quality or shelf-life to the processed foods.

Packaging is an integral part of the processing and preservation of foods and can influence many of these factors. It can influence physical and chemical changes, including migration of chemicals into foods. The flavour, colour, texture as well as moisture and oxygen transfer is influenced by packaging. The effects of temperature changes and light can be modified by packaging materials.

Types of Packaging

Early forms of packaging ranged from gourds to sea shells to animal skin. Later came pottery, cloth and wooden containers. These packages were created to facilitate transportation and trade.

Utilizing modern technology, today's society has created an overwhelming number of new packages containing a multitude of food products. A modern food package has many functions, its main purpose being to physically protect the product during transport. The package also acts as a barrier against potential spoilage agents, which vary with the food product. For example, milk is sensitive to light; therefore, a package that provides a light barrier is necessary. The milk carton is ideal for that. Other foods like potato chips are sensitive to air because the oxygen in the air causes rancidity. The bags containing potato chips are made of materials with oxygen-barrier properties. Practically all foods should be protected from filth, microorganisms, moisture and objectionable odours. We rely on the package to offer that protection.

Aside from protecting the food, the package serves as a vehicle through which the manufacturer can communicate with the consumer. Nutritional information ingredients and often recipes are found on a food label. The package is also utilized as a marketing tool designed to attract the attention at the store. This makes printability an important property of a package.

The food industry utilizes four basic packaging materials: metal, plant matter (paper and wood), glass and plastic. A number of basic packaging materials are often combined to give a suitable package. The fruit drink box is an example where plastic, paper and metal are combined in a laminate to give an ideal package. This concept can be easily seen in your peanut butter jar. The main package containing the food (primary package) is made of glass (or plastic), the lid is made of metal lined with plastic, and the label is made of paper.

Each basic packaging material has advantages and disadvantages. Metal is strong and a good overall barrier, but heavy and prone to corrosion. Paper is economical and has good printing properties; however, it is not strong and it absorbs water. Glass is transparent, which allows the consumer to see the product, but breakable. Plastics are versatile but often expensive. Therefore, combining the basic materials works well in most cases. So, for a product like milk, which is an essential food for children and young adults and therefore cannot be very expensive, paper makes a good economical material. It also provides a good printing surface. However, since paper absorbs water, it will gain moisture from the milk, get weaker and fail, thereby exposing the milk to spoilage factors. It may even break and waste the product. When a thin layer of a plastic called polyethylene is utilized to line the inside of the milk carton, it serves as a barrier to moisture and makes an economical, functional package.

After making a food product and placing it in the appropriate package, a number of these individual packages must be placed in a large container to facilitate shipment. These larger containers are called secondary packages. The paperboard box is a very common secondary package. Plastics also can serve as secondary packages. The milk case in which a number of milk cartons are delivered to the supermarket is a good example.

Driving the Choice: Preservation, Protection and Safety

The main factors that influence the decision of food packaging materials are preservation, protection and safety. Packaging must protect the contents from physical damage and from external contamination *e.g.,* microbiological contamination. It must preserve the quality of the contents, whether for short shelf-life of some days or for the extended shelf-life of several months. Food and water packaging must equally protect the nutritional value and taste of the contents.

More importantly, packaging must be safe and have no risk of affecting health. The intrinsically inert nature of polyethylene and polypropylene help make them ideal candidates for safe packaging and they easily comply with relevant national and international regulations.

Upgrading of Food Packaging

Packaging accounts for an increasing share of the costs of the food processing industry in modern days. In 2016, food processing was a ~$45 billion market when measured by revenues. This sub-sector comprises machinery for activities that include processing, grading, sorting, heating, and milling. Rising from 4 per cent in 1947 to 10 per cent in 1987, now in 2016, food packaging had a ~$16 billion market comprising equipment machinery (*e.g.*, preparing, combining, aligning) and packaging machinery (filling, cartoning/wrapping, case-packing, palletizing) out of total~$45 billion market for food processing and it shall grow more in the times to come. Despite new materials which have reduced packaging weight, the total and relative costs of food and beverage packaging are increasing. Packaging costs encompass packaging material, packing operations, distribution, and the expense of inappropriate packaging. On average, the cost of packaging materials represents about one-fifth of material costs, however, in 10 out of 40 food industry sectors, packaging costs exceed the costs of the edible food stuff ingredients.

In most countries packaging continues to be driven by consumer demand, with regulatory bodies playing a limited role. In Europe, however, legislation and taxes on packaging have been established to encourage reduced packaging waste and more sustainable packaging practices.

Upgrading of traditional food packaging technologies in many cases, introduces exogenous factors, *i.e.*, the importation of technology from abroad. Whether or not adapted to local circumstances, the use of imported packaging technologies in many developing countries remains restricted to modern technologies; even when these are locally developed, they are more complex to use, repair, and maintain. They are also expensive and tend to rely on imported components and non-renewable sources of energy.

The Image of Packages

Packaging and packages have a bad image, in some part thanks to environmental activists. Probably glass and metal are the least harmful thought of and plastics and composite materials the most.

The bad image of packaging, however, does not translate into consumer hostility at point of sale. Products and not packages are bought. Packages are not noticed during purchase/transport/use of the product, but the minute the product is consumed, and the package had fulfilled its function, it turned waste. At that minute it is already seen as an environmental burden, something what is "polluting the environment".

It is convenient to fill the shopping trolley with goods, then stuff it all into plastic bags, place them into the boot of the car, thus get them all safe home. One seldom realises that it is exactly thanks to packaging that it is possible. Indeed it is more annoying to handle goods with no, or just light package. It is much easier is to carry home goods in rigid plastic bottle, or in aseptic box. But the very same plastic bottle or aseptic box is seen as nuisance, environmentally unfriendly when it ends up in the waste-basket. Then suddenly one's environmental consciousness arise, and thinks of what could the industry, retailers or government do about "all those" packages.

In the same time, the same consumer buys French apples, kiwi from New Zealand, orange from Israel, drinks Colombian coffee and Chinese tea, seasons the food with spices from India or Africa, uses car, takes baths, uses dishwashers, washing machines, all of what means tremendous amount of (drinking) water consumption. At public places one uses paper towels. The thing, however, what, through the media, seems the most conceived as a harm done to the environment for consuming resources, is the plastic bottle, and aseptic box in the waste basket. One may feel somewhat relieved to find containers for collecting them separately. During the pilgrimage to these containers the citizen may feel that he has done something good for the environment's sake, and does not necessarily want to know whether it is really good or not. One would probably argue that food and drinks, toiletry and cleaning products are needed. To get them, however, we need packaging as well.

We cannot discuss food packaging without discussing the effects of packaging waste on the environment. Clearly, recycling is a sound approach. However, the problem often lies in feasibility of collection, separation and purification of the consumer's disposed food packages. This mode of recycling is called post-consumer recycling. While it offers a logistic challenge, recycling is gaining in popularity, and the packaging industry is cooperating in that effort. Aluminum cans are the most recycled container at this time. Plastic recycling is increasing, yet most plastic is recycled during manufacturing of the containers; not as post-consumer recycling. For example, trimmings from plastic bottles are reground and reprocessed into new ones.

The plastics industry is helping to facilitate consumer recycling by identifying the type of plastic from which the container is made. A number from 1 to 7 is placed within the recycling logo on the container's bottom. For example, 1 refers to PET (Polyethylene Terephthalate), the plastic used for the large 2 litre soft drink bottles. Plastics have the advantage of being light. This helps to conserve fuel during transport and also reduces the amount of package waste.

There are many interesting packaging concepts being explored by the industry to keep up with the changing life style of the consumer and new technologies. Many professionals are involved in designing and manufacturing the modern package. Today's package is designed with the consumer's safety and convenience in mind. Examples are microwaveable popcorn packages, squeezable ketchup bottles and the tamper-proof milk bottle cap.

The Influence of Packaging on the Environment

Packaging in the modern world has a considerable impact on the environment. Food packaging makes up two fifths of the household waste. Life-cycle assessment for new packaging materials, such as nanobased and bio-based materials, is essential for their success. This is a comprehensive analysis of materials from production to disposal to determine their environmental impact. It involves evaluation of the materials and energy usage in product manufacture and use and evaluation of the type and amount of waste generated.

Uniform criteria or a universal model with standard inputs and outputs is needed to evaluate packaging materials. Analysis of the performance and sustainability of new materials, especially biobased materials, is also required to document their advantages over conventional petroleum-based materials. Investigations into the compatibility of new materials with food products are important.

Environmental Concerns that Impact Material Choices

Whether considering a full life cycle analysis or the packaging conversion process alone, polyethylene and polypropylene are materials that favourably address environmental concerns such as energy consumption, climate protection, and water conservation, compared to competing and traditional packaging materials.

Polyethylene and polypropylene intrinsically offer low densities which mean low packaging unit weight. This results in an important reduction of material used and minimises waste at end of life. The materials also significantly lower fuel consumption for transport and contributes to lowering our CO_2 emissions. In addition, advanced polyolefins are recyclable or can be turned back into energy to substitute fossil fuels in clean energy recovery processes.

Reducing the Environmental Impact of Packaging Materials

The food companies are also concerned that the products are designed to be eco-friendly. In particular, efforts are underway to reduce the weight of packaging materials (which turn into waste after consumption), adopt materials with less environmental impact, and collect and recycle used containers. In Japan, with the Containers and Packaging Recycling Law instituted in 1995, municipalities began to separate waste and supermarkets placed several collection boxes at their storefronts to recycle packaging materials. Public efforts to recycle packaging materials are thus, very important for the environmental health.

Review and Change of Food Packaging Materials

Food packaging materials play a primary role in protecting the contents from ultraviolet rays, germs in the air, shocks during transportation, *etc.*, while they turn into waste after consumption. Of non-industrial waste comprised primarily of household garbage, waste packaging materials account for some 60 per cent of the total volume, and 20-30 per cent of the total weight, according to some estimates. Besides, they need to be collected and treated separately as they are usually made of several materials, a factor that makes the recycling of packaging materials rather difficult.

For this reason, the companies have begun to change the packaging materials for frozen foods in February 2002, from composite materials (aluminized polypropylene or nylon) to polypropylene. The design and strength of inner trays were also reviewed for weight reduction purposes. All these efforts are designed to simplify the recycling of waste packaging materials and reduce their environmental impact (new packaging materials, when incinerated, produce less CO_2 and hazardous gases). The food companies are now committed to providing customers with products designed to be eco-friendly, thereby fulfilling corporate social responsibility.

Recycle Plaza JB (located in Saitama City, Japan), an intermediate treatment plant for used beverage containers, has been in operation since May 2003. This plant was jointly established by JT and its affiliate, Japan Beverage, a vending machine operator. These beverage vending machines are installed in various locations such as office buildings, factories and along roadsides, the disposal of used containers had always been a major concern to the company. According to Mr. Toru Sunaga, manager of the environmental promotion division, Japan Beverage Inc., they felt that they were responsible for taking care of the entire process, from sales of products to collection and recycling of used containers – which resulted in the establishment of Recycle Plaza JB, an intermediate treatment plant for used containers.

Recycle Plaza JB is capable of treating 64 tons of used containers a day, which include aluminum cans, steel cans, paper cups, and brick packs collected by Japan Beverage's 38 sales offices in the Kanto area. It is the first of its kind managed by a vending machine operator, recycling a variety of beverage containers. He reported that they are, and will be, promoting recycling efforts on a company- wide basis, based on the awareness that paying attention to the environment is a critical theme that should be addressed not only by companies but also by each individual.

Going Green on Fast Food Packaging

With the hectic lifestyle of today, more and more people have less and less time to walk around and take in the scenery, let alone cook their own food. A lot of people rely on the existence of a burger joint or some other variation of fast food chain that can be located around every turn of each street corner. Upon receiving and paying for their ordered food, few people take the time to notice how their food has been packaged.

Fast food packaging is something that has been talked about and debated about by various environmentalists, primarily due to the type of materials that are being used as utensils and containers. Fast food joints usually make use of plastic spoons and forks, Styrofoam or polystyrene thermal insulation food containers, plastic or Styrofoam cups, plastic straws and stirrers, greaseproof and sulphite bags and burger wraps. For potato fries, cardboard containers or paper containers are used.

As can already be observed, fast food packaging consists of disposable items. Some environmentalists would argue that these disposable items, particularly the ones made of polystyrene thermal insulation material can be harmful to the earth if not properly disposed or recycled.

Although plastic is recyclable, most of these materials usually end up in dumps and never really decays creating more garbage. Fast food conglomerates argue that the use of disposable products is environmentally friendly or at the very least acceptable since less water is used. With these disposable fast food packagings, dishwashers would not need to be working around the clock saving more energy and water.

There are some fast food joints these days that prefer using recycled paperbags for take aways or take-out meals. This is a fairly good move since these items are biodegradable. Cardboard boxes are also being used by some fast food chains as opposed to using Styrofoam containers when serving their burgers. Another environment-friendly alternative implemented by top fastfood chains include the use of unbleached paperbags as opposed to bleached paperbags or plastic for take away food.

The progress of making fast food packaging more environment-friendly is rather sluggish, and the great amount of polystyrene wastes that are accumulating is increasingly becoming alarming because of the pollution they produce. There are some countries that encourage their citizens and the fast food chains to do their share of saving the environment by proper trash disposal and recycling. If more fast food chains and consumers would care about efficient and environment-friendly fast food packaging, then perhaps the progress of revolutionizing the packaging system would become faster.

Eco-friendly Consumer Choices

There aren't perfect solutions, short of buying foods only in their whole, natural form.

Given the culture we live in, we inevitably purchase packaged and throw-away items.

Yet, as non-renewable resources dwindle and technology advances, new possibilities are emerging.

PCC merchandisers have considered several novel ideas for containers and packaging with an eye toward more sustainable options. They're looking at containers made from palm fiber, a waste product of the palm oil process, and they're investigating new "food films" made from non-GE eucalyptus pulp as an alternative to plastic wrap. Both products are recyclable, biodegradable and can compost in own backyard. In the meantime, for shoppers who buy throwaway products, PCC offers bioplastic utensils made from non-GE corn and paper products made from 100 per cent recycled paper and no chlorine. When reviewing new items, merchandisers always give preference to products with less packaging.

Retailers, however, can't alter the marketplace alone. Consumers need to take responsibility, too, and support more sustainable solutions:

1. Bring your own glass or reusable containers for daily foods.

2. Re-use paper or plastic bags for bulk flour, nuts, *etc.*
3. Use long-lasting canvas bags for packing groceries to take home.
4. For fresh produce, don't use a bag since nature provides a natural skin as a "wrapper" and all produce can be packed safely at the register for transport home.
5. Call ahead and ask for butcher paper instead of plastic wrap for meat.
6. Choose products without packaging or reduced packaging, such as berries in fiberboard instead of plastic boxes.

The old maxim "Reduce, reuse, recycle" still holds. Yet recycling and reusing a product really starts with buying products that can be re-used. This is called "precycling." After all, when we throw something away, there is no away. To be a sustainable retailer, we need the consumers to make sustainable choices.

The food packaging industry is committed to developing products and implementing practices that can contribute to a cleaner, safer and better world.

Sustainable Packaging to Protect the Environment

Sustainability of packaging means:

1. To strengthen or support
2. To keep something going
3. Over time or continuously

It is the concept of using technology for meeting the needs of 'today' without in any way compromising the interests of future generations or the 'tomorrow'. The requirements of sustainable technology are:

1. Must not impose any burdens on the environment or material resources, renewable or non-renewable.
2. Must not cause deterioration in their quality or availability.

Sustainable packaging is integrating environmental criteria in the design process of a product-packaging combination. This means that besides the normal criteria, such as marketing, functional, economic and technical criteria, also environmental criteria are taken into account. The scope of sustainable packaging is the complete life-cycle of the product-packaging combination. This can result in improving the current packaging system, but also in developing a completely new product-packaging system for the circular economy or even packaging solutions with a positive impact.

Therefore, the overall objective of sustainable food packaging is to integrate functional and/or innovative materials into packaging that can improve economic and environmental health. Recyclable materials can also be employed together with the use of lifecycle assessments and lifecycle inventories to minimize the ecological footprint and environmental impact of the resulting packaging. To address the above mentioned concerns, we should look into the issues to achieve the objective of sustainable packaging as it has the following advantages:

1. It preserves natural resources like non-renewable including fossil fuels and minerals. Packaging material should prevent depletion of non-renewable resources.
2. Sustainable packaging is beneficial, safe and healthy for individuals and communities throughout its life cycle. It meets the market criteria for performance and cost.
3. It can save other scarce resources like energy. It should be physically designed in such a way as to optimise materials and energy. Packaging material should conserve energy efficiency. It should be sourced, manufactured, transported and recycled using renewable energy.
4. Packaging should be eco-friendly. It should prevent pollution of air and water, global warming, and not have an adverse effect on forest cover. The material to be used should be such that it preserves the quality of environment. It maximises the use of renewable recycled source materials and is manufactured using clean production technologies and best practices. It is made from materials healthy in all probable end of life scenarios.
5. The packaging material should not generate but reduce waste. Use renewable and recyclable materials. Use cleaner and greener processes and should avoid or reduce greenable emissions. It should be effectively recovered and utilised in biological and/or industrial cradle to cradle cycles.

In response to the need of sustainable packaging, the Granting society with low environmental innovative packaging (GLOPACK), a European 2020 innovation action is set on developing new generation bio-based and fully degradable composite food-trays made of biopolymers and fibres from agro-wastes. The bio-wastes used are residues from the processed fruits and vegetables industry for producing biopolyesters for use as the matrix. The lignocellulosic fillers are solid residues from the wheat industry which are blended to the matrix and processed using hot melt extrusion to produce the composites.

Food packaging is generally viewed as an unnecessary cost. It is considered as an additional economic and environmental cost instead of rightfully being seen as an added value for the reduction of food loss by improving shelf life. It is reported that at least 20 per cent food waste is a result of errors or misinterpretation of food date labels, and this shows the importance of the packaging in achieving several functions. Packaging material and the product to be packaged should be considered as one system from the design phase of the material to match the properties of the food. Sustainable of green packaging development using edible or biodegradable materials and their composites, plant extracts, and nanomaterials can help to bring down the environmental impacts of food packaging.

Chapter 22

The Vision for Future Packaging

Lessons in Packaging Learned from Nature

Perfect packaging exists in nature - examples include the banana and the egg, together with the many smart materials and systems that control plant and biological functions. Learning from Nature to solve engineering problems (biomimetics) is not new, although the term itself is. Animals and plants have evolved many successful structural and functional mechanisms and increasing our study of biological materials and systems will almost certainly yield promising engineering concepts applicable to smart packaging. Packaging in the future could indeed be 'smart by name, smart by Nature!'

The vision of the future of packaging, according to the recently published Foresight report 'Materials: Shaping Our Society', is one in which the package will increasingly operate as a smart system incorporating both smart and conventional materials, adding value and benefits across the packaging supply chain. For smart materials to be adopted in packaging, they need to be inexpensive relative to the value of the product, reliable, accurate, reproducible in their range of operation, and environmentally benign and food contact safe.

Smartness in Packaging

Traditional food packaging is meant for protection, communication, convenience and containment (Paine, 1991; Robertson, 2006). The package is used to protect the product from the deteriorative effects of external environmental conditionals like heat, light, presence or absence of moisture, pressure, microorganisms, gaseous emissions and so on. It also provides the consumer with the greater ease of use and time saving convenience and contain product of various size and shapes (Yam *et al.,* 2005; Marsh and Bugusu, 2007). The key safety objective for traditional packaging materials which comes in contact with food is to be inert as possible. Traditional food packages are passive barriers designed to delay the adverse effects of the environment on the food product. Modern food packages, instead, besides the passive role of containment, protection, and marketing of the product are characterized by an active function that allows packages to play a dynamic role in food preservation (during processing and storage), retaining the safety and quality of food throughout the distribution chain. In other terms, the key safety objective for traditional packaging materials coming in contact with food is to be as inert as possible; whereas the innovative packaging systems concepts are based on the useful interaction between packaging and the environment inside, to provide active protection to the food. Many terms are used to describe innovative packaging technologies such as "active", "interactive", "smart", "clever", "intelligent", "indicators", *etc.* These terms often lack in a clear definition and are interchangeable in some literature.

Bioengineered food packaging materials are closely related with active, intelligent, and smart packaging. A packaging material that provides an active function apart from inert containment and protection to the product is referred to as active packaging. Intelligent and smart packaging usually incorporates the capacity to sense or quantify the attributes of the produce, the inner environment of the package, or the transport atmosphere. The main difference is that intelligent packaging does not act directly on food with any action other than monitoring the condition of the packaged product, while active packaging acts on the environment surrounding food to increase the shelf-life. Therefore, active packaging is the component that takes some action, while intelligent packaging is the component that senses and shares the information. Therefore, it is important to differentiate their meanings. Intelligent and active packaging can, almost inevitably, work in synergy to create what is called a "smart" packaging (Biji *et al.,* 2015).

"Smart packaging" is an umbrella term that describes packaging that uses technology to do things that go beyond just housing a product. Smart packaging is usually embedded with technology that can sense or provide information about the product or brand. Smart packaging describes a large category of packaging that leverages technology to provide functionality that goes beyond simply housing a product.

In principle, "Smart Packaging" describes packaging with extended functions. There are usually two types of smart packaging - active packaging and intelligent packaging. 'Smartness' in packaging is a broad term that covers a number of functionalities, depending on the product being packaged, including food, beverage, pharmaceutical, household products *etc.* Examples of current and future functional 'smartness' would be in packages that:

1. Retain integrity and actively prevent food spoilage (shelf-life)
2. Enhance product attributes (*e.g.,* look, taste, flavour, aroma *etc.*)

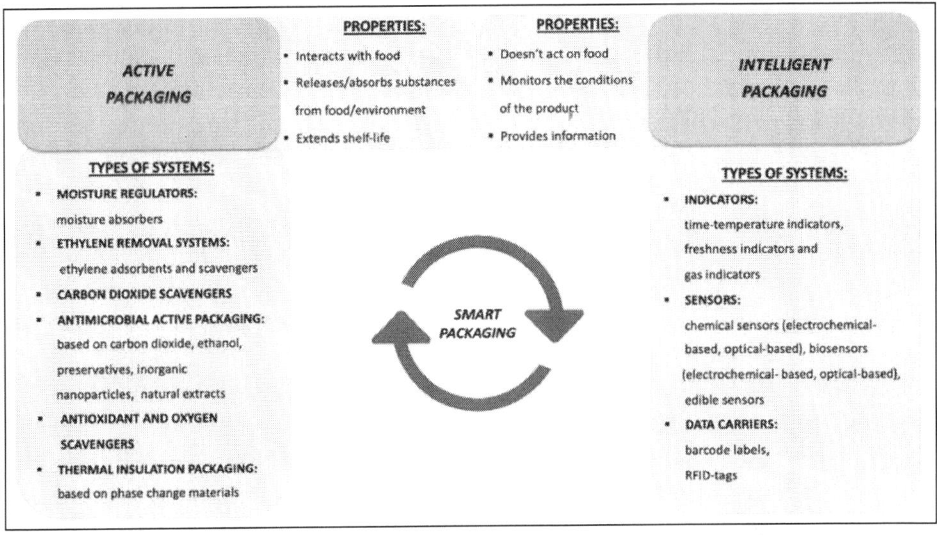

3. Respond actively to changes in product or package environment
4. Communicate product information, product history or condition to user
5. Assist with opening and indicate seal Integrity
6. Confirm product authenticity, and act to counter theft.

Active Packaging

There is an important distinction between package functions that are smart/ intelligent, and those that become active in response to a triggering event, for example, filling, exposure to UV, release of pressure *etc.* and then continue until the process is exhausted. Some smart packaging already exists commercially and many other active and intelligent packaging concepts are under development (Table 22.1).

Table 22.1: Smart Packaging Under Development

Active	*Intelligent*
Oxygen scavenging	Time-temperature history
Anti-microbial	Microbial growth indicators
Ethylene scavenging	Light protection (photochromic)
Heating/cooling	Physical shock indicators
Odour and flavour absorbing/releasing	Leakage, microbial spoilage indicating
Moisture absorbing	

Active packaging can be defined as a system in which the product, package and the environment interact in a positive way to extend the shelf life or to achieve some characteristics. It has also been defined as a type of packaging that changes the condition of the packaging to extend shelf life or to improve safety or sensory properties while maintaining the quality of the packaged food (Ahvenainen, 2003). The goal of active packaging is to enhance the preservation of food in the package and prolonging shelf life involves application of various strategies like temperature control, oxygen removal, moisture control, addition of chemicals such as salt, sugar, carbon dioxide or natural acids or a combination of these with effective packaging. These developments in active packaging have led to advances in many areas including delayed oxidation in muscle foods, controlled respiration rate in horticultural products, microbial growth and moisture migration in dried products. In addition, active packaging also manipulates the selectivity to modify the atmospheric concentration of gaseous compounds inside the package by coating, micro perforation, lamination, co extrusion, or polymer blending. For example, active packaging may release antimicrobial agents to prevent bacterial growth when it senses that a product's freshness is being compromised. A good example of active packaging is the highly successful foam-producing 'widget' in a metal can of beer. Some beers packaged in plastic bottles contain oxygen absorbers in the screw caps that can extend the beer's shelf life from three to six months. Another is the oxygen scavenging MAR technology patented by CMB Packaging Technology (now Crown Cork and Seal). Film packaging with ethylene absorbers can absorb the ripening hormone ethylene while the food is being stored to ensure a longer shelf life. Table 22.2 shows some examples of commercially available active food packaging for each of the categories.

Oxygen Scavengers

The most widely used active packaging technology for foods today are oxygen scavengers. Oxygen facilitates the growth of aerobic microbes, off flavor and odour development, colour changes and nutritional losses and overall shelf life stability of muscle. Therefore, control of oxygen levels in food packages is important to limit the rate of such spoilage reactions in food. The oxygen absorbers are designed to reduce oxygen levels to less than 100 ppm in package headspace. The commercially available oxygen scavengers utilize one or more of the following technologies: iron powder oxidation, ascorbic acid oxidation, photosensitive dye oxidation, enzyme oxidation, saturated fatty acid oxidation, immobilized yeast on solid material *etc.* Majority of currently available oxygen scavengers are based on iron powder oxidation in the form of small sachets containing various iron based powders containing as assortment of catalysts. The chemical substances react with the water supplied by the food to produce a reactive hydrated metallic reducing agent that scavenges oxygen within the food package.

Table 22.2: Commercially available Food Active Packaging

Commercial Name	*Principle*	*Application*	*Materials and Forms*
Activ-Film™ (www.csptechnologies.com)	Moisture absorber	Fruit and vegetables	Low-density polyethylene (LDPE) film
Tenderpac[fi] (www.sealpacinternational.com)	Moisture absorber	Meat products	Polyethylene terephthalate (PET) tray
PEAKfresh™ (www.peakfresh.com)	Ethylene scavenger	Fruit and vegetables	Film impregnated with a natural mineral
BIOPAC (www.biopac.com.au)	Ethylene scavenger	Fresh products	Sachet in porous material mixed with potassium permanganate
ATCO[fi] (www.emcotechnologies.co.uk)	Carbon dioxide absorber	Fresh products	Film—bags
SANDRY[fi] (www.hengsan.com)	Carbon dioxide absorber	Fruit, coffee, fermented food	Sachets
McAirlaid's CO_2Pad (www.mcairlaids.net)	Carbon dioxide emitter	Fish, meat and fruit products	Cellulose-based pads
FreshPax[fi] (www.multisorb.com)	Carbon dioxide emitter	Processed and pre-cooked food	Packets and films realize with food grade materials
Celox™ (www.grace.com)	Oxygen scavenger	Beverages	Cans sealants and closure coatings
$ZERO_2$ (www.ipl-plastics.com)	Oxygen scavenger	Fresh products	Multilayer film fused to injection-molded containers
Biomaster[fi] (www.biomasterprotected.com)	Antimicrobial	Chilled and frozen products	Cool bags
Food-touch[fi] (www.microbeguard.com)	Antimicrobial	All food products	Various forms of paper products
ATOX (www.artibal.com)	Antioxidant	Cereal products	Film coating containing oregano essential oils
Pure Temp (www.puretemp.com)	Phase Change Materials	Frozen food, cold storage	Palm oil, coconut oil and soybean oil-based
Green Box (www.greenbox.it)	Phase Change Materials	Perishable products	Vegetable oil-based

Source: Drago *et al.*, 2020. Innovations in Smart Packaging Concepts for Food: An Extensive Review. Foods, 9: 1628.

Antimicrobial Active Packaging

Food shelf-life is strongly affected by the presence of microorganisms. Contamination could occur at every stage of the supply chain, whenever food is exposed to the environment and its outcomes can range from simple alterations of sensory features to serious health hazards for consumers. Antimicrobial packing is a form of active packaging in which the packaging acts to reduce, inhibit or retard the growth of microorganisms that may be present in the packaged food or packaging material itself (Appendini and Hotchkiss 2002). To control undesirable microorganisms on foods, antimicrobial substances can be incorporated in or coated onto food packaging materials. The major potential food applications for antimicrobial agents include meat, poultry, bread, cheese, fruits and vegetables.

Antimicrobial package material can be classified into two types: those containing antimicrobial agents that migrate to the surface of the packaging material, and those are effective against surface microbes without migration of the active agent to the foods. Antimicrobial enzymes like lactoperoxidase and Lactoferrin, antimicrobial peptides like megainins, cecropines, defensins, natural phenols like hydroquinons and catechins, fatty acid esters, antioxidant phenolic, antibiotics and metals like copper are incorporated into polymers. Packaging systems that release volatile antimicrobials include chlorine dioxide, sulphur dioxide, carbon dioxide and ethanol. In this system, polymer need not be directly in contact with the food. Here the antimicrobial agents are directly incorporated into the polymer or into the carriers that may be extruded or coated into the packaging materials. Some antimicrobial packaging uses covalently immobilized antibiotics or fungicides to suppress the growth of microbes (Brody *et al.,* 2002). Examples of antimicrobials with functional groups are peptides, enzymes, polyamines and organic acids. Some polymers like chitosan are inherently antimicrobial and are used in films and coatings. Antimicrobial edible coatings and films prepared from polysaccharides, proteins and lipids have variety of advantages like biodegradability, edibility, biocompatibility, aesthetic appearance, barrier properties *etc.*

Moisture Absorbers

Moisture absorbers are active non-migratory packaging, working as controllers of excess moisture. High content of water into the package may be provided by several causes, such as trapping during packaging process, release due to temperature fluctuations, respiration of fresh products, low vapor permeability of the package, *etc.* Excess water reduces food shelf-life, since promotes microbial proliferation inducing undesired changes in food quality, particularly in dry food.

The main purpose of moisture absorbers is the reduction of food water activity by using hygroscopic substrates or substances, providing an environment less suitable for mold, yeast and bacteria growth and spoilage. Some packaging materials used as passive system are inherently moisture scavengers due to their hygroscopicity Moisture scavengers can be defined as "relative humidity controllers" if they provide a reduction of humidity in the headspace, while "moisture removers" absorb liquids exuded from food and are usually placed on the bottom of fresh products and meat packages. Examples of moisture absorber commercially available are absorbent pads, sheets, trays and blankets, as summarized in Table 22.3.

Table 22.3: Main Features of Commercially available Systems Working as Moisture Regulators

Structures		*Desiccants*	
		Inorganics-based	*Organic-based*
Sachets	Systems where moisture scavenger materials are enclosed into a small porous bag.	Bentonite Calcium chloride Calcium oxide Calcium sulfate Molecular sieves Natural clays Silica gel	Cellulose Fructose Modified starch Sorbitol
Pads, blankets, sheets, trays:	Mostly composed of porous materials, foamed and perforated sheets and moisture superabsorbent materials.		
Labels	Systems composed of adhesive dessiccant labels, suitable for low-level humidity systems.		

Source: Drago *et al.,* 2020. Innovations in Smart Packaging Concepts for Food: An Extensive Review. Foods, 9: 1628.

Ethylene Scavengers

Ethylene is a natural plant growth hormone which accelerates respiration of fruits and vegetables, induces fruit ripening, fruit softening and senescence even at low concentration. It causes yellowing of vegetables, russet spotting on lettuce and has detrimental impact on shelf life of many fruits and vegetables. The reduction of ethylene in the package atmosphere by active systems can slow undesired effects on these products. Ethylene removal can be carried out by ethylene absorbers, which physically absorb and hold its molecules, and scavengers, acting by chemical reactions.

Flavour or Odour Absorbers and Releasers

The volatile compounds that accumulate inside the package as a result of food degradation such as aldehydes, amines and sulfides can be selectively scavenged. Flavour scavengers prevent the cross contamination of pungent odour while transportation of mixed loads. The package consists of an odour impermeable plastic like polyethylene tetraphthalate (PET) or polyethylene of suitable thickness together with a port to allow for the passage of respiratory gases and a sachet made from a mixture of charcoal and nickel to absorb odour. Volatile amines formed due to the protein breakdown in fish muscle can be removed by incorporating acidic compounds like citric acid in polymers. Use of high barrier packaging materials can also prevent the absorption of other non-food odours like taints.

Antioxidants

They are widely used in food to improve the oxidation stability of the food to prolong the shelf life. Antioxidants are incorporated in packaging films as a source of antioxidants in some foods since the increased consumer demand for reduced antioxidants and additives in foods. Antioxidant incorporation can also stabilize the polymer inorder to protect the films from degradation. Incorporating natural antioxidants like Vitamin C, and E on packaging films can reduce oxidative reactions like development of rancid odour and colour changes in fatty fishes. Vitamin E is also safe and effective for low to medium water activity cereal and snack food products.

RFID Tags in Packaging: Smart Packaging

Packaging technology could possibly realize its most influential innovations through smart packaging. Smart packaging encompasses a wide range of technologies including films that indicate product freshness, as well as security inks used to protect against counterfeiting in the food and pharmaceutical sector. Perhaps the most significant technological advancement currently evolving is radio frequency identification (RFID). Radiofrequency identification (RFID) is an automatic identification technology that uses wireless sensors to identify items and gather data without human intervention. An RFID is based on tags and readers. Most RFID tags store some sort of identification number based on which reader can retrieve information about the ID number from a database and acts upon it accordingly. RFID tags could be used on food packaging to perform relatively straightforward tasks, such as allowing cashiers in supermarkets to tally all of a customer's purchases at once or alerting consumers if products have reached their expiration dates.

RFID tags are of two categories; passive and active. Passive tags rely on the power supplied by the reader. When radio waves from the reader are encountered by a passive RFID tag, the coiled antenna within the tag forms a magnetic field. The tag draws energy from it and sends the information encoded in the tags memory. Semi passive RFID tags use battery to maintain memory in the tag or power the electronics that enable the tag to modulate the Electromagnetic waves emitted by the reader antenna. Active RFID tags are powered by an internal battery, used to run the microchips circuitry and to broadcast a signal to the reader. The use of RFID tags in packaging is increasing for all end-markets; however, a majority of current activity can be seen in the food and beverage and pharmaceutical sectors. RFID will play a crucial role in packaging for the foreseeable future.

RFID tags are controversial because they can transmit information even after a product leaves the supermarket. Privacy advocates are concerned that marketers will have even greater access to data on consumer behavior. Wal-Mart in the US and TESCO in the UK have already tested RFID tagging on some products in some stores.

The tagging of food packages will mean that food can be monitored from farm to fork during processing, while in transit, in restaurants or on supermarket shelves and eventually, even after the consumer buys it. Coupled with nanosensors, those same packages can be monitored for pathogens, temperature changes, leakages *etc.* (Scott, 2002).

RFID has been successfully applied to traceability control and supply chain management processes because of its ability to identify, categorize and manage the flow of goods. Studies indicate that RFID is more advanced than the zebra black and white paper; the barcode system for food traceability. It provides supply chain visibility, which enables fast automated processes at the supply chain visibility which enables fast and automated processes at supply chain level such as exception management and information sharing.

Wal-Mart is two years into the world's most ambitious effort to implement RFID technology and is making significant headway with its Next Generation logistics initiative. The company kicked-off their RFID program in 2004 with suppliers that included Gillette, Hewlett-Packard, Johnson and Johnson, Kimberly-Clark, Kraft Foods, Nestlé Purina PetCare, Procter and Gamble, and Unilever, and has already begun to realize a return on its investment with improved sales of promotional display items and a sizable reduction in product stock-outs.

It is estimated that the market for smart technology in packaging should grow to nearly $5 billion by 2011 and reach $14 billion by 2014. As electronic printing technologies improve and smart packaging technologies become more economically viable, interoperable RFID products will allow for improved inventory management, logistics, and supply chain management on a global scale.

Intelligent Packaging

Intelligent materials and articles means materials and articles which monitor the condition of packaged food or the environment surrounding the food. The concept of internal migration of preservatives to food and the communication function of the package to facilitate decision making are related with intelligent packing. In 'intelligent' packaging, the package function switches on and off in response to changing external/internal conditions and can include a communication to the customer or end user as to the status of the product. Intelligent packaging systems provide the user with information on the conditions of the food or its environment (temperature, pH). It is an extension of the communication function of traditional packaging and communicates to the consumer based on its ability to detect, sense and record the changes in the products environment.

In the literature there are many definitions of the concept of intelligent packaging. A simple definition of intelligent packaging is 'packaging which senses and informs', and nowhere does this generate a more potent vision than within the smart home of the future (Butler, 2001). European Commission defines intelligent packaging as "materials and articles that monitor the condition of packaged food or the environment surrounding the food". Intelligent packaging has also been defined as science and technology that uses the communication function to facilitate decision making to extend shelf-life, improve safety, guarantee quality, provide information, and warn of any problems by monitoring changes in the internal and external environment of the packages. More in details, intelligent packaging is a packaging system that uses the internal (*e.g.*, metabolites) or external (*e.g.*, temperature) package environment as "information" to monitor the status of product quality, in order to improve not only product safety, but also to track the product for automatic identification and traceability. The development of new intelligent packaging capable of continuously providing information on the conditions of both the food product and the packaging integrity, allows for a safer and more efficient supply chain, avoiding unnecessary transports and logistics and reducing food waste.

In contrary to active components, intelligent components do not have the intention to release their constituents into the food. The intelligent packaging can also contribute to improving Hazard Analysis and Critical Control Points' (HACCP) and Quality Analysis and Critical Control Points' (QACCP) systems (Heising *et al.,* 2014), which are developed to onsite detection of unsafe food, identify potential health hazards and establish strategies to reduce or to eliminate their occurrence. It also helps to identify processes that strongly affect the quality attributes and efficiently improve the final food quality (Vanderroost *et al,.* 2014). Basically, there are three intelligent systems; sensors, indicators and radiofrequency identification (RFID) systems

Indicators

Indicators can be defined as substances that indicate the presence, absence or concentration of another substance or the degree of reaction between two or more substance by means of a characteristic change, especially in colour.

The main function of indicators is to transmit to the consumer information related to the presence or absence of a specific substance, to indicate whether a reaction is taking place between two or more components or to monitor the concentration of a certain substance. This information is translated into signals usually in the form of immediate visual changes (*e.g.*, different colour intensities or diffusion of a dye along a straight path), providing qualitative or semiquantitative information. In most of the cases, the basic requirement of an indicator is that these changes in colour or intensity are irreversible. The indicators represent a very consistent class of intelligent packaging; the most used classification, is based on the type of variable controlled, so they can be grouped into three macro categories: time-temperature indicators, freshness indicators, and gas indicators. For a more general classification, the indicators can be classified as external or internal, based on their positioning on the packaging.

Freshness Indicator

Freshness indicators are intelligent devices that allow the monitoring of the quality of food products during storage and transport. Freshness indicators provide the product quality information resulting from microbial growth or chemical changes within a food product. Freshness indicators, unlike temperature indicators, provide direct information on the quality of the product, analyzing the chemical reactions of food deterioration due to target microorganisms. Changes in the concentration of metabolites such as glucose, organic acids, ethanol, carbon dioxide, biogenic amines, volatile nitrogen compounds or sulfur derivatives are indicators of microbial growth and their presence can therefore be exploited in the freshness indicators. In most cases, freshness indicators are based on the use of dyes sensitive to pH variations caused by the deterioration of the product, which lead to the visible change in the colour of the indicator. These indicators find applications in various products including fresh food, fruit, and seafood.

Time-Temperature Indicator

Temperature is a fundamental factor in monitoring and determining the shelf-life of a food product, especially for perishable products. One of the main challenges of the perishable food industry is to ensure the quality of food by protecting it from unwanted temperature fluctuations, a challenge that is not simple at all. Studies report that poorly managed temperature in perishable food logistics can cause up to 35 per cent loss of products. To meet these needs, time-temperature indicators (TTI) provide valuable support. Time-temperature indicators find application majorly for temperature-sensitive foods such as chilled and frozen products. Furthermore, TTIs have also been applied to control the pasteurization and sterilization process. There are three basic types of TTI available in market: critical temperature indicators, partial history indicators and full history indicators. Basically, TTIs are made of small, self-adhesive labels attached onto containers or individual consumer packages, so as to be clearly visible and interpretable. Currently available commercial TTIs are diffusion, enzymatic and polymer based systems. These systems must represent time-temperature dependent changes in an easily measurable manner that can be quickly correlated with the extent of deterioration and the remaining shelf-life of the food.

Integrity Indicator

A leak indicator to package ensures package integrity throughout the production and distribution chain. Visual oxygen indicators with redox dyes change its colour with changes in oxygen concentration. Disadvantage of such system is that the device should be highly sensitive and the residual oxygen in the package are susceptible to indicators. Oxygen comes through leakage may also be consumed by the natural microbes present in the food.

Gas Indicator

The activity of the food, the nature of the package, and the environmental conditions to which the package is subjected lead to the variation of the gas composition in the headspace of the package. Gas indicators are used to monitor these changes and are also often used to evaluate the effectiveness of active packaging components (*e.g.*, O_2 and CO_2 scavengers) or to detect the occurrence of leaks across the packaging, and therefore, to detect package integrity, which is an essential requirement, especially when the food is stored in a modified atmosphere packaging (MAP). The most well-known gas indicators are used to control oxygen and carbon dioxide concentrations. Oxygen is responsible for the microbial and biochemical deterioration of food and is therefore removed from the packaging and replaced with gas such as nitrogen or using oxygen scavenger. However, during transport and storage, oxygen may be able to penetrate the package due to poor sealing, defects, or damage. For this reason, the use of visible oxygen indicators becomes essential to quickly and easily verify the presence of oxygen inside the package, without using professional equipment or laboratory analysis. The most commonly used oxygen indicator is a colorimetric indicator based on a redox dye, generally methylene blue and a reducing agent, such as glucose, in an alkaline solution such as sodium or potassium hydroxide.

The other predominant class of gas indicators is related to CO_2. The microbial growth responsible for the deterioration of food usually begins immediately after packaging, and therefore there is a production of CO_2 due to the metabolic activity of microorganisms. The rate of carbon dioxide production depends on the type of food, time, and conditions of storage, and on packaging materials. Carbon dioxide is also used, in combination with oxygen and nitrogen, in protective atmosphere technology to inhibit microbial metabolism over time. Again, most of the carbon dioxide indicators are colorimetric labels, whose colour changes as a function of the pH of the product.

Gas indicators, mainly in the form of labels or printed directly on the packaging films, are positioned inside the package to monitor changes in the internal gas composition, thus providing a strategy for monitoring the quality and safety of food products.

Sensors

The approach to sensors is made difficult by the fact that the terms indicators and sensors are often used indiscriminately. In fact, in the literature, the term sensor is often referred to a device capable of detecting a specific analyte, a definition which includes both sensors and indicators. Actually, they represent two very distinct classes of devices. A sensor can be defined as a device used to detect, locate or quantify energy or matter giving a signal for the detection or measurement of a physical or chemical property to which the device responds. A sensor is a device that responds to a chemical, biological, or physical property by providing a quantifiable signal proportional to the measurement. The most popular traditional sensors are designed to measure temperature, humidity, pH, and light exposure. In addition to these, the need to monitor food quality and packaging integrity has led researchers to place a growing interest in applying disposable and advanced sensors (such as edible sensors) for intelligent packaging. Sensor usually consists of four main components: the first part is a receptor (the sensitive part of the sensor), which generally consists of a selective coating that acts as a sampling area that is able to detect the presence, activity, composition, or the concentration of specific chemical analyte by surface adsorption, with consequent modification of a certain property of the coating. This change is typically detected by the transducer, which is the second part of the sensor, converting the signal change into an output signal. If the transducer requires external power for measurement, the transducer is called "active", if not, it is called "passive". Finally, the sensor is completed by a part of electronics for signal processing and a signal display unit. The integration of these disposable into food packaging is not simple or immediate; in fact, there are very few products currently on the market. In particular, extensive research efforts are still needed to study the potentially dangerous effects of the chemical and biological components present in the sensor on foods, to avoid any migration of these components, and also to avoid potential changes in taste and consistency of the food product. Disposable sensors for food packaging should also satisfy the following requirements: selectivity for the target species; sensitivity to changes in target species concentrations; quick response times; small size (miniaturization); low construction cost; use inexpensive, sustainable or biodegradable materials; be simple to use; operate without or with an affordable, portable instrument.

Sensors intended for food packaging applications can be classified into two macro categories: chemical sensors and biosensors. Each of these two categories can in turn be divided into subclasses based on the method for signal transduction: electrochemical; optical; mechanical; magnetic; thermometric, and microgravimetric.

Biosensor

Biosensors are used to detect, record and transmit information pertaining to biological reactions. Biosensors contain bioreceptors and transducers. The bioreceptor recognizes the target analyte and the transducer converts biochemical signals into quantifiable electronic response. The bioreceptors may be either organic or biological materials like enzyme, hormone, nucleic acid, antigen, microbes *etc.* The transducers may be of optical, acoustic or electrochemical. Food Sectinel System® (SIRA Technologies Inc.) is a commercial biosensor developed to detect the food pathogens. Specific antibodies are attached to the membrane forming part of the sensor or the barcode. The pathogens cause localized dark bar formation results the barcode unreadable (Yam *et al.,* 2005). Biosensors for the detection of xanthine, (adenine nucleotide degradation product in animal tissue) was developed by Arvanitoyannis and Stratakos (2012) by immobilization of xanthine oxide onto the electrodes made of materials such as platinum, silver and pencil graphite (Devi *et al.,* 2013).

Gas Sensor

Gas sensors are used for detecting the presence of gaseous analyte in the package. It includes oxygen sensors, carbon dioxide sensors, water vapour sensor, ethanol sensor, metal oxide semiconductor field effect transistors, organic conducting polymers and piezoelectric crystal sensors *etc.* Optochemical sensors are used to detect the quality of products by sensing gas analyte such as hydrogen sulphide, carbon dioxide and volatile amines (Wolfbeis and List 1995). The optochemical sensing methods are of three types including fluorescence based system using a pH sensitive indicator, absorption based colourimetric sensing and energy transfer approach using phase fluorimetric detection (Neurater *et al.,* 1999; Mills *et al.,* 1992).pH sensitive dyes can be used to develop sensors for the detection of basic volatile amines in fish, meat and poultry. Indicators based on methyl red/cellulose membrane, curcumin/bacterial cellulose membrane respond through visible colour changes to volatile amines released during fish spoilage (Kuswandi *et al.,* 2014).

Chemical Sensor

The chemical sensor or the receptor is a chemical selective coating capable of detecting the presence, activity, composition, concentration of particular chemical or gas through surface adsorption. Presence of particular chemicals are being observed and converted into signals by transducer. Transducers are of either active or passive depends on the external power requirement for measurement. Carbon nano materials like nanoparticles, grapheme, graphite, nano fibers and nanotubes are applied in chemical sensors because of their excellent electrical and mechanical properties along with the high specific surface area (Vanderroost *et al.,* 2014). Nano based sensors can be used to detect pathogens, chemical contaminants, spoilage, product tampering, track ingredients or products through the processing chain (De-Azeredo 2009; Liu *et al.,* 2007).

Electronic Nose

Nose system consist of an array of either chemical or biosensors with partial specificity and statistical methods enabling the recognition of simple of complex flavour, odour or savour. Electronic nose system was proved to be successful in the quality evaluation of fresh yellowfin tuna and vacuum-packed beef (Dobrucka and Cierpiszewski 2014). The aroma emitted by fruits and vegetable canindicate the quality of marketed products. Rajamaki *et al.* (2004) studied the quality of modified atmosphere packed broiler chicken cuts using electronic nose. The electronic nose results were compared with those obtained by microbiological, sensory and head space gas composition analysis. The e-nose could clearly distinguish the chicken packages with deterioration from fresh packages.

Printed Electronics

Printed electronics is an emerging technology on flexible substrates using electrically functional inks. The unique properties of printed electronic sensors include light weight, bendable, rollable, portable and foldable. Possibility of creating sensors on a variety of substrates each shaped and individually tailored to operate uniquely. The flexible printed chemical sensors contain a receptor printed on top of a printed transducer.

Intelligent Tamper-Proof Packaging

Knowing whether a package has been tampered with is equally important to consumers. Tamper evidence technologies that cannot easily be replicated, *e.g.,* based on optically variable films or gas sensing dyes, involving irreversible colour changes, will become more widespread and cost-effective for disposable packaging of commodity items. Piezoelectric polymeric materials might be incorporated into package construction so that the package changes colour at a certain stress threshold. In this way, a 'self bruising' closure on a bottle or jar might indicate that attempts had been made to open it.

Easier to Open Packaging

Easier to open packaging will be a paramount feature of future packaging. The focus will be on better design (size, shape, *etc.*) and the optimum use of materials, to produce easy to open packages consistent with the strength capabilities of an ageing population. Developments in low peel-force adhesives and structures, even smart packages that are self-opening, *e.g.,* based on shape memory alloys (the metal 'rubber' band), can be envisaged.

Quality Assurance Using Intelligent Labels

Another important need is for consumer security assurance, particularly for perishable food products. The question as to whether, for example, a chilled ready-meal is safe to use or consume is currently answered by 'best by' date stamping. However, this does not take into account whether the product has inadvertently been exposed to elevated temperatures during storage or transportation. In the future, microbial growth and temperature-time visual indicators based on physical, chemical or enzymatic activity in the food will give a clear, accurate and unambiguous indication of product quality, safety and shelf-life condition. As an example, COX Technologies has developed a colour indicating tag that is attached as a small adhesive label to the outside of packaging film, which monitors the freshness of seafood products. A barb on the backside of the tag penetrates the packaging film and allows the passage of volatile amines, generated by spoilage of the seafood. These are wicked passed a chemical sensor that turns FreshTag progressively bright pink as the seafood ages.

Intelligent Packaging for Fresh Fruit and Vegetables

Fresh-cut produce continues to be one of the fastest growing segments of food retailing and while conventional film packaging is suitable for lettuce and prepared salads, it cannot cope with the high respiration rates of pre-cut vegetables and fruit, leading to early product deterioration. In the USA, novel breatheable polymer films are already in commercial use for fresh-cut vegetables and fruit. Landec Corporation supplies Intellipac packaging films that are acrylic side-chain crystallisable polymers tailored to change phase reversibly at various temperatures from 0-68°C. As the side-chain components melt, gas permeation increases dramatically, and by further tailoring the package and materials of construction, it is possible to by tune the carbon dioxide to oxygen permeation ratios for particular products. The final package is 'smart' because it automatically regulates oxygen ingress and carbon dioxide egress by transpiration according to the prevailing temperature. In this way, an optimum atmosphere is maintained around the product during storage and distribution, extending freshness and allowing shipping of higher quality products to the consumer.

Self-Heating and Self-Chilling Packaging

Improved convenience is a value-added function that customers are likely to pay extra for as lifestyles change. Self-heating packages, for soup and coffee, for example, and self-cooling containers for beer and soft drinks have been under active development for more than a decade, but have yet to achieve commercial status. However, Crown Cork and Seal is pioneering the development of a self-chilling beverage can in conjunction with Tempra Technologies and development is nearing completion. The Crown/Tempra technology uses the latent heat of evaporating water to produce the cooling effect. The water is bound in a gel layer coating a separate container within the beverage can, and is in close thermal contact with the beverage. The consumer twists the base of the can to open a valve, exposing the water to the desiccant held in a separate, evacuated external chamber. This initiates evaporation of the water at room temperature. The unit has been designed to meet a target specification set by major beverage customers cooling 300ml of beverage in a 355ml can by 16.7°C in three minutes. This performance level has been achieved in laboratory tests and working samples are currently undergoing focus group trials with customers.

Thermochromic Labelling

Give a self-heating or self-cooling container a sensor to tell the consumer it is at the correct temperature and the package becomes 'smart' (such packaging is currently under development). The most common use a thermochromic ink dot to indicate the product is at the correct serving temperature following refrigeration or microwave heating. Plastic containers of pouring syrup for pancakes can be purchased in the USA that are labelled with a thermochromic ink dot to indicate that the syrup is at the right temperature following microwave heating. Similar examples can be found on supermarket shelves with beer bottle labels that incorporate thermochromic-based designs to inform the consumer when a refrigerated beer is cold enough to drink.

Factors that Aid the Growth of Intelligent Packaging

Consumer and societal factors are likely to drive the adoption of smart packaging in the future. The growing need for information on packaging will mean there has to be a step change in providing this information. Consumers increasingly need to know what ingredients or components are in the product and how the product should be stored and used. Intelligent labelling and printing, for example, will be capable of communicating directly to the customer via thin film devices providing sound and visual information, either in response to touch, motion or some other means of scanning or activation. Voice-activated safety and disposal instructions contained on household and pharmaceutical products will be used to tell the consumer how they should be disposed of after consumption - information that can be directly used in the recycling industry to help sort packaging materials from waste streams. Drug delivery systems in smart packaging will be programmed to communicate patient information back to healthcare centres.

Possible Concerns over Intelligent Packaging

When it comes to the environment, consumer attitudes towards packaging are generally confused and contradictory. There are increasing concerns about the amount of waste created by packaging, but the growth of more elaborate and attractive packaging is being driven by consumers to fuel our desire for convenience and feed our lifestyle choices. Ultimately, future consumers could react negatively to the perception of increased waste and lack of recyclability of disposable smart packages. The perception of extra cost and complexity, and the possible mistrust/ confusion of technology - for example, if there is both a date stamp and a visual indicator on a food pack, which does the customer take note of - are further factors that could slow widespread market introduction of smart packages. The overall acceptance barriers to smart packaging can be summed up as:

1. Extra cost - can it be absorbed/passed on to consumer?
2. Unreliability of indicating devices showing either food to be safe when it is not (potential liability?) or food to be unsafe when it is (increased spoilage stock loss)
3. Food safety and regulatory issues - *e.g.* possible migration issues of complex packaging materials into product
4. Recycling features and environmental regulations.

What Areas will Benefit from Intelligent Packaging First?

Cost issues will probably mean that early adopters of smart packaging are likely to be in non-commodity products, *e.g.,* pharmaceuticals, health and beauty, and packaging that plays a part in lifestyle and leisure activities. A further consideration is the need for education to reassure the consumer of package safety, and ensure against incorrect operation and mistrust of smart technology. The successful adoption of smart packaging concepts in the future must create advantages for the whole of the supply chain.

Smart Packaging Concepts for Pharmaceuticals

Smart packaging concepts that improve case of use could include 'dial-a-dose' smart caps and closures that allow the safe dispensing of exact controlled quantities of product, *e.g.* pharmaceuticals, cleaning materials, and other potentially hazardous materials. Already a prescription drug bottle with bottle cap alarm is available - it beeps to alert users when it is time to take the medication, and it displays how many times the bottle has been opened and the intervals between openings. The bottle can be connected via a modem to the healthcare centre for the automatic transmission of drug usage and, if necessary, provide feedback to the patient if not in compliance. Eventually, programmed skin patches using smart gels that rely on changes in skin properties to trigger drug delivery could replace conventional pill-taking medication.

Bibliography

1. Ahmed E. and Bird, K. 2007. Nanoscale particles to block UV light in plastic packaging. *Nanowerk LLC* Report.
2. Ahvenainen, R. 2003. Novel Food Packaging Techniques; CRC Press: Boca Raton, FL, USA.
3. Alfadul S.M.1. and Elneshwy, A.A. 2010. Use of nanotechnology in food processing, packaging and safety – review. AFJAND. 10 (6) :2719-2739.
4. Allaoui, S., Aboura, Z., and Benzeggagh, M.L. 2009. Effects of the environmental conditions on the mechanical behaviour of the corrugated cardboard. Composites Science and Technology 69: 104–110.
5. Altaf, U., Kanojia, V. and Rouf. A. 2018. Novel packaging technology for food industry. J. Pharmacognosy and Phytochemist. 7(1): 1618-1625.
6. Anonymous (2004). Down on the Farm: the Impact of Nano-Scale Technologies on Food and Agriculture, *etc.* Group Report.
7. Arvanitoyannis I.S., Stratakos, A.C. (2012) Application of modified atmosphere packaging and active/smart technologies to red meat and poultry: A review. Food Bioprocess Tech 5(5):1423–1446.
8. Asadi, G. and Mousavi, M. Application of Nanotechnology in Food Packaging. Available: http://iufost.edpsciences.org
9. Bijii, K.B., Ravishankar, C.N., Mohan, C.O. and Gopal, T.K.S. 2015. Smart packaging systems for food applications: a review. J. Food Sci. Technol. 52 (10): 6125–6135.

10. Brody, A.L. (1989). Modified atmosphere packaging of Seafoods. In: *Controlled/ Modified Atmosphere Vacuum Packaging of Foods,* A.L. Brody (Ed.), Food and Nutrition Press, Inc., Trumbull, Connecticut, Chap. 4.
11. Brody A.L., Strupinsky, E.R., Kline, L.R.2002. Active packaging for food applications. Boca Raton: CRC press.
12. Brody, A.L., Bugusu, B., Han, J.H., Sand, C.K. and McHugh, T.H. 2008. Innovative food packaging solutions. J. Food Sci.: 73 (8): R 107-116.
13. Brooks, J. (1929). Biochem. J., **23**: 1391.
14. Brooks, J. (1931). Proc. Royal Soc., London, Ser. B., **109**: 35.
15. Brooks, J. (1937). Proc. Royal Soc., London, Ser. B., **123**: 368.
16. Brooks, J. (1938). Food Res., **3**: 75.
17. Butler, O.D., Bratzler, L.J. and Mallman, W.L (1953). Food Technol., **7**: 397.
18. Butler, P. 2001. Smart Packaging - Intelligent Packaging for Food, Beverages, Pharmaceuticals and Household Products. Materials World, 9 (3): 11-13.
19. Chang,W.S., Ventsel, E., Krauthammer,T. and John, J. 2005. "Bending behaviour of corrugated-core sandwich plates" Composite Structures 70 (2005) 81–89.
20. Clark, D.S. and C.P. Lentz (1969). Can. Inst. Food Sci. Technol. J., **2**: 72.
21. Clark, D.S. and C.P. Lentz (1972). Can. Inst. Food Sci: Technol. J., **5**: 175.
22. Clark, D.S. and C.P. Lentz (1973). Can. Inst. Food Sci. Technol. J., **6**: 194.
23. Clark, D.S., C.P. Lentz and L.A. Roth (1976). Can. Inst. Food Sci. Technol. J., **11**: 363.
24. Cole, A.B. (1986). Reciprocal Meat Conference Proc., **39**: 106.
25. Coles, R,. Mcdowell, D. and Kirwan, M.J. 2003. Food Packaging Technology. Blackwell Publishing Ltd.
26. Dainty, R.H., Shaw, B.G. and Roberts, T.A. (1983). Microbial and chemical changes in chill-stored red meats. In: Food Microbiology: Advances and Prospects, T.A. Roberts and F.A. Skinner (Eds.), Academic Press, London, England, p. 151.
27. Daniels, J.A., Krishnamurth, R. and Rizvi, S.S.H. (1985). J. Food Protect., **48**: 532.
28. Dainelli, D.; Gontard, N.; Spyropoulos, D.; Zondervan-van den Beuken, E.; Tobback, P. 2008. Active and intelligent food packaging: Legal aspects and safety concerns. Trends Food Sci. Technol., 19, S103–S112.
29. Das, R. (2018). Flexible Packaging vs Rigid Packaging: Which one should you use? Retrieved from: https://bizongo.com/blog/flexible-packaging-vs-rigid-packaging/

30. Dawson, L.E. (1987). Packaging of processed poultry. In: The Microbiology of Poultry Meat Products, F.E. Cunningham and N.A. Cox (Eds.), Academic Press, Inc., London, Chap. 7.

31. De-Azeredo H.M.C. 2009. Nano composites for food packaging applications. Food Res Int., 42:1240–1253.

32. Devi R, Yadav S, Nehra R, Yadav S and Pundir, C.S. 2013. Electrochemical biosensor based on gold coated iron nanoparticles/chitosan composite bound xanthine oxidase for detection of xanthine in fish meat. J Food Eng.;115(2):207–214.

33. Denton, J.H. and F.A. Gardner (1981). J. Food Sci., **47**: 214.

34. Dimitriyevic, M. *et al.* (2015). Safety aspects of nanotechnology applications in food packaging. Procedia Food Science. **5**. 57-60.

35. Dobrucka R and Cierpiszewski R.2014. Active and Intelligent Packaging Food Research and Development- A Review. Pol J Food Nutr Sci.,64(1):7–15.

36. Donald, A. 2004. Food for thought. Nature Materials. 3: 578-581.

37. Eriksson, I., Lidbrandt, O. and Westermark, U. 1988.Wood Science and Technology 22, 251.

38. Enfors, S.O. and Molin, G. (1980). Can. J. Microbiol., **27**: 15.

39. Enfors, S.O. and Molin, G. (1981). J. Appl. Bacteriol., **48**: 409.

40. Enfors, S.O. and Molin, G. (1981). The effect of different gases on the activity of microorganisms. In: Psychrotrophic Microorganisms in Spoilage and Pathogenicity, T.A. Roberts, G. Hobbs, J.H.B. Christian and N. Skovgaard (Eds.), Academic Press, London, England, Chap. 34.

41. European Commission Regulation No. 450/2009 of 29 May 2009 on active and intelligent materials and articles intended to come into contact with food. Off. J. Eur. Union 2009, 135, 3–11.

42. Fabech, B., Hellstrøm, T., Henrysdotter, G., Hjulmand-Lassen, M., Nilsson, J., Rüdinger, L.; Sipiläinen-Malm, T., Solli, E.;,Svensson, K. and Thorkelsson, Á.E. *et al.,* 2000. Active and Intelligent Food Packaging—A Nordic Report on the Legislative Aspects; TemaNord: Stockholm, Sweden; ISBN 9289305207.

43. Farber, J.M. (1991). J. Food Protect., **54**: 58.

44. Fengel, D. and G. Wegener, Wood, Chemistry, Ultrastructure, Reactions, (Walther de Gruyter, New York, 1984).

45. Finne, G. (1981). Principles of controlled atmosphere extension of food shelf life. In: Proc. Int. Conf. Control. Allnos., *CAP*'84, Chicago, Illinois, p. 37.

46. Gardner, G.A. (1983). Microbial spoilage of cured meats. In: Food Microbiology: Advances and Prospects, T.A. Roberts and F.A. Skinner (Eds.), Academic Press, London, England, p. 179.

47. Genigeorgis, C. (1985). Int. J. Food Microbiol., **1**: 237.

48. Gentilini, C., Nobile, L., Keith and Seffen, A. 2009. Numerical analysis of morphing corrugated plates. Procedia Engineering 1:79–82.

49. Georgala, D. and Davidson, C.M. (1970). British Patent, **1**: 199, 998.

50. George, P. and C.J. Stratmann (1952). *Biochem. J.,* **51**: 418.

51. Geueke, B. *et al.* (2018). Food packaging in the circular economy: Overview of chemical safety aspects for commonly used materials. Journal of Cleaner Production. **193**. 491-505.

52. Gibbs, P.A. and Patterson, J.T. (1977). Some preliminary findings on the microbiology of vacuum- and gas-packaged chicken portions. In: *The Quality of Poultry Meat,* S. Scholtyssek (Ed.), Universitat Hohenheim, Stuttgart, p. 198.

53. Gill, C.O. (1982). Microbial interaction with meats. In: *Meat Microbiology,* M.H. Brown (Ed.), Applied Science Publishers Ltd., London, England, Chap. 7.

54. Gill, C.O. (1988). Meat Sci., **22**: 65.

55. Gill, C.O. (1989). Brit. Food J., **91(7)**: 11.

56. Gill, C.O. (1990). Food Control J., **74**.

57. Gill, C.O. and Tan, K.H. (1980). Appl. Environ. Microbiol., **39**: 317.

58. Gill, C.O. and Penney, N. (1985). Meat Sci., **14**: 43.

59. Gill, C.O. and Penney, N. (1986). Meat Sci., **18**: 41.

60. Gill, C.O., Harrison, C.L. and Penney, N. (1990). Int. J. Food Micro., **11**: 151.

61. Grau, F.H. (1983). J. Food Sci., **48**: 326.

62. Grayhurst, P. and Girling, P.J. (2011). Packaging of food in glass containers. In: Coles, R., Kirwan, M. (Eds.), Food and Beverage Packaging Technology, second ed. Place: John Wiley and Sons Ltd, UK.

63. Grigore, M.E. 2017. Methods of Recycling, Properties and Applications of Recycled Thermoplastic Polymer-Review. Recycling, 2, 24: 1-11.

64. Gokularaman S.1., Stalin Cruz A.1., Pragalyaashree1, M.M., Nishadh, A. 2017. Nanotechnology Approach in Food Packaging – Review. J. Pharm. Sci. and Res. 9 (10), 1743-1749.

65. Goyal, S. and Goyal, G.K. 2012. Nanotechnology in food packaging – a critical review. Russian Journal of Agricultural and Socio-Economic Sciences, No. 10 (10): 14-24.

66. Han, J.H., Ho, C.H.L. and Rodrigues, E.T. 2005. Intelligent Packaging; Elsevier: Amsterdam, The Netherlands: ISBN 9780123116321.

67. Harte, B.R. (1987). Packaging of restructured meats. In: Advances in Meat Research, A.M. Pearson and T.R.

68. Dutson (Eds.), Elsevier Applied Science Publishers, Essex, England, Chap. 13.

69. Hauschild, A.H.W. and Simonsen, B. (1985). J. Food Protect., **48**: 997.

70. Hayakawa, K., Henig, Y.S. and Gilbert, S.G. (1975). J. Food Sci., **40**: 186.

71. Henig, Y.S. and Gilbert, S.G. (1975). J. Food Sci., **40**: 1033.

72. Herner, R.C. (1987). High CO_2 effects on plant organs. In: Postharvest Physiology of Vegetables, J. Weichmann (Ed.), Marcel Dekker, Inc., New York, Chap. 11.

73. Heising, J.K., Dekker M., Bartels P.V. and Van Boeke,l M.A. 2014. Monitoring the quality of perishable foods: opportunities for intelligent packaging. Crit Rev Food Sci Nutr.;54(5):645–654. doi: 10.1080/10408398.2011.600477.

74. Hicks, A. 2002. Minimum Packaging Technology for Processed Foods: Environmental Considerations. AU J.T. 6(2): 89-94.

75. Higuchi, T. Ed., Biosynthesis and Biodegradation of Wood Components, (Academic Press, Inc., New York, 1985).

76. Hintlian, C.B. and Hotchkiss, J.H. (1986). Food Technol., **40(12)**: 70.

77. Hirose, K., Harte, B.R., Giacin, J.R., Miltz, J. and Stine, C. (1988). Sorption of d-limonene by sealant films and effect on mechanical properties. In: Food and Packaging Interactions, J.H. Hotchklss (Ed.), ACS Symp. Ser. #365, American Chemical Society, Washington, D.C., Chap. 3.

78. Hobbs, G. (1983). Microbial spoilage of fish. In: Food Microbiology: Advances and Prospects, T.A. Roberts and F.A. Skinner (Eds.), Academic Press, London, England, p. 217.

79. Holdsworth, S.D. (1983). *The Preservation of Fruit and Vegetable Food Products.* The Macmillan Press Ltd, London, England.

80. Hogan, S.A. and Kerry, J.P. 2008. Smart packaging of meat and poultry products. In: Kerry J, Butler P, editors. Smart packaging technologies for fast moving consumer goods. West Sussex, England: John Wiley and Sons Ltd; pp. 33–59.

81. Huxsoll, C.C. and. Bolin, H.R (1989). Food Technol., **43**: 124.

82. Isaksson,P. and Gradin, P. 2009. Shear buckling in the core of a corrugated board structure. Composite Structures 88: 610–614.

83. Isenberg, M.F.R. (1979). Hort. Rev., **1**: 337.

84. Jackson, J.M. and Shinn, B.M. (1979). Fundamentals of Food Canning Technology. AVI Publishing Company, Inc., Westport, Connecticut.

85. Jurin, V. and Karel, M. (1963). Food Technol., **17(6)**: 104.

86. Kader, A.A. (1980). Food Technol., **34(3)**: 51.

87. Kader, A.A. (1986). Food Technol., **40(5)**: 99.

88. Kader, A.A. (1987). Respiration and gas exchange of vegetables. In: *Postharvest Physiology of Vegetables, J.* Weichmann (Ed.), Marcel Dekker, Inc., New York and Basel, Chap. 3.

89. Kader, A.A., D. Zagory and E.L. Kerbel (1989). *CRC Crit. Rev. Food Sci. Nutr.*, **28**: 1.

90. Kader. A.A. (1985). Postharvest and biology. In: Postharvest Technology of Horticultural Crops. A.A. Kader, R.F. Kasmire, F.G. Mitchell, M.S. Reid, N.F. Somer and J.F. Thompson, Cooperative Extension University of California Division of Agriculture and Natural Resources Special Publication 3311, Chap. 2.

91. Karel, M. and Go, J. (1964). Modern Pack., **37(6)**: 123.

92. Karel. M. (1975). Protective packaging of foods. In: Principles of Food Science Part II: Physical Principles of Food Preservation, M. Karel. O.R. Fennema and D.B. Lund, Marcel Dekker, Inc., New York, Chap. 12.

93. Kawada, K. (1982). Use of polymeric films to extend postharvest life and improve marketability of fruits and vegetables-UNIPACK: individually wrapped storage of tomatoes, oriental persimmons and grapefruit. In Proceedings of the Third National Controlled Atmosphere Research Conference on Controlled Atmospheres for Storage and Transport of Perishable Agricultural Commodities, D.G. Richardson and M. Meheriuk (Eds.), Timber Press, Beaverton, Oregon, p. 87.

94. King, A.D and H.R. Bolin (1989). Food Technol., **43**: 132.

95. Kirwan, M. (2003). Paper and paperboard packaging. In: Coles R., Kirwan M. (Eds.) Food Packaging Technology. 241-281. CRC Press.

96. Kramer, A., T. Solomos, F. Wheaton, A. Puri, S. Sirivichaya, Y., Lotem, M., Fowke and Ehrman, L. (1980). Food Technol., **34**: 65.

97. Kress, G., Winkler, M. 2010. Corrugated laminate homogenization model. Composite Structures 92 795–810.

98. Kuswandi, B., Jayus, R.A., Oktaviana, R., Abdullah, A. and Heng, LY. 2014. A novel on-package sticker sensor based on methyl red for real-time monitoring of broiler chicken cut freshness. Packag Technol Sci. 2014;27(1):69–81.

99. Moon, K., Kwon, Mi., Eun-Bin, Lee, B. and Kim, C.Y. 2020. Recent Trends in Controlling the Enzymatic Browning of Fruit and Vegetable Products -Review. Molecules, 25:1-15.

100. Labuza, T.P. (1984). Moisture Sorption: Practical Aspects of Isotherm Measurement and Use. American Association of Cereal Chemists, St Paul. Minnesota.

101. Lakin, W.D. (1987). J. Packag. Tech., **1**: 82.

102. Lee, D.S., Haggar, P.E., Lee, J. and Yam, K.L. (1991). J. Food Sci., **56**: 1580.

103. Lee, J.J.L. (1987). The design of controlled or modified packaging systems for fresh produce. In: Food Product: Package Compatibility, J.I. Gray, B.R. Harte and J. Miltz (Eds.), Technomic Publishing Co., Inc., Pennsylvania, p. 157.

104. Liew, K.M., Peng, L.X. and Kitipornchai, S. 2007. Nonlinear analysis of corrugated plates using a FSDT and a meshfree method. Comput. Methods Appl. Mech. Engrg. 196:2358–2376.

105. Lipton, W.J. (1975). Controlled atmospheres for fresh vegetables and fruits why and when. In: Postharvest Biology and Handling of Fruits and Vegetables, N.F. Haard and D.K. Salunke (Eds.), AVI Publishing Co. Inc, Wesport, Connecticut, p. 130.

106. Liu Y, Chakrabartty, S. and Alocilja E.2007. Fundamental building blocks for molecular bio-wire based forward error-correcting biosensors. Nanotechnology.18:1–6.

107. Lougheed, E.C. (1987). Hort Sci., **22**: 791.

108. Majid, I. *et al.* (2016). Novel food packaging technologies: Innovations and future prospective. Journal of the Saudi Society of Agricultural Sciences. **17**. 454–462.

109. Mandal, P. K., Choi, K., Min, Sang-Gi and Lee, Chi-Ho. 2009. Application of Nanotechnology in Food Packaging: An Overview. Korean J. Food Sci. Ani. Resour. 29 (4): 403~408.

110. Mannapperuma, J.D., Zagory, D. Singh, R.P. and Kader, A.A. (1989). Design of polyme packages for modified atmosphere storage of fresh produce. Proceedings 5th Controlled Atmosphere Research Conference, Vol. 2, Wenatchee, Washington, p. 225.

111. Mannheim, C.H., J. Miltz and Letzter, A.J. (1987). J. Food Sci., **52**: 737.

112. Meena, M.K., Chetti, M.B., Nawalagatti, C.M. and Chandra Naik, M. 2017. Adv Plants Agric Res. 7(1):221225.

113. Marsh, K. and Bugusu, B. (2007). Food Packaging – Role, Materials, and Environmental Issues. Institute of Food Technologists. **72**. Nr.3. 39-55.

114. Mathew, R. and Dorothy Jaganathan, D. 2017. Packaging and storage practices of meat. Global. J. Bio. Agri. Health Sci., 6 (1):32-40.

115. Metlitskii, L. V., Sal'Kova, *e.g.*, Volkind, N.L. Bondarev, V.I. and Yanyuk, V. (1983). Controlled Atmosphere Storage of Fruits. Amerind Publishing Co. Pvt. Ltd., New Delhi, India, Chap. VI.

116. Mills A, Qing Chang Q, Mc Murray N. 1992. Equilibrium studies on colorimetric plastic film sensors for carbon dioxide. Anal Chem.: 64:1383–1389.

117. Mohanty, F. and Swain, S. K. (2017). Bionanocomposites for Food Packaging Applications. Nanotechnology Applications in Food. 363-379.

118. Moraru, C., Panchapakesan,C., Huang Q. and Takhistov, P., 2003. Nanotechnology: A new frontier in food science. Food Technol. 57: 25-27.

119. Müller, P.; Schmid, M. 2019. Intelligent packaging in the food sector: A brief overview. Foods, 8.

120. Myers, R.A. (1989). Food Technol., **43**: 129.

121. Nelson, P.E. and Tressler, D.K. (1980). Fruit and Vegetable Juice Processing Technology, 3rd edn., AVI Publishing Co. Inc., Westport, Connecticut.

122. Neurater G., Klimant, I, Wolfbeis, O.S. 1999. Microsecond life time based optical carbon dioxide sensor using luminescence resonance energy transfer. Analytica Chimica Acta.; 382:67–75.

123. Obiora B. Ezeudu 1, Jonah C., Agunwamba, Ezeasor, I. C. and Madu C.N. 2018. Sustainable Production and Consumption of Paper and Paper Products in Nigeria: A Review. Resources 8: 53.

124. Qbst, J.R. 1990. In: Caulfield, D.F.; Passaretti, J.D.; Sobczynski, S.F., eds. Materials interactions relevant to the pulp, paper, and wood industries: Proceedings, Materials Research Society symposium; 11 1990 April 18-20; San Francisco, CA. Pittsburgh, PA: Materials Research Society; 1990: 11-20. Vol. 197.

125. Pantastico, E.B., Chattopadhyay, T.K. and Subramanyam, H. (1975). Storage and commercial operations. In: Postharvest Physiology, Handling and Utilization of Tropical and Subtropical Fruits and Vegetables, E.B. Pantastico (Ed.), AVI Publishing Co. Ltd., Connecticut, Chap. 16.

126. Peleg, K. (1966). Automatic wrapping and packaging of citrus. M.Sc. Thesis, Technion Haifa, Israel.

127. Peleg, K. (1985). Retail produce packaging. In: Produce Handling, Packaging and Distribution, AVI Publishing Co. Inc., Wesport, Connecticut, Chap. 11.

128. Pitarresi, G., Carruthers, J.J., Robinson, A.M., Torre, G., Kenny, J.M., Ingleton, S. Velecela, O. and Found, M.S. 2007 A comparative evaluation of crash worthy composite sandwich structures. Composite Structures 78 (2007) 34–44.

129. Peng, L.X., Liew, K.M. and Kitiporncha, S. 2007. Analysis of stiffened corrugated plates based on the FSDT via the mesh-free method" International Journal of Mechanical Sciences 49 (2007) 364–378.

130. Powrie, W.D. (1988). Optimization of parameters to prolong retention of prepackaged fruit. In: Proceedings of the Fourth International Conference on Controlled Atmosphere Packaging CAP'88, Scotland Business Research, Inc., New Jersey, p. 245.

131. Powrie, W.D. and Skura, B.J. (1991). Modified atmosphere packaging of fruits and vegetables. In: Modified Atmosphere Packaging of Food, B. Ooraikul and M.E. Stiles (Eds.), Ellis Horwood Ltd., Chichester, England, Chap. 7.

132. Prince, T.A. (1989). Modified atmosphere packaging of horticultural commodities. In: Controlled/Modified Atmosphere/Vacuum Packaging of Foods, A.L. Brody (Ed.), Food and Nutrition Press, Trumbull, Connecticut, Chap. 5.

133. Rajamäki, T., Alatomi, H., Titvanen, T., Skyttä, E., Smolander, M., and Ahvenainen, R.2004. Application of an electronic nose for quality assessment of modified atmosphere packaged poultry meat. Food Contr.; 17:5–13.

134. Realini, C.E. and Marcos, B. 2014. Active and intelligent packaging systems for a modern society. Meat Sci.; 98:404–419.

135. Reany, J., Joachim, Grenestedt, L. 2009. Corrugated skin in a foam core sandwich panel. Composite Structures 89: 345–355.

136. Reid, M.S. (1985). Ethylene in postharvest technology. In: Postharvest Technology of Horticultural Crops, A.A. Kader, R.F. Kasmire, F.G. Mitchell, M.S. Reid, N.F. Somer and J.F. Thompson, Cooperative Extension University of California Division of Agriculture and Natural Resources Special Publication 3311, Chap. 13.

137. Reitman, N. and Doppelt, B. 2007. Packaging and Plastics: What's a Consumer to Do ? PCC Natural Markets' Newsletter: The Sound Consumer.

138. Risch, S. J. (2009). Food Packaging History and Innovations. Journal of Agricultural and Food Chemistry, 57(18), 8089–8092.

139. Risse, L.A., and Miller, W.R. (1986). J. Plastic Film and Sheeting, **2**: 163.

140. Robertson, G. L. (Ed.). (1993). Food packaging: Principles and practice. New York, NY: Marcel Dekker.

141. Rydz, J. *et al.* (2018). Present and Future of Biodegradable Polymers for Food Packaging Applications. Biopolymers for Food Design. 431-467.

142. Salunke, D.K. and Desai, B.B. (1984). Postharvest Biotechnology of Vegetables, CRC Press, Inc., Florida, Vols. I and II.

143. Sarkanen, K.V. and Ludwig, C.H. Eds. 1971. Lignins, Occurrence, Formation, Structure and Reactions, (Wiley-Interscience, New York, 1971).

144. Sarwar, M.H., Muhammad Farhan Sarwar, M.F., Sarwar, M., Qadri, N.A. and Moghal, S. 2013. The importance of cereals (Poaceae: Gramineae) nutrition in human health: A review. J. Cereals Oilseeds.4(3): 32-35.

145. Schaefer, D. and Cheung, W. M. (2018). Smart Packaging: Opportunities and Challenges. Procedia CIRP. **72**, 1022-1027.

146. Scott, A.(2002). BASF takes big steps in small tech, focusing on nanomaterials. Chemical Week.164,1-45.

147. Siracusa, V. and Rosa, M. D. (2018). Sustainable Packaging. In Sustainable Food Systems from Agriculture to Industry. 275-307.

148. Smock, R.M. (1979). Hort. Rev., **1**: 301.

149. Talbi,N., A., Batti, A., Ayad, R. and Guo, Y.Q. 2009. An analytical homogenization model for finite element modeling of corrugated cardboard. Composite Structures 88: 280–289.

150. Thomas Szemplenski. 2012. The U.S. markets for aseptic packaging from: Handbook of Aseptic Processing and Packaging CRC Press.

151. Tian,Y. S. and Lu, T.J. 2005. "Optimal design of compression corrugated panels" Thin-Walled Structures 43 (2005) 477–498.

152. Tolle, W.E. (1971). Variables affecting film permeability requirements for modified-atmosphere storage of apples. *U.S. Dept. Agr. Tech. Bull.,* No. 1422.

153. Troller, J.A. and J.H.B. Christian (1978). Water Activity and Foods, Academic Press Inc., New York.

154. Vanderroost, M.; Ragaert, P.; Devlieghere, F.; De Meulenaer, B.2014. Intelligent food packaging: The next generation. Trends Food Sci. Technol., 39, 47–62.

155. Veeraju, P. and Karel, M. (1966). Mod. Pack., **40(2)**: 168.

156. Wang, Y., Weissmüller, J., and Duan, H.L.2010. "Mechanics of corrugated surfaces" Journal of the Mechanics and Physics of Solids accepted date 12 July 2010.

157. Westermark, U. 1985. Wood Science and Technology 19, 323.

158. Winkler, M., and Kress, G. 2010 "Deformation limits for corrugated cross-ply laminates" Composite Structures 92 (2010) 1458–1468.

159. Wolfbeis OS, List H (1995) Method for quality control of packaged organic substances and packaging material for use with this method. US Patent 5407829

160. Weichmann, J. (1987). Low oxygen effects. In: Postharvest Physiology of Vegetables, J., Weichmann (Ed.), Marcel Dekker, Inc., New York and Basel, Chap. 10.

161. Yam, K.L., Takhistov, P.T., and Miltz, J. 2005. Intelligent packaging: concepts and applications. J Food Sci.; 70:1–10.

162. Yang, C.C. and Chinnan, M.S. (1988). J. Food Sci., **53**: 869.

163. Zagory, D., and Kader, A.A. (1988). Food Technol., **42(9)**: 70.

Index

C

D

E

F

G

H

I

J

K

L

M

N

O

P

Q

R

S

T

U

V

W